BOOKS BY
CHARLES B. BREED
AND
GEORGE L. HOSMER

THE PRINCIPLES AND PRACTICE OF SURVEYING

Volume I. Elementary Surveying.

Eleventh Edition. 751 pages, 324 cuts.
Flexible Binding, Pocket Size.

Volume II. Higher Surveying.

Eighth Edition. 543 pages, 237 cuts.
Flexible Binding, Pocket Size.

BY CHARLES B. BREED

SURVEYING

Third Edition. 481 pages, 219 cuts.
Flexible Binding, Pocket Size.

THE PRINCIPLES AND PRACTICE OF

SURVEYING

VOLUME I: ELEMENTARY SURVEYING

by

CHARLES B. BREED
Late Professor of Civil Engineering

and

GEORGE L. HOSMER
Late Professor of Geodesy

AND REVISED BY

DR. W. FAIG, P. Eng.
Associate Professor of Surveying Engineering
University of New Brunswick

ASSISTED BY

BR. B. AUSTIN BARRY, FSC, RE.
Professor of Civil Engineering
Manhattan College

ELEVENTH EDITION

NEW YORK

JOHN WILEY & SONS, INC.

LONDON

PRINTED IN U.S.A.

5/77

PREFACE TO THE ELEVENTH EDITION

This edition attempts to preserve the basic good of the initial text and of the several succeeding revisions that have constantly infused new life into the work and maintained its modernity. Rather extensive changes have nonetheless been made.

Specifically this revision brings together into any one place all that pertains to any single subject (e.g., distance measurement, computation). A new numbering system for articles and illustrations is introduced, which should prove most convenient. Logarithmic computation has at last been abandoned, in acknowledgment of the pocket calculator on the modern scene. And the attempt is begun to ease into the values of SI (Système Internationale).

The shift has been made from probable error to standard deviation (Chapter I); all angular measurement has been consolidated (Chapter II), with a notable emphasis on modern instruments; EDM in distance measurement is given full play (Chapter III); topographic mapping procedures supplement the usual stadia procedures (Chapter VII); computing and computational tools are covered in a fresh manner (Chapter XIII); and, finally, emphasis is focused on coördinates in surveying computation (Chapter XV).

The text is updated and modernized, as surveying itself has been in these past years. The attitude engendered throughout is that the student (and the practitioner) is not learning a set of prescribed procedures and successive steps, but rather is coming to grips with an enticingly vibrant and ever new subject matter unfolding before him.

PREFACE TO THE FIRST EDITION

In the preparation of this volume, it has been the authors' chief purpose to produce a text-book which shall include the essentials of a comprehensive knowledge of practical surveying and at the same time be adapted to the use of teachers and students in technical schools. In this book, which is essentially an elementary treatise, such subjects as stadia, plane table, hydrographic and geodetic surveying, are entirely omitted, these subjects being left for a later volume.

Considerable stress is laid upon the practical side of surveying. The attempt is made not only to give the student a thorough training in the fundamental principles and in approved methods of surveying, computing, and plotting, but also to impress upon him the importance of accuracy and precision in all of his work in the field and the drafting-room. In carrying out this purpose it has seemed necessary to lay particular stress upon some points which to the experienced engineer or the advanced student may appear too obvious to require explanation, but which teaching experience has shown to be most helpful to the beginner. The most common errors and mistakes have therefore been pointed out and numerous methods of checking have been explained. Every effort has been made to inculcate right methods even in minor details, and for this purpose a large number of examples from actual practice have been introduced.

In arranging the subject matter of the work, the four parts are presented in what appears to be a logical sequence. First, the use, adjustment, and care of instruments are taken up; then the next three parts, surveying methods, computations, and plotting, are taken in the order in which they are met in the daily practice of the surveyor. To show more clearly the steps in the process, the notes which are used as illustrations in surveying methods are calculated in the computation section, and are treated again under the methods of plotting, finally appearing as a completed plan.

While the authors recognize fully their indebtedness to those who have preceded them in this field, they hope that they have made some useful contributions of their own to the treatment of the subject. Thus in the section on Surveying Methods, many practical suggestions have been inserted which they have found of value in their own work and which, so far as they are aware, now appear in a text-book for the first time. On the subject of Computations, much emphasis is laid upon the proper use of significant figures and the arrangement of the work, matters which heretofore have not been adequately treated in books on surveying. The section on Plotting contains many hints referring particularly to surveying drafting, which are not given in the published books on drawing and lettering. It is hoped also that the complete set of original illustrations which have been introduced throughout the book will aid materially in making the text clear.

A comprehensive cross-reference system giving the page as well as the article number has been adopted: this, together with the complete index at the end of the book and the many practical hints throughout the volume, will, it is hoped, render it useful to the practical surveyor as a reference book.

The authors desire to acknowledge their indebtedness to their various associates in the teaching and engineering professions who have kindly responded to requests for information and assisted in the preparation of this work, particularly to Blamey Stevens, M. Sc., of Ellamar, Alaska, who supplied the entire chapter on Mining Surveying. They are also under obligations for the use of electrotype plates of tables: to W. H. Searles for Tables IV, V, and VI; to Professor J. C. Nagle for Tables II and III; and to Professor Daniel Carhart for Table I; all of these plates were furnished by John Wiley & Sons. The authors are under special obligation to Professors C. F. Allen, A. G. Robbins, and C. W. Doten of the Massachusetts Institute of Technology, and to H. K. Barrows, Engineer U. S. Geological Survey, who have read the entire manuscript and who have offered many valuable suggestions in preparing the work for the press. The authors also desire to express their appreciation of the excellent work of W. L. Vennard, who made the drawings for illustrations.

No pains have been spared to eliminate all errors, but the authors cannot hope that their efforts in this line have been completely successful, and they will consider it a favor if their attention is called to any which may be found.

C. B. B.
G. L. H.

BOSTON, MASS., *September*, 1906.

CONTENTS

PART I

USE, ADJUSTMENT AND CARE OF INSTRUMENTS

Chapter I—Introduction to Surveying—Adjustment of Errors

Chapter II—Angular Measurements

CHAPTER III—MEASUREMENT OF DISTANCES

TAPING.

Chapter IV—Measurement of Difference of Elevation

PART II

SURVEYING METHODS

CHAPTER V—LAND SURVEYING

CHAPTER VI—TRAVERSE LINES—LOCATION OF BUILDINGS—MISCELLANEOUS SURVEYING PROBLEMS

Chapter VII—Topographical Surveys

Chapter VIII—Observations for Meridian and Latitude

OBSERVATIONS AND MERIDIAN.

ADJUSTMENTS OF THE SOLAR ATTACHMENT.

OBSERVATIONS FOR LATITUDE.

Chapter IX—Leveling

CHAPTER X—CITY URBAN SURVEYING

LINES AND GRADES—CIRCULAR CURVES—SURVEY OF CITY LOTS, BUILDINGS AND PARTY WALLS.

ESTABLISHING AND STAKING OUT CITY LINES AND GRADES.

CURVES.

CHAPTER XI—MINE SURVEYING

Chapter XII—Photogrammetry

PART III

COMPUTATIONS

Chapter XIII—Computational Tools and Their Use in Surveying

Chapter XIV—Survey Computations

CHAPTER XV—SIMPLE COÖRDINATE CALCULATIONS, TRAVERSES

PART IV

PLOTTING

Chapter XVI—Drafting Instruments and Materials—Reproducing Plans

ENGINEERING DRAFTING INSTRUMENTS.

DRAWING PAPERS.

REPRODUCING PLANS.

Chapter XVII—Methods of Plotting

METHODS OF PLOTTING TRAVERSES.

THE PRINCIPLES AND PRACTICE OF SURVEYING

PART I

USE, ADJUSTMENT, AND CARE OF INSTRUMENTS

CHAPTER I

INTRODUCTION TO SURVEYING

ADJUSTMENT OF ERRORS

1.1. INTRODUCTION.—Surveying is the science or art of making such measurements as are necessary to determine the relative position of points above, on, or beneath the surface of the earth, or to establish such points. Surveys that cover so small a part of the earth's surface that its curvature need not be taken into account are called "*plane surveys.*" Surveys in which the earth's curvature must be taken into account are called "*geodetic surveys.*"

1.2. KINDS OF SURVEYS.—The common types of surveys are known as cadastral, topographic, engineering, construction, route, underground, aerial, photogrammetric, and hydrographic.

Cadastral surveys (land surveys) are surveys which are made for setting land boundaries on the ground and determining areas of land parcels; they are also used for locating and identifying property lines.

Topographic surveys are those made for discovering the shape of the ground and plotting it on a map along with natural and artificial features.

Engineering or *construction* surveys are surveys of the sites where construction of an engineering nature is to be undertaken.

Route surveys may also be properly classed under engineering surveys. They comprise surveys of linear construction such as railroads, highways, canals, transmission lines and include the layout of lines and grades for these projects.

Underground surveys consist of surveys for locating the workings below the earth's surface, as in mines, tunnels, and aqueducts.

Aerial surveys utilize photographs taken from an airplane. Measurements are obtained from the photographs or from three-dimensional projections of stereo-pairs of photographs. The field which deals with these measurements is called *photogrammetry*.

Hydrographic surveys are those made for determining the shape of the bottom of lakes, rivers, harbors, and oceans. They also include the measurement of flow of water in streams, the estimation of water resources, and the positioning of sounding vessels on the water surface.

1.3. DUTIES OF THE SURVEYOR.—*The land surveyor* should be expert in subdividing land, retracing old boundaries, analyzing evidence as to the legality of boundaries, identifying boundaries, and making surveys of properties. He furnishes accurate descriptions of boundaries and areas which are required in preparing land titles. Such titles are usually drawn by an attorney called a conveyancer. The surveyor does not have the legal authority to determine property boundaries, but his expert knowledge and advice will often enable him to bring owners of adjoining land to an agreement as to boundaries and thereby avoid adjudication. For this reason he must know and be able to apply the legal principles governing competency and weight of evidence as to boundaries.

A large amount of a nation's wealth is either in land itself or in properties which are permanently fixed to the land. It is the responsibility of both the surveyor and the conveyancer to protect the public in the purchases and sales of lands.

The topographic surveyor uses the control points established by government or private surveys and fills in the details of the earth's surface and physical features, such as hills, valleys, streams, roads, railroads, buildings, and fences. For small surveys additional control is usually provided by a transit and tape survey which, wherever practicable, is tied in to one or more government survey points. Presently most topographic information is obtained by means of photogrammetry.

The *engineering* or *construction* surveyor is responsible for an accurate determination of the terrain upon which engineering works are to be built. The design of the structure depends in many respects upon the information shown on the surveyor's plans. He has a further responsibility in physically laying out on the ground the design elements of the proposed construction. An error in the survey which is not discovered until the construction work begins may result in very costly changes in design and construction. The surveyor also measures and computes the pay quantities of the different classifications of construction work designated by the contract.

The geodetic surveyor's field of interest includes practical astronomy, terrestrial magnetism, oceanography, gravity, the figure of the earth, and seismology. His service to the ordinary surveyor lies in locating, with high precision, survey stations for horizontal control, and "bench marks" (permanent points of elevation) for vertical control of detailed surveys.

1.4. Sources of Surveying Information—Local Sources.—The engineering or the public works departments in counties, cities, and towns have on file the official planned or actual alignment of streets, as also the determined values and descriptions of permanent monuments (of fixed and known horizontal location) and of bench marks (of fixed and known vertical location). In metropolitan districts there are usually special offices from which may be obtained many engineering and surveying data relating to parks, sewage works, water supply, waterfront improvements, and rapid transit facilities.

The registry of deeds will furnish a description of every parcel of land that has ever been conveyed. Although these descriptions are often not sufficiently definite to permit plotting of a plan, they do often, though not always, suffice to identify boundaries. It is, therefore, often necessary to inspect the original survey records for detailed information. With the proper and increasing use of coordinates this problem should eventually disappear.

1.5. State Surveys.—State governments keep records of boundary surveys for the state and its subdivisions (counties and towns), which in most states are in the custodial domain of the counties. In most states the Federal government has established a plane coordinate system covering the entire state, and has published the coordinates of many control points which are useful to the surveyor; some state geodetic survey bureaus have extended and improved the Federal work. State highway departments have plans of all state highway locations and definitions of rights-of-way in which these highways lie. The larger cities may use the state plane-coordinate system, but most have local coordinate systems, often predating and unrelated to the state system.

1.6. FEDERAL SURVEYS.—Several U. S. Federal surveying and mapping agencies have a wide range of information useful to the surveyor. Typically, these are:

National Oceanic and Atmospheric Administration
Defense Mapping Agency
United States Geological Survey
Bureau of Land Management
Bureau of Reclamation
United States Forest Service
Corps of Engineers, Department of the Army
Soil Conservation Service
Federal Highway Administration
International Boundary Commission
Tennessee Valley Authority
National Parks Service
National Bureau of Standards
Naval Oceanographic Office

The best source of information about maps, charts, photography etc. in the U. S. A. is:

Public Inquiries Office
U. S. Geological Survey, National Centre
12201 Sunrise Valley Drive, Reston, Virginia 22092

The National Oceanic and Atmospheric Administration through its National Ocean Survey (formerly Coast and Geodetic Survey) has available for public use data obtained from various type surveys in the fields of geodesy, gravity, astronomy, hydrography, oceanography, and marine technology. The National Ocean Survey (NOS) also provides charts for the safety of marine and air navigation.

Of particular interest to federal, state, and local surveyors are the publications listing the horizontal and vertical control data produced by the NOS through its National Geodetic Survey which establishes, improves, and maintains the basic national horizontal and vertical networks of geodetic control.

The U. S. Geological Survey has available for public use data obtained by its Topographic, Geologic, Water Resources and Conservation Divisions. Among its publications are:

The Geologic Atlas of the United States. The atlas is a series of folios which contain descriptions of, and maps showing, the geology of various regions of the country.

Topographic Maps of the United States. These maps are printed on paper about 16½ by 20 inches and are published on the following scales: 1/250000, 1/62500, 1/31680, and 1/24000.

Water Supply Papers give measurements of stream flow and descriptions of streams and gaging stations.

In addition to the above, many special maps and publications on geology, mineral sources, topography, hydrography, and hydrology are available.

The Bureau of Land Management (Public Land Surveys) is in charge of the survey of the public lands of the United States. A manual specifying in detail the regulations and methods of carrying on the public land surveys, maps showing the locations of the surveys that have been made, and data concerning the actual surveying, may be procured from this office.

The Bureau of Standards is engaged in tests and comparisons of weights and measures; in the establishment and maintenance of various standards and units of measurement; in the development of measuring instruments and methods of determining physical constants and the properties of materials. Its activities cover a wide range of subjects in the fields of physics, chemistry, engineering, and commercial practice. For a small fee surveyors may have their tapes standardized by the Bureau.

The Corps of Engineers (U. S. Army) makes surveys in connection with navigable waters, over which it has jurisdiction. At its district offices will be found maps defining pier heads, harbor lines, channel lines, depths of water, and bridge clearances in the harbors and rivers of its districts.

The U. S. Lake Survey has established bench marks in all harbors of the Great Lakes, the elevations of which are based on the Lake Datum. It also has records of the stages of the Lakes since about 1860.

The Mississippi River Commission of the Dept. of the Army has made extensive surveys of the Mississippi Basin and has published topographic quadrangle sheets on the scale of about 1 inch = 1 mile. The Commission has available also airplane photographs, triangulation, and bench-mark data. It keeps a record of channel lines and depths, river stages and discharge, and it maintains guides for navigation.

Definitions of Surveying and Associated Terms published by American Society of Civil Engineers in Manual of Engineering Practice—1972 gives definitions of terms used in surveying as well as a listing of agencies from which maps may be obtained and a bibliography of publications on surveying.

BASIC UNITS FOR SURVEYING USE

1.7. THE ENGLISH SYSTEM.—The basic unit of length used within most English speaking countries is the foot. The definition of a foot is based on the international metre (see section 1.9). Although in general practice, parts of a foot are expressed in inches, the surveying profession has adopted a decimal system based on the foot, and the tapes are therefore graduated into tenths and hundredths of a foot.

It is frequently necessary in construction work to convert decimals of a foot into inches and fractions, but the inch is not commonly used in surveying fieldwork. The following table gives relations which enable the surveyor to convert rapidly from one unit to the other.

TABLE 1.7.1

DECIMALS OF FOOT IN INCHES

Decimal of Foot		Inches	Decimal of Foot		Inches
.01	=	⅛−	.25	=	3 (exact)
.08	=	1−	.50	=	6 (exact)
.17	=	2+	.75	=	9 (exact)

Decimals of a foot can be easily converted mentally into inches by use of the equivalents in the above table, for example, 0.47 ft. = 0.50 − 0.03 = 6″ − ⅜″ = 5⅝″.

Other English units, applied occasionally in surveying are:

1 yard	=	3 ft
1 rod, perch or pole	=	16.5 ft.
1 chain (Gunther's chain)	=	66 ft.
1 mile	=	5280 ft. = 1760 yards
	=	320 rods = 80 chains

The basic unit for angular measurement is the degree of arc (1°), with 360° describing a full circle. This angular unit can also be expressed in minutes of arc (1′) or seconds of arc (1″) according to the following relations.

1° = 60′ = 3600″

01′ = 60″

Areas are expressed either in square feet or acres or square chains.

1 acre = 43,560 sq. ft. = 10 sq. chains.

1.8. The International System of Units (SI-System).—The basic unit of length for practically all measurements is the *metre;* even the foot is presently defined in terms of a metre. Originally the metre was defined as being the 1/10 millionth part of a meridianal earth quadrant and standardized in terms of the distance between markings on a metal bar at certain conditions. This bar is located in Paris, France and called the "International Prototype Metre." However, the accuracy of this standard is insufficient for modern precision measurements, so the International metre was redefined in 1960 as 1 650 763.73 wave lengths of the orange-red light of krypton 88.

As the SI system is a strictly decimal system, decimal multiples of SI units are denoted with certain Greek and Latin prefixes and then used as measuring units.

These prefixes are in table 1.8.1. Greek prefixes indicate values larger than the unit and Latin prefixes indicate values smaller than the unit.

Applying the prefixes of table 1.8.1 to the metre, one sees that 1/100 of a metre equals 1 centimetre (1 cm), that 1/1000 of

TABLE 1.8.1

PREFIXES IN SI-SYSTEM

Multiplying Factor		Prefix	Symbol
10^{12} =	1 000 000 000 000	tera	T
10^{9} =	1 000 000 000	giga	G
10^{6} =	1 000 000	mega	M
10^{3} =	1 000	kilo	k
10^{2} =	100	hecto	h
10^{1} =	10	deca	da
10^{0} =	1		
10^{-1} =	0.1	deci	d
10^{-2} =	0.01	centi	c
10^{-3} =	0.001	milli	m
10^{-6} =	0.000 001	micro	μ
10^{-9} =	0.000 000 001	nano	n
10^{-12} =	0.000 000 000 001	pico	p
10^{-15} =	0.000 000 000 000 001	femto	f
10^{-18} =	0.000 000 000 000 000 001	alto	a

a metre equals 1 millimetre (1 mm), and that 1000 metres are 1 kilometre (1 km). These are the commonly used units, and it is not expected that the others will have great usage.

The angular unit in the SI system is the radian (rad) which equals $180°/\pi$. The well known units degree, minute, and second of arc are universally permissible within the SI system and expressed in radians as follows:

$$1° = (\pi/180) \text{ radians}$$

$$01' = \frac{1°}{60} = \frac{1}{60}\left(\frac{\pi}{180}\right) \text{ radians} = \frac{\pi}{10{,}800} \text{ radians}$$

$$01'' = \frac{01'}{60} = \frac{1}{3600}\left(\frac{\pi}{180}\right) \text{ radians} = \frac{\pi}{648{,}000} \text{ radians}$$

Areas are expressed in square meters (m^2) or square kilometers (km^2). For practical purposes, two other units are used in land surveying, namely

1 are (1 a) = 100 m^2

1 hectare (1 ha) = 100 a = 10000 m^2

1.9. Conversion from English to SI-Units and Vice Versa.—Even with the approaching change from the English system to the SI system, the surveying profession will be required to convert frequently between these systems. Therefore, a few basic conversion factors are listed. It has to be kept in mind, that measurements in the United States conducted prior to 1959, when the foot was redefined, are based on the definition of 1893, which is smaller by 2 parts per million and is denoted as the "U. S. Survey Foot." For all plane surveys, however, this discrepancy can be neglected. The *present* conversion factors, all by the *new* definition, are:

1 inch = 25.4 millimetres (exactly)
= 2.54 centimetres (exactly)
1 foot = 0.3048 metre (exactly)
1 yard = 0.9144 metre (exactly)
1 mile = 1.609344 kilometres (exactly)
1 metre = 39.37007874 inches
= 3.280839895 feet
= 1.093613298 yards
1 square foot = 0.09290304 square metres (exactly)
1 acre = 4046.856 4224 square metres
1 square metre = 10.76391042 square feet
1 are = 1076.391042 square feet
1 hectare = 107639.1042 square feet
= 2.471053815 acres

The *old* relationship (prior to 1960) is retained as the U. S. Survey Foot. The basic relations to the SI-system are as follows:

1 metre = 39.37 inches (exactly) = 3.280833333 feet

With these values, all others are also slightly different, namely:

1 inch = 25.4000508 millimetres
= 2.54000508 centimetres
1 foot = 0.3048006095 metres (the U. S. Survey Foot)
1 yard = 0.9144018288 metres
1 mile = 1.609347219 kilometres
1 metre = 39.37 inches (exactly) = 3.280833333 feet
= 1.093611111 yards
1 square foot = 0.0929034116 square metres
1 acre = 4046.87261 square metre
1 square metre = 10.76386736 square feet
1 acre = 1076.386736 square feet

1 hectare = 107638.6736 square feet
= 2.47104393 acres

BASIC ERROR THEORY

1.10. PRECISION AND ADJUSTMENT OF ERRORS OF MEASUREMENTS.—While the cost of any survey is influenced by the prevailing physical difficulties such as obstructions along the measured lines, distance from the survey office and weather conditions, it is also greatly influenced by the precision required.* The degree of precision sought should depend upon the value of the land, present purpose of the survey and its future use. Many surveys are made with a precision not justified by these conditions; on the other hand, they are also made where an increase in precision would have paid well. Herein lies a phase of the art of surveying, namely, **intelligent selection of degree of precision for every one of the processes that comprise the entire survey.** It should be much higher for the basic survey lines than for the details which are to be attached to these lines by measurements.

As in all other engineering fields, surveying accuracy can be increased by increased efforts in terms of man-power and/or equipment and procedures. It is essential to know the accuracy requirements before conducting a survey and then select the most suitable approach. A knowledge of the errors that occur during measurements, and of their propagation, is essential for determining the accuracy of survey results.

The surveyor should distinguish carefully between errors which are of such a nature that they tend to balance each other and those which continually accumulate. The latter are by far the more serious. Suppose that a line 5000 ft. long is measured with a steel tape which is supposed to be correct but is really 0.01 ft. too long but that the error (uncertainty) in measuring a tape-length is, say, 0.02 ft., which may of course be a + or a −

*Various government agencies have developed specifications for the precision required for different types of surveys. Three designations are used: First order, second order and third order, in sequence of decreasing precision requirements. See "Specifications for Geodetic Control Surveys" prepared by the U. S. Bureau of the Budget in cooperation with governmental agencies concerned with surveys (excerpts of it are added in the Appendix).

error. This latter is an "accidental" error. There will then be 50 tape-lengths in the 5000-ft. line. A study of the laws governing the distribution of accidental errors (Method of Least Squares) shows that in such a case as this the number of errors that will probably remain uncompensated equals the **square root of the total number of opportunities for error,** i.e., in the long run this would be true. Hence the total number of such uncompensated errors left in the measurement of the line is $\sqrt{50} = 7$; and $7 \times \pm 0.02 = \pm 0.14$ ft., which is the error in the total line due to inaccuracies in marking tape-lengths. Since the error due to erroneous lengths of tape increases directly as the number of tape-lengths, and since these errors are not compensating, the total error in the line due to the fact that the tape is 0.01 ft. too long is $50 \times 0.01 = 0.50$ ft. The small (0.01) **cumulative** (systematic or constant) error is therefore seen to have far greater effect than the larger (0.02) **compensating** (accidental or random) error.

The usual cumulative errors in taping are those which are due to error in tape-length, temperature, and others. The amounts of these may be computed by the formulas given in articles 3.25 to 3.32. No attempt should be made at distributing compensating errors until the effect of cumulative errors has first been applied to all measurements.

The following articles deal primarily with the precision and adjustment of distances; the same principles apply to the adjustment of angles or of level circuits. See Art. 5.18, for a discussion of the accuracy of angle measurements; and Art. 9.26 for distributing errors in a level circuit.

1.11. Probability of Errors.—The theory of probability is based upon the following assumptions relative to the occurrences of errors:

1. Errors of small magnitude are more frequent than errors of large magnitude.
2. Positive and negative errors of equal magnitude are equally likely to occur.
3. The probability of very large errors occurring is small.
4. The mean of an infinite number of observations is the true value.

The purposes of the adjustment of observation are: first, to derive the best attainable value of a set of observations; and second, to provide results free from inherent discrepancies. It must be understood that the results of such adjustments are not the true values, but are the best values derivable from the given observations, as, for example, a simple mean obtained from a series of observations. In these adjustments, only accidental or compensating errors are considered; accumulative (systematic) errors may be still present in the adjusted results.

1.12. Adjustment of Accidental Errors.*—Assume that corrections for all accumulative errors have been made. Then, if the measurements are taken under like conditions, the most probable value (adjusted result) of a set of corrected observations is a simple mean; and this adjusted result should have usually the same number of significant figures as are present in any single observation. The following measurements were made of a line *A* to *B*.

(1) 615.42 ft., (2) 615.36 ft., (3) 615.44 ft.

If the mean, 615.4067 feet, were carried to four decimal places and published as such, a false impression of the precision of the mean measurement would be obtained; 615.41 ft. gives a more nearly correct idea of the degree of precision attained.

It does occur, nevertheless, that a measurement taken a very large number of times can be thus refined enough to be expressed with one more digit.

If n direct observations $M_1, M_2, M_3, \ldots M_n$ are made of the value of a quantity, M, and all are taken under like conditions,

---

*For fuller treatment of the theory of errors and the adjustment of measurements see Vol. II, Chap. 10, and the following references.

"The Adjustment of Observations," T. W. Wright and J. F. Hayford, D. Van Nostrand, 1906; "An Elementary Treatise on Precision of Measurement," W. S. Franklin, Franklin and Charles, Lancaster, Pa., 1925; "Precision of Measurements and Graphical Methods," H. M. Goodwin, McGraw-Hill Book Co., 1920; "Practical Least Squares," O. M. Leland, McGraw-Hill Book Co., 1920; "Introduction to Theory of Error," Yardley Beers, Addison-Wesley Publishing Co., 1953; "Surveying for Civil Engineers," Philip Kissam, Chapters 21 and 22, McGraw-Hill Book Co., 1956; B. A. Barry "Engineering Measurements," Wiley & Sons, 1964.

the best attainable value, M_0, is the arithmetical mean of the observations.

$$M_0 = \frac{M_1 + M_2 + M_3 + \cdots M_n}{n} = \frac{\sum M}{n}$$

The *Residual* (v) of an observation is the difference between an observed value of a measured quantity and the value of the mean.

There are several ways of denoting errors, the most common one being the mean square error, or standard deviation (σ). This value indicates that 67% of all errors of a series are smaller or equal to it while the remaining ones are larger. The 2σ indicates a 90% confidence region, and the 3σ a 99% confidence region.

From the Method of Least Squares, the following formulae may be derived:

The standard deviation of an observation, $\sigma = \sqrt{\dfrac{\sum v^2}{(n-1)}}$

and the standard deviation of the mean, $\sigma_0 = \sqrt{\dfrac{\sum v^2}{m(n-1)}}$

From the above measurements of the line to B, we have

	M	v	v^2
1.	615.42 ft.	+0.01	0.0001
2.	615.36 ft.	−0.05	0.0025
3.	615.44 ft.	+0.03	0.0009
Mean	615.41 ft.	$\sum v = -0.01$	$\sum v^2 = 0.0035$

While $\sum v$ should equal zero, the values in this example have only been carried to hundredths, and the $\sum v = 0.01$ is a small discrepancy due to rounding off the mean values.

$$\sigma = \sqrt{\frac{0.0035}{2}} = 0.04 \text{ ft.}$$

$$\sigma_0 = \sqrt{\frac{0.0035}{3 \times 2}} = \pm\frac{0.04}{\sqrt{3}} = \pm 0.03 \text{ ft.}$$

Final expression for length A to B is 615.41 ±0.03 ft., which is the best available value of the length and its standard deviation. The degree of precision is commonly expressed as a fraction (with unity in the numerator) as follows:

$$\text{Precision} = \frac{0.03}{615.41} = \frac{1}{20{,}000}$$

Usually one cannot expect that three measurements will give great statistical confidence in the result: the theory demands an infinity of measurements, a practical impossibility. One really depends upon having carefully used a standard procedure to make the measurements and when they seem to cluster properly about a mean one gains the confidence afforded by the procedure employed in measuring. (See Art. 1-13.)

As a further note, the standard deviation (σ) can be understood as the deviation from the mean that is likely to occur in the *next* measurement, if one more were to be made. And the standard error (σ_0) or standard deviation of the mean or root mean square error (for obvious reasons) can be regarded as the deviation from the mean that is likely to occur in another *series*. Really it is also properly construed as the likely deviation from the *true* value.

In reality, since we cannot know the true value, we can never know the true error in our measurement. But the use of statistical reasoning points out that in a multitude of careful measurements (having all systematic or cumulative errors corrected) we can expect that the mean of our result is bound to be close to the true value. And the σ_0 value gives us an idea of how close.

The acceptance of the probability theory gives one, therefore, the confidence to believe that residuals, or variations from the mean of the measurements, behave like and are of the same magnitude as errors, or variations from the true value of the quantity being measured. Having this confidence enables one to speak, frequently enough, of "sigma errors" or "errors" and to assign values or magnitudes to them, although the best that one can ever know is the magnitude of a residual. But it is permissible thus to speak of the $\sigma-$ error, the $2\sigma-$ error, the

$3\sigma-$ error, etc. One generally assumes that a standard deviation (σ or σ_0) means a one-sigma error.

Another error value which is used is the probable error (r). It is an error such that one-half the errors of the series are greater than it and the other half are less than it; that is, the probability of making an error greater than *r* is just equal to the probability of making an error less than *r*.

The numerical conversion from the one-sigma error to the probable is rather simple, namely

$$r = 0.6745\sigma$$

or, for most practical purposes: $r = \frac{2}{3}\sigma$

Whenever any quantity is made up of the algebraic sum of several other independent quantities each being subject to accidental errors, the combined standard deviation can be found by the relation

$$\sigma = \sqrt{\sigma_1^2 + \sigma_2^2 + \cdots}$$

An illustration of this would be a distance of 400 ft. made up of short sections, say, each 100-ft. tape-length long. If the accidental error in measuring a tape-length is ±0.008 ft., then for the total distance

$$\sigma = \sqrt{0.008^2 + 0.008^2 + 0.008^2 + 0.008^2} = 0.016 \text{ ft.}$$

This principle was stated in words in Art. 1.10.

In combining two or more sets of measurements to obtain a more accurate value, it is logical to give greater weight to that set which had a small error than to a set that had a larger error, and it is customary to apply the weight inversely proportional to the square of the error. For example, if the standard deviation in one set was ±.01 and in another it was ±.03, nine times the weight would be given to the first as compared with the second precision measurement.

The standard deviation of the final value may be computed from the formula

$$\sigma = \sqrt{\frac{w_1^2\sigma_1^2 + w_2^2\sigma_2^2 + w_3^2\sigma_3^2 + w_4^2\sigma_4^2}{(\sum w)^2}}$$

in which σ is the standard deviation of each set and w is the corresponding weight.

The weighted mean (M_0) is found from the relation

$$M_0 = \frac{w_1M_1 + w_2M_2 + w_3M_3 + w_4M_4}{w_1 + w_2 + w_3 + w_4} = \frac{\sum w \times M}{\sum w}$$

and the check on the calculation of the weighted mean is the fact that $\sum w \times v = 0$

The example of Fig. 1.12.1 shows weighting of measurements and precision values for a line AB taken on four different days.

	M	σ	Weighting of Measurements*	w	$w \times M$ for decimal part only	$w^2\sigma^2$
I	615.41	± 0.02 ft.	$\frac{M \times .0036}{.0004} = 9M$	9	3.69	0.0324
II	615.40	± 0.03	$\frac{M \times .0036}{.0009} = 4M$	4	1.60	0.0144
III	615.42	± 0.03	$\frac{M \times .0036}{.0009} = 4M$	4	1.68	0.0144
IV	615.41	± 0.02	$\frac{M \times .0036}{.0004} = 9M$	9	3.69	0.0324
			Totals	26	10.66	0.0936

$$\text{Weighted } M_0 = 615 + \frac{10.66}{26} = 615.41 \text{ ft.}$$

$$\sigma = \sqrt{\frac{w_1^2\sigma_1^2 + w_2^2\sigma_2^2 + w_3^2\sigma_3^2 + w_4^2\sigma_4^2}{(\sum w)^2}} = \sqrt{\frac{0.0936}{(26)^2}} = 0.01 \text{ ft.}$$

Final expression for A to B: 615.41 ± 0.01 ft.

$$\text{Precision} = \frac{0.01}{615.41} = \frac{1}{62{,}000}$$

FIG. 1.12.1 WEIGHTING OF MEASUREMENT.

If all the standard deviations are equal, however, the final value may be obtained by finding a simple mean of the obser-

*In the above problem the inverse square of each standard deviation is multiplied by such a factor (0.0036) as will give weights of integral whole numbers. Thus the weight given to the first set is $\frac{1}{0.0004} \times 0.0036 = 9$, and for the second the weight is $\frac{1}{0.0009} \times 0.0036 = 4$, etc. The factor 0.0036 is the least common multiple of the squares of the standard deviations.

vations, and by finding the standard deviation of this mean. The latter may be found by dividing the standard deviation of one determination by the square root of the number of precision values.

Thus if all the standard deviations of the above series had been 0.02 ft., the standard deviation of the final value would be $\frac{0.02}{\sqrt{4}}$ or 0.01 ft.

It will be observed that the precision reached in a series of sets of measurements of the same distance is more refined than the precision reached in one set.

In the case illustrated above four sets of observations gave a standard deviation of one-half the standard deviation for one set of observations. Since the standard deviation of the mean varies inversely as the square root of the number of observations, additional measurements beyond a certain number not only have little effect on the resultant standard deviation, but also greatly increase the cost of the survey.

Errors may not always remain accumulative or compensating, i.e., an accidental error may become a systematic error and *vice versa.* There is in reality no fixed boundary between the systematic and the accidental errors. Every accidental error has some cause, and if the cause were perfectly understood and the amount and sign could be determined, it would cease to be an accidental error, but would be classed as systematic.

However, when a systematic error is compensated, it very likely cannot be exactly compensated, thus a slight error remains of unknown magnitude and of unknown sign. This remaining error is, of course, defined as an accidental error. But though both magnitude and sign are unknown, they do not vary and therefore cause a cumulative error that through ignorance can only be treated as accidental. This can, however, result in unexplainable results sometimes.

On the other hand, errors which are either constant or systematic may be brought into the accidental class, or at least made partially to obey the law of accidental error, by so varying the conditions, instruments, etc., that the sign of the error is frequently reversed. If a tape is 0.01 ft. different from the standard, this produces a constant error in the result of a measurement. If, however, we use several different tapes, some of which are 0.01 ft. too long and others 0.01 ft. too short, this error may be positive or negative in any one case. In the long run these different errors tend to compensate each other like accidental errors.

1.13. Accuracy and Precision.—In determining the length of a line, or, in fact, in making any observation (linear or angular) it is customary to repeat the measurements. In plane surveying these repetitions are not always for the purpose of obtaining a more precise result, but for insuring accuracy within the limits of the precision desired. **Precision** implies **refinement of measurements such as would produce closeness in agreement** between several measurements, whereas **accuracy** implies **correctness** or closeness to the truth. Freedom from mistakes or carelessness is also essential. Precision is of no significance unless accuracy is also obtained.

An example may be cited in which mistakes have been introduced; a distance is measured three different times with close agreement to the nearest .01 ft.; the results were 5280.16, 5282.38 and 5281.47 ft. These three measurements were precise because they were observed to within the nearest hundredth of a foot, yet they were not accurate because they had different values in the whole foot digit. In other words, this was a precise but an inaccurate survey.

Now, suppose the measurements had been made to the nearest foot only and the three results were 5280, 5282 and 5281 ft., indicating reasonably accurate work for the precision attempted, because it was obviously free from error, but it was not a very precise series because the closeness of agreement required for the three independent measurements was 1 ft.

A surveyor may attempt to obtain a precision which the method he employs does not warrant. In such a case, after he has determined the mean of his measurements, he must use the figures only within the degree of precision that is indicated by the agreement between the independent measurements, which will clearly indicate to him the degree of precision probably obtained by the method employed.

1.14. Field Notes.—The best and most accurate survey is of no value if it is not properly recorded. This recording is performed on the job, i.e., in the field. Since all subsequent presentations and evaluations can only be based on this original field record, only the original observations, not derived quantities, are recorded. It is essential that the field notes be *neat, complete, legible, clear, accurate* and *unbiased.*

The frequent use of sketches is suggested. A hard pencil gives sharper lines and a neater and clearer appearance. Any wrong entry is not to be erased but to be crossed out in order to avoid misinterpretation at a later date. The use of ink pens is not recommended, since a few rain drops can ruin a day's work or more in this case.

It is essential that all entries be clearly labeled and that the names or observer and recorder, the date, weather conditions and any other pertinent data necessary for subsequent interpretation and evaluation be included. When sketches are used, an arrow pointing north should be added for easier orientation.

Any records copied from a piece of paper or written from memory into the fieldbook after the fieldwork has been completed is dangerous and bad practice, since it frequently leads to incorrect entries. Although at times loose leaf field records are preferred, it is general practice to use a fieldbook, whose hard cover and pocket size are specially designed for outdoor use. Since such a fieldbook may contain records for different projects, an updated index on one of the first pages is necessary. A complete address might also be helpful, especially in case of loss.

CHAPTER II

ANGULAR MEASUREMENTS

INSTRUMENTATION

2.1. The Surveyor's Compass.—The surveyor's compass (Fig. 2.1.1) is an instrument for determining the horizontal direction of a line with reference to the direction of a magnetic needle. The needle is balanced at its center on a *pivot* so that it swings freely in a horizontal plane. The pivot is at the center of a horizontal circle which is graduated to degrees and half-degrees, and numbered from two opposite zero-points each way to 90°. The zero-points are marked with the letters N and S, and the 90° points are marked E and W. The circle is covered with a glass plate to protect the needle and the graduations, the part enclosed being known as the *compass-box*. A screw is provided for raising the needle from the pivot by means of a lever. The needle always should be raised when the compass is lifted or carried, to prevent dulling the pivot-point; a dull pivot-point is a fruitful source of error. Both the circle and the pivot are secured to a brass frame, on which are two vertical sights so placed that the plane through them also passes through the two zero-points of the circle. This frame rests on a tripod and is fastened to it by means of a ball-and-socket joint. On the frame are two spirit levels at right angles to each other, which afford a means of leveling the instrument. This ball-and-socket joint is connected with the frame by means of a spindle which allows the compass-head to be revolved in a horizontal plane, and to be clamped in any position.

The magnetic needle possesses the property of pointing in a fixed direction, namely, the *Magnetic Meridian*. The horizontal angle between the direction of this meridian and of any other line may be determined by means of the graduated circle, and this angle is called the *Magnetic Bearing* of the line, or simply its *Bearing*. If the bearings of two lines are known the angle between them may be computed. Bearings are counted from 0° to 90°, the 0° being either at the N or the S point and the 90° either at the E or the W point. The quadrant

FIG. 2.1.1. THE SURVEYOR'S COMPASS.

in which a bearing falls is designated by the letters N.E., S.E., S.W., or N.W. For example, if a line makes an angle of 20° with the meridian and is in the southeast quadrant its bearing is written S 20° E.

2.2. The Pocket Compass.—The *pocket compass* is a small hand instrument for obtaining roughly the bearing of a line. There are two kinds, the *plain* and the *prismatic*. The former is much like the surveyor's compass, except that it has no sights. In the prismatic compass the graduations, instead of being on the compass-box, are on a card which is fastened to the needle (like a mariner's compass) and which moves with it. This compass is provided with two short sights and the bearing can be read, by means of a prism, at the same instant that the compass is sighted along the line.

2.3. METHOD OF TAKING A MAGNETIC BEARING.—The surveyor's compass is set up (and leveled) at some point on the line whose bearing is desired. The needle is let down onto the pivot, and the compass sights pointed approximately along the line. While looking through the two sights the surveyor turns the compass-box so that they point exactly at a lining pole or other object marking a point on the line. The glass should be tapped lightly over the end of the needle to be sure that the latter is free to move. If it appears to cling to the glass this may be due to the glass being electrified, which condition can be removed at once by placing the moistened finger on the glass. The position of the end of the needle is then read on the circle and recorded. Bearings are usually read to the nearest quarter of a degree although it is possible to estimate somewhat closer.

Since the needle stands still and the box turns under it, the letters E and W on the box must be reversed from their natural position so that the direct reading of the needle will give not only the angle but also the proper quadrant. Reference to Fig. 2.3.1 will show the following rule to be correct. **When the north point of the compass-box is toward the point whose bearing is desired, read the north end of the needle.** When the south point of the box is toward the point, read the south end of the needle. If a bearing of the line is taken looking in the opposite direction it is called the *reverse bearing*. **Reverse bearings should be taken.**

Since iron or steel near the instrument affects the direction of the needle, great care should be taken that the tape, axe, or marking pins are not left near the compass. Small pieces of iron on the person, such as keys or iron buttons, also produce a noticeable effect on the needle. Electric currents are a great source of disturbance to the needle and in cities, where electricity is so common, the compass is practically useless. Vehicles standing nearby or passing will attract the needle.

In reading the compass-needle, the surveyor should take

care to read the farther end of the needle, always looking **along** the needle, not across it. By looking at the needle sidewise it is possible to make it **appear** to coincide with a graduation which is really at one side of it. This error is called *parallax.*

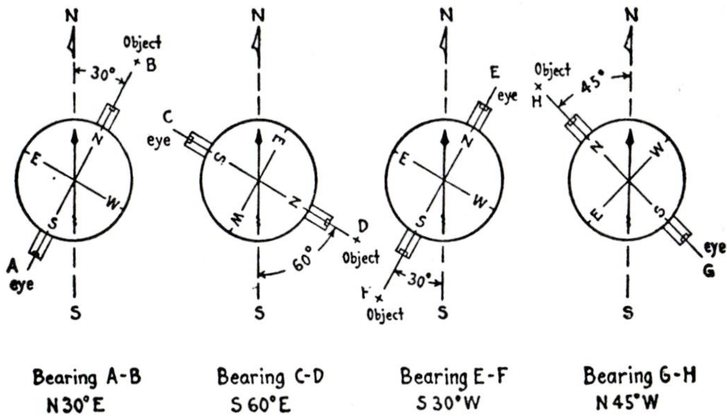

FIG. 2.3.1. ILLUSTRATING METHOD OF READING BEARINGS.

2.4. THE EARTH'S MAGNETISM.—Dip of the Needle.—The earth is a great magnet. On account of its magnetic influence a permanent magnet, such as a compass-needle, when freely suspended will take a definite direction depending upon the direction of the lines of magnetic force at any given **place** and **time.** If the needle is perfectly balanced before it is magnetized it will, after being magnetized, dip toward the pole. In the northern hemisphere the end of the needle toward the north pole points downward, the inclination to the horizon being slight in low (magnetic) latitudes and great near the magnetic pole. In order to counteract this dipping a small weight, usually a fine brass or silver wire, is placed on the higher end of the needle at such a point that the needle assumes a horizontal position.

2.5. DECLINATION OF THE NEEDLE.—The direction which the needle assumes after the counterweight is in position is called the magnetic meridian and this rarely coincides with the true meridian. The angle which the needle makes with the true meridian is called the *declination of the needle.* When the north end of the needle points east of the true, or geographical, north the declination is called *east:* when the north end of the needle points west of true north is has a *west* declination.

2.6. Variations in Declination.—The needle does not constantly point in the same direction. Changes in the value of the declination are called *variations of the declination.** The principal variations are known as the *Secular, Daily, Annual,* and *Irregular.*

The *Secular Variation* is a long, extremely slow swing. It is probably periodic in character but its period covers so many years that the nature of it is not thoroughly understood. The following table shows the amount of secular variation as observed in Massachusetts during two centuries. It cannot, of course, be predicted.

TABLE 2.6.1†

DECLINATIONS OF NEEDLE IN EASTERN MASSACHUSETTS
(Latitude 42°N, Longitude 70°W)
(prepared by Solid Earth Data Services Division, NOAA)

Year	Declination	Year	Declination	Year	Declination
1750	7°38′W	1840	9°38′W	1930	15°33′W
1760	7°17′W	1850	10°24′W	1940	15°51′W
1770	7°04′W	1860	11°07′W	1945	15°56′W
1780	7°01′W	1870	11°39′W	1950	15°52′W
1790	7°05′W	1880	12°08′W	1955	15°52′W
1800	7°23′W	1890	12°31′W	1960	15°54′W
1810	7°47′W	1900	13°02′W	1965	15°53′W
1820	8°17′W	1910	14°00′W	1970	15°53′W
1830	8°55′W	1920	14°49′W	1975	15°49′W

The *Daily Variation* consists of a swing which averages about 7 minutes of arc from its extreme easterly position at about 8 A.M. to its most westerly position at about 1:30 P.M. It is in its mean position at about 10 A.M. and at 5 or 6 P.M. The amount of daily variation is from 3 to 12 minutes according to the season and the locality.

The *Annual Variation* is a periodic fluctuation during 12 months, independent of the secular variation and is so small

*The *Declination* is usually called *Variation* by navigators.

†Table prepared by Geomagnetism Division, Coast and Geodetic Survey Environmental Science Services Administration.

(about one minute a year) that it need not be considered in surveying work.

Irregular Variations in the declination are caused chiefly by magnetic storms. These variations are uncertain in character and unpredictable. They are, however, usually observed whenever there is a display of the Aurora Borealis. Such storms often cause variations of from 10 to 20 minutes in the United States, and even greater variations in higher latitudes.

2.7. Isogonic Chart.—If lines are drawn on a map so as to join all places where the declination of the needle **is the same at a given time,** the result will be what is called an *isogonic chart.* (See Fig. 2.7.1.) Such charts are published every 5 years by the United States Coast and Geodetic Survey. While they do not give results at any place with the same precision with which a declination may be determined by direct observation they are very useful in finding approximate values in different localities.

In the isogonic chart of the United States, Fig. 8, the full lines are isogonic lines for each whole degree of declination and the dashed lines show annual rates of change in declination. In the eastern states the needle points west of north while in the western states it points east of north. The agonic line of no declination is shown by heavy solid line, which passes (in 1965) through Georgia, Kentucky, Michigan and Ontario. In 1965 the declinations were decreasing west of the agonic line and increasing east of it, except for an area north of the no-change (double-dash) line in the Quebec and New Brunswick area where they were decreasing.

2.8. Observations for Declination.—For any survey where the value of the present declination is important, it should be found in the field by determining the true bearing of a line by observation on Polaris or on the sun, and comparing that true bearing with the observed magnetic bearing. The value found at one place may be considerably different from that at a place only a few miles distant. The method of finding the meridian by observation on the Pole-Star is described in Art. 8.3. The solar observation for meridian is described in Art. 8.9.

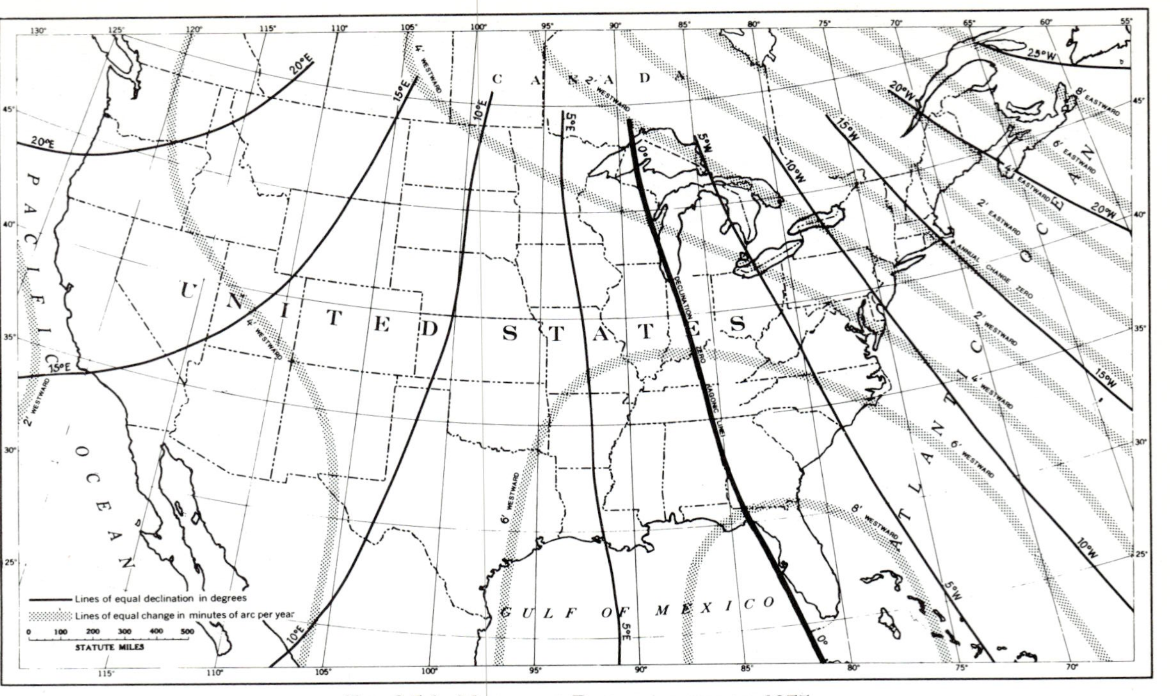

FIG. 2.7.1. MAGNETIC DECLINATION FOR 1975.
(Derived from USGS Map 1-911)

2.9. Detecting Local Attraction of the Needle.—As the needle is always affected by masses of iron near the compass it is important that the bearings in any survey should be checked. This is most readily done by taking the bearing of any line from both its ends or from intermediate points on the line. If the two bearings agree it is probable that there is no local magnetic disturbance. If the two do not agree it remains to discover which is correct.

In Fig. 2.9.1 suppose that the compass is at *A* and that the bearing of *AB* is N 50°¼ E, and with the compass at *B* the bearing *BA* is found to be S 49° W. It is evident that there is

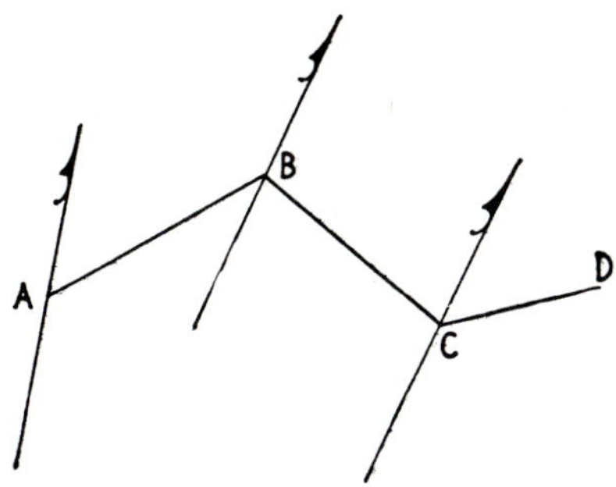

Fig. 2.9.1. Diagram Illustrating Local Attraction at *A*.

local attraction at one or both points. In order to ascertain the correct magnetic bearing, turn the compass toward a point *C* which is apparently free from magnetic disturbance, and observe the bearing of *BC*, which is, say, S 72° E. Now move the compass to *C* and observe the bearing *CB*. If this is N 72° W it indicates that there is no local attraction at *C* or *B*, hence S 49° W is the correct bearing of line *BA*, and there is 1°¼ error in all bearings taken at *A*. If the bearings of *BC* and *CB* had not agreed it would have been necessary to take the bearing and reverse bearing of a new line *CD*. This process is continued until a line is found whose bearing and reverse bearing differ by exactly 180°. In order to be certain, however, that these latter bearings are really free from attraction several other stations should be occupied. Since local attraction drags the needle a **fixed** amount from the magnetic meridian it follows that **the angles at any one point computed from the bearings are not affected by local attraction.**

2.10. Calculating Angles from Bearings.—In calculating the angle between two lines it is necessary only to remember that the bearing is always counted from the meridian, either N or S, toward the E and W points. In Fig. 2.10.1

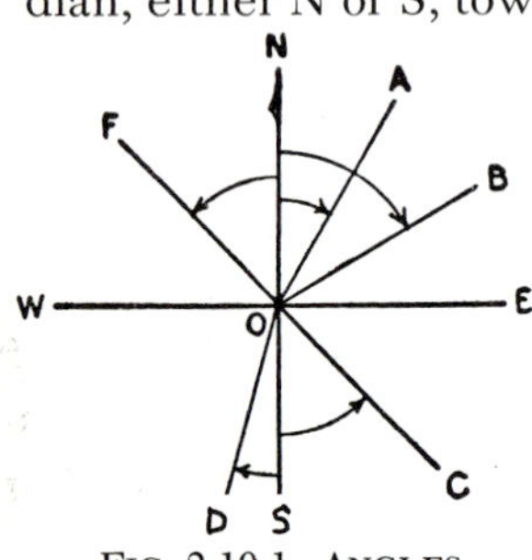

Fig. 2.10.1. Angles Marked with Arcs are Bearings.

AOB = difference of bearings.
AOC = 180° — sum of bearings.
AOD = 180° — difference of bearings.
AOF = sum of bearings.

For example: if the magnetic bearing of OA = N 35° 15′ E and that of OD = S 18° 30′ W, the angle AOD = 180° − (35° 15′ − 18° 30′) = 163° 15′.

For a discussion of the method of calculating bearings see Art. 15.4.

2.11. Remarks on the Use of the Compass.—Great accuracy cannot be expected from the magnetic compass, and it is no longer used on surveys of importance. It is, however, quite important that the surveyor of the present day should understand the instrument, its peculiarities and its limitations. (1) The subdivision of nearly all of the land in the older states was made by means of the compass. Since the surveyor who is making re-surveys must re-trace the old lines as nearly as possible he should know the changes in the declination and the possibilities of the compass as a surveying instrument. (2) Calculations for area, rectangular coördinates, etc., from bearings, are the same as those used when the compass was employed for measuring the bearings. These calculations are better understood if the surveyor is familiar with the compass. (3) In spite of its lack of precision the compass is still much used for obtaining rough checks on angles or azimuths taken with the transit.

2.12. Common Sources of ERROR in Compass Work.—

1. Iron or steel, or electric current, near compass.
2. Parallax in reading needle.

2.13. Common MISTAKES.—

1. Reading wrong end of needle.
2. Not letting needle down on pivot.
3. Reading the wrong side of the 10th degree, viz., reading 61° instead of 59°.

THE THEODOLITE OR TRANSIT

2.14. General Description of a Theodolite.—The surveying instrument designed for measuring horizontal and vertical angles is the theodolite, commonly called the transit. The basic parts are shown schematically in Fig. 2.14.1, being principally the base and leveling head (tribrach), the lower plate and outer hollow spindle, the upper plate (alidade) with standards, and the telescope.

The tribrach is attached to the tripod head or surveying pillar and leveled with the aid of foot screws. Base and tribrach are connected solidly and act as one piece. The lower plate fits with the outer spindle into the base and carries the horizontal circle. When clamped to the inner spindle, it rotates with the upper plate; when clamped to the base it remains stationary with the base. The inner spindle carries the upper plate and fits into the outer spindle. This upper plate, or upper part (alidade) rotates around the vertical axis (azimuth axis) with respect to the base. It consists basically of the plate level, the telescope standards, and the vertical circle. The standards bear the telescope which is rotatable around the horizontal axis (elevation axis).

All angular measurements are based on three principal axes, namely: COLLIMATION AXIS, AZIMUTH AXIS and ELEVATION AXIS, as shown in Fig. 2.14.2. While the latter two are physical axes around which the telescope can rotate, the collimation axis extends in the direction of the telescope, representing the line of sight. Its inclination from a horizontal plane is measured as vertical angle, while the angle between its projection onto a horizontal plane and a given reference direction (the zero-direction) is measured as horizontal direction. The difference between two (horizontal) directions is called a horizontal angle.

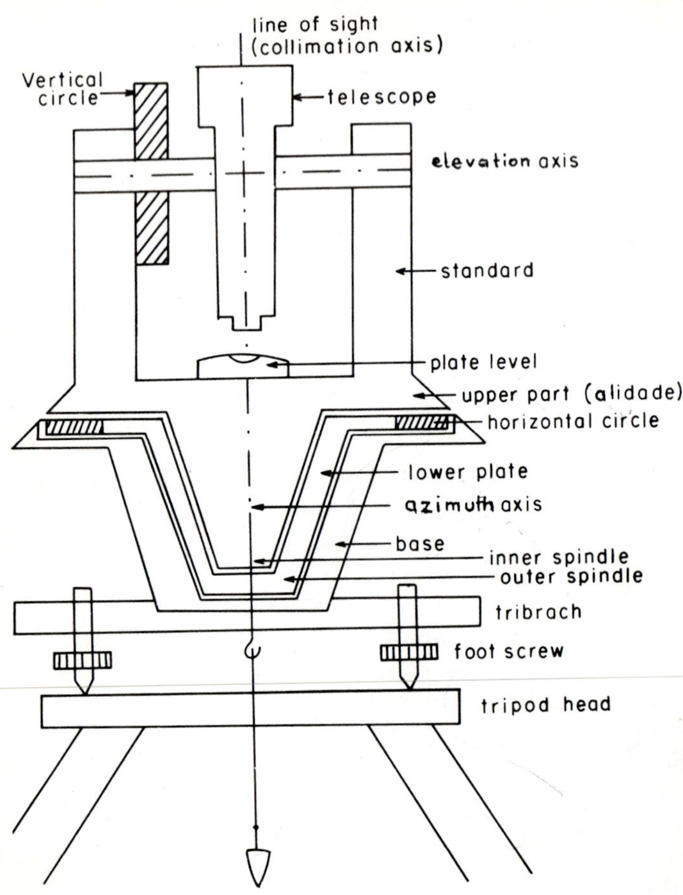

FIG. 2.14.1. SCHEMATIC SECTION OF A THEODOLITE.

2.15. CLASSIFICATION OF THEODOLITES.—As mentioned before, accuracy can be increased with proper, more costly instrumentation. For angular measurements, the limiting factor is the degree of graduation of the circle and the efficiency of the reading mechanism. For highly accurate work, other factors, such as telescope resolution must also be considered.

The engineer's transit is the simplest type of theodolite, equipped usually with graduated metal circles and having

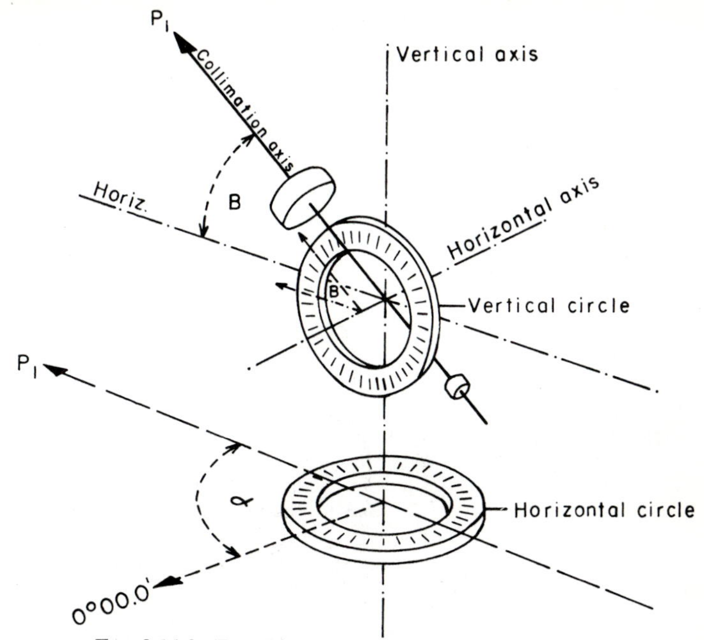

FIG. 2.14.2. THE TRIAXIAL SYSTEM OF A THEODOLITE.
α = horizontal direction
β = vertical angle

verniers to subdivide the least circle division. Recently, with the perfecting of accurate glass circles, theodolites can be built with much smaller divisions and be easily read without eyestrain. This is because light can be directed right through the glass circle and magnified greatly as contrasted with the lesser brightness possible using only light reflected from an illuminated metal circle.

It is obvious that the smaller the circle the finer the graduation for the same least count. Therefore, high precision surveys demand instruments which are large and heavy. On the other hand, many practical surveys where medium to low accuracy is sufficient permit the use of smaller, simpler, and cheaper theodolites. According to the achievable accuracy, theodolites can be classified in four categories as shown in table 2.15.1.

TABLE 2.15.1

THEODOLITE CLASSIFICATION

Category	Reading direct	Reading estimated	Reading Device
1 (small theodolite)	10′	01′ to 20″	Vernier
2 (engineer's theodolite)	01′	10″ to 06″	scale microscope
3 (precision theodolite)	01″–02″	00.5″	optical micrometer
4 (high precision theod.)	01″	00.1″	optical micrometer

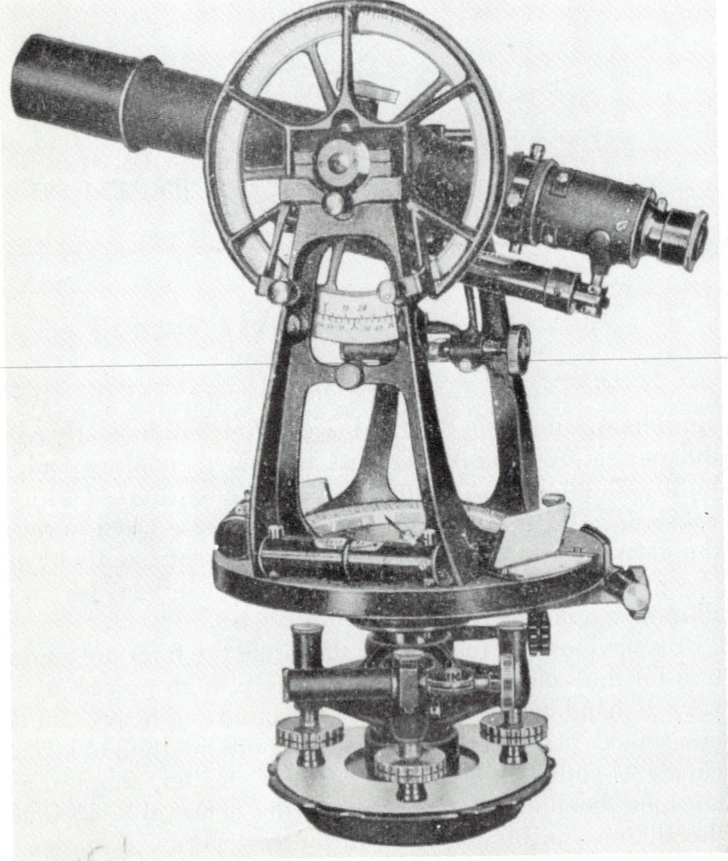

FIG. 2.15.1. ENGINEER'S TRANSIT.

FIG. 2.15.2. ENGINEER'S THEODOLITE, AN INSTRUMENT OF CATEGORY II.

(Courtesy, Kern Instruments, Inc.)

Instruments of category 1 are used for simple construction and local surveys. All transits belong to this category.

Engineers' theodolites (category 2) are used for property surveys, small scale traversing, and simple trigonometric work, while precision theodolites (category 3) are primarily used for 2nd and 3rd order triangulation and large traverses.

A high precision theodolite (category 4) is needed for 1st order triangulation, astronomical observations and precision engineering surveys, such as deformation measurements.

In table 2.15.2 most of the available theodolites are listed, according to the company which manufactures and/or distributes them. Due to constantly changing developments, this list has to be considered as an overview rather than a complete set.

FIG. 2.15.3. PRECISION THEODOLITE, AN INSTRUMENT OF CATEGORY III.
(Courtesy of Wild, Heerbrugg)

TABLE 2.15.2

COMMON THEODOLITES, CLASSIFIED IN CATEGORIES

Category	Designation of Instrument	Manufacturer/Distributor
1	GAVEC, TEKAT	Breithaupt, W. Germany
	Transits 6060, 6140, 6170	Dietzgen–Franke, U.S.A.
	FT5	Fennel, W. Germany
	Transits 62–20, –30, 112 Gurley Nikon #5	Gurely, U.S.A.
	Paragon Transits	Keuffel & Esser, U.S.A.
	Transits 10C, 550, 625	Lietz, U.S.A.
	TOM	Mashpriboritorg, U.S.S.R.
	Transit TR303	Path (David White/Path), U.S.A.
	STN 0	Slom, France
	Transits No. 10, BT20, GT60	Sokkisha, Japan
	V11	Vickers, U.K.
	T0	Wild, Switzerland
	Theo 080	Zeiss (Jena)/Zena Co. U.S.A.
	Th5	Zeiss (Oberkochen)/Keuffel & Esser, U.S.A.
2	Tts	Askania, W. Germany
	TEAUT	Breithaupt, W. Germany
	6020–A1, –A6	Dietzgen–Franke, U.S.A.
	Tatha FT1A, Natha FTS	Fennel, W. Germany
	Gurley–Nikon NT–2	Gurley, U.S.A.
	K1-A, DKM-1	Kern, Switzerland
	TM10C, 20C, T60C	Lietz, U.S.A.
	TT4, OTIII	Mashpriboritorg, U.S.S.R.
	Te–C1, –D1, –D2, –E6	MOM, Hungary
	T–202	Path (David White/Path), U.S.A.
	Microptic 1	Rank, U.K.
	4149-A, 4150-NE	Salmoiraghi, Italy
	STN–27	Slom, France
	TM10, 20	Sokkisha, Japan
	V22	Vickers, U.K.
	T16	Wild, Switzerland
	Theo 020A	Zeiss (Jena)/Zena Co., U.S.A.
	Th3, 4	Zeiss (Oberkochen)/Keuffel & Esser, U.S.A.

TABLE 2.15.2 (continued)

COMMON THEODOLITES, CLASSIFIED IN CATEGORIES

Category	Designation of Instrument	Manufacturer/Distributor
3	TU	Askania, W. Germany
	6020–A2	Dietzgen–Franke, U.S.A.
	Sethe FT2N	Fennel, W. Germany
	DKM–24	Kern, Switzerland
	TM1	Lietz, U.S.A.
	TB–1	Mashpriboritorg, U.S.S.R.
	Te–B3	MOM, Hungary
	Microptic 2	Rank, U.K.
	4200–A	Salmoiraghi, Italy
	Travistock 2	Vickers, U.K.
	T2	Wild, Switzerland
	Theo 010A	Zeiss (Jena)/Zena Co., U.S.A.
	Th2	Zeiss (Oberkochen)/Keuffel & Esser, U.S.A.
4	DKM-3	Kern, Switzerland
	OT-02	Mashpriboritorg, U.S.S.R.
	Microptic 3	Rank, U.K.
	Geod. Travi.	Vickers, U.K.
	T3	Wild, Switzerland
	Theo 002	Zeiss (Jena)/Zena Co., U.S.A.

2.16. The Telescope.—The essential parts of the telescope are the *objective*, the *cross-hairs*, and the *eyepiece*. (See Figs. 2.17.1 and 2.18.1.)

The line of sight, or *line of collimation*, is the straight line drawn through the optical center of the objective and the point of intersection of the cross-hairs. When light from any point falls upon the objective, the rays from it are bent and brought to focus at a single point called the *image*. The cross-hairs are placed in the telescope tube near where the image is found, as shown in Fig. 2.17.1 where the fine lines representing rays of light intersect.

2.17. External Focusing Telescope.—In this type (Fig. 2.17.1) the objective lens is fixed in a tube, which slides inside the main tube, moved by means of a rack-and-pinion screw so as to bring the plane of the image of the object into coincidence with the plane of the cross-hairs. The instrument is so constructed that the motion of this tube is **parallel** to the line of sight. The eyepiece is simply a microscope for viewing the image and the cross-hairs. The adjustment of the eyepiece and the objective, so that the cross-hairs and the image can be seen clearly at the same time, is called *focusing*.

2.18. Internal Focusing Telescopes.—Internal focusing telescopes are similar to those previously described with the exception that an additional lens is placed between the cross-hairs and the

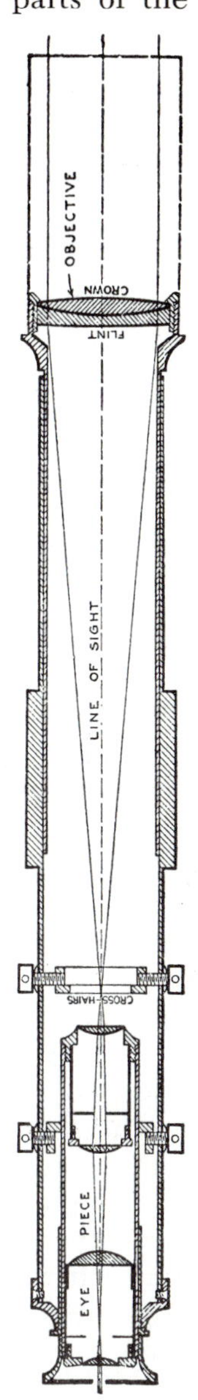

Fig. 2.17.1. Longitudinal Section of an External Focusing Transit Telescope (Erecting Eyepiece). Light rays meet at Cross-hairs; in Eyepiece they become inverted and again erected before reaching the eye.

objective (Fig. 2.18.1). This lens is usually a concave (negative) lens. In this type of telescope both the objective and the cross-hairs are fixed in position in the telescope tube. The negative lens is assembled in a tube which fits inside the tele-

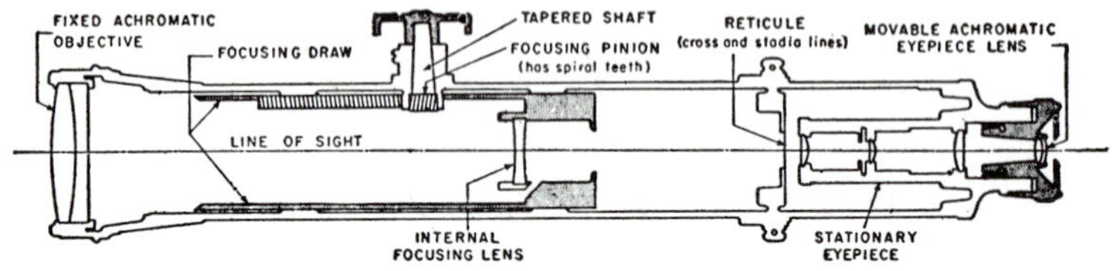

FIG. 2.18.1. SECTION OF INTERNAL FOCUSING TRANSIT TELESCOPE.

scope. The telescope is focused by moving this interior tube back and forth by means of a rack and pinion to bring the image into the plane of the cross-hairs. This arrangement has three advantages: (1) the barrel of the telescope is in one piece, closed at one end by the objective and at the other by the eyepiece and therefore there is less likelihood of dust or moisture getting into the telescope; (2) the telescope is better balanced on its horizontal axis when sighting short distances because the moving parts are nearer to the telescope axis; (3) the stadia constant (Art. 3.38) is negligible.

In addition to external and internal focusing telescopes, there are also prismatic and mirror telescopes (Fig. 2.18.2 shows schematically the mirror telescope of the Kern DKM-3 theodolite). This allows a reduction in the length of the telescope and therefore permits a more compact design of a theodolite.

2.19. The Objective.—The objective might consist of a simple bi-convex lens, like that shown in Fig. 2.19.1, which is formed by the intersection of two spheres. The line OO' joining the centers of the two spheres is called the *optical axis.* If rays parallel to the optical axis fall on the lens those near the edge of the lens are bent, or refracted, more than those near the center, so that all the rays are brought to a focus (nearly) at a point F on the optical axis called the *principal focus*. If light falls on the lens from any direction there is always one of the rays such as AC or BD which passes through the lens without

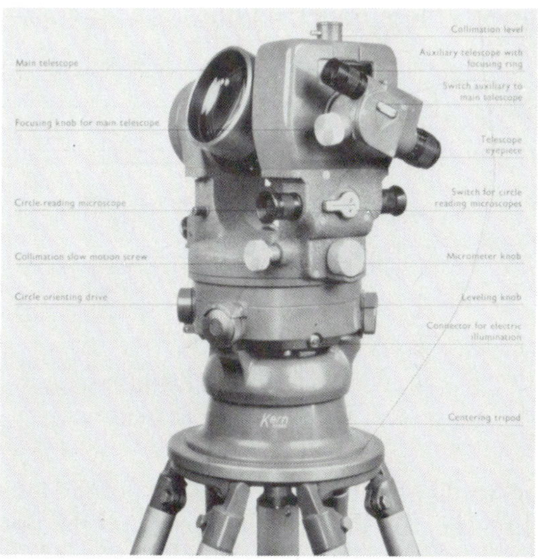

FIG. 2.18.2. HIGH PRECISION THEODOLITE, AN INSTRUMENT OF CATEGORY IV.
(Courtesy, Kern Instruments, Inc.)

permanent deviation, i.e., it emerges from the other side of the lens parallel to its original direction. All such rays intersect at a point x on the optical axis which is called the *optical center*. It should be noted that the vertex of the angle formed by rays from A and B is not point x but a point (m) nearer to the surface of the lens. This is called the "Nodal Point." Rays from C and D intersect at (n) the other nodal point, to the left of x. The planes through these points perpendicular to the optical axis are called "Nodal Planes."

A simple bi-convex lens does not make the best objective because the rays do not all come to a focus at **exactly** the same point. This causes indistinctness (spherical aberration) and also color (chromatic aberration) in the field of view, particularly near the edges. This difficulty is overcome by using a combination of two or more lenses of "crown" and of "flint" glass (having different indexes of refraction) as shown in Fig. 2.17.1; this arrangement, also called achromatic lens, very nearly corrects these imperfections.

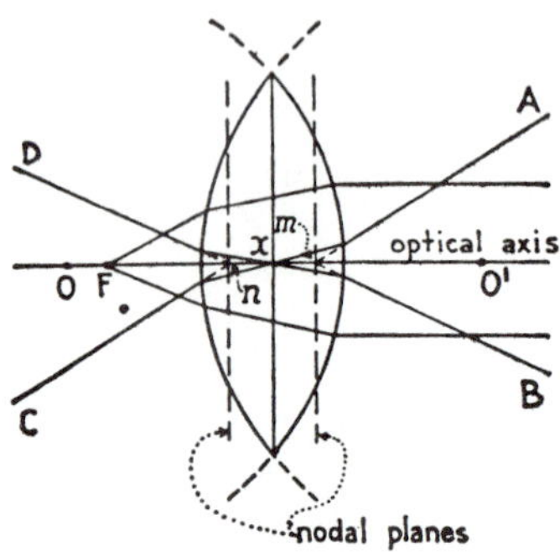

FIG. 2.19.1. BI-CONVEX LENS.

The position of the image of any point is located on a straight line (nearly) through the point and the optical center; it will be seen therefore that the image formed by the objective is inverted.

2.20. Cross-Hairs.—In older theodolites, the cross-hairs consist of two very fine spider threads stretched across a metallic ring at right angles to each other and fastened by means of shellac. The cross-hair ring, diaphragm, or reticule (Fig. 2.20.1) is held in place by four capstan-headed screws

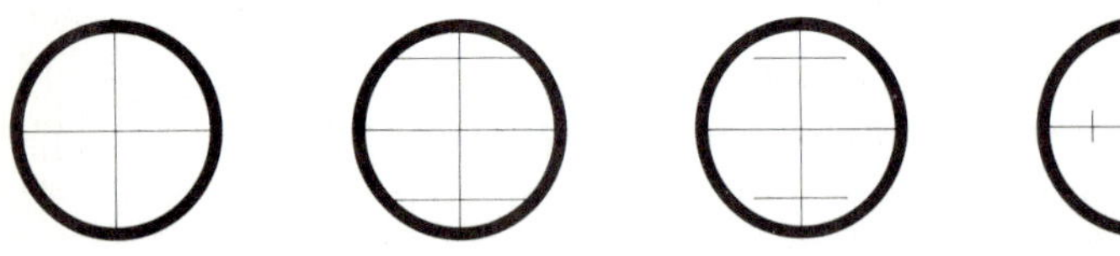

FIG. 2.20.1. COMMON RETICULE PATTERNS FOR THEODOLITES.

which permit of its being moved vertically and horizontally in the telescope tube. The holes in the tube through which the screws pass are elongated enough to allow a slight rotary motion of the ring in adjusting. A disadvantage of the spider thread is that it may slacken when moist. It requires some skill to mount it with just the right tension. Fine platinum wires and tungsten wires are sometimes used. In fact in all modern theodolites the cross-hairs consist of fine lines etched onto a thin glass plate.

2.21. Line of Sight.—When changing the focus, the cross-hair has to be moved with respect to the objective lens by moving either the objective lens or the internal focusing lens. For all practical purposes, the line of sight (collimation axis) is defined by the centers of lens and cross-hair (actually, the collimation axis is the locus on which the center of the cross-hair is imaged, independent of the focus setting). If multiple horizontal cross-hairs are used separated by a constant amount Δs, they will always include the same angle $\Delta\alpha$. This principle is utilized for stadia measurements (see Art. 3.10).

An advantage of the glass reticule is that the stadia hairs may be marked by short lines in the middle of the field of view, and so not easily be confused with the horizontal hair which is ruled all the way across. Disadvantages of the glass type are the slight loss of light caused by the glass and the possibility of dust or film collecting on the glass and fogging the image.

2.22. Eyepiece.—The eyepiece may be either of two kinds, that which shows an inverted image or that which shows an erect image (Fig. 2.17.1). In the inverting type all objects appear to be upside down and those on the right appear to be on the left. An erecting eyepiece requires two more lenses than the inverting eyepiece, which add to its length and also absorb light; yet the erecting eyepiece is generally used on American transits. With the same length of telescope, however, a greater magnifying power and a clearer definition of the image can be obtained by the use of the inverting eyepiece. Instruments of European design usually have relatively short telescopes with optical systems which give an inverted image.

2.23. Magnifying Power.—The magnifying power of a telescope is the amount by which an object is increased in apparent size. It is equal to $\dfrac{\tan \frac{1}{2}A}{\tan \frac{1}{2}a}$ $\left(\text{or nearly equal to } \dfrac{A}{a}\right)$, A being the angle subtended by an object as seen through the telescope and a the angle as seen by the unaided eye.

The magnifying power may be measured in two ways. (1) The dimensions on a graduated rod will appear magnified when viewed through a telescope. If, with one eye at the telescope, the rod is viewed directly with the other eye it will be noticed that one space as viewed through the telescope will appear to cover a certain number of spaces as seen with the naked eye. This number is approximately the magnifying power of the telescope; it is expressed as so many "diameters."

(2) Viewed through a telescope wrong-end-to, an object is reduced in apparent size in the same ratio that it is magnified when seen through the telescope in the usual manner. Measure with a transit some small angle A between two distant points and then place the telescope to be tested in front of the transit, with its objective close to the objective of the transit. By careful focusing and sighting with the transit through the telecope to be tested, the two distant stations will be observed in the field of view.

Measure this small angle a between the stations with the transit. Then the Magnifying Power $= \dfrac{\tan \frac{1}{2}A}{\tan \frac{1}{2}a}$. The magnifying power of the ordinary transit telescope is between twenty and thirty.

2.24. Resolution of the Telescope.—Resolution is a measure of indicating how far two objects have to be separated in order to be recognized as separate objects. Like magnification, it is an instrumental constant for a given telescope, depending primarily on the opening (aperture) of the objective lens. For most theodolite telescopes the resolution amounts to a few seconds of arc. This means that two objects viewed through a telescope with a resolution of 3″ have to be approximately 2 inches apart in order to be recognized as separate objects, because of

$$\mathrm{R} = \frac{3'' \cdot 3 \times 10^6 \text{ cm}}{206265''} \approx 4.4 \text{ cm} = 1.7 \text{ inches}$$

2.25. Field of View.—The field of view is the angular space that can be seen at one time through the telescope. It is the angle subtended at the optical center of the objective by the opening in the eyepiece. In the ordinary transit this angle is about one degree, but in some instruments it is considerably more.

When the magnification is great, the field of view is narrowed. In some of the more precise theodolites the field of view is so narrow that pointing is difficult because the instrument man cannot find the target. To assist, for example in the Kern DKM 3, an auxiliary telescope of lesser magnification is incorporated to enable the operator to recognize enough terrain features to aim the instrument. The main telescope is then used for the precise sighting.

2.26. Circle Reading Devices.—As indicated earlier, the accuracy of reading the circle is the limiting factor for the accuracy of angular measurements. There are several different types of circle reading devices which are discussed in the subsequent articles. The simplest one, the vernier, is used for simple instruments such as transits and is therefore explained in more detail than the others.

2.27. Principle of the Vernier.—The vernier is a device for determining the fractional part of the smallest division of a scale more accurately than it can be estimated by eye. The principle of the vernier is illustrated in Fig. 2.28.1, which shows a scale that is to be read to tenths of its smallest division. A distance equal to 9 spaces on the scale is made to represent the *full* length of the vernier, and this is then divided into 10 equal parts. Consequently one space on the vernier equals 9/10 of a space on the scale; hence *ab* equals 1/10 of a scale space, *cd* equals 2/10, etc. On the vernier there is an "index" mark, which points to the actual scale reading. If the vernier in Fig. 2.28.1 is now raised until *a* coincides with *b*, the reading becomes 3.01. If the vernier is moved until *c* coincides with *d*, the reading becomes 3.02, as shown in Fig. 2.28.2. Referring to Fig. 2.28.3 the procedure to follow in reading the scale by use of the vernier is: first, follow up the scale in the direction of increasing numbers and find the nearest scale reading **below** the vernier index, which is 3.1 in Fig. 2.28.3, then follow up the vernier in the **same direction** in which the scale was read until the line on the vernier is found which coincides with a line on the scale. In this instance it is 6. The reading, therefore, is 3.16. It will be observed that it is

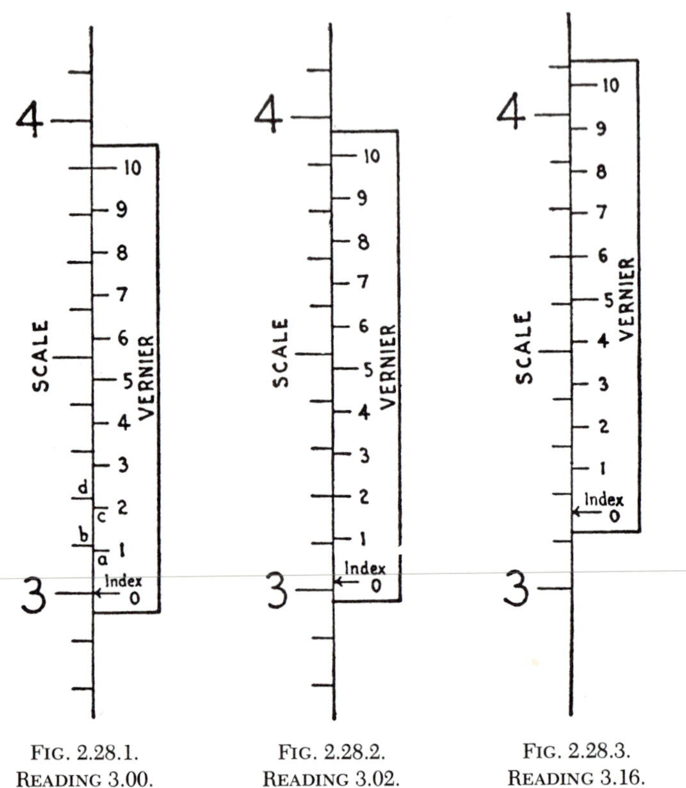

FIG. 2.28.1. READING 3.00.

FIG. 2.28.2. READING 3.02.

FIG. 2.28.3. READING 3.16.

impossible to have more than one coincidence. The kind of vernier just described is often used in reading level rod targets (Art. 4.4).

2.28. CIRCLE GRADUATIONS AND VERNIERS OF TRANSITS.—The horizontal circles of modern transits are divided into spaces of half-degrees (30 minute), thirds of a degree (20 minute), or sixths of a degree (10 minute). The circle graduations are usually numbered at 10° intervals continuously from 0° to 360°, in both directions from 0°. The inside row of figures increases in a clockwise direction, and the outside row in a counterclockwise direction. Verniers are provided for reading the angle closer than the smallest circle division, as explained in Art. 2.27. The graduation of the vernier in every case depends upon the subdivision of the circle. For example, to read to 01′ on a circle graduated in half degrees (30′) spaces, the space between each line on the vernier should be 29/30 of the 30′ arc space on the circle; that is, an arc composed of 29 subdivisions of 30′ each (14° 30′ on the circle) is subdivided into 30 equal parts to obtain the space between division lines on the vernier. Therefore, one division on the vernier is 01′ less in angular measurement than one division on the circle.

Fig. 2.28.4 shows a 01′ vernier set with its zero (index) opposite the zero (360°) of the circle, ready for measuring an angle. The lines on the vernier on both sides of zero fail to

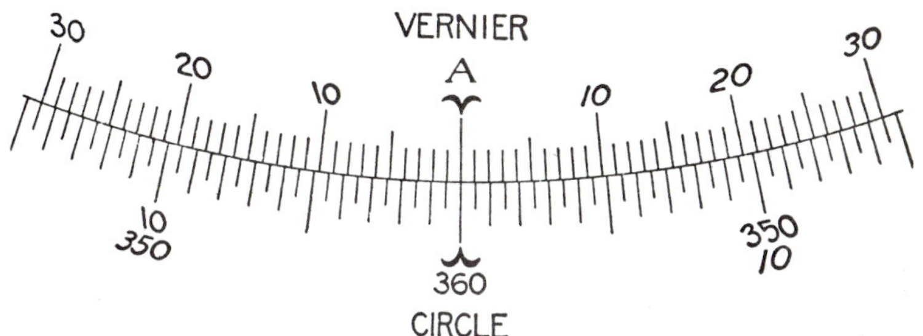

Fig. 2.28.4. One-Minute Vernier Set at 0°.

match the lines on the circle by 01′; the second lines on the vernier on both sides of zero fail to match the lines on the circle by 02′, and so on. In setting exactly at 0°, this can best be accomplished by noting that the vernier lines on both sides of the zero fail to match the circle divisions by the **same amount,** i.e., they fail to match symmetrically.

In measuring horizontal angles, the outer circle remains fixed in position while the inner circle (with the vernier) moves with the telescope and thus travels along the stationary graduated (outer) circle. If the telescope is turned to the right (clockwise), the zero of the vernier moves clockwise around the circle; if the telescope is turned to the left (counterclockwise), the zero (index) of the vernier moves counterclockwise.

The arrow at the zero of the vernier always points to the reading on the circle. In Fig. 2.28.5 one can see that the reading on the outer row is approximately 9¼°. The exact reading is obtained by reading the degrees and half-degrees on the cir-

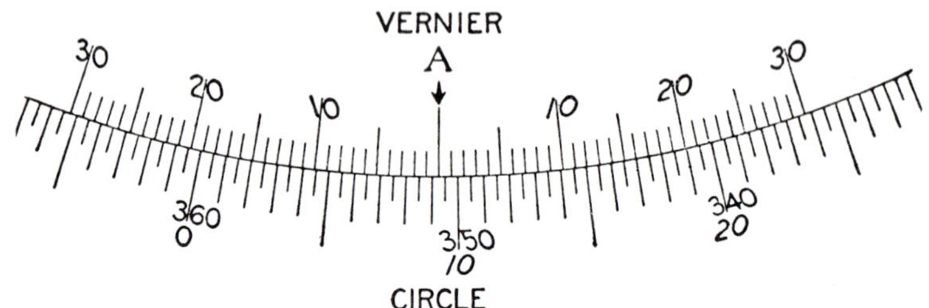

FIG. 2.28.5. READING OUTER ROW OF FIGURES 9° 16′ (angle measured counterclockwise), READING OF INNER ROW, 350° 44′ (angle measured clockwise).

cle, and then crossing over to the vernier and passing along it in the **same direction** in which the angle has been turned until the graduation on the vernier is found that coincides with a graduation on the circle. The reading on the vernier at this coincidence line is the number of minutes to be added to the degrees and half-degrees that were read from the circle, i.e., 9° 00′ + 16′ = 9° 16′ in Fig. 2.28.5. A reading glass is commonly employed for reading the vernier.

Notice that in both Figs. 2.28.4 and 2.28.5 there are two verniers, one running in each direction from the common zero-point; this is called a *double vernier.* Double verniers are used in order that the circle may be read either clockwise or counterclockwise, depending upon which way the telescope moved in describing the angle being measured.

If the circle is divided into 20′ divisions and it is desired to read the angle to the nearest ½ minute, or 30″, the vernier is so constructed that 39 of the circle divisions are divided into 40 equal parts. Thus each vernier division is 39/40 of one of the circle divisions, or 1/40 less than a circle division; and since each circle division represents 20′ of angle, 1/40 ×20′ = ½′, or

30″ (Fig. 2.28.6). In a similar manner a vernier may be devised for any type of circle graduation.

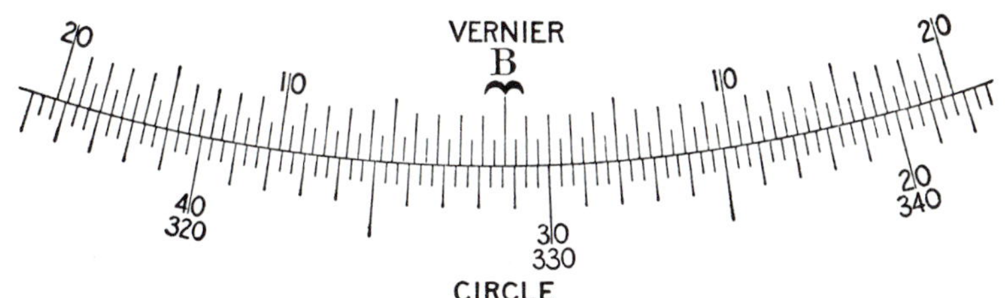

FIG. 2.28.6. READING OF INNER ROW, 31° 17′ 30″.
READING OF OUTER ROW, 328° 42′ 30″.

Fig. 2.28.7 shows a circle divided into 10′ spaces and a vernier in which 59 divisions of the circle are divided into 60 parts to make a vernier reading to 10″. This is a *single vernier*

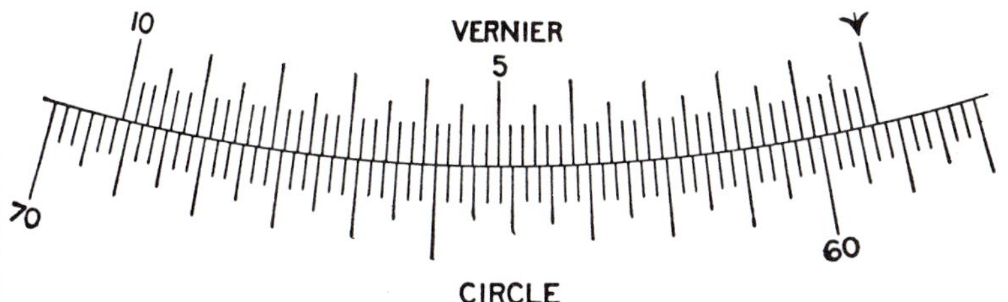

FIG. 2.28.7. SINGLE VERNIER. READING 59° 15′ 50″.

such as might be found on an instrument used for triangulation. It can be read in only one direction.

Fig. 2.28.8 illustrates a *folded vernier* which is sometimes used on transits or plane table alidades where there is not enough space for a double vernier. The folded vernier is read like the ordinary 1′ vernier except that if a coincidence is not reached by passing along the vernier in the direction in which the circle is numbered, it is necessary to go to the other end of the vernier and continue in the same direction, toward the center, until the coincidence is found.

2.29. Scale Microscope.—Transits and simpler theodolites (categories 1 and 2) use the simplest optical reading device, a scale microscope. In a separate eyepiece, which is usually located alongside the telescope eyepiece, a small segment of

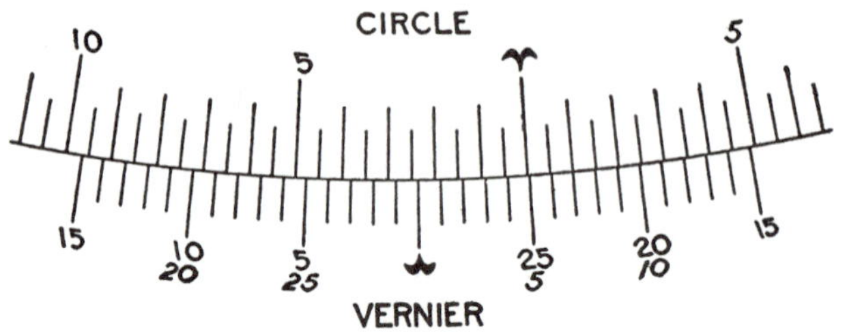

FIG. 2.28.8. FOLDED VERNIER. READING 2° 23′.

the circle is imaged. Much as with a cross-hair, a reading line is superimposed onto this image. To facilitate still easier reading estimation, usually a fine graduated scale supplements this line (see Figs. 2.29.1 and 2.29.2). The image is very bright,

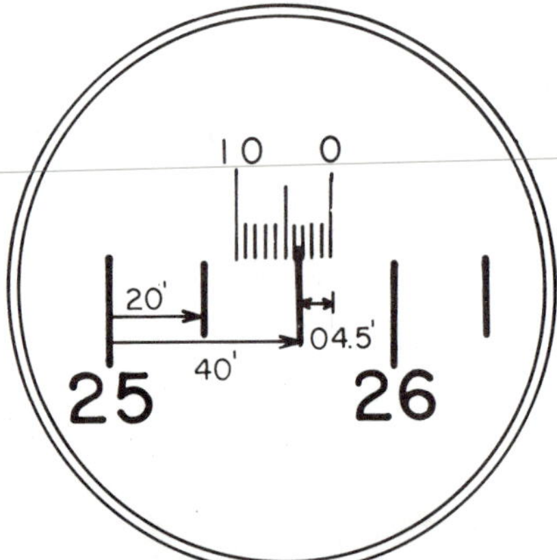

FIG. 2.29.1. SIMPLE SCALE MICROSCOPE.
READING 25° 44′.5.

despite the high magnification, because daylight is directed through the glass circle by the use of a mirror.

The real advantage of all optical reading devices is the capability of greater precision with the higher brightness/

magnification ratio possible through the use of transparent circles. As an added bonus, the observer is not inconvenienced by having to change positions when reading the circle,

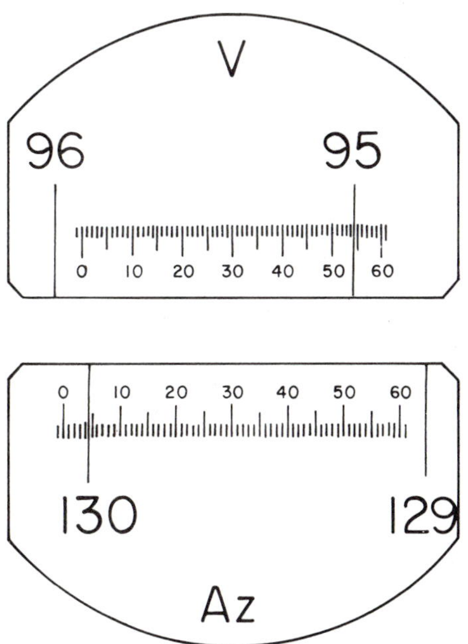

FIG. 2.29.2. SCALE MICROSCOPE OF THE CATEGORY II THEODOLITE.
(Courtesy of Wild, Heerbrugg)

since the circle reading eye piece is right beside the sighting telescope.

2.30. Optical Micrometer.—All of the precise theodolites (categories 3 and 4) utilize the principle of the optical micrometer for circle reading. (See Figs. 2.30.1 and 2.30.2.) By optical means, two diametrically opposite circle positions are superimposed into one image. Rather than estimating fractions of the circle subdivision, one measures these by a micrometer. The instrument man rotates a coincidence knob to shift the two opposite images optically until their respective lines appear to coincide. The strip value can then be read directly at

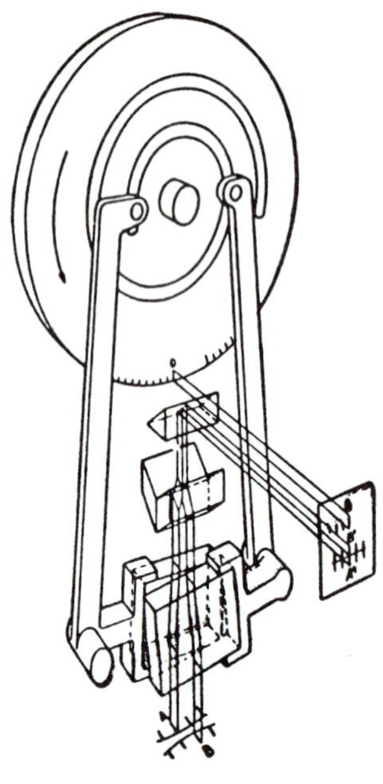

FIG. 2.30.1 OPTICAL MICROMETER.

the micrometer drum (see Fig. 2.30.3). Both the coinciding and reading requires experience. In order to reduce this, newer instruments separate coincidence and reading completely as also shown in Fig. 2.30.3.

2.31. Automatic Readout.—Newest developments in certain of the high precision theodolites tend toward full automation of the angular readings. In this case the circle is digitally coded and then photographed instead of being read. The angular value is then automatically decoded and transferred directly onto punched paper tape for processing in an electronic computer. Some experimentation has also been done with direct digitizing of angular information punched directly onto paper tape in the field.

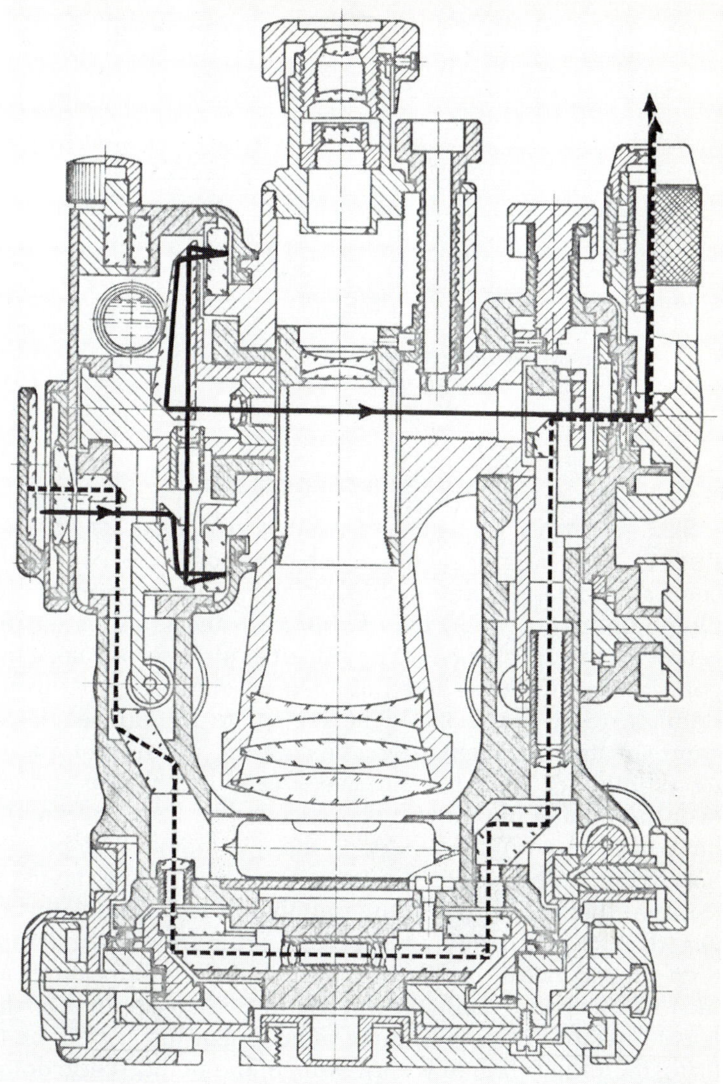

Fig. 2.30.2. Path of Light Rays in an Optical-Reading Theodolite Showing How Opposite Sides of the Circles Are Simultaneously Read and Averaged.

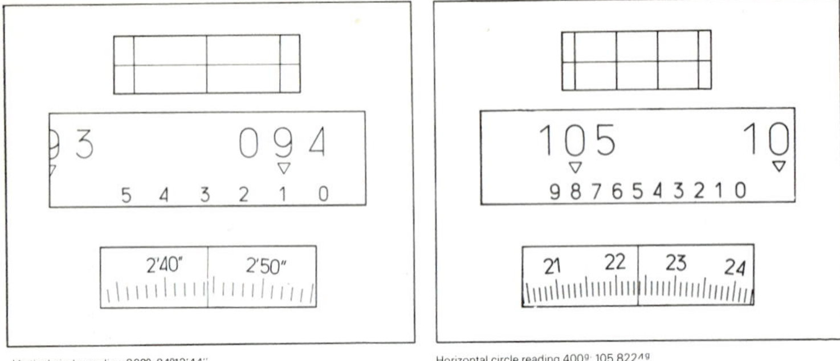

Vertical circle reading 360°: 94°12'44"

Horizontal circle reading 400^g: 105.8224^g

FIG. 2.30.3. OPTICAL MICROMETER READING OF A CATEGORY III THEODOLITE.

(Courtesy of Wild, Heerbrugg)

USE OF THE THEODOLITE

2.32. SETTING UP AND CENTERING THE TRANSIT OVER A POINT.—The simplest and most common device used to center a transit or theodolite above a given point is the plumb-bob, suspended by a string from the hook under the center of the instrument. The transit is first set up near the point, with the legs spread at such an angle as will insure stability. (See Fig. 2.32.1) The foot plate should be placed approximately horizontal by adjusting the position of one or more of the legs. For the purpose of centering the instrument over the point a plumb-bob is suspended by a string from the hook under, and in the center of, the instrument. The plumb-bob string is tied with a slip knot, or some slipping device is used, so that the height of the plumb-bob may be readily adjusted. Without changing the general set-up the instrument is then moved bodily until the plumb-bob is within about an inch of the point. Sufficient pressure is then applied to the legs to insure firmness in the ground and at the same time to bring the plumb-bob to within ⅛-in. of the point. While doing this care should be taken to see that the foot plate remains nearly horizontal, thereby requiring little turning of the leveling screws to make the plates level. The final centering of the plumb-bob over the point is made by means of the shifting

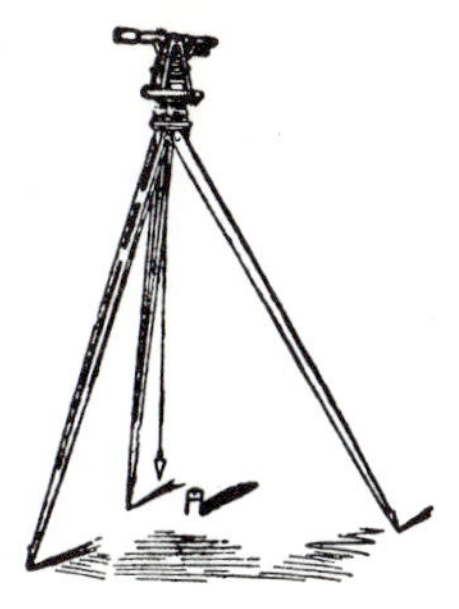

1st. Set instrument on ground near point, with legs spread to give convenient height. Without reference to point, move legs so that plates are nearly level. Tripod leg screws should be firm but not binding.

2nd. Pick up instrument bodily without disturbing relative position of legs and head.

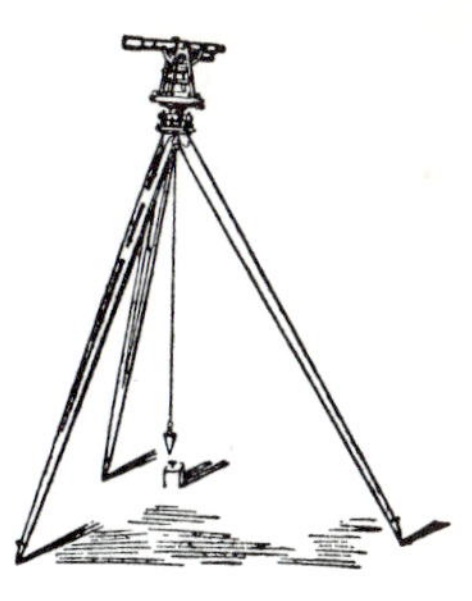

3rd. Place down again with plumb-bob within 1″ of point and bring bob to within $\frac{1}{4}''$ to $\frac{1}{2}''$ of point by shifting legs slightly.

4th. Swinging one or more legs in arc of circle will approximately level head without changing materially the position of bob.

5th. Press each leg firmly into ground about same amount, watching position of bob so that it will finally return to within $\frac{1}{8}''$ to $\frac{1}{4}''$ of the tack. Plates should still be nearly level.

6th. Bring bob exactly over tack by loosening two *adjacent* leveling screws and shifting head. Then tighten leveling screws and proceed to level the plates.

FIG. 2.32.1. STEPS IN SETTING UP A TRANSIT.

head of the transit. By loosening two adjacent leveling screws the transit head with the attached plumb-bob is lowered a slight amount, and in this position it may be moved about a quarter of an inch in any direction from its center. When the plumb-bob is over the point, the leveling screws are tightened just enough so that they have a firm bearing. The foregoing steps are illustrated in Fig. 2.32.1.

While the above-mentioned six steps may be necessary for the beginner, the first two may be performed by grasping two legs of the tripod as shown in Fig. 2.32.2. Place the third leg on the ground at such a point, with respect to the stake, that when the other two legs are allowed to touch the ground the foot plate will be level and the height of the telescope will be convenient for the transitman and, at the same time, will allow sufficient spread of the legs to keep the instrument stable. It obviously requires practice to perform this step effectively.

FIG. 2.32.2. FIRST STEP IN SETTING UP TRANSIT. (ALTERNATE METHOD.)

When setting up the transit on sloping ground the usual practice is to place one leg of the tripod up, and the other two legs down, the slope.

Many tripods have telescopic legs, an advantage, especially in hilly terrain. By slightly changing the length of one of the legs, a small shift similar to step 3 in Fig. 2.32.1 can be obtained. This also leads to a small tilt of the tripod head, which normally can be overcome during the levelling process.

2.33. Other Centering Devices and Their Use.—There are other devices which have the advantage of not being influenced by wind. The most common one is the optical plummet, which consists basically of a 45° prism diverting the line of sight by 90°, thus enabling the observer to look down in the vertical axis of the theodolite (see Fig. 2.33.1) and see the point on the ground. When the theodolite is leveled, the line of sight is vertical. A cross-hair or centering circle then indi-

FIG. 2.33.1. PRINCIPLE OF THE OPTICAL PLUMMET IN THEODOLITE. ALSO VISIBLE IS THE OPTICAL TRAIN OF PRISMS AND LENSES FOR READING THE VERTICAL AND HORIZONTAL CIRCLES.

(Courtesy of Wild, Heerbrugg)

cates the point where the vertical axis hits the terrain. By shifting the theodolite horizontally, the vertical axis is shifted so as to be centered above the desired point. Often,

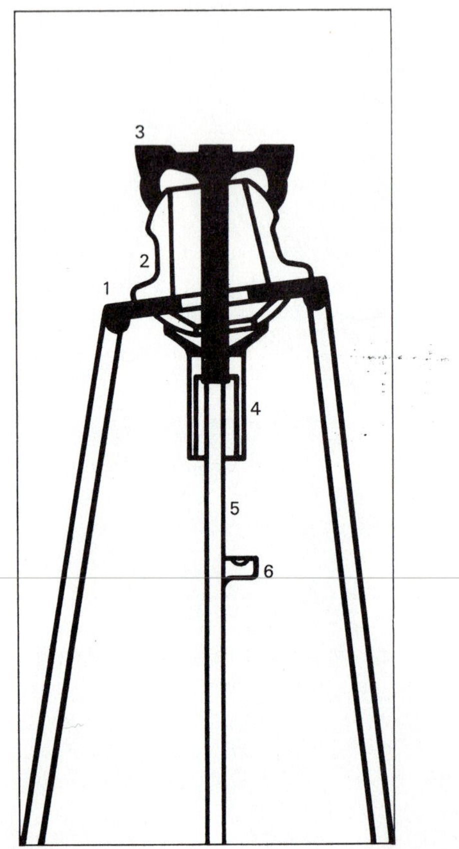

FIG. 2.33.2. THE KERN CENTERING TRIPOD.

(Courtesy, Kern Instruments, Inc.)

the optical plummet is used for fine centering only after the rough centering has been done with the plumb-bob.

The Kern centering tripod is unique due to the fact that the simple operation of the centering automatically establishes the coarse leveling of the instrument. Consequently, the centering tripod provides a simplified method of operation that results in a speed of set up not found in any other tripod.

Construction

The construction principle of the centering tripods is shown in Fig. 2.33.2. The tripod plate (1) does not carry the instrument directly but rather the shifting tripod head (2) which has a spherical zone on top. The instrument support plate (3) rests on the spherical zone. The support plate is rigidly attached at right angles to the telescoping centering rod (5). The centering rod carries an adjustable bullseye level (6) that can be checked by reversal. The instrument support plate and the tripod head are clamped to the tripod plate with the clamping grip (4).

Operation

The following simple method of operation is made possible by the construction features described above: the tripod is set over the ground point and the point of the centering rod inserted into the center of the station mark. Then by shortening or extending the legs, the centering rod is brought close to the vertical as indicated by the rough centering of the bullseye level.

The tripod head is then shifted on the tripod plate until the bullseye level is accurately centered. This makes the centering rod vertical and the instrument support plate horizontal. When the clamping grip is tightened the instrument may be placed on the tripod. The instrument is automatically centered over the ground point with a mean error of ±0.5 mm and is level within about 1′. A fraction of a turn of the leveling knobs is now sufficient to center the plate level of the instrument.

In this way the observer can set up quickly and easily on all kinds of terrain. Compared to the conventional tripod, set up time is cut at least 50% and where the terrain is difficult or the observer inexperienced, a much greater time saving is realized.

2.34. Leveling the Transit.—In this context, we have to distinguish between theodolites with three foot screws, which are sufficient for leveling an instrument, and instruments (mainly transits) with four foot screws, an arrangement for symmetry having one redundant supporting screw. Since the procedures vary, both will be discussed here.

To level a tranist equipped with four leveling screws, turn the instrument about its azimuth axis until one plate bubble is lined up parallel to a line through two diagonally opposite level screws. The second bubble will then be parallel to the other pair of leveling screws. Grasp a pair of opposite screws between the thumbs and the forefingers and center the bubble approximately. This is done by **uniformly** turning this pair of leveling screws in such a way as to move the thumbs either toward or away from each other, as shown in Fig. 2.34.1, thereby tightening one screw by the **same** amount that the other is loosened. This tilts the leveling head while at the same time sustaining a definite support for it on both screws. The screws should bear firmly on the foot plate at all times, but they should never be allowed to bind. Similarly, the other bubble is centered by means of the other pair of screws. Alternate this process until **both** bubbles are in the center. Then observe the position of the plumb-bob; if it has moved off the point, reset by means of the shifting head and again level. Should one of the screws bind, the other should be turned faster, and if this does not release the tight screw, one of the screws of the other pair should be loosened slightly. When using the leveling screws it is convenient to note that the bubble moves in the direction of the left thumb.

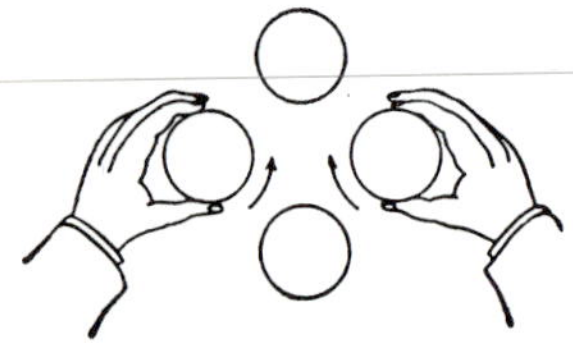

FIG. 2.34.1. SHOWING HOW TO MANIPULATE FINGERS WHEN LEVELING.

After both bubbles have been centered in one position, turn the plate about the vertical axis through an angle of 180°. If the bubbles do not remain central, it indicates that the adjustment of the spirit level is imperfect, but the instrument can still be leveled (without the necessity of adjusting the plate levels) by moving the leveling screws until **each bubble moves half-way back to the center of its tube.** If this is done correctly, the plates will be truly horizontal and the bubble will then stay in the **same place in the tube** (not in the center), as the plates are turned slowly around the vertical axis.

It is always possible to level a transit by means of the long bubble attached to the telescope, and since this bubble is more sensitive than the plate bubbles, the result will be more accurate. The telescope is first turned so that it is over one pair of opposite leveling screws, the vertical arc reading zero, and the bubble is brought to the center of its tube by means of the leveling screws. The telescope is then reversed 180° about the vertical axis (without touching the clamp or the tangent screw on the standard). If the bubble is no longer central it should be moved half-way back by means of the tangent screw, the other half by means of the leveling screws. By additional trials the adjustment is perfected so that the telescope may be reversed without causing any change in the position of the bubble. The adjustment over the other pair of screws is made in the same way. When the leveling is completed it should be possible to turn the telescope in any azimuth without any change in the position of the bubble. It will save time if the transit is first leveled as accurately as possible with the plate bubble.

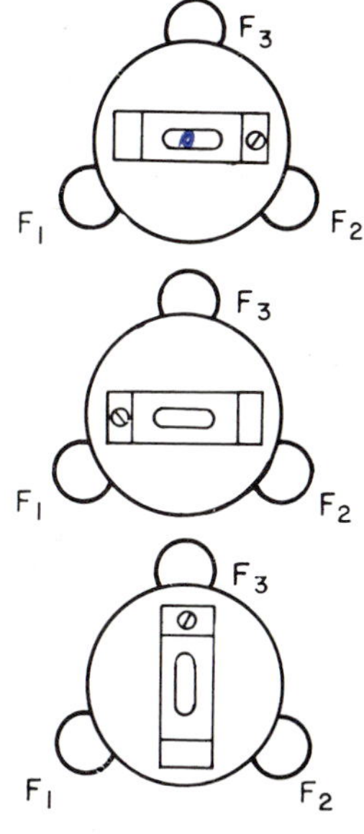

FIG. 2.35.1. LEVELING WITH 3 FOOT SCREWS.

2.35. Leveling a Theodolite.—For three foot screws, one plate level is used. As a first step, the theodolite is rotated until the plate level is parallel to two foot screws F_1 and F_2 (see Fig. 2.35.1). Then the bubble is centered using either one or both foot screws (F_1 and F_2). Then the theodolite is rotated by 180°. If the axis of the bubble were perpendicular to the azimuth axis of the instrument (vertical axis), the bubble remains centered. If the bubble does move, it means that the azimuth axis is inclined by a small amount. This is eliminated by bringing the bubble halfway back, using the foot screws. This new position (truly level position) of the bubble indicates the position in which the azimuth axis is vertical. One must either remember to use it at this off-center position or adjust the bubble to true center position at this time (see also Art. 2.34).

Now the theodolite is rotated by 90° and the instrument is leveled using the third foot screw (F_3) only, by bringing the bubble to make it read level. With the instrument thus leveled in two perpendicular directions, it is level for any direction. However, as it is difficult to obtain exact parallelism between plate level and foot screws, the procedure is to be repeated for checking and small refinements found necessary.

2.36. To Measure a Horizontal Angle.—Basically, the difference between the readings obtained from sighting to two targets constitutes a horizontal angle. If the left target can be sighted with a circle reading of zero, the reading for the right target will immediately be the horizontal angle. Although in a few theodolites (e.g. Wild T-2) the horizontal circle can be directly rotated by any desired amount, usually an indirect approach of alternately clamping and loosening the lower plate (carrying the horizontal circle) against the upper plate (carrying the telescope) has to be used. This procedure is best explained with an example, in this case using a vernier transit.

Different reading and/or clamping devices as found on other theodolites simplify the procedure somewhat.

After setting the instrument up over the point, first set the zero of one of the verniers opposite the zero of the circle. This is done by turning the two plates (with both clamps loose) until the two zeros are nearly opposite, clamping the plates firmly together with the **upper clamp**, and then bringing the two into exact coincidence by means of the tangent screw which goes with the **upper clamp.** If a line on the vernier is coincident with a line on the circle then the two adjacent lines on the vernier will fail to coincide with the corresponding lines on the circle by **equal amounts** (Art. 2.28). Hence the coincidence of any line on the vernier with a line on the circle can be more accurately judged by examining also the adjacent divisions and noting that they are symmetrical with respect to the coincident lines. A pocket magnifier, or "reading glass," is generally used for setting and reading the vernier. **Never touch the clamp after a setting has been made by means of the tangent screw.** In setting with the tangent screw it is better to do this by a **right-hand turn,** i.e., by turning the screw in the direction which **compresses** the spring against which it works. If the screw needs to be turned back, instead of turning it to the exact setting turn it back too far and then bring it up to the accurate setting with a right-hand motion, thereby insuring a firm bearing of the spring against the screw. The two plates which are now clamped in proper position are free to turn together about the vertical axis. Turn to the first object and point the telescope at it approximately by looking over the top of the telescope. When turning the instrument so as to sight the first point it is good practice to touch the lower plate only. Focus the telescope by moving the eyepiece until the cross-hairs are distinct and then moving the objective until the image is distinct. It is sometimes convenient to point the telescope at the object when focusing the cross-hairs so that they can be readily seen.* Test for parallax by moving the eye slightly from one side to the other. Move the telescope until the vertical cross-hair is very nearly on the point. It is better to use that part of the cross-hair which is near the center of the field of view. Clamp the lower plate by means of the **lower clamp,** and set exactly on the point by the **lower** tangent screw. The line of sight is now fixed on the first object. To measure the angle loosen the **upper clamp,** turn the telescope to the second point, and focus the objective if necessary. Set nearly on the point, clamp the **upper** plate, and set the vertical cross-hair exactly on the point by means of the **upper** tangent screw. The angle is then read by **using that vernier which was set at 0°.**

---

*If the eyepiece is focused on the cross-hairs with the telescope pointing at the sky, as is frequently done, they will be found to be nearly in focus when looking at the object; but for accurate work the eyepiece should be focused on the cross-hairs when the objective is in focus on the object.

The tangent screws should **not** be used to move the plates over large angles. Acquire the habit of setting closely by hand and using the tangent screw for slight motions only.

2.37. To Measure an Angle by Repetition.—This method is used to utilize telescope resolution and high pointing accuracy even with instruments with low precision reading devices (e.g. vernier or scale microscope). Rather than reading a single angle, the angular measurement is repeated several times, and is added mechanically along the horizontal circle. The final reading represents a multiple of the angle, which is obtained by division with the proper number. The least count is thereby reduced according to the same ratio. In this case, it is convenient to start with a zero reading. The procedure is again explained for a vernier transit. The first point is set with zero reading as explained in Art. 2.36. Then, after the first angle has been measured leave the two plates clamped together, loosen the **lower clamp** and turn back to the first point. Set on the first point, using the **lower clamp** and **its** tangent screw. Then loosen the upper clamp and set on the second point, using the **upper clamp** and **its** tangent screw, thus adding another angle, equal to the first one, to the reading on the circle. Repeat this operation, say, six times. The total angle divided by six will give a more precise result than the first reading. Suppose that the angle is actually 18° 12′ 08″; if a "one-minute" instrument is being used it is impossible to read the 08″ on the vernier, so the reading will be 18° 12′. Each repetition will add 08″ (nearly) and after the 6th repetition, the amount will be 48″ which will be read as 1′. After the sixth pointing the total angle will then be read 109° 13′ which divided by 6 gives 18° 12′ 10″, a result which is correct to the nearest 10″. (See notes below.)

Stn.	Target	No. of Repetitions	Circle Reading	Final Angle
1	A	—	0°–00′	
	B	1	(18°–12′)	
		6	109°–13′ ÷ 6 =	18°–12′ 10″

To eliminate errors in the adjustment of the transit the above process should be repeated with the instrument reversed and the mean of the two values used. (See Art. 2.54). It is customary to take only the first and sixth readings, but as a check against mistakes it is well for the beginner to examine the vernier reading after each repetition and see that ½ the second reading, ⅓ the third, etc., nearly equals the first reading.

Repetition has also the advantage of eliminating, to a great extent, errors of graduation. If an angle is about 60° and is repeated 6 times it will cover a whole circumference. If there are systematic errors in the graduations the result is nearly free from them. The effect of accidental, or irregular, errors of graduation is decreased in proportion to the number of repetitions. In the best modern instruments, however, the errors of graduation seldom exceed a few seconds.

Little is gained by making a very large number of repetitions as there are systematic errors introduced by the action of the clamps, and the accuracy apparently gained is really lost on this account. Three repetitions with the telescope normal and three with the telescope inverted are sufficient for anything but very exact work.

Division by 6 is easy in the sexagesimal system, which also accounts for favoring the 3D–3R routine—three direct and three reverse angles, totaling 6. Dividing 109° by 6 gives 18° with 1 to carry: this 1 is placed in the tens place of the minutes column. Dividing 13′ by 6 gives 2 with 1 to carry: this 1 is placed in the tens place of the seconds column.

It is desirable that as little time as possible should elapse between pointings, as the instrument cannot be relied upon to remain perfectly still. As a matter of fact it is vibrating and "creeping" nearly all the time from numerous causes. For example, when the instrument is set up on frozen ground, it will quickly change its position on account of the unequal settlement of the tripod legs. Changes of temperature, causing expansion or contraction of the metal of the instrument, and the effect of wind introduce errors. The more rapidly the measurements can be made, consistent with careful manipulation, the better the results will be. If the transit is set up on shaky ground the transitman should avoid walking around his instrument.

2.38. "Doubling" the Angle.—Repetition is useful not only to secure precision, but also as a check against mistakes. If a mistake is made on the first reading of an angle the vernier, on the second reading, falls in a new place on the circle so that the mistake is not likely to be repeated. It is common practice to repeat, or "double," all important angles and divide the second reading by 2 simply as a check on the first reading.

2.39. To Measure Directions in Sets.—Theodolites with optical micrometers are matched as far as pointing and reading accuracy is concerned. Therefore, the need for repetition facility does not exist. It is still desirable for accuracy reasons, however, to make multiple measurements of an angle and spread these over the circumference of the horizontal circle. This can be done by measuring directions in sets. In this case, more than one angle at a time can be measured, which saves time in triangulation work. The procedure can be stated in steps:

1. Carefully center and level instrument over station.
2. Sight in direct mode to first target and read horizontal circle which is **set** close to zero. (It is generally impossible to set a precise theodolite exactly on zero, nor is it needed or desirable).
3. Sight in direct mode to next target and read horizontal circle; sight and read circle on third target, on fourth etc., continuing until the last target.
4. Invert telescope and sight again to last target, then read horizontal circle.
5. Now sight in inverted mode to the other targets in reversed order and read horizontal circle each time. This consitutes one set.
6. Before starting a new set, change the horizontal circle reading by approximately 180°/n, where n is the number of sets planned.
7. Repeat steps 2 to 5, and 6, as necessary.

An example is given in table 2.39.1. Columns 1, 2 and 3 describe the station, set number and targets. Columns 4, 5 contain the observed directions, and columns 6–11 contain computed values. The simple mean (column 6) is obtained by

TABLE 2.39.1

EXAMPLE OF DIRECTION MEASUREMENT IN SETS:

Stn.	Set	Target	Readings: Direct ° ′ ″	Readings: Reverse ° ′ ″	Simple Mean ° ′ ″	Reduced Mean ° ′ ″	Accuracy: v	Accuracy: v′	Accuracy: $\sum v'^2$	Mean of Sets ° ′ ″
A	1	B	0 05 39	180 06 06	0 05 52.5	0 00 00	—	+0.3	0.09	0 00 00
		C	266 55 48	86 56 17	266 56 02.5	266 50 10	−0.8	−0.5	0.25	266 50 09.2
		D	308 20 51	128 21 05	308 21 05	308 15 12.5	—	+0.3	0.09	308 15 12.5
		E	310 49 54	130 50 26	310 50 10	310 44 17.5	+1.7	+2.0	4.00	310 44 19.2
		F	317 18 15	137 18 47	317 18 31	317 12 38.5	−2.5	−2.2	4.84	317 12 36
			150 27	152 55	151 41	122 18.5	−1.6	−0.1		
			151 41 (Check No. 1)		−(29 22.5)		/5			
					122 18.5 (Check No. 2)		= −0.3			
	2	B	60 09 14	240 09 39	60 09 26.5	0 00 00	—	−1.0	1.00	
		C	326 59 19	146 59 48	326 59 33.5	266 50 07	+2.2	+1.2	1.44	
		D	8 24 23	188 24 53	8 24 38	308 15 11.5	+1.0	0.0	0.00	
		E	10 53 30	190 54 01	10 53 45.5	310 44 19	+0.2	−0.8	0.64	
		F	17 21 50	199 22 12	17 21 01	317 12 34.5	+1.5	+0.5	0.25	
			168 16	170 33	169 24.5	122 12	+4.9	−0.1		
			169 24.5 (Check No. 1)		−(47 12.5)		/5			
					122 12 (Check No. 2)		= +1.0			
	3	B	120 08 51	300 09 18	120 09 04.5	0 00 00	—	+0.6	0.36	
		C	26 59 03	206 59 27	26 59 15	266 50 10.5	−1.3	−0.7	0.49	
		D	68 24 04	248 24 32	68 24 18	308 15 13.5	−1.0	−0.4	0.16	
		E	70 53 10	250 53 41	70 53 25.5	310 44 21	−1.8	−1.2	1.44	
		F	77 21 26	257 21 53	77 21 39.5	317 12 35	+1.0	+1.6	2.56	
			166 34	168 51	167 42.5	122 20	−3.1	−0.1		
			167 42.5 (Check No. 1)		−(45 22.5)		/5			
					122 20 (Check No. 2)		= −0.6		17.61	

$$\sigma = \sqrt{\frac{17.61}{2 \times 4}} = 1''.5 \qquad \sigma_M = \sqrt{\frac{17.61}{3 \times 2 \times 4}} \times 0''.9$$

using the degree value of the direct reading and combining it with the mean of the minute and second values of direct and reverse readings. It would not make any difference if the degrees were included into the mean, because all values would differ only by 90°. The reduced mean (column 7) is obtained by reducing all values of the simple mean by the amount required to make the first value zero. The final mean of sets (column 11) represents the arithmetic mean of each reduced mean value of the direction to each target. All these columns are computed first, to obtain the mean values for the three sets.

The v, v′ and $\sum(v')^2$ columns are optional. They are used to obtain the accuracy of the directions. The standard deviation of a single observation is

$$\sigma = \sqrt{\frac{\sum v'^2}{(s-1)(n-1)}}$$

where n is the number of rays and s the number of sets. For subsequent use, the standard deviation of the final mean is needed, which is

$$\sigma_M = \sqrt{\frac{\sum v'^2}{s(s-1)(n-1)}}$$

The v-values are the differences between the final value (mean of sets) and the respective reduced mean, while the v′-values are normalized by shifting the v-values of each set such that their algebraic sum equals zero (as nearly as can be done). This takes care of the fact that the first direction is not better than any other. There are two sum checks which aid the notekeeper in detecting blunders when he performs the meaning. These checks should be completed before leaving the station in order to avoid costly resurveys.

No. 1: The simple mean is checked by utilizing the fact that the sum of the mean (column 6) is equal to the mean of the sums (columns 4 & 5).

No. 2: The reduced mean is checked by subtracting n-times the first reading from the sum of the simple means, which has to result in the sum of the reduced means.

The mean of sets is checked by balancing of the v-values,

which have to add up to zero for each direction. Here, they almost do, although the effect of round-off errors is noticeable.

2.40 To Measure a Vertical Angle.—In surveying, a vertical angle is the angle in a vertical plane measured from the horizontal plane through the axis of the telescope to the point in question. When the point sighted is above the horizontal plane, the vertical angle is called an angle of elevation, or positive angle; when the point sighted is below the horizontal plane, the angle is called an angle of depression, or negative angle. In the notes the former is preceded by a + sign, and the latter by a − sign.

While a few transits are equipped with only a vertical arc, the vast majority of theodolites and most transits have a full vertical circle. It carries the zero reading either at the horizontal or at the top vertical position. In the latter case, a horizontal line will read either 90° (direct) or 270° (reverse). With this arrangement, values read on the vertical circle are called zenith distances. Although a few sophisticated theodolites have an automatic gravity actuated index, most instruments have an index bubble which has to be centered before each reading in order to refer it to the horizontal plane.

To measure a vertical angle to a point, the transit is carefully set up and leveled. The telescope is directed to the object and when it is observed in the telescope, the vertical motion is clamped. By means of the vertical motion tangent screw the middle horizontal cross-hair is set exactly on the point. The vertical arc reading is the correct vertical angle, provided that the instrument is in adjustment and has been carefully leveled. The conditions that must be fulfilled are the following: the axis of the telescope spirit level and the line of sight must be parallel; when the plate bubbles are centered, the plate must be truly horizontal; and the vertical arc must read 0° when the telescope bubble is in the center of the tube.

If the first condition is not fulfilled it is impossible to obtain a correct vertical angle by means of the vertical arc only. With a full vertical circle this error may be eliminated by observing the vertical angle first with the telescope direct, then with it reversed, and by taking the average of the two readings.

If the last two conditions are not fulfilled because the theodolite is slightly out of adjustment, it is possible to determine a correction by means of which the observed vertical angle may be rectified. This correction is called the *index correction.*

The easiest way to determine the index correction is by adding the direct and reverse readings. The difference to 360° is twice the index correction. By applying half of this difference to the vertical angle the index correction is satisfied and the vertical angle is correct. (See notes below.)

Stn.	Target	Direct Reverse	Index corr.	Corrected Vertical Angle
A	B	15° 23′ 6″	+11″	15° 23′ 27″
		344° 36′ 22″	+11″	344° 36′ 38″
		359° 59′ 38″		360 00 00
		360° 00 00		
		Diff.: +22		

It should be noted that the index correction should remain either constant for a particular instrument or change continuously with time due to differential heating of the index bubble. Any large jumps might indicate blunders.

Since the measurement of vertical angles requires this additional correction, and since the readings cannot be spread over the whole circle, the vertical angles are usually less precise than the horizontal angles measured with the same instrument.

2.41. Using the Theodolite as a Level.—Level the plates, and after revolving through 180° about the vertical axis, bring each plate bubble **half-way back** to the mid-position. Now bring the telescope bubble to the center of its scale, using the clamp and slow-motion screw of the vertical arc. Turn the telescope toward the rod and read it where the horizontal cross-hair cuts it **at the instant that the telescope bubble is centered.** Care must be exercised when leveling to use the middle cross-hair, not one of the stadia hairs.

Approximate leveling may be done by measuring the vertical angle and the inclined distance from the horizontal axis of the telescope. The vertical height which the point sighted is above (or below) the axis of the transit equals Slope Distance × Sin Vertical Angle. The slope distance may be measured by tape or it may be obtained by stadia, as explained in Art. 3.10. The vertical angle method is not capable of the same degree of precision as is the method of direct leveling, because vertical arcs usually read only to one minute. At 300 ft. an arc of 01′ subtends about 0.10 ft. In direct leveling, a rod-reading to 0.01 ft. is easily obtained at this distance.

2.42. Common Sources of ERROR in Transit Work.—Even though the transit is in adjustment, or errors of adjustment have been eliminated by proper manipulation of the transit, the following sources of error may be present:

1. Eccentricity of circle (i.e., imperfectly centered), and errors of graduation.
2. Changes due to temperature and wind.
3. Uneven settling of tripod.
4. Poor focusing (parallax).
5. Inaccurate setting over point.
6. Irregular refraction of atmosphere

2.43. Common MISTAKES in Transit Work.

1. Reading in the wrong direction from the vernier index on a double vernier.
2. Reading the vernier opposite the one which was set.
3. Reading the circle incorrectly, e.g., reading 39° for 41°. If the angle is nearly 90°, reading the wrong side of the 90° point, e.g., 89° for 91°.
4. Using the wrong tangent screw.
5. Neglecting to add 30′ to a reading; for example, 26° 18′ for 26° 48′.
6. Where the circle is graduated in both directions from 0° to 360°, reading wrong circle scale near 180°.

Many errors in angle readings can be avoided if the degrees and minutes are estimated mentally before the precise vernier reading is taken.

2.44. Care of the Theodolite.—There are certain precautions which every instrument man should take in handling the transit; the following are the most important. The clamps at the top of the tripod legs should be kept sufficiently tightened so that there will be no apparent looseness. When being transported in a vehicle, the transit should be carried in its box. When being taken out of the box, it should be lifted by the foot plate, not by the telescope axis. In the field, it should be carried from place to place on the shoulder, with the tripod legs pointing forward; but in entering buildings the head of the transit should be held forward to avoid striking it against door frames. It should not be jolted at any time and especial care should be taken to avoid jolting it when it is in a horizontal position. It is well to center the leveling screws and the transit head just before picking the transit up to go to the next station. In rainy weather the telescope should not be left pointed upward so that water can collect on the object glass; the moisture may leak down into the barrel of the telescope and cause the cross-hairs to break. The metal parts of the instrument should be wiped dry before it is put away. Care should be taken to avoid scratching the object glass when cleaning it. A camel's hair brush should be used first, then an old soft and clean piece of linen, moistened with alcohol to dissolve grease. It should not be rubbed, but should be wiped gently and brushed again with the camel's hair brush. A special lens paper is available which may be used. The vertical arc should never be touched with the fingers, as it will tarnish.

Force should never be used in tightening any clamps or screws; a definitely firm, but not severe, tightening is all that ever is necessary. When placing the transit in the box, especial care should be taken to make sure that the telescope does not touch the sides of the box. The telescope is clamped in an approximately horizontal position, the upper motion is also clamped, but the lower clamp is left loose so that the telescope can be guided into its position in the box. It is then clamped so that the telescope cannot swing against the side of the box while being transported.

2.45. PRECAUTIONS IN THE USE OF THE THEODOLITE.—In the preceding text several sources of error and also precaution against mistakes have been mentioned, but in order that the beginner may appreciate the importance of handling the instrument carefully he should make the following simple tests.

1. Set the transit up with the three points of the tripod rather near together so that the instrument will be high and unstable. Sight the cross-hair on some definite object, such as the tip of a church spire, so that the slightest motion can be seen. Take one tripod leg between the thumb and forefinger and twist it strongly; at the same time look through the telescope and observe the effect.
2. Press the tripod leg laterally and observe the effect on the level attached to the telescope; center the bubble before testing.
3. Step on the ground about 1 or 2 inches from the foot of one of the tripod legs and observe the effect on the line of sight.
4. Breathe on one end of the level vial and observe the motion of the bubble.
5. Press laterally on the eyepiece and observe the effect on the line of sight.

These motions, plainly seen in such tests, are really going on all the time, even if they are not readily apparent to the observer, and show the necessity for careful and skillful manipulation. The overcoat dragging over the tripod, or a hand carelessly resting on the tripod, are common sources of error in transit work.

Before picking up the transit **center the movable head, bring the leveling screws back to their mid positions, loosen the lower clamp, and turn the telescope either up or down.**

ADJUSTMENTS OF THE THEODOLITE

2.46. Errors in Adjustment of the Theodolite.—As a result of handling in the field, changes in temperature, and jarring in transportation, the transit is likely to get out of adjustment.

In general, there are two kinds of adjustments to the transit: shop adjustments, which can be made only in the factory; and field adjustments, which can be made readily by the instrument man in the field. Only the latter will be explained in this chapter.

The more common errors of adjustment that may be encountered, and their effects on measurements with the engineer's transit, are tabulated below:

1. **Plate Bubbles.** Axes of plate bubbles may not be perpendicular to the vertical axis of the transit.

   Effect. When bubbles are centered, the horizontal plate will lie in a plane which is inclined to the horizontal. This condition affects the accuracy of horizontal and vertical angles and the lining in of objects, especially when the points sighted are at different elevations.
2. **Cross-Hairs.** Vertical hair may not be perpendicular to the horizontal axis of the telescope because the ring holding the cross-hairs has rotated from its correct position.

   Effect. Objects appear to move off the vertical hair when the telescope is raised or depressed.
3. **Cross-Hairs.** The line of sight as defined by the vertical cross-hair may not be perpendicular to the horizontal axis of the telescope.

   Effect. Introduced errors in horizontal angles when the objects sighted are at elevations above or below the instrument. Also causes angular errors in prolonging a straight line by a single reversal of the telescope.
4. **Standards.** The horizontal axis of the telescope may not be truly level, i.e., exactly perpendicular to the vertical axis of the instrument.

   Effect. Introduces errors in horizontal angles when objects sighted are above or below the horizontal, and in plumbing vertical objects with the transit.
5. **Telescope Bubble.** The axis of the telescope bubble and the line of sight may not be parallel.

   Effect. When using transit as a level, differences in elevation are in error when sight distances are unequal.
6. **Vernier of Vertical Arc. (Index Error.)** The vernier of the vertical circle may be displaced, causing the zero of the vernier and the zero of the vertical arc to fail to coincide

when the telescope is leveled by means of the telescope level.

Effect. Introduces an error in the measurement of vertical angles.

In the following articles, the tests for adjustment and the methods of adjusting the instrument are described in detail. When making tests and adjustments, the transit should be set up in the shade on firm level ground with the tripod leg screws firm but not binding. Certain of the adjusting screws are of the capstan-headed type; they are turned by inserting a small steel pin through holes in their heads. The adjusting pins should fit snugly in order to avoid burring of the holes. Excessive force should not be applied in turning the screws, for there is danger of breaking off the heads of the screws.

Adjustments must be made in the order listed above, so that the effect of one adjustment may not be reflected in the others.

If adjustment errors are small, satisfactory measurements may still be made provided certain precautions are taken (Art. 2.54).

2.47. ADJUSTMENT OF THE PLATE BUBBLES.—To adjust the Plate Levels so that Each lies in a Plane Perpendicular to the Vertical Axis of the Instrument.—Set up the transit and bring the bubbles to the centers of their respective tubes. Fig. 2.47.1a shows one bubble after this first step. Next, turn the instrument 180° about its vertical axis and observe whether the bubble remains in the center. In Fig. 2.47.1b it has moved from the center toward the high end of the tube. To adjust, turn the capstan-headed screws on the bubble-tube casing by means of an adjusting pin, thus raising or lowering one end until the bubble moves half-way back to its center, as in Fig. 2.47.1c. Perform these three steps with each bubble separately. Through this third step the leveling screws have not been moved. If the ajustment has been carefully made the axes of the bubbles are now perpendicular to the vertical axis, or parallel to the horizontal plate. The bubbles are now brought to their central positions by means of the leveling screws; this process will make the vertical axis truly vertical and the horizontal plate truly horizontal, as in Fig. 2.47.1d.

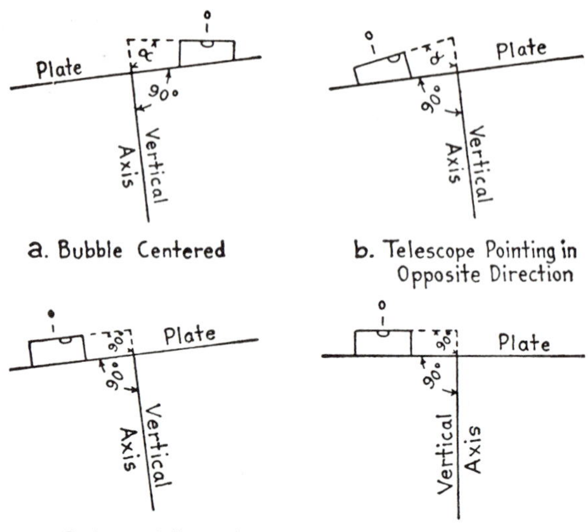

FIG. 2.47.1. ADJUSTMENT OF PLATE BUBBLES.

2.48. ADJUSTMENT OF THE CROSS-HAIRS.—1st. To put the Vertical Cross-Hair in a Plane Perpendicular to the Horizontal Axis.—Sight the vertical hair on some well-defined point, and, leaving both plates clamped, rotate the telescope slightly about the horizontal axis (see Fig. 2.48.1).

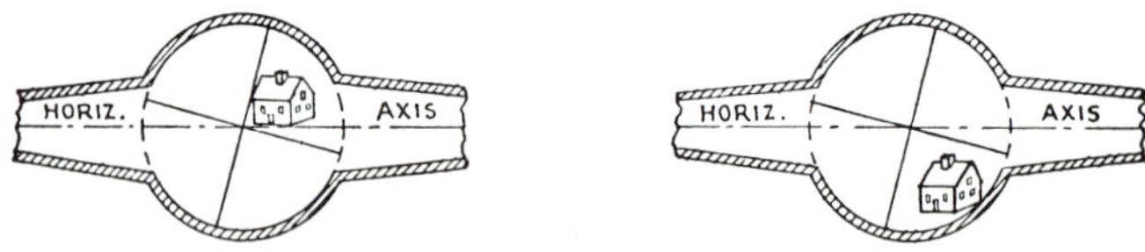

FIG. 2.48.1. ADJUSTMENT OF THE VERTICAL CROSS-HAIR (FIRST PART).

The point should appear to travel up or down on the vertical cross-hair throughout its length. If it does not, loosen the screws holding the cross-hair ring, and, by tapping lightly on one of the screws, rotate the ring until the above condition is satisfied. Tighten the screws and proceed with the next adjustment.

2.49. 2nd. To make the Line of Sight Perpendicular to the Horizontal Axis.—(See Fig. 2.49.1.) Set the theodolite at *A*. Level it, clamp both plates, and sight accurately on *B* which is approximately at the same level as *A*. Reverse the telescope

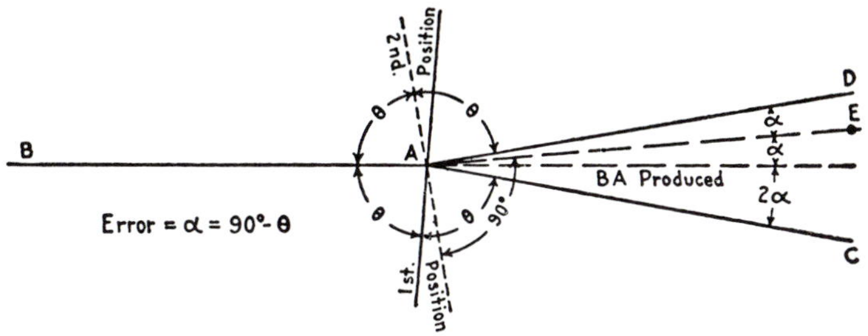

FIG. 2.49.1. ADJUSTMENT OF THE VERTICAL CROSS-HAIR (SECOND PART).

and set *C* in line with the vertical cross-hair. *B*, *A*, and *C* should be in a straight line. To test this, turn the instrument about the **vertical** axis until *B* is again sighted. Clamp the plate, reverse the telescope, and observe if point *C* is in line. If not, set point *D* in line with the cross-hair just to one side of point *C*; then the cross-hair ring must be moved until the vertical hair appears to have moved to point *E*, which is set at **one-fourth** the distance from *D* toward *C*, since the angular error of adjustment, α, subtends one-fourth of *CD* (Fig. 2.49.1).

The cross-hair ring is moved by loosening the screw on one side of the telescope tube and tightening the opposite screw. If *D* falls to the **left** of *C* then the cross-hair ring should be moved to the **left**; but if the transit has an erecting eyepiece the cross-hair will **appear** to move to the **right** when viewed through the telescope. If the transit has an inverting eyepiece the cross-hair appears to move in the same direction in which the cross-hair is actually moved.

The process of reversal should be repeated until no further adjustment is required. When finally adjusted, the screws should hold the ring firmly but without straining it.

2.50. ADJUSTMENT OF THE STANDARDS.—To make the Horizontal Axis of the Telescope Perpendicular to the Vertical

Axis of the Instrument.—(See Fig. 2.50.1.) Set up the transit and sight the vertical cross-hair on a high point *A*, such as the top of a church steeple. Lower the telescope and set a point *B* in line, on the same level as the telescope. Reverse the telescope, turn the instrument about its vertical axis, and sight on *B*. Raise the telescope until the point *A* is visible and see if the cross-hair comes on *A*. If not, note point *C* in line and at the same height as *A*. Then half the distance from *C* to *A* is the error of adjustment. Loosen the screws in the pivot cap and raise or lower the adjustable end of the horizontal axis by means of the capstan-headed screw under the end of the axis. Repeat the test until the high and low points are both on the cross-hair in either the direct or reversed positions of the transit. The adjusting screw should be brought into position by a right-hand turn, otherwise the block on which the horizontal axis rests may stick and not follow the screw. The cap screws should then be tightened just enough to avoid looseness of the bearing.

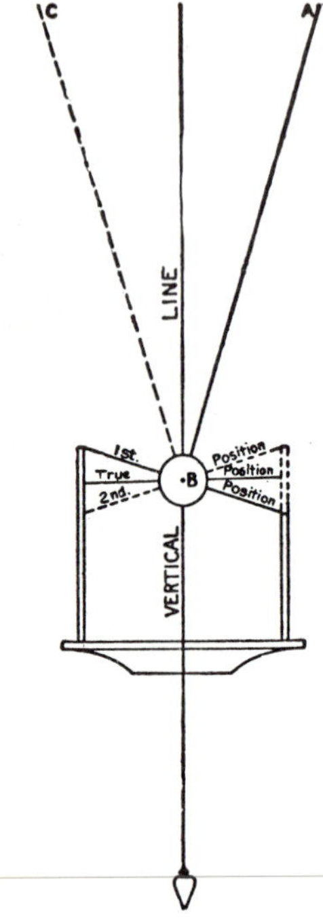

FIG. 2.50.1. ADJUSTMENT OF THE STANDARDS.

2.51. Adjustment of the Telescope Bubble.—This is adjusted by the *"peg" method*, or *direct method*, as explained in Art. 4.40. After the "peg" test has been completed the true difference in elevation between the two turning points becomes known. Knowing the height of the transit above one of the turning points, and the true difference in elevation, a rod reading (at the other turning point) is then computed to give a point at the same elevation as the center of the instrument. This establishes a true horizontal line through the center of the instrument. After the telescope has been inclined (by means of the tangent screw) so that the horizontal hair is set at the computed reading, the bubble is brought to its zero position on the scale by means of the adjusting screw on the level case.

Adjusting the bubble will throw out the adjustment of the vernier of the vertical arc; this necessitates a readjustment of the vernier. If the error is but slight this may be avoided by moving the cross-hairs instead of the bubble. That is, center the bubble, and then move the upper and lower adjusting screws of the diaphragm until the horizontal hair shows the proper reading on the rod. It is not advisable to move the cross-hair ring very far, however, as an error will be introduced when very short sights are taken.

2.52. Adjustment of the Auxiliary Level on the Vernier of the Vertical Arc.—(See Art. 2.40) **To adjust the Level so that it is in the Center of the Tube when the Line of Sight is Level and the Vernier reads 0°.**—If there is no level attached to the telescope the auxiliary level is adjusted by the "peg method" (Art 4.40). The bubble is first brought to the center of the tube by means of its tangent screw. Then the telescope is moved until the vernier of the vertical arc reads 0°. The instrument is then in condition to be used as a leveling instrument and is adjusted by the "peg method."

When the telescope is provided with an attached level the auxiliary level could be adjusted by comparing it with the telescope level as follows. Level the telescope by means of its attached level, make the vernier read 0° by means of the tangent screw of the vernier, and then bring the bubble of the auxiliary level to the center by means of its adjusting screws.

2.53. Adjustments of the Vernier of the Vertical Arc or Circle.—To make the Vernier read 0° when the Telescope Bubble is in the Center of Its Tube.—Level the instrument first by means of the plate levels, and then by means of the long bubble on the telescope, so that it will remain central when the transit is revolved. Adjust the distance between the vernier and the circle so that the 30′ lines coincide with the circle divisions. To make the vernier zero coincide with the circle zero loosen the screws holding the circle or arc and, by tapping the spokes lightly, perfect this adjustment. Then tighten the screws.

2.54. How to Eliminate the Effect of Errors of Adjustment in the Theodolite.—Errors of adjustment in the plate bubble may be avoided by leveling up and reversing as when adjusting. Then, instead of altering the adjustment, simply move the bubble half-way back by means of the **leveling screws.** This makes the vertical axis truly vertical. Then the bubbles should remain in the same parts of their respective tubes as the instrument revolves about its vertical axis.

Errors of the line of sight and errors of the horizontal axis are eliminated by using the instrument with the telescope in the direct and then in the reversed position and taking the mean of the results whether the work is measuring angles or running straight lines.

Errors of eccentricity of the circle are completely eliminated by reading the two opposite verniers and taking the mean.

Errors of graduation of the circle are nearly eliminated by reading the angle in different parts of the circle or by measuring the angle by repetition.

In table 2.54.1 the errors are listed together with the procedure used to minimize them. The remaining influence is also indicated.

TABLE 2.54.1

THEODOLITE ERRORS & MEASURING PROCEDURES FOR THE MINIMIZATION

Error	Procedure	Effect on error
1) Collimation Error (collimation axis not perpendicular to elevation axis)	Measure in direct and reverse telescope positions and take mean value	elimination
2) Elevation Axis Error (not perpendicular to vertical axis)	Measure in direct and reverse telescope positions and take mean value	elimination
3) Azimuth Axis Error	Level as precisely as possible	influence increases with inclined sights

TABLE 2.54.1 (cont.)

THEODOLITE ERRORS & MEASURING PROCEDURES FOR THE MINIMIZATION

Error	Procedure	Effect on error
4) Eccentricities (of telescope or circle)	Measure in direct and reverse telescope positions and take mean value	elimination
5) Circle Subdivision (not uniformly divided)	Spread readings over entire circle, either by repetition or when measuring direction in sets	reduction to a negligible amount
6) Targeting or pointing error	Repeat n times	reduction $\sigma = \frac{\sigma_0}{\sqrt{n}}$
7) Reading error	Repeat n times	reduction by $\sigma = \frac{\sigma_0}{\sqrt{n}}$
8) Centering error	Check plummet, work carefully	negligible for long sighting distances.

CHAPTER III

MEASUREMENT OF DISTANCES

3.1. Horizontal Distances.—In surveying, the distance between two points means the **horizontal** distance. If the boundary lines of a parcel of land are along sloping surfaces of the ground, the distances recorded in a deed will be the horizontal distances between the corners, and not the slope distances along the ground. In determining the distance between two points the horizontal distance is measured either directly or inclined measurements are made and subsequently reduced to the corresponding horizontal distance. Distances plotted on engineer's plans are always horizontal distances.

3.2. Chains.—Old surveys were sometimes made with the *Surveyor's* (or *Gunter's*) *Chain*, which was 66 ft. in length and divided into 100 links, each link 7.92 in. long. Later, the *Engineer's Chain*, which was 100 ft. long, came into use; this had 100 links, each 1 ft. long. Both of these chains are obsolete, but the surveyor may have occasion to use old plans on which the distances are expressed in chains. (See footnote, Art. 5.27.)

The following table shows the simple relation which the Gunter's (or Surveyor's) Chain bears to the mile and acres.

1 Gunter's Chain	= 4 Rods = 66 feet = 100 Links.
1 Mile	= 80 Chains.
1 Acre	= 10 Square Chains = 43,560 square feet.

3.3. Steel Tapes.—For accurate measurements steel tapes are used most commonly. The common lengths are 50 ft., 100 ft., and 200 ft., although longer tapes are available. Some steel tapes are graduated throughout their length in feet, tenths, and hundredths. Tapes graduated in metric units are available in lengths of 25, 30, and 50 metres. Fig. 3.3.1 shows some of the current types of tapes and their graduations. On some tapes the zero point is at the extreme end of the brass loop; on some it is at the extreme end of the steel ribbon; on others it is marked on the steel ribbon a short distance from the end.

3.4. Metallic tapes, sometimes called woven tapes, are cloth tapes with fine tinsel twisted around each thread to prevent excessive stretching. They are usually graduated into feet, tenths, and half-tenths and are made in lengths of 25 ft., 50 ft.,

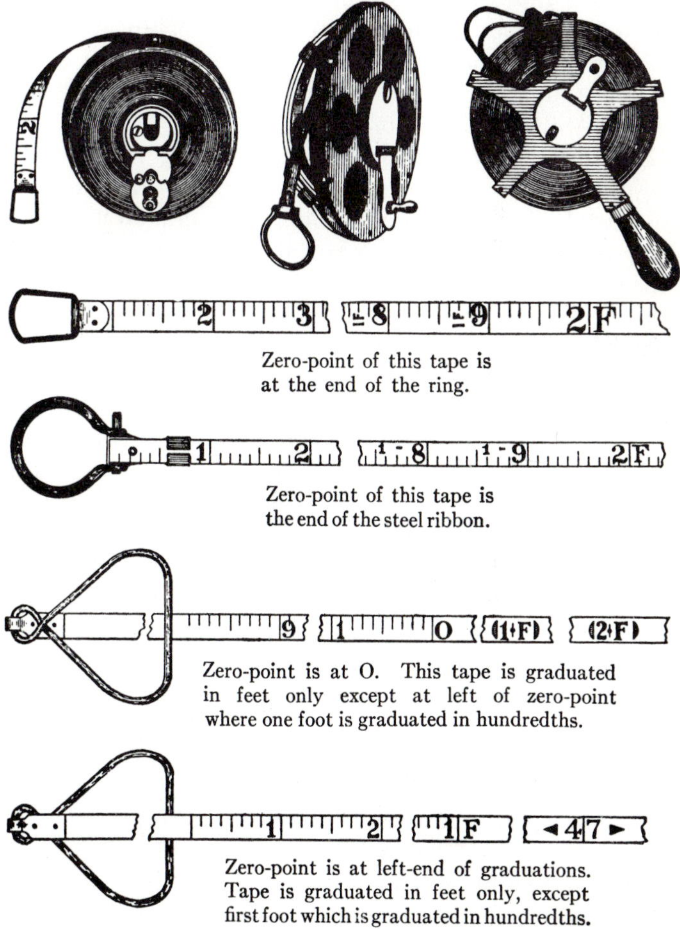

Zero-point of this tape is at the end of the ring.
Zero-point of this tape is the end of the steel ribbon.
Zero-point is at O. This tape is graduated in feet only except at left of zero-point where one foot is graduated in hundredths.
Zero-point is at left-end of graduations. Tape is graduated in feet only, except first foot which is graduated in hundredths.

FIG. 3.3.1. COMMON TYPES OF STEEL TAPES.

and 100 ft. They are used for short measurements which are not required nearer than 0.1 ft., such as measurements of buildings for plotting purposes or measurements for topography. When precise results are required, a steel tape should be used.

3.5. Cloth tapes, without woven wires, stretch so much as to be practically useless in surveying.

3.6. Invar tapes are made of a nickel steel alloy having a very small thermal coefficient of expansion. They are used for surveys requiring a high degree of precision. For a given change in temperature, the change in length of these tapes is about one-thirtieth that of ordinary steel tape.

3.7. Other equipment needed for taping includes note book, plumb-bobs, range poles, set of taping pins (surveyor's arrows), stakes and nails for setting points, crayon for marking stakes, axe or hammer for driving stakes and cutting tools for clearing lines.

3.8. The *odometer* is an instrument attached to the wheel of a vehicle to measure its revolutions, which, when multiplied by the circumference, gives slope length of distance covered.

This is a rather inaccurate instrument and of little use in surveying. A variation, known as the measuring wheel sometimes finds handy use being pushed along pavement for quantity measurements or for rough reconnaissance.

3.9. MEASURING HORIZONTAL DISTANCES.—Two common methods for measuring the horizontal distance between two points with a tape are (1) making the measurement by holding the tape horizontally; (2) measuring the slope or inclined distance between the points. By the first method the horizontal distance is obtained directly. By the second method the slope distance is later reduced to the horizontal distance; and for this purpose either the difference in elevation of the points or the vertical angle between the line connecting these points and the horizontal line must be known or determined.

In measuring long distances it is important to keep a correct count of the number of tape lengths. This may be done in the following manner, using taping pins. A taping pin, or surveyor's arrow, is made of metal, about 3/16 in. in diameter and 14 in. in length, pointed at one end with an eye at the other.

One man, the head-tapeman, takes the forward end of the tape and ten marking pins and goes ahead along the line to be measured, while the rear-tapeman, with one pin, takes his position at the stake marking the beginning of the line. The rear-tapeman, with his eye over the point, places the head-tapeman in line with some object, such as a lining or range pole which marks the other end of the line to be measured. When the head-tapeman is nearly in line he takes a pin and, standing to one side of the line, holds it upright on the ground a foot or so short of the end of the tape and the rear-tapeman motions him to the right or left until his pin is on the line. When the head-tapeman has the pin in line he stretches the tape taut, seeing that there are no kinks and that no obstructions cause bends in the tape. The rear-tapeman at the same time holds the zero-point of the tape at his pin and when he calls out, "All right here," the head-tapeman, stretching the tape past his line pin, removes this line pin, places it at the end graduation of the tape, and presses it vertically into the ground. When the tapemen are experienced the pin may be set for both line and distance at the same time. When the pin is in place the head-tapeman calls, "All right," the rear-tapeman takes the pin left at his end of the line and they proceed to the next tape-length. The pin that the rear-tapeman has is a record of the first tape-length. Just before reaching the second pin the rear-tapeman calls out, "Tape," to give the head-tapeman warning that he has nearly reached a tape-length. The process of lining in the head-tapeman and measuring a tape-length is then repeated. After the third pin has been stuck in the ground the rear-tapeman pulls out the second pin; in this way the number of pins the rear-tapeman holds is a record of the number of tape-lengths measured. There is always one pin in the ground which marks the distance but is not counted. When 10 tape-lengths have been measured the head-tapeman will be out of pins and calls to the rear-tapeman, who brings forward 10 pins. The pins are then counted by **both men.**

In most engineering surveys the lines are comparatively short, and the intermediate (100-ft.) points on the lines are needed for side measurements, such as ties, offsets, etc. In all such surveys the accuracy is comparatively high and it is necessary to employ exact means of marking the points. Sometimes **stakes** are driven into the ground and **tacks** or **pencil marks** used to mark the points. A short galvanized **nail** with a large head, pressed into the ground so that the center of the head is in the proper position, makes a good temporary mark, but of course is easily lost. In measuring on the surfaces of hard roads heavy steel **spikes** are used for permanent marks.

All measurements should be checked by repeating the measurement, and in such a way as not to use the same intermediate points on both of the measurements.

In measuring with the tape some prefer to make a series of measurements between points set in the ground a little less than 100 ft. apart, summing up the partial measurements when the end of the line is reached. This guards against the mistake of omitting a whole tape-length. Another advantage is that it is easier to read the distance to a fixed point than to set a point accurately at the end of the tape; this is especially true in measurements where plumbing is necessary. This method takes less time than the usual method, but it is not applicable when it is necessary to mark the 100-ft. points on the line.

In measuring a long line much time can be saved if the head-tapeman will **pace** the tape-length and then place himself very nearly in the line by means of objects which he knows to be on line as, for example, the instrument, a pole, or the last pin. In all surveys the accuracy should be frequently tested by the application of rough check measurements, and these are readily determined by pacing. If much pacing is to be done it will be found more accurate and certainly less tiresome to assume one's natural gait, determining by actual trial the number of steps per 100 ft., and not attempt to take steps exactly 3 ft. long.

3.10. HORIZONTAL MEASUREMENT ON SLOPING GROUND WITH A TAPE.—If the ground is level, the tape is stretched on the ground and is therefore supported horizontally when the measurements are taken. But if the terrain is not horizontal or if obstructions such as bushes are on line, the tape must be held horizontally by supporting it at one or at both ends when the measurement is being made, and the distance must be transferred to the ground by means of a plumb-bob, as in Figs. 3.10.1 and 3.10.2. One end of the tape must be held on the stake, as in Figs. 3.10.1 and 3.10.2; plumbing at both ends is ordinarily not permitted.

When measuring horizontally the tape must be held as nearly level as is possible by estimation, ordinarily without special instruments to determine that it is horizontal. For ordinary measurements the pull on the tape (tension) is judged by the feeling on the hand; for precise work it is measured with a spring balance. Although some random error is likely to be introduced in transferring the measurement from the tape to the stake by means of the plumb-bob, a greater systematic error is likely to occur by failing to hold the two ends of the tape at exactly the same level. **Holding a tape level requires much practice.** Level lines are present in the outside finish of buildings and on the shore of water surfaces; when available these level lines should be used as guides in holding the two ends of the tape level. Another guide for keeping the tape level is to be sure that the plumb-bob string is parallel to tape divisions when the tape is held on edge. **The importance of having the two ends of the tape at the same level cannot be overstated.**

3.11. "Breaking Tape."—In taping a line by the usual method of holding the tape horizontally, it will be frequently found that on steep ground the length of the line must be made up of a series of horizontal measurements, all less than a full tape-length. To illustrate this method, assume that the horizontal distance *AB* (Fig. 3.11.1) is required. The tape is laid on the ground, the zero end being in the direction of *B* and the 100-ft. end at *A*. The head-tapeman takes a position about on line, and so selected that the rear-tapeman will not have to plumb higher than his chest to make the tape level. Under the direction of the rear-tapeman, the head-tapeman sets a pin or temporary mark on line. The tape is now held

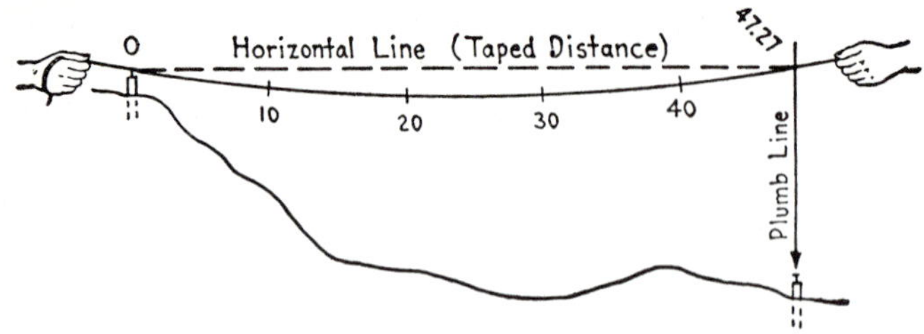

FIG. 3.10.1. HORIZONTAL DISTANCE MEASURED DIRECTLY BY PLUMBING. SUFFICIENT PULL ON TAPE GIVEN TO ELIMINATE EFFECT OF SAG.

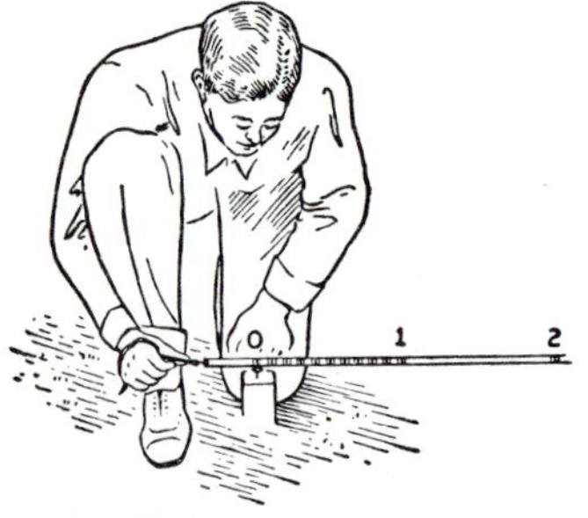

A. Holding zero end of tape on a tack in top of stake.

B. Plumbing over tack in stake and reading tape graduations which are on observer's side of tape.

FIG. 3.10.2. THESE MEN ARE SHOWN IN PROPER POSITIONS PERFORMING THE TAPING INDICATED IN FIG. 3.10.1.

horizontally through the line mark and the necessary tension is gradually applied, the rear-tapeman plumbing the 100-ft. mark on the tape over A. When the rear-tapeman calls "O.K.", the head-tapeman inserts a pin or nail in the ground at some full foot, as point *a* (say at the 60-ft. mark). It is good practice to repeat this measurement as a check. When the head-tapeman calls "all right," the rear-tapeman drops the tape and advances to point *a*, the head-tapeman taking a position at the zero end of the tape. In a similar manner point *b* is set, the rear-tapeman

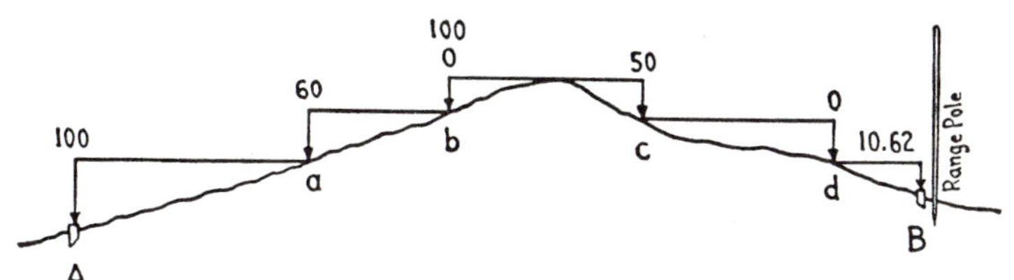

FIG. 3.11.1. MEASURING HORIZONTAL DISTANCES BY USE OF PLUMBING.

plumbing the 60-ft. tape mark over point *a* and the head-tapeman setting point *b* at the zero end of the tape. The tape is now pulled ahead until the 100-ft. mark is at *b*. Similarly, points *c* and *d* are set. It will be noted that in measuring *bc* both tapemen have to plumb, but in setting points *a* and *d* only the head-tapeman plumbs. When point *d* has been set the measured horizontal distance Ad = 100 ft.; the distance dB still remains to be measured. The tape is now drawn ahead, the head-tapeman plumbing the zero of the tape over *B*; if the tape is graduated throughout its length to hundredths, the rear-tapeman reads 10.62. The total horizontal distance AB =210.62 ft.

If, however, only the end foot is graduated to hundredths, the rear-tapeman holds on a full foot-mark (in this case 11 ft.) at *d* so that the graduated end foot will come over *B*. In this case the rear-tapeman calls "11 ft." The head-tapeman actually reads on the tape 0.38, which he mentally subtracts from 11 ft. The head-tapeman therefore records 10.62 ft.

Measuring a line in the above manner is called "breaking tape." By this method it is an easy matter to keep account of the number of 100-ft. lengths measured. Of course, the line could have been measured by taping each individual distance, as *Ab*, *bc*,etc., the sum giving the length *AB*.

To check the measurement, the line is taped from *B* to *A*, using different intermediate points from those used on the forward measurement.

3.12. Suggestions on Taping.—It is easier to measure downhill than uphill. In the former case the rear-tapeman holds the tape steady at the point on the ground while the head-tapeman applies a uniform pull on it, plumbs exactly over the point, and reads the tape. The head-tapeman, who is doing the plumbing, can keep his balance as he applies the pull because he has the whole process so much under his control.

In taping uphill the process is more difficult because in this case the rear-tapeman must hold himself in perfect balance, keep his plumb-bob steady and exactly over the point, while the head-tapeman is applying a steady pull on the tape, which tends to pull the rear-tapeman off balance.

In making tape measurements, both head-tapeman and rear-tapeman should take such positions as will enable them to apply tension to the tape **gradually** without losing their balance and so causing irregular or sudden pull on the tape. The positions which the men should take will depend on the method used in taking the measurement. If, for example, the tape is supported throughout its length, both men should take a position with one knee on the ground, as in Fig. 3.10.2-A. If one end of the tape is on the ground and the other end is being plumbed, one man must be in a standing posture, as in Fig. 3.10.2-B. If both ends are plumbed, both men must be standing in perfect balance. In any event, each should retain a firm grip on the tape and should increase tension gradually rather than in jerks. They should stand so that they can look squarely across the tape graduations and at the point to which they are measuring; at the same time they should keep far enough off the line being measured for the transitman to sight past them.

3.13. Slope Measurements.—When taping "on the slope" the measurement may be made (at the right tension) **directly between the two points** and the distance read from the tape; or the tape may be stretched directly **from a tack in a stake to the horizontal axis of the transit telescope.** The slope distance being known, the horizontal distance may be calculated if either the difference in height of the two ends of the tape or the angle of slope of the tape is known.

Taping bucks are convenient for slope taping. These are low tripods with a head in which a small hole is drilled for use as a taping reference. Measurements are made directly from transit axis to a buck or between bucks, thus avoiding the necessity for plumbing. See Vol. II, Art. 1-26.

If the slope distance is measured directly between two points and the difference in their elevations is known, the horizontal distance can be obtained as follows. In the right triangle formed by the slope distance s, the difference in elevation of the ends of the tape h, and the horizontal distance d, $d = \sqrt{s^2 - h^2}$. Because the solution of this expression is rather awkward, a shorter approximate expression is commonly used which gives **the difference between slope distance and horizontal distance.**

$$s - d = \frac{h^3}{2s}$$

The proof of this relation is demonstrated from Fig. 3.13.1 wherein:

$$s^2 - d^2 = h^2$$
$$(s + d)(s - d) = h^2$$

assuming $s = d$ and applying this to first parentheses only,

$$2s(s - d) = h^2 \text{ (approximately)}$$
$$s - d = \frac{h^2}{2s} \text{ (approximately)}$$

Similarly, $s = d = \frac{h^2}{2d}$ (approximately)

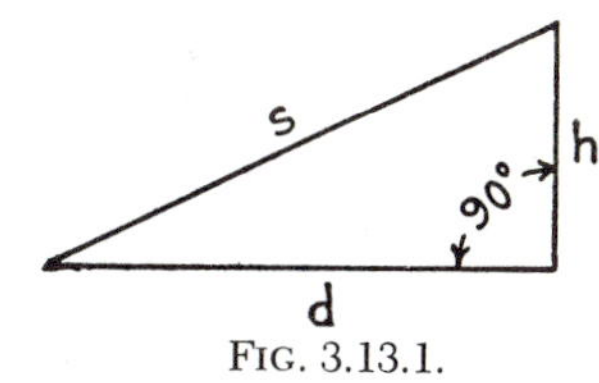

FIG. 3.13.1.

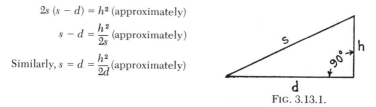

The ordinary slide rule is suitable for this computation.

It is evident that the smaller h is in comparison with the other two sides, the more exact will be the results obtained by this formula. This formula is correct to the nearest 0.01 ft. even when $h = 14$ ft. and $s = 100$ ft., or when $h = 30$ ft. and $s = 300$ ft. Greater precision may be obtained from the following expanded form:

$$s - d = \frac{h^2}{2s} + \frac{h^4}{8s^3}$$

If the angle of slope v of the tape is measured with the transit, for instance, the horizontal distance will be $s \cos v$ (Fig. 3.13.2). But here again it is easier to compute the **difference** in length between the slope and the horizontal distance.

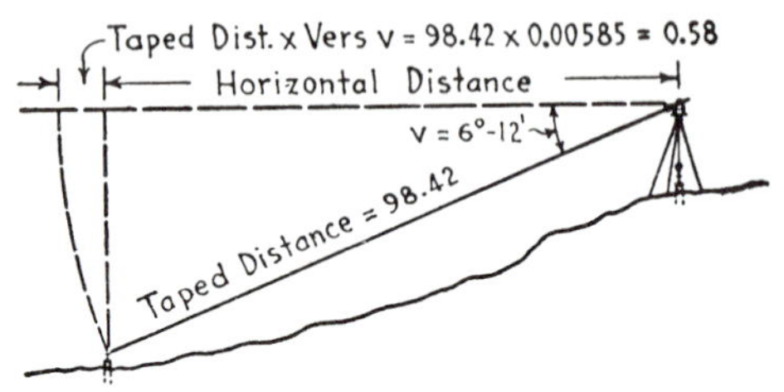

FIG. 3.13.2. SLOPE DISTANCE AND VERTICAL ANGLE MEASURED IN FIELD. HORIZONTAL DISTANCE COMPUTED BY VERS SINE METHOD.

Instead of computing $d = s \cos v$, compute $s - d = s \text{ vers } v$* and obtain the result by slide rule or by simple arithmetic. When the angle v is smaller than 10°, the difference between the inclined and the horizontal distance is small; for example, if the slope distance is 98.42 and the vertical angle is 6° 12′, the horizontal distance is 98.42 − 98.42 × vers 6° 12′ = 98.42 − (98.42 × .00585) = 98.42 − 0.58 = 97.84 ft. (Fig. 3.13.2). Generally speaking, a slide rule calculation of $s = d$ is as accurate as a 5-place trigonometric calculation of d would be if the formuls $s \cos v$ were used.

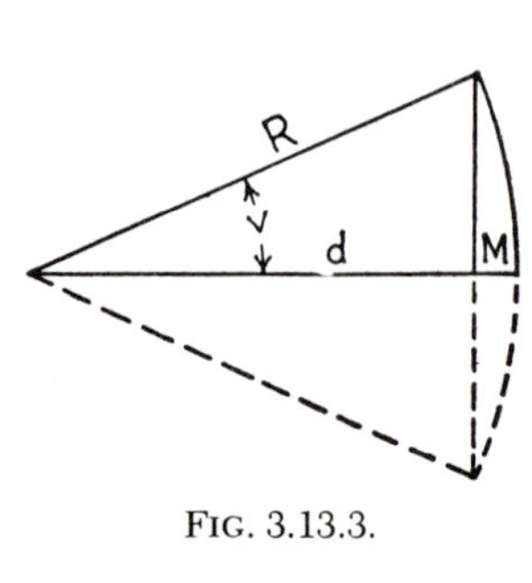

FIG. 3.13.3.

*In most texts on Trigonometry vers v is expressed as being $1 = \cos v$, which is correct; but it is not in this form best suited to the surveyor. He customarily senses each trigonometric function of an angle as the ratio of one side of a right triangle divided by another. It is therefore advisable to think of vers v as the ratio of the distance M divided by the radius R. In Fig. 3.13.3, vers $v = \frac{M}{R}$, and it is this ratio which should be visualized and kept in mind. It can readily be shown that, since $M = R - d$, the vers $v = \frac{R-d}{R} = 1 - \frac{d}{R} = 1 - \cos v$.

3.14. Common Mistakes in Taping

1. Failure to observe the correct position of the zero-point of the tape:—(1) End of ring; (2) end of steel band; (3) on steel band.
2. Omitting a whole tape-length.
3. Transposing figures, e.g., 63.56 for 63.65 (mental); or reading tape upside down, e.g., 6 for 9, or 86 for 98.
4. Reading wrong foot-mark, as 63.56 for 62.56.
5. Subtracting incorrectly when using tape graduated on first foot only.
6. Tape not stretched straight.

Most of these blunders are detected by repeating the distance measurement from the other end. It is, however, essential that great care is being taken during the measurements to avoid such blunders. A taping crew has a certain responsibility and a systematic, clear and distinct communication between tapeman and note keeper is just as important as familiarity with the particular tape and its graduation. Maintaining a straight path for the tape is essential, since bending the tape around trees, bushes or boulders will cause the measured distance to be too great.

3.15. Sources of ERROR in the Measurement of Lines.—The principal sources of tape measurement errors are

1. Erroneous length of tape
2. Variation in temperature
3. Incorrect tension
4. Sag
5. Wind
6. Incorrect alignment—Horizontal and Vertical
7. Pinsetting, reading, plumbing

3.16. Testing Length of Tape.—The U. S. Bureau of Standards, for a small fee, will test a steel tape, giving its length at standard temperature (68° F. = 20° C.) when supported its full length and subjected to a 10-pound pull. A certificate, giving the result of such a test, will be furnished provided that the tape is made of a single piece of metal ribbon and that none of the graduations are on pieces of solder, on sleeves attached to the tape, on wire loops, spring balances, tension handles, or on other attachments which are liable to be detached or to become changed in shape. If these requirements are not met, a "*report*" will be furnished, but not a "*certificate.*" For an additional fee the Bureau will test the tape when supported at the ends. A copy of a typical tape certificate is in Fig. 3.16.1.

It is good practice to have the U. S. Bureau of Standards test, for instance, a **new** 100-ft. tape and a **new** 300-ft. tape and submit a full report, based on test made with different amounts of tension both when the tapes are supported throughout their length and when they are suspended only at their ends. Thereafter these tapes **should never be taken into the field** but should be kept in the office to serve as available standards with which the tapes used in the daily work can be frequently compared.

It is obvious that in making precise measurements the tape should be used under the conditions for which it was standardized; however this seldom can be done. Consequently, the corrections must be computed and applied.

3.17. Incorrect Length of Tape. Tapes that are either too long or too short will obviously give incorrect measurements of distance. In either case **the error is cumulative.** If the error per tape-length is known, a correction can be computed which, when applied to the observed measured distance, will give the correct distance between the points. The corrections may be + or −, depending on whether the tape is too long or too short; it is obviously **important to apply it in the proper manner.** If the tape is too long it will not go as many times as it should between the two points, and the observed distance therefore will be too short; the correction should be added to obtain the true length. If the tape is too short it will go more times than it should between the points, and the observed distance will be too long; the correction should be subtracted to obtain the true length.

Suppose, for example, that the distance is known to be exactly 500 ft. between points A and B and that the line AB is

Form 979

U. S. DEPARTMENT OF COMMERCE
WASHINGTON

National Bureau of Standards
Certificate

FOR
100-Foot Steel Tape
NBS No. 8272

Maker's Identification Mark
The Lufkin Rule Co.

SUBMITTED BY

Massachusetts Institute of Technology,
Department of Civil & Sanitary Engineering,
Camp Technology,
East Machias, Maine

This tape has been compared with the standards of the United States. It complies with the specifications for a standard tape, and the intervals indicated have the following lengths at 68° Fahrenheit (20° centigrade under the conditions given below:

Supported on a horizontal flat surface:

Tension	Interval	Length
10 pounds	(0 to 100 feet)	100.002 feet
15 pounds	(0 to 100 feet)	100.005 feet
20 pounds	(0 to 100 feet)	100.008 feet

Supported at the 0 and 100-foot points:

Tension	Interval	Length
10 pounds	(0 to 100 feet)	99.876 feet
15 pounds	(0 to 100 feet)	99.949 feet
20 pounds	(0 to 100 feet)	99.976 feet

See Note 3(a) on the reverse side of this certificate.

For the Director,

Lewis V. Judson
Lewis V. Judson,
Chief, Length Section,
Metrology Division

Test No. II-1/113327
Test completed: June 25, 1947

The comparisons of this tape with the United States Bench Standard were made at a temperature of 75°Fahrenheit and in reducing to 68° Fahrenheit (20° centigrade), the coefficient of expansion of the tape is assumed to be 0.00000645 per degree Fahrenheit (0.0000116 per degree centigrade).

11—8242 U.S. GOVERNMENT PRINTING OFFICE

3(a) Unless otherwise stated, the comparisons of this tape with the Bench Standard have been made at the center of the lines on the edge to which the shortest graduations are ruled. If all the graduations extend entirely across the tape, the ends farthest from the observer when the zero of the tape is at his left hand are used.

FIG. 3.16.1. TAPE CERTIFICATE.

to be measured with a 100-ft. tape which is 0.02 ft. short; i.e., its length is actually 99.98 ft. as the standard tape would be if the temperature were at about 30° F. In measuring AB with this tape, its full length will obviously be used five times and each length will be recorded as 100 ft., although the tape will fail to reach the point B by 0.10 ft., because each time it is laid down it will cover only a distance of 99.98 ft. Consequently, the length of AB that this erroneous tape will give is 500.10 ft. To obtain the correct length, 5 × 0.002 ft. = 0.10 ft. should be subtracted from the recorded length, giving 500.00 ft.

Steel tapes are not easily broken when used properly. In unreeling a tape, the reel (or handle) should be held in such a position that the tape will unwind in a direction **tangent to the reel.** Many tapes may be removed from their handles; when unreeling and removing the tape, care should be used to avoid breaking off the end ring to which the handle is attached.

In placing the tape in the reel, the 100-ft. end should be attached to the reel in such a way that the graduations will face the inside of the coil. If the reel is held in the left hand, the 100-ft. end of the tape should be attached to the reel with the graduations up; the winding handle of the reel should be turned clockwise, using the right hand.

Some surveyors remove the tape from the reel at the beginning of the day's work and replace it when the day's work is done. Others do not replace the tape, but loop it in one hand in about 5-ft. lengths so that it forms a figure 8, and then give the 8 a twist that turns it into an O shape as the tape is placed in the kit-bag, or hung on a hook for the night.

Sharp bends and kinks in the tape should be avoided; tapes having kinks in them are easily broken when tension is applied. It is not advisable to allow even rubber-tired vehicles to run over tapes.

3.18. Temperature—Effect, Variation, and Application.—Steel tapes are of standard length when at a temperature of 68° F. If the length of the tape is known at 68° F. and a measurement of a line is made when the temperature is 30° F., it is evident that the tape will be shorter at 30° F. than at 68° F.; hence a correction must be applied to the observed length of the line to obtain its true length. If the measurement is made when the temperature is 90° F., it is obvious that the tape will be longer than at 68° F. If the temperature varies during the measurement of the line, the error is systematic so long as observed temperatures are consistently below 68° F. or above 68° F. If, during the measurement, some of the observed temperatures are above and some below 68° F., the errors tend to be random.

In computing the change in length of a steel tape caused by a change in temperature, we must know the "*thermal coefficient of expansion*" of the tape. The value adopted for steel tapes is 0.00000645 per 1° F., which is the change per unit length of the tape caused by a change in temperature of 1° F. Thus, the change in the length of a tape 1 ft. long for a change of temperature of 1° F. will be 0.00000645 ft. For a 100-ft. tape, the change will be for 1° F., $100 \times 0.00000645 = 0.000645$ ft. For 15° F. change in temperature, the change in length of a 100-ft. tape will be $15 \times 0.000645 = 0.0097$ ft., or approximately 0.001 ft.; a value which the surveyor **should always keep in mind.**

In determining the temperature only high-grade thermometers should be used. Especially constructed thermometers that may be attached to the tape by clips are available. Two of these are commonly used, one placed at, or near, each end. Sometimes the thermometers are placed on the ground. For best results the measurements should be made on cloudy days when the temperature is about constant or changing slowly; on sunny days there may be considerable difference between the temperature of the tape and of the air.

Example: A line measured with a steel tape under a tension of 10 lbs. was found to be 896.72 ft. long when the mean temperature was 36° F. If the tape under a tension of 10 lbs. at 68° F. was 100 ft. long, compute the correct length of the measured line. The change in temperature $68° - 36° = 32°$. The correction per tape-length of 100 ft. is $32° (0.00000645)(100) = 0.021$ ft. Since there are 8.9672 tape-lengths, the total correction to the measured distance is $8.9672 \times 0.021 = 0.19$ ft. Therefore, the corrected distance, since the tape is too short, is $896.72 - 0.19 = 896.53$ ft.

3.19. Tension.—The standard tension for tapes 100 ft. or less in length when supported throughout their length on a level surface is 10 lbs. It is evident that a different tension applied to the tape, say, 20 lbs., will cause the tape to stretch; hence its length will be changed. Light steel tapes will stretch between 0.01 and 0.02 ft. in 100 ft. if the tension is increased to 20 pounds. Heavier tapes will stretch much less than this. The increase in length caused by an increased tension is sometimes computed by the formula

$$C_p = \frac{L(t - t_0)}{SE},$$

in which C_p is the correction per tape-length in feet

L is the length of the tape in feet
t is the applied tension in pounds
t_0 is the standard tension in pounds
S is the cross-sectional area in square inches
E is the modulus of elasticity of the steel in pounds per square inch (about 28,000,000 to 30,000,000).

The increase in length that is caused by an increase in tension may be measured directly for any tape by varing the tension by means of a spring balance and noting the resulting change in length. Fig. 3.19.1 shows tension handle with spring balance which can be clipped to end of tape.

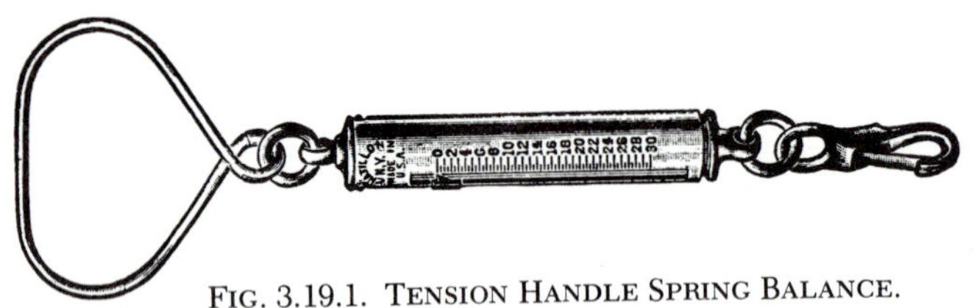

FIG. 3.19.1. TENSION HANDLE SPRING BALANCE.
(Courtesy, Keuffel & Esser Company)

In ordinary surveying work, experienced tapemen can estimate the tension sufficiently close to come within the required precision. It is only on precise work that the spring balance is used in applying tension.

3.20. Sag.—A tape is of **standard length when it is supported throughout its length and under a definite tension.** Suppose that while the same tension (10 lbs.) is being applied a measurement is made when the tape is supported only at its ends. It is evident that this tape will hang in a "catenary" curve and that the horizontal distance between the end points will be less than the tape reading. The amount of this shortening caused by sag depends on the weight of the tape, the distance between supports, and the tension applied. It may be computed by the formula

$$C_s = \frac{w^2 l^3}{24t^2}$$

in which C_s is the correction between supports in feet
w is the weight of the tape in pounds per foot of tape
l is the distance between supports in feet
t is the tension in pounds.

A practical way of determining the effect of the sag for any given tape is, as follows: Select a horizontal piece of ground near a vertical wall of a wooden building where the tape may be stretched out on the level ground, supported throughout its entire length. Drive a nail A firmly into the wall about waist high. The head of the nail should be 2 or 3 in. from the wall and a definite point should be marked on the nail-head. By means of a plumb-bob, set point A' vertically under A, flush with the surface of the ground. Mark A' carefully to indicate the exact point. Next, stretch the tape out on the ground along the wall. Hold the zero end on A', apply a 10-pound tension with a spring balance at the other end of the tape, and set B' at the same distance from the face of the wall as A'. Indicate by a mark on B' the exact position of the 100-ft. mark on the tape. By means of a plumb-bob, set a nail B vertically over B' and **at the same height** as A. Mark the exact point B carefully. With a 10-pound tension on the tape, measure the distance AB along the face of the building, the tape being unsupported in this case. The difference in the readings $A'B'$ and AB is the shortening because of sag. If the pull on the tape is now increased until the 0 and 100-ft. marks on the tape coincide exactly with the elevated points A and B respectively, the additional tension applied is the amount required to eliminate the effect of sag.

It should be kept in mind that the distances $A'B'$ and AB are not necessarily exactly 100 ft. apart, for the reason that the graduated length of the tape may be slightly in error. If the tape has been compared with a standard and its correct length is known, the distance $A'B'$ may be laid out exactly 100 ft., and a temperature correction may be applied, if necessary. (See Art. 3.18.) It is advisable to lay out AB and $A'B'$ in a shady location where the temperature is likely to change but little while comparisons are being made.

3.21. WIND.—Wind will blow the tape off line especially when plumbing, thus causing the measured distance between the points to be **too great.** The effect is equivalent to the sag, with a horizontal wind force component acting on the tape.

3.22. INCORRECT ALIGNMENT.—When a line is longer than a tape-length it must be measured in parts and the intermediate points must be "lined in" either by eye or by the transit. If the distance only is required, it is usually possible to line the intermediate points in so closely by eye that the resulting error in length is negligible. The magnitude of the error may be computed by the approximate formula $s - d = \frac{h^2}{2s}$. For example, if one end of the tape is 1 ft. off line in a 100-ft. tape-length the error will be $(1)^2 \div 200 = 0.005$ ft. In 40 ft. a 1-ft. error in alignment will be $(1)^2 \div 80 = 0.012$ ft. Fig. 3.22.1 illustrates the amount of error introduced by holding the tape 1 ft. off line for different distances and shows clearly the **importance of correct alignment when the distances are short.** "Lin-

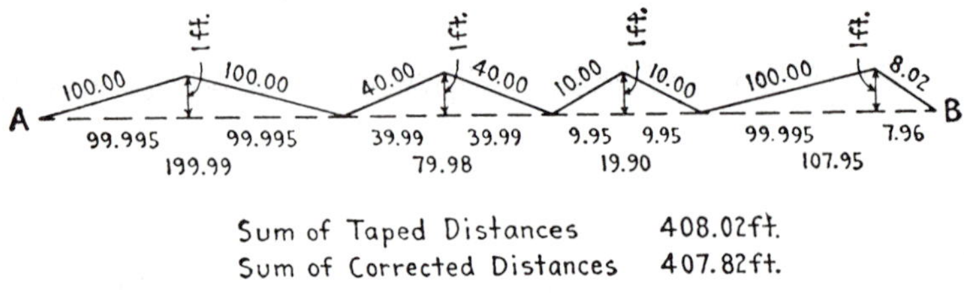

FIG. 3.22.1. ERRORS IN MEASUREMENT OF DISTANCES DUE TO CARELESS ALIGNMENT OF INTERMEDIATE POINTS. NOTE THE LARGE ERROR WHEN TAPE-LENGTHS ARE SHORT.

ing in" can ordinarily be done by eye much closer than within one foot, usually within 0.2 or 0.3 ft.

By the same method the error in length caused by failing to hold the tape horizontal may be computed.

It will be observed that **errors in horizontal and vertical alignment are systematic.**

3.23. PINSETTING, READING, PLUMBING.—It is obvious that carelessness in these operations will cause errors such that the length of each segment might be affected. Errors of this kind are random, and although they cannot be completely avoided, they can be reduced by proper care to a small amount.

3.24. CARE OF STEEL TAPE.—The tape should be kept in the reel when it is not being used, and it should be cleaned at the end of the day. A damp or muddy tape should be cleaned, and wiped first with a dry cloth and then with an oiled one; a tape that is only dusty may be wiped with a dry or slightly oiled cloth. If the tape is taken care of in this manner, it will not get rusty. If, however, through neglect it should get rusty, the spots can be removed with steel wool or dry Portland cement, and the tape polished off with an oily cloth. After it has been reeled, it should be placed carefully, not thrown, in the bag.

3.25. THE PARALLACTIC TRIANGLE.—Indirect optical distance measurements utilize geometric relations in the parallactic triangle (see Fig. 3.25.1).

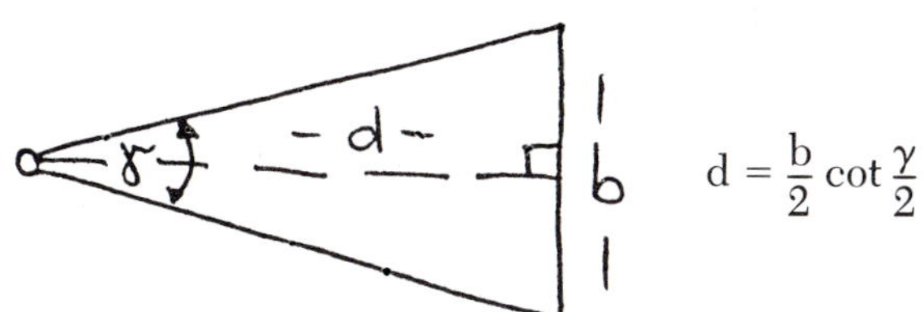

FIG. 3.25.1. THE PARALLACTIC TRIANGLE.

With d being the unknown distance, there are two possibilities for its indirect determination, namely:

a) keep γ constant and measure b, as with ordinary stadia method or with the variable base range finder

b) keep b constant and measure γ, as with the subtense bar or with the fixed base range finder.

3.26. The Variable Base Rangefinder.—This instrument uses principle (a). In this case, the base is at the observation end. Two small telescopes which are mounted such that their respective lines of sight intersect at a fixed angle can be set onto a target by moving them along the base. The instrument gives a low accuracy and is mainly used for simple military ranging.

3.27. Fixed Base Rangefinders.—In simplicity, cost, precision and application equivalent to the variable base rangefinders, these instruments use two telescopes mounted on a fixed base. By rotation of one or both of them, the parallactic angle changes. These changes are directly translated into distances which can be read on the rotation knob.

3.28. The Stadia.—Distances can be measured very rapidly by the stadia method. In this method the distance is obtained by sighting the telescope of a theodolite on a graduated rod held at the other end of the line, and observing the interval on the rod included between two special horizontal cross-hairs (stadia-hairs) set in the telescope. From the observed interval, the distance from the instrument to the rod is readily computed.

Although this approach is rather simple and fast, there are some drawbacks, namely the limited accuracy due to the direct reading of the rod and the fact that the distances obtained are slope distances. Further instrument development has dealt with these problems which led to rather complicated instruments which are mentioned here for completeness' sake.

Once useful double image tacheometers (e.g. Zeiss Redta, Wild RDH) which apply an optical vernier system to increase the accuracy, have given way to simple electro-optical distance measuring devices. (See 3.39)

Self reducing tacheometers (e.g. Wild RDS, Kern DR–KV) use curves instead of horizontal crosshairs, thereby varying the effective angle as a function of telescope inclination. The stadia distances are then automatically reduced to horizontal distances, a fact which makes these instruments very useful for conventional mapping.

3.29. Stadia Method of Measuring Distances.—The stadia method of locating points is one in which distances are measured by observing through the telescope of a transit the space, on a graduated rod, included between two horizontal hairs called *stadia hairs.* If the rod is held at different distances from the telescope different intervals on the rod are included between the stadia hairs, these intercepted spaces on the rod depending upon the distances from the rod to the instrument, so that the intercepted space is a **measure** of the distance to the rod.

Owing to the fact that in making a stadia measurement the intervening country does not have to be traversed, as is necessary when making a tape measurement, distances can be taken across ravines and water surfaces, and over rough as readily as over smooth ground. This gives the stadia a great advantage over the tape in point of speed. Another advantage of this method over that of tape measurements is that the chief errors of stadia measurements are random while those of tape measurements are mostly systematic. Furthermore, the accuracy of the stadia measurements is not diminished in rough country, so that the results obtained by this method are, under some conditions, as accurate as tape measurements. While surveys of property boundaries ordinarily demand the use of transit and tape, still the stadia method is well adapted to the survey of cheap land, such as marsh or timberland. In many instances the boundaries of such properties are so uncertain that the latter method is amply accurate. In highway or railroad surveys preliminary plans must often be prepared in a short time; for this purpose the stadia method is well adapted. Furthermore, where an accurate tape and transit survey is being made it is often desirable to locate also certain physical features whose precise location is not required; these may properly be located by stadia measurements, and the use of stadia for locating such details is common. As a means of obtaining a rough check upon steel tape measurements the stadia is of especial value. The lengths of all lines may be read quickly by stadia, and mistakes of a foot or more in taping readily detected. It is also useful for taking sights across a traverse, or "cut-off lines," for the purpose of dividing the traverse into parts which may be checked independently. The stadia method is also especially adapted to topographical surveying.

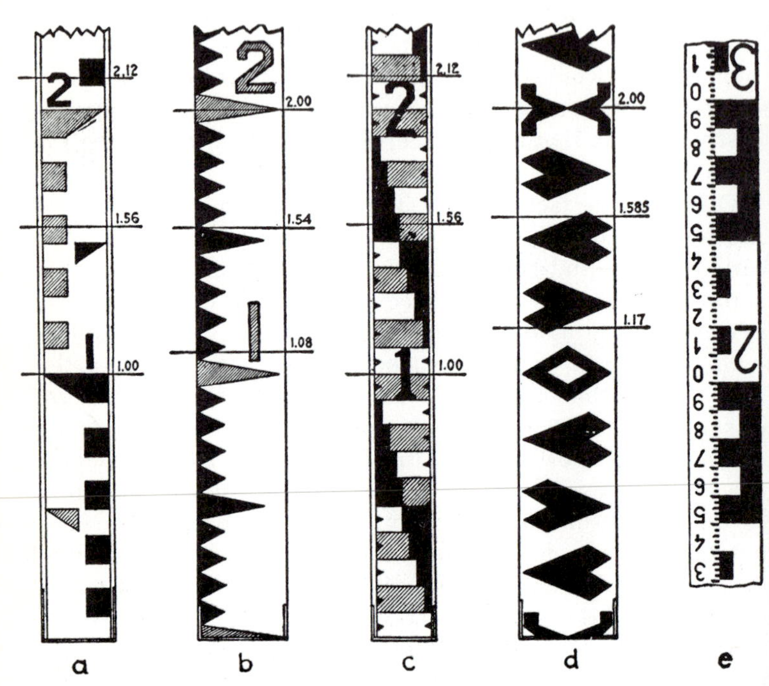

FIG. 3.30.1. STADIA RODS.

3.30. Stadia Rods.—There are many kinds of stadia rods, differing both as to the construction of the rod itself and the style of diagram used to mark the graduations. The diagrams for ordinary work should be simple, so that long distances may be read quickly. Complicated diagrams are to be avoided except where all of the sights are short and where greater precision is desired than is usually required in stadia work. In the rods shown in Fig. 3.30.1 the diagrams are made so that the

Leveling and Stadia Rod No. 5 E
of wood, single folding type, length 3 m or 4 m, with upright figures. For use with the GK0-E, GK0-A, GK1-A, GK23-E levels and the RK planetable equipment

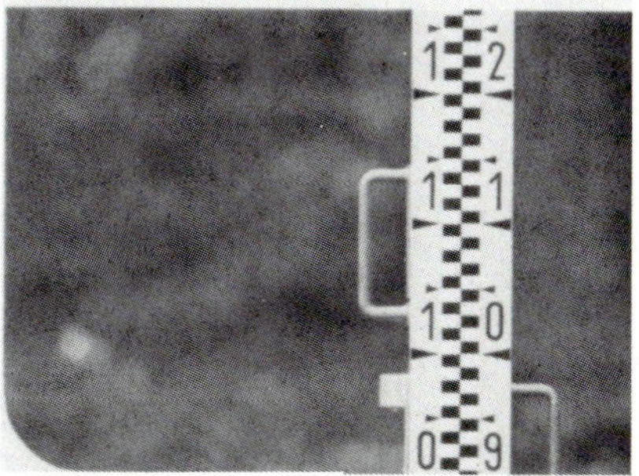

eveling and Stadia Rod No. 3
wood, single folding type, ngth 3 m or 4 m. For use with the K0, GK1, GK23 levels and e K1-RA tachymeter theodolite

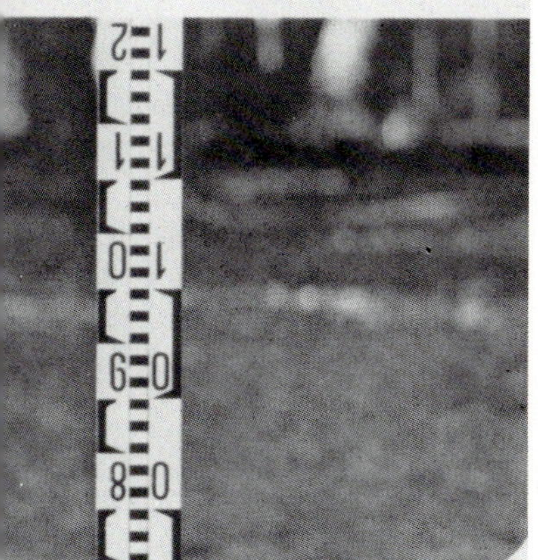

Topographic Rod No. 7
of wood, single folding type, length 3 m or 4 m, with upright figures. For use with the RK planetable equipment

FIG. 3.30.2. ALTERNATIVE STADIA RODS.

0.05-ft. or 0.10-ft. spaces can be distinguished easily and the hundredths of a foot estimated.

The rods on which these graduations are painted consist of wooden strips from 3 to 5 inches wide and 10 to 15 ft. in length. For convenience in carrying the rods they are usually made in two sections joined by hinges, but sometimes the two parts are separate and are clamped together when in use.

In Fig. 3.30.1 the rod (a) is particularly useful for reading distances up to about 500 ft. Rod (b) is readable up to about 1000 ft., and rod (c) is good for both short and medium distances. Rod (d) is particularly applicable to distant readings. Rod (e) is a combination Philadelphia and stadia rod with the graduations inverted for viewing through inverted telescopes, such as found on many foreign instruments. The Philadelphia leveling rod, Fig. 3.30.1, may be used for taking stadia readings up to a distance of about 300 feet. In Figure 3.30.2 alternative rods are illustrated.

3.31. Principle of the Stadia.—The fundamental principle upon which the stadia method depends is the simple geometric proposition that in two similar triangles homologous sides are proportional. According to Fig. 3.31.1 This proposition yields

$$i : a = s : (D - C)$$

Which leads to

$$\frac{1}{a} = \frac{s}{i\,(D-C)}$$

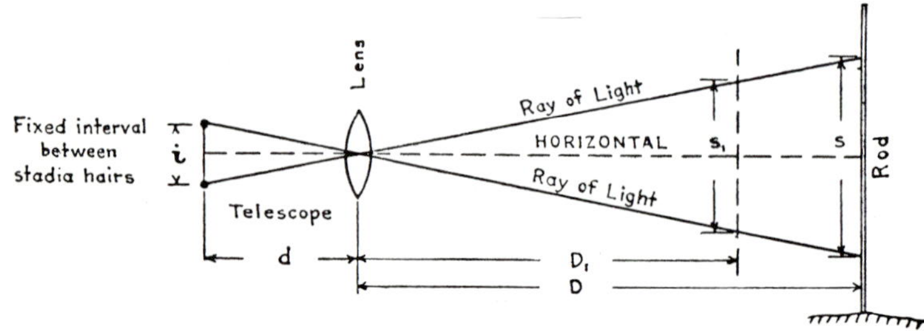

FIG. 3.31.1. DIAGRAMMATIC REPRESENTATION OF PRINCIPLE OF THE STADIA.

a, however, is the image distance for the stadia lens, and therefore has to satisfy the general lens equation:

$$\frac{1}{D-C}+\frac{1}{a}=\frac{1}{F}$$

By substitution we obtain:

$$\frac{1}{D-C}+\frac{s}{i\,(D-C)}=\frac{1}{F}$$

and

$$\frac{i+s}{i\,(D-C)}=\frac{1}{F}$$

$$Fi + Fs = iD - iC$$

$$D = \left(\frac{F}{i}\right) s + F + C$$

Since F, i and C are constant for any given instrument, a direct relation between rod reading and distance is obtained. Using a given focal length for the objective lens, the other quantities are dimensioned such that the evaluation is simplified.

With $\frac{F}{i} = k$

the formula becomes

$$D = k \cdot s + (F + C)$$

For most theodolites k equals 100 and (F + C) is negligibly small. This means that the stadia distance equals 100 times the rod section, it also shows clearly, that a reading accuracy of 1/100th of a foot at the rod results in an accuracy for the distance of one foot.

3.32. Formulas for Inclined Sights.—In practice it is customary to hold the rod plumb rather than perpendicular to the line of sight, because the former position can be judged easily and accurately, whereas it is not easy to determine when the rod is perpendicular to the line of sight and often is difficult to maintain it in that position. On inclined sights, when the rod is plumb, the vertical and horizontal distances evidently cannot be found simply by solving a right triangle. In Fig. 3.32.1 let *AB* be the intercept on the rod when it is held vertical, *A′B′*

the intercept when the rod is perpendicular to the line of sight, i.e., $A'B'$ is perpendicular to CO. In the triangle AOA', $\angle O = \angle \alpha$, the measured vertical angle; $\angle A' = 90° + m$; and $\angle A = 90° - (\alpha + m)$, m being half the angle between the

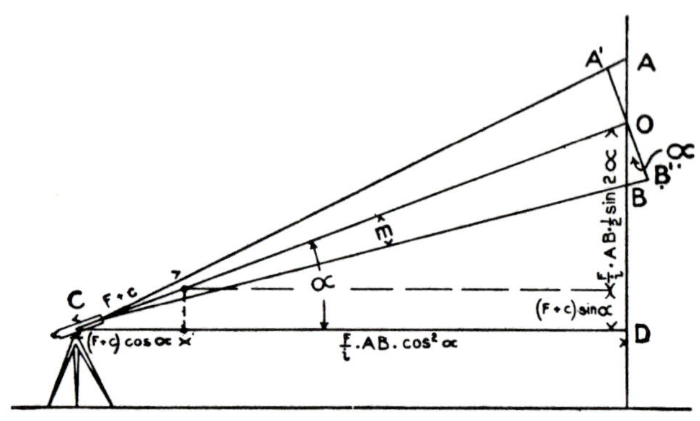

FIG. 3.32.1.

stadia hairs. In the triangle BOB', $\angle O = \angle \alpha$; $\angle B' = 90° - m$; and $\angle B = 90° - (\alpha - m)$.

Then
$$\frac{AO}{A'O} = \frac{\sin(90° + m)}{\sin\{90° - (\alpha + m)\}}$$

and
$$\frac{BO}{B'O} = \frac{\sin(90° - m)}{\sin\{90° - (\alpha - m)\}}$$

$$AO + OB = AB = \tfrac{1}{2}A'B'\left\{\frac{\cos m}{\cos(\alpha + m)} + \frac{\cos m}{\cos(\alpha - m)}\right\}$$

from which may be obtained

$$A'B' = AB\cos\alpha - AB\frac{\sin^2\alpha}{\cos\alpha}\tan^2 m$$

The value of the second term is very small; for $AB = 15.00$ ft., $\alpha = 45°$, and $m = \tan^{-1} 0.005$, this term is only 0.0002 ft., and hence it may always be neglected. In other words it is sufficiently accurate to regard the angles A' and B' as right angles.

$$\therefore A'B' = AB\cos\alpha$$

The difference in elevation between the center of the instrument and the point O on the rod is derived as follows:

$$DO = CO \sin \alpha$$

$$= \left\{\frac{F}{i}A'B' + (F + c)\right\} \sin \alpha$$

$$= \frac{F}{i}AB \sin \alpha \cos \alpha + (F + c) \sin \alpha$$

$$= \mathrm{k} \cdot \mathrm{s} \cdot \tfrac{1}{2} \sin 2\alpha + (F + c) \sin \alpha \qquad [A]$$

For the horizontal distance from the transit point to the rod we have

$$CD = CO \cos \alpha$$

$$= \left\{\frac{F}{i}A'B' + (F + c)\right\} \cos \alpha$$

$$= \frac{F}{i}AB \cos^2 \alpha + (F + c) \cos \alpha$$

$$= \mathrm{k} \cdot \mathrm{s} \cdot \cos^2 \alpha + (F + c) \cos \alpha \qquad [B]$$

3.33. Application of Stadia Formulas.—Most of the stadia tables, diagrams and stadia slide rules are based upon these two formulas [A] and [B]. Table A in Figure 7.33.1 gives values for the horizontal correction per 1.00 ft, rod interval based on k = 100 (the usual stadia hair interval).

The difference in elevation between the axis of the telescope and the point on the rod where the middle hair cuts it is called the *vertical height*, and is obtained by multiplying the rod interval by the number in Table B. But that gives only the first term of equation [A]. This is sufficient when internal focusing instruments are being used. With external focusing transits the vertical height represented by the second term is obtained as follows:

The error in DO is almost negligible if $(F + c)$ is first added to $100 \times AB$ which is equivalent to adding 0.01 ft. to the rod interval, and then obtaining the vertical height from the table as before. Hence the process of finding the vertical height,

DO, involves merely adding 0.01 ft. to the rod interval and multiplying the result by the tabular amount (Table B).

Table A provides the amount the horizontal distance is less than $100 \times AB$. This is not equation [B]; it is the value of $100\ AB \sin^2 \alpha$. To allow for the constant $(F + c)$ its amount (1 ft.) is added to $100 \times AB$ and from this sum the horizontal correction is subtracted in obtaining the true horizontal distance.

Strictly, the way in which the second factor in equations [A] and [B] have been used is not precise, but wholly in keeping with the fact that the rod interval is not read closer than 0.01 ft. corresponding to 1 ft. in horizontal distance.

3.34. SUBTENSE BAR.—This device consists of two targets supported on each end of a metal bar or tube mounted horizontally on a tripod. Distances are obtained by measuring angles subtended by the targets. In Fig. 3.34.1 the targets are

FIG. 3.34.1. SUBTENSE BAR.
(Courtesy, Kern Instruments, Inc.)

maintained exactly two meters apart by means of an invar wire under tension. This represents one of the major applications of principle (b). The other one is stereophotogrammetry where distances to object points imaged on two overlapping photographs are obtained by using angles measured or derived from measurements and the fixed photobase. This simple principle is fundamental for a whole scientific field, photogrammetry, which is briefly described in a separate chapter.

Subtense bars are manufactured by most of the major European instrument firms and can be used with any theodolite. As the subtense bar is precisely 2 metres long, the formula for indirect distance measurement is reduced to

$$D\left[\text{metres}\right] = \cot\frac{\gamma}{2}$$

The use of good cotan tables will evaluate the distance in metres, with division by 3.280839895 to give the result in feet. Tables of distances in feet for the subtended angle are also available from the manufacturer. Since nothing is required except a firm support for the theodolite and for the subtense bar, it follows that distance can be measured across streams, chasms, busy highways, swamps, etc.—even from atop a building to another building or to the ground—as long as there is intervisibility between the stations. An additional advantage of the subtense bar method is that the distance obtained is always the horizontal distance (not the slope distance), regardless of any difference of elevation. The reason for this "automatic" reduction is that the theodolite measures a horizontal angle between two vertical planes formed by the azimuth axis of the theodolite and the respective ends of the subtense bar. Thus, any inclination of the line of sight has no effect.

The accuracy of the distance is a function of the accuracy of the fixed 2-meter length of the subtense bar and the accuracy of the angle measured. It can be stated, closely enough, that the relative error in distance equals the sum of the relative errors in base and angle, thus

$$\frac{\Delta D}{D} = \frac{\Delta b}{b} + \frac{\Delta\alpha}{\alpha}$$

Example: To achieve a relative accuracy of 1:5,000 for a distance of 100 metres, the difference ΔD cannot exceed ± 2 cm. Since 1:5,000 is the result of an addition, a relative accuracy of 1:10,000 is required for both parameters b and γ. For the base, this means 0.2 mm, which is easily maintained by the manufacturer. The parallactic angle γ is about $1° 08' 45''$ for 100 m, which means

$$\Delta\gamma'' = 1.146° \times \frac{1}{10{,}000} \times 60 \times 60 = \pm 00.4 \text{ seconds of arc.}$$

This is a surprisingly high accuracy requirement. Assuming an average 0.05 second of arc and 1:10,000, then 75 metres represents the maximum distance that can be measured directly. By making 8 to 10 repetitions with an engineer's theodolite or measuring 3 sets with a precision theodolite such angular accuracy will result.

For longer distances, it is advisable to set up the subtense bar between the stations. Beyond 150 metres, the use of the subtense bar equipment becomes cumbersome and often impractical. It is, however, used in the so-called trigonometric traversing, where a base is established by subtense bar perpendicular to a distance to be measured. Then the parallactic angle is measured to both end points of the base while the subtense bar principle is again applied.

3.35. Electronic Distance Measurement.—These distance measurements are accomplished by transmitting beams of electromagnetic waves between instruments located at the ends of the line to be measured. The beams are modulated precisely so that by comparing the transmitted and received beams, at one or both ends of the line, the distance can be determined. Accurate distance determination depends upon accurate knowledge of the velocity of the particular electromagnetic radiation being used, involving corrections for temperature, atmosphere pressure, and relative humidity. These instruments generally require that the ends of the line be intervisible and occupiable. The distances measured are the slope length between sending and returning components of the instruments. The instruments are constructed so that they may be operated by a surveyor after a short period of instruction and without much knowledge of the electronic features. Failures of the electronic features are referred to the manufacturer or an electronic engineer.

The advantage of the electronic method is that long distances can be measured with high accuracy in much less time and expense than by tape. Except for the necessity of line of sight between stations, the measurements are independent of roughness of terrain. Distance may be measured easily where taping is difficult if not impossible.

A number of electronic distance measuring instruments are available. The electronics of these instruments and the details of their operation are too complex to be described fully in this text. A detailed description of the instruments and their use may be obtained from manuals prepared by their manufacturers.

3.36. Instruments Utilizing Radio Waves.—These instrument systems use the longer waves (several hundred metres of wave length) over long ranges (several hundred miles), yielding a limited absolute accuracy. Their main uses are in hydrographic and oceanographic surveying and in navigation. For such applications the long range is essential, while the accuracy requirements are rather relaxed. It has to be kept in mind, however, that an error of ±100 cm over a distance of 1000 km still represents a relative accuracy of 1:10,000. Some systems of this type are: Cubic Autotope, Decca, Shoran, Hiram, Loran and Omega.

3.37. Microwave Instruments.—Utilizing wave lengths of a few centimetres, these instruments are capable of reaching absolute accuracies of better than ±2 centimetres over distances of up to 100 km. This means distance measuring accuracy over very long distances equalling or bettering the best taping accuracy.

An instrument system consists of two units, the master and the slave (or remote) unit.

The master unit is located at one end of the line and the remote unit is located at the other end. Each unit transmits a modulated beam which is independent of the beam being received from the other unit. This beam may be visualized as constructing a series of tapelengths placed end to end traveling through space at the velocity of the micro-waves. The beam being received at each end of the line is compared with that being transmitted. The result of this comparison at the remote station is relayed to the master station. The two com-

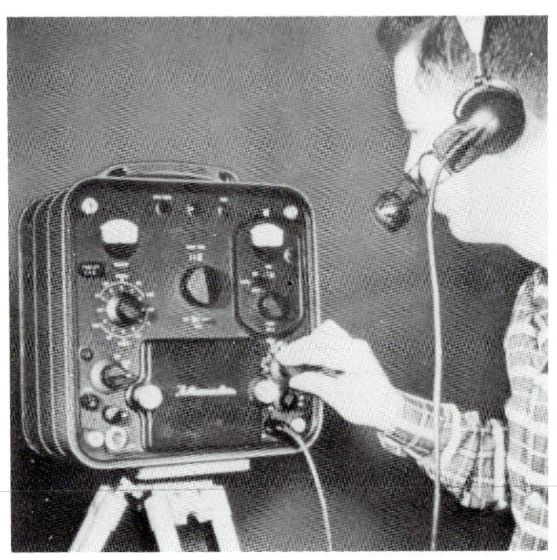

FIG. 3.37.1. TELLUROMETER MODEL MRA-3.
(Courtesy, Tellurometer, Inc.)

parisons are then combined to provide the measurement data.

The uncertainty of the meteorological conditions between the instruments may be a source of error when measuring over long distances. Measurements may be successfully made in haze, mist and light rain, and through light brush, but there should be no intervening terrain. Sights over large flat surfaces, such as bodies of water and highways tend to reflect the micro-wave radiation creating more than one pathlength between the instruments. Elevating the instruments on towers frequently reduces the effect of deflected waves. Another device for measuring over the terrain is by the use of a helicopter spotted over a point by an optical "hover sight." A rotating beacon on the aircraft marks the target. In table 3.37.1 some of the micro-wave systems are listed together with their measuring range and the achievable accuracy.

TABLE 3.37.1

MICROWAVE INSTRUMENTS

Instrument	Manufacturer	Range	Reading Accuracy
Electrotape	Cubic, U.S.A.	10 m to 50 km	1 cm.
Distameter	Ertel, W. Germany	50 m to 100 km	2 cm.
MRA2	Tellurometer, S. Africa	100 m to 150 km	5 cm.
MRA3		30 m to 80 km	15 cm.
MRA4		50 m to 50 km	3 mm.
MRA101		100 m to 50 km	1.5 cm.
Distomat DI50	Wild, Switzerland	50 m to 150 km	2 cm.

3.38. Instruments utilizing visible light.—The use of electro magnetic waves in the visible light range has led to an additional increase in accuracy, mainly because these extremely short wavelengths are less affected by atmospheric conditions. However, the range is considerably reduced, especially for daylight operations (e.g. 5 km for geodimeter 6). At night, and with a high powered light source, such as a laser, the range can be significantly increased but will hardly be as long as for microwave instruments.

Because of the nature of light waves, a remote transmission unit is not needed and an optical reflector takes its place. For practical purposes the reflector consists of one or more retroprisms which return light rays to the source, even if not placed exactly perpendicular to the direction of the rays.

There are basically two instrument types using radiation of this type, namely the Geodimeter and the Mekometer instruments, the latter manufactured by Kern, Switzerland. While the latter are relatively unknown, it should be mentioned, that they represent the most precise EDM instruments for distances exceeding one mile.

The geodimeters are in world wide use. There are several

FIG. 3.38.1. DISTANCE MEASURING DEVICE USING LIGHT.
(Courtesy, AGA Lidingö, Sweden)

models, indicating progress in development as well as instruments for different applications and ranges, such as geodimeter Model 8, which is equipped with a laser light source. All are using the same principle of comparing modulated light returned from the reflector with the transmitted beam. The use of three different frequencies leads to an accurate distance determination.

3.39. Instruments utilizing infrared radiation.—Although operating with approximately double the wavelength of the visible light type instruments, the accuracy of infrared equipment remains practically as good. There is not much difference in the operating procedure; however, due to the lower power transmission of the infrared diode, the range is limited to approximately one mile. Instruments of this type (see table 3.39.1) are usually compact, small and light-weight, and equipped with direct numerical readout. They are mostly used as a substitute for the tape, as their precision matches that of careful taping, even over short distances.

TABLE 3.39.1

SOME EDM INSTRUMENTS USING INFRARED RADIATION.

Instrument	Manufacturer	Range	Resolution	Internal Accuracy
HP 3800	Hewlett-Packard, U.S.A.	3 km.	2 mm.	0.5 cm.
DM 500	Kern, Switzerland	½ km.	2 mm.	0.6 cm.
DM 2000	Kern, Switzerland	2½ km.	1 mm.	0.4 cm.
MA 100	Tellurometer, S. Africa	2 km.	0.1 mm.	1.5 mm.
Disomat DI-10	Wild, Switzerland	1 km.	1 cm.	1 cm.
EOK 2000	Zeiss Jena, E. Germany	2 km.	1 mm.	1 cm.
SM11	Zeiss Oberkochen, W. Germany	1 km.	1 mm.	1 cm.

3.40. Laser EDM Instruments.—There are several instruments which use laser light instead of ordinary light. This means basically an extension of the range while maintaining all basic characteristics of the instrument (e.g. Laser Geodimeter).

3.41. Combined EDM-Angular Measuring Instruments.—While many EDM instruments can be directly interchanged with theodolites (forced centering), or attached to theodolites (e.g. Wild DI10, Kern DM500) a few combined instruments are on the market. These instruments are sometimes called electronic tacheómeters. The more popular ones are the Hewlett-Packard Total Station, Zeiss (Oberkochen) Regelta, distributed by Keuffel and Esser, and the Geodimeter 700.

FIG. 3.39.1. DISTANCE MEASURING DEVICE USING INFRARED LIGHT.
(Courtesy, Hewlett-Packard)

3.42. EDM-Measurements.—Depending on the instruments used, the procedures vary and cannot therefore be described here. Each equipment is provided with an operating manual by the manufacturers. Usually the procedures are rather simple and clear and require little or no previous training. Further, direct readout and/or storage facilities reduce blunders to a great extent.

CHAPTER IV

MEASUREMENT OF DIFFERENCE OF ELEVATION

4.1. Level Surface.—A level surface is a **curved** surface which at every point is perpendicular to the direction of gravity at that point, such, for example, as the surface of still water. Any line of sight which is perpendicular to the direction of gravity at a given point is therefore tangent to the level surface at that point and is called a *horizontal line*.

4.2. The Spirit Level.—In nearly all instruments the direction of gravity is determined by means of either a plumb-line or a spirit level. A spirit level is a glass tube, the inside of which is ground to a circular curve longitudinally, and nearly filled with a liquid such as alcohol or ether, leaving enough space to form a bubble (Fig. 4.2.1). The grinding is usually

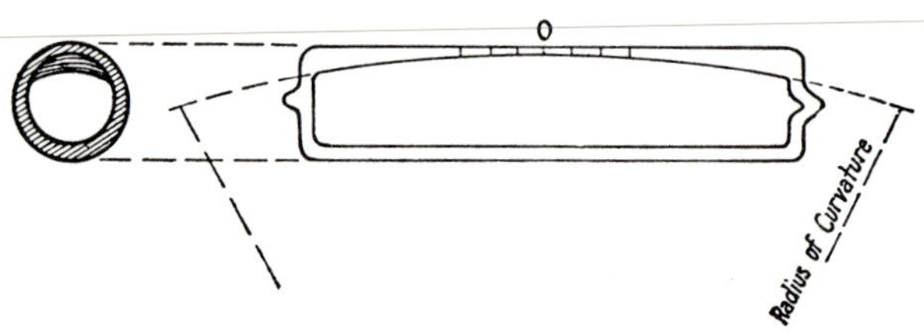

Fig. 4.2.1. Level Vial.

done only on the inside upper surface of the tube. The radius of the curve varies according to the use which is to be made of the level; a very short radius makes a slow-moving bubble while a long radius makes a very sensitive bubble. It is important that the curve should be exactly circular so that equal distances on the tube subtend equal angles at the center of curvature. The level is provided with a scale of equal parts, consisting of lines etched on the glass vial. A point near the middle of the tube is selected as the *zero-point* and the graduations are considered as numbered both ways from that point. The straight line tangent to the curve at the zero-point of the scale is called the *axis of the bubble*. The position of the bubble in the tube is determined by noting the positions of both ends. The bubble will change its length with changes in temperature, consequently the reading of one end is not sufficient to determine the position of the bubble. On account of the action of gravity on the liquid the bubble will always move toward the higher end of the tube; hence, when the bubble is **central** the axis of the tube is **horizontal.**

4.3. Angular Value of One Division of the Level Tube.—The angular value of one division of a level tube is the angle, usually expressed in seconds, through which the axis of the tube must be tilted to cause the bubble to move over the length of one division on the scale. The simplest way of finding this in the field consists in moving the bubble over several divisions on the scale by means of the leveling screws and observing the space on a rod passed over by the horizontal cross-hair, the rod being placed at a known distance from the instrument. The space on the rod divided by the distance to the rod gives the natural tangent of the angle through which the line of sight has moved. Since the angle is very small its value in seconds of arc may be obtained by dividing its tangent by the tangent (or sine) of one second.* Dividing the angle just found by the number of divisions of the scale passed over by the bubble in the experiment gives the average number of seconds corresponding to a single division.

For very precise leveling one would need an instrument with a bubble sensitivity of 0.25 second of arc or better. Lower order leveling might demand a 0.5 second sensitivity. For ordinary construction leveling a bubble of 10 to 20 seconds may well suffice (see also Art. 4.14).

4.4. Levels.—There are several instrument types used for the direct determination of differences of elevation such as the

*Instead of dividing by tan 1″ (= arc 1″ = sin 1″) it is often more convenient to multiply by 206 264.8, which is the reciprocal of arc 1″, and therefore the number of seconds in one radian.

Dumby Level, the *Tilting Level,* the *Self-Leveling* or *Automatic Level,* and the *Hand Level.*

All of these instruments, except the self-leveling level, are so constructed that the line of sight is horizontal when the bubble of the attached spirit level is in the middle of its tube. In the self-leveling instrument the line of sight is horizontal when it is perpendicular to a line that is defined by a pendulum device which hangs vertically in the direction of gravity.

4.5. THE DUMPY LEVEL.—In the dumpy level (Fig. 4.5.1) the telescope, the vertical supports, the horizontal bar and the

FIG. 4.5.1. THE DUMPY LEVEL.
(Courtesy, W. & L.E. Gurley)

vertical spindle are all made in one casting or else the parts are fastened together rigidly. The spirit level is fastened to the horizontal bar and can be adjusted in the vertical plane; there is no other adjustable part except the cross-hair ring.

The dumpy level is rigidly constructed, which means that the parallelism between the bubble axis and the line of sight should remain in adjustment. Nevertheless, it is essential to perform a peg test periodically to check this adjustment. It is good engineering practice to do this at the beginning of each job, particularly if the level has not been used for some time.

4.6. PRECISE TILTING LEVEL.—While it is possible to obtain a fairly high precision with the dumpy level, there are certain disadvantages in using it for longer lines. With the ordinary dumpy level it is not convenient for the observer to view the level bubble from the observing position. Should the bubble be not centered, he must shift from the observing position in order to center it. In changing his position the instrumentman not only loses time but may even cause a slight change in the elevation of the instrument. Furthermore even the longer time period required, if not the observer's movement around the tripod could cause the instrument to sink, especially when set up on marshy, sandy or frozen ground, or asphalt pavement. To overcome these difficulties, a level with a built-in tilting mechanism is used.

The precise tilting level (such as that of Fig. 4.6.1) may have four level screws, but more frequently three. A spirit level is

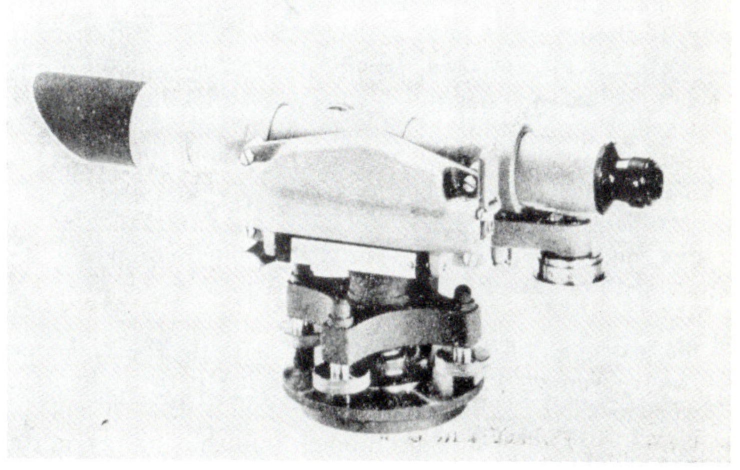

FIG. 4.6.1. PRECISE TILTING LEVEL.
(Courtesy, Keuffel & Esser)

mounted alongside the telescope and is viewed from the end by means of prisms that appear to split the bubble and make it look like Fig. 4.6.2.

The Precise Tilting Level shown in Fig. 4.6.1 is usually supported on a base by means of three leveling screws. A spirit level equipped with an air chamber is mounted by brackets on the left side of and close to the telescope; thus it is alongside the line of collimation. By means of a prismatic reading device attached above and to the left of the level vial casing, the level bubble is made to appear to be split longitudinally and also transversely so that it looks like Fig. 4.6.2. The left half of Fig. 4.6.2 shows half of one end of the bubble and the right half shows half of the other end of the bubble, hence when the bubble is in the central position the two half-ends will appear to coincide. Without changing his position the observer can view the position of the bubble image with his left eye and the rod and cross-hairs with his right.

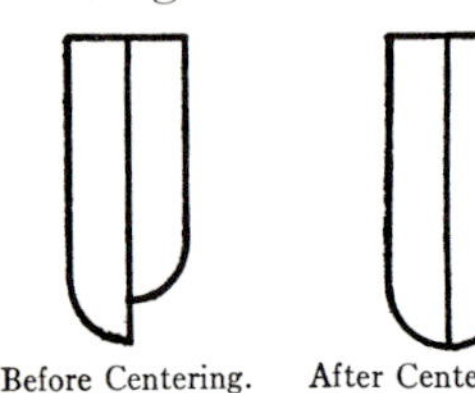

FIG. 4.6.2. BUBBLE AS SEEN IN PRISMS.

By means of a micrometer screw bearing at a point directly beneath the eyepiece and connected by an arm of the cradle supporting the telescope, the telescope may be raised and lowered (tilted) about a horizontal axis directly above the vertical axis, thus insuring no change of the height of the telescope. To prevent undue wear, a cam is provided for lifting the telescope off the micrometer screw head when observations are not being taken. The micrometer screw is graduated for setting the telescope at any desired slight inclination from the level position in setting grade points on a slope. (See Art. 9.19.)

The instrument is first made approximately level by the use of a small circular (or universal) level located to the right of the telescope. The telescope is then pointed at the rod and the ends of the level bubble are brought into coincidence in the prism by rotating the micrometer screw. At the moment the bubble ends are in coincidence all three hairs are read, the mean giving the final rod reading. It is important not to disturb the leveling screws between observations at a given set-up, since, with the usual three-leveling-screw base, motion of any one screw raises or lowers the height of the telescope.

4.7. Engineers and Precision Levels of Newer Design.—In Europe and elsewhere the expertise in designing and manufacturing optical instruments has led to the development of numerous levels which embody several unique features: prisms for viewing the bubble, split bubbles, shorter telescopes, prism telescopes, coated lenses to improve light transmission, three-screw (vs. four-screw) leveling heads, tilting screws, optical micrometers, etc. The effort has produced compact and lightweight instruments. As an example, one is illustrated in Fig. 4.7.1.

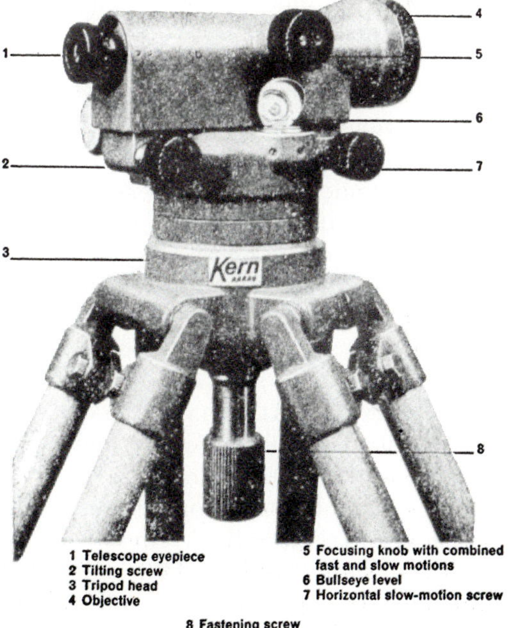

Fig. 4.7.1. Kern GK 23 Precise Level.
(Courtesy, Kern Instruments, Inc.)

The base of this instrument has no footscrews. It consists of a hollow cylinder, the inner surface of which is ground to a spherical surface of the tripod head. Preliminary leveling is accomplished by shifting the instrument on the spherical surface until the circular level is centered. Then the fastening screw is tightened to hold the instrument firmly in position. By elimination of footscrews the instrument can be prepared for operation very quickly and with improved stability of the connection of the instrument to the tripod. The level vial is completely enclosed and sealed with a protective glass to avoid disturbing external influences such as sun and wind.

The tilting screw has a fine micrometer thread which provides excellent control for bubble centering.

Coincidence prisms and an optical train produce an image of the bubble of the telescope level in the plane of the telescope reticule. Thus, the image of the rod and the bubble are observed simultaneously in the telescope eyepiece.

The short-barreled telescope has a large aperture and provides a sharp image with rich contrast. Focusing is internal so the telescope does not vary in length.

The GK 23-C level has a horizontal circle of glass graduated either in degree or grade intervals, 360° or 400^g. The circle is read by means of a microscope, the eyepiece of which is next to the telescope eyepiece.

All control screws can be operated conveniently with the right hand. Only the slightest movement of the head is required to observe the circular and telescope levels, the rod image and the circle image.

4.8. SELF-LEVELING OR AUTOMATIC LEVEL.—In the self-leveling type of instrument an optical system is employed to establish a precise level line of sight. This system replaces the sensitive spirit level bubble used for that purpose in the conventional type levels. In the self-leveling level illustrated in Fig. 4.8.1 the line of sight is reflected through a chain of three prisms within the telescope. The two end prisms are fixed while the middle one, known as the compensating prism, is suspended by wires from the roof of the telescope and is free to swing as a pendulum acting under the force of gravity. A

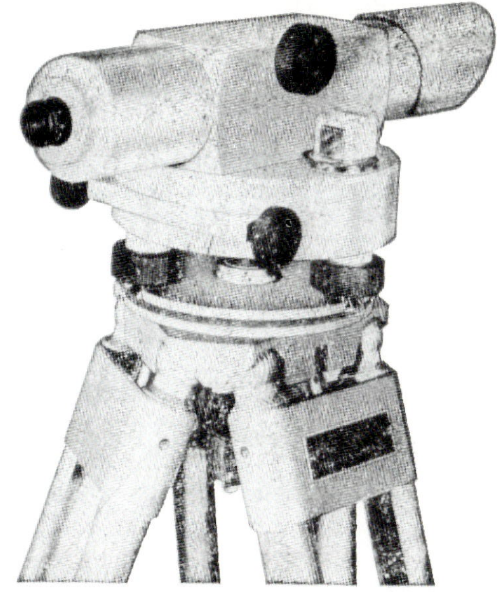

FIG. 4.8.1. ZEISS Ni2 SELF-LEVELING LEVEL.
(Courtesy, Keuffel & Esser)

line of sight coming from an object enters the first fixed prism, is reflected downward through the suspended prism, and then upward through the third (fixed) prism from which it is reflected to the cross-hair and the eyepiece (Fig. 4.8.2). So long as the pendulum prism hangs freely, a level line of sight is maintained regardless of whether or not the telescope tube is truly level.

The principle of the compensating element is shown diagrammatically in Fig. 4.8.3b. For simplicity the damping device has been omitted. In the upper diagram the telescope axis is level and the level line of sight is parallel to it. In the lower diagram the telescope tube is shown slightly tilted. The suspended prism has now swung under the force of gravity to a new position such that the level line of sight, although not parallel to the telescope axis, is still transmitted through the chain of prisms to the cross-hair and to the eyepiece.

In setting up the instrument it is necessary to level it only approximately using the circular bubble, to be sure that the pendulum prism hangs freely. Thereafter the line of sight will automatically seek the level position on every setting. The swinging prism is brought quickly to rest by an air damping device (Fig. 4.8.2) consisting of a shallow drum which

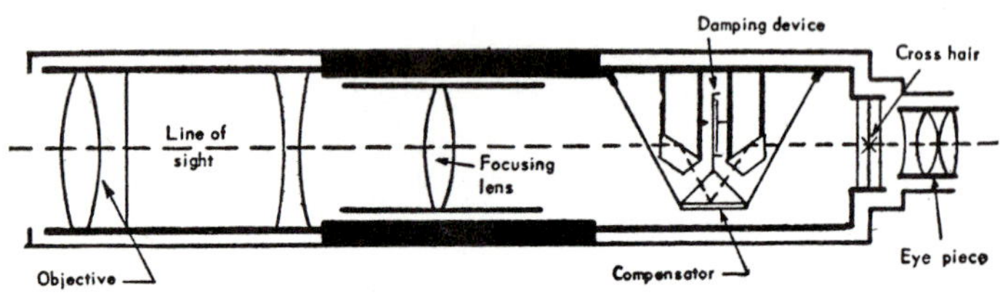

FIG. 4.8.2. LONGITUDINAL SECTION THROUGH ZEISS Ni2 TELESCOPE.

swings with the compensating prism against a close-fitting piston disc attached to the fixed prism support. A right-angle prism mounted over the circular bubble (Fig. 4.8.1) allows the instrumentman to center that bubble from his sighting position where he normally stands behind the telescope. The optical system gives an erect image. A low- and high-gear focusing screw permits rapid and precise focusing.

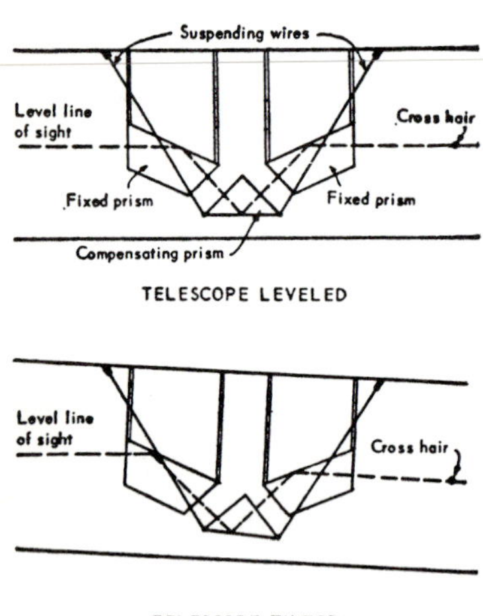

FIG. 4.8.3. SCHEMATIC DIAGRAMS OF COMPENSATOR IN ZEISS Ni2 SELF-LEVELING LEVEL.

Leveling work is greatly speeded up by the elimination of the necessity for fine adjustment of the bubble with the foot screws and by reducing the time spent by the operator in walking around the instrument.

In addition to the center horizontal cross-hair or "wire," the self-leveling level has two short-length stadia wires placed equidistant above and below the middle one. For precise leveling all three wires may be read and an average taken. For this purpose a rod graduated in yards and thousandths of a yard is convenient, since the sum of the three rod readings is then automatically the mean of the three readings in feet. When using such a rod the stadia interval should be 0.3 per 100 instead of the usual 1.0 per 100 so that the rod intercept in yards times 1000 will give the distance in feet.

Other types of automatic levels employ the same principle of the suspended prism, but they differ in their construction. The Filotecnica Salmoiraghi, MOM and Zena automatic levels have a pendulum device suspended in a vertical tube with eyepiece at bottom of tube and objective at top giving the instruments the appearance of a periscope.

4.9. PLAN-PARALLEL PLATE AND MICROMETER.— Generally the cross-hair image will intersect a leveling rod between two graduations. The difference to the next lowest graduation mark is estimated in order to increase the reading precision. This visual linear interpolation, however, can only provide a limited improvement. Based on a similar principle

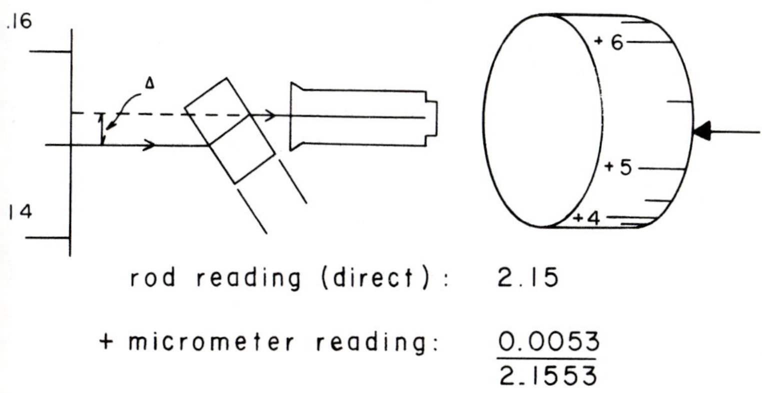

FIG. 4.9.1. PLAN-PARALLEL PLATE AND MICROMETER.

as the optical micrometer in theodolites (see Art. 2.30) the line of sight is therefore shifted up or down until the cross-hair coincides with a graduation mark. This is accomplished by rotating a plan-parallel glass plate placed in front of the objective lens (see Fig. 4.9.1). The vertical shift is a function of the angular rotation and can be read very precisely on a micrometer drum. All geodetic levels for high precision work have this plan-parallel plate and micrometer attachment, which is either part of the instrument (e.g. Wild N-3, Zeiss Ni-1) or can be attached to some instruments (e.g. Zeiss Ni-2), thus upgrading regular precision levels to geodetic standards. This applied to both tilting and automatic levels.

Construction and dumpy levels (e.g. Wild N-1, Gurley) are not accurate enough to warrant such an attachment.

4.10. THE HAND LEVEL.—The hand level, used for approximate leveling only, is a metal tube with a plain glass cover at each end and with a spirit level fastened to the top of the tube (Fig. 4.10.1). When looking through the tube, the bubble and

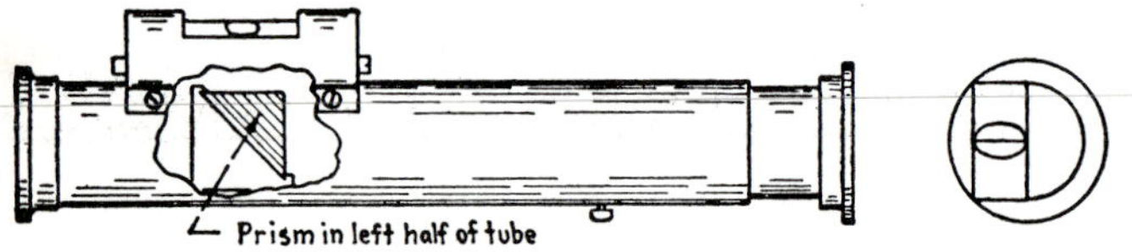

FIG. 4.10.1. LOCKE HAND LEVEL.

the cross-wire (reflected by a prism) are seen in the left-half side of the field of view, and the landscape is seen in the right-half side. In order that the bubble and the landscape may be seen simultaneously, a half-lens is placed in a sliding tube to permit focusing on the bubble. The instrument is held by hand at the eye and the farther end is raised or lowered until the bubble is in the center of the tube, at which instant the horizontal wire will intercept the rod, or any other object, at a point which is at the elevation of the observer's eye. The cross-wire which defines the level line may be adjusted by means of two opposing screws.

4.11. The Clinometer (Abney Hand Level).—The clinometer (Fig. 4.11.1) may be used for leveling or for measuring vertical angles. When the vernier index is set to 0° on the vertical scale, the instrument may be used for leveling in the

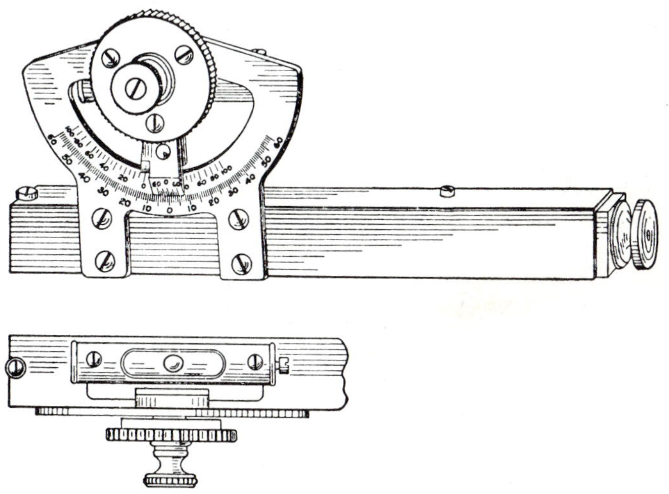

FIG. 4.11.1. ABNEY HAND LEVEL AND CLINOMETER.

same manner as in the Locke hand level (Art. 4.10). For measuring vertical angles the horizontal hair is set on the object. While still looking through the tube at the object, the control screw to which the vernier and bubble are rigidly attached is turned until the bubble appears to be bisected by the cross-hair. The angle is then read on the vertical arc.

The upper scale on the arc reads in per cent of grade and the lower scale in degrees of vertical angle.

4.12. LEVELING RODS.—Leveling rods are classified as *Target* rods and *Self-Reading* rods. Target rods are read by the rodman, whereas self-reading rods are read directly by the levelman as he looks through the telescope. Some rods may be used either as a target or as a self-reading rod.

Most leveling rods are graduated in hundredths of feet, but some are graduated in half-tenths only. Self-reading rods may be read by estimation closer than the smallest division, i.e., readings on rods graduated to hundredths may be read to half-hundredths. The target used on some rods carries a vernier by means of which readings may be made to thousandths of feet.

The rod most commonly used is the Philadelphia rod. This rod is both a target and a self-reading rod. It is made of wood and may be obtained as a single strip of wood or as two or more strips which slide one over the other; the latter is called an extension rod. Although these rods may be obtained in various lengths, the one most commonly used is 12 to 13 ft. long; when extended, its parts are held in position by clamps. Referring to Fig. 4.12.1, which shows several types of Philadelphia rods, A is a one-piece rod 12½ ft. long and is graduated throughout its length into hundredths of feet, zero being at the bottom of the rod; B shows this rod in more detail. The face of the rod is painted white and on it the graduations, 0.01 ft. wide, are painted in black, and are spaced 0.01 ft. apart. Thus each alternate black and white space is 0.01 ft. high; the tops of the black graduations are the even hundredths. The large numbers, usually in red, indicate the foot-marks; and the smaller numbers, usually in black, the tenths of feet.

When used as a self-reading rod the levelman observes the point where the horizontal wire intersects the face of the rod. In Fig. 4.12.1 two separate readings are shown. The lower one is 4.240 ft., and by estimation the upper one is 4.545 ft.

Fig. 4.12.1C shows the front face of a two-piece Philadelphia rod with target attached at the 7-ft. mark. When this rod is used as a target rod and the readings are less than 7 ft., the target is moved up or down at the direction of the levelman until it is bisected by the horizontal cross-hair of the level. It is then clamped to the rod, as shown in the lower half of Fig. 4.12.1D, at the point where the reading is 5.386. The reading is obtained by means of a vernier (sometimes by a scale divided into 20ths of a foot) which is attached to the target. For readings higher than 7 ft. the center line of the target (zero of target vernier) is clamped at the 7-ft. mark, and the rear strip of the rod is raised until the target is bisected by the horizontal

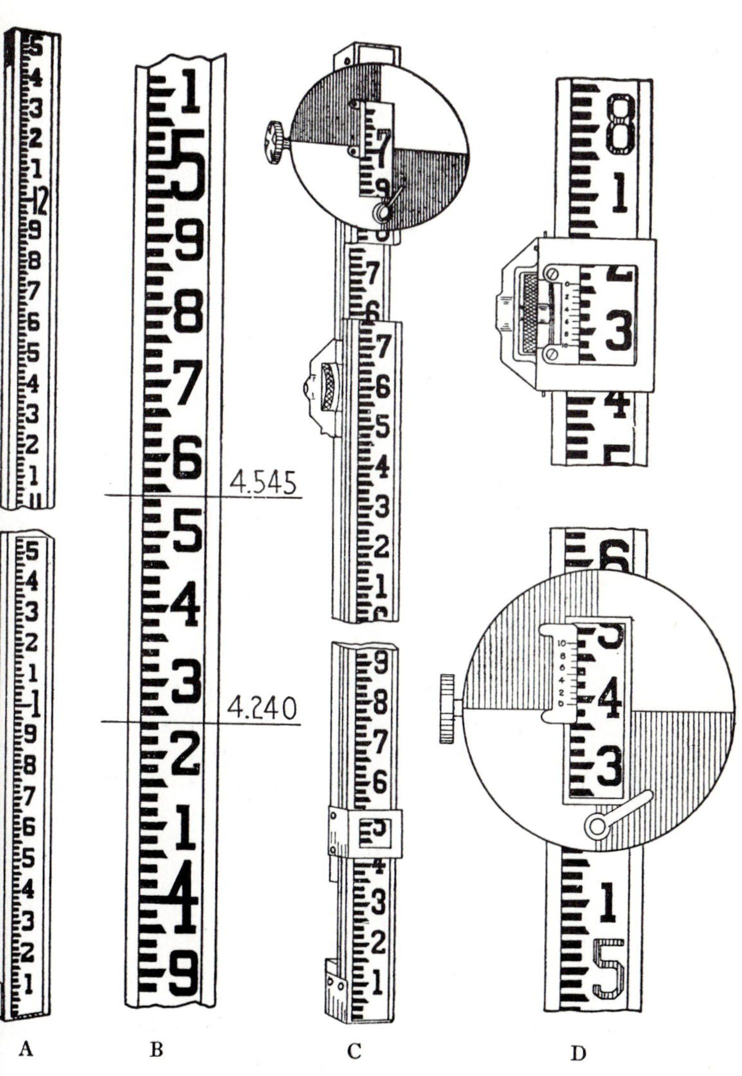

FIG. 4.12.1. LEVELING RODS.

cross-hair. The two strips are then clamped together and the reading is obtained by reading the graduations and vernier on the rear face of the rod, as shown in the upper half of Fig. 4.12.1D, where the reading is 8.238. This collapsible rod may also be used as a self-reading rod; it is then extended to its full length for readings greater than 7 ft. When the rod is fully extended, the graduations on the front of the rod read continuously from the bottom to the top of the rod.

The *Architect's rod* is graduated in eighth inches by fine lines; target reads to 1/64 inch. The rod is 5 ft. 6½ inches long when closed, and about 10 ft. long when extended. The face is ¾ inch wide.

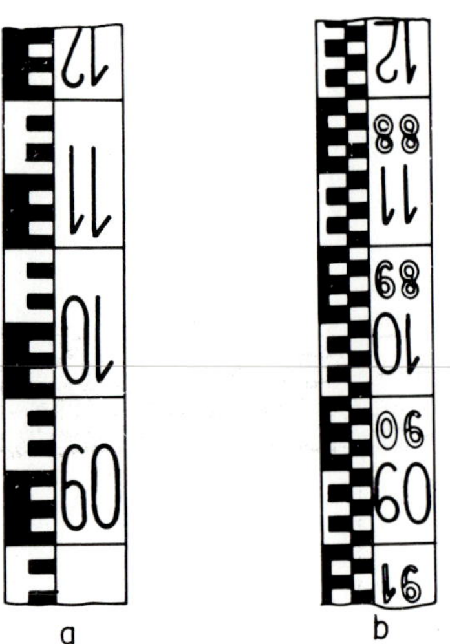

FIG. 4.12.2. METRIC LEVELING RODS.
(For use with Inverting Telescope.)

There are numerous metric rods on the market. Although many of them have a line-type graduation, there are others with area graduation. The most common ones are the E-rod and the checker-board rod (see Fig. 4.12.2 A and B). Also for some uses yard rods exist.

4.13. Advantages of the Self-Reading Rod.—Experience shows that as precise results may be obtained with a self-reading rod as with a target rod. The target will be found most useful, however, in dense woods, in dark places, in high winds, for exceptionally long sights, and for setting several points on the same straight grade line.

The exactness to which readings are made depends on the requirements of the project. For some work, the nearest tenth of a foot is sufficient, while in other kinds of work the nearest hundredth is required. To obtain the latter precision in the final result, a self-reading rod should be read to the nearest 0.005 ft. on intermediate readings; if a target rod is used it is customary to record the reading to the nearest 0.001 ft.

Under favorable conditions a self-reading rod graduated to hundredths can be read by estimation to the nearest half-hundredth at a distance of 300 ft. from the instrument.

The value of one division of the ordinary level bubble is about 20″; this corresponds to 0.03 ft. in 300 ft. The distance between the graduations on the level vial is about 0.1 in. It is reasonable to assume that the bubble is not centered closer than a fifth of a division (1⁄50 of an inch) at the instant the rod-reading is taken. At 300 ft. this discrepancy corresponds to a rod interval of 0.006 ft; at 200 ft., 0.004 ft.; and at 100 ft., to 0.002 ft. The error in reading a rod by estimation to the nearest 0.005 is 0.0025 ft., which is the error to expect in reading a self-reading rod. When using a target rod, values read and recorded to the nearest 0.001 ft. often give a false impression of the precision actually obtained.

4.14. The Tape—or Direct Elevation Rod.—The direct elevation rod is a ten-foot endless loop of steel, graduated in feet and hundredths, mounted on an extensible rod in such a fashion that the loop can be moved along and then fixed temporarily for a sequence of rod readings. The loop is correctly fixed as follows. The rod is held on a point of known elevation (say, 186.50) and the loop is moved up or down until the levelman reads the elevation value on the rod (in this case he will read 6.50). The rodman tightens the clamp that will hold the tape loop fast to the rod at one point.

Subsequently, when the rod is held at an unknown point, the instrumentman will read the elevation of that point directly. For example, if he reads 8.00 on the rod he records the elevation as 188.00; for 4.50 he writes 184.50; for 0.50, he writes 190.50 (judging by eye that it should not be 180.50); etc. Until the instrument is moved, the rod gives elevations correctly. When the level is set up in a new location, the rod loop clamp must be loosened and the tape loop reset while sighting back on a point of known elevation before more leveling can be done.

The direct elevation rod is particularly convenient for profile leveling, for leveling over a gridded area for cross-sections, for many operations in building construction, and for any case where the quick result is desired without the extra subtraction chore.

Two such rods exist, one of which is the Lenker Rod.

4.15. Flexible or Pocket Leveling Rod.—This consists of a specially prepared strip of woven fabric graduated the same as any other self-reading rod. When in use it may be tacked to a light board of convenient size. For transportation it may be rolled up and packed in a very small space. These rods are not so reliable as wooden rods because they are subject to stretching and shrinkage, but where accuracy is not of primary importance they have a decided advantage over the heavy rods for work requiring much traveling.

4.16. Geodetic Level Rod.—The rods used by the U. S. National Oceanic and Atmospheric Administration and the U. S. Geological Survey are of the non-extensible pattern. Those used by the NOAA are about 3.3 m. long and are made of flat pieces of wood on which are painted the meter and decimeter graduations, and to which are attached strips of invar on which are painted the centimeter divisions. The Geological Survey rod is graduated in yards and hundredths; otherwise it is similar to the NOAA rod. Each rod is provided with a spirit level for plumbing, and with a centrigrade thermometer. The foot of the rod consists of a metal piece, with a flat end, connected rigidly (at the top) with the invar strip.

4.17. Attachments to the Rod for Plumbing.—In accurate work it is necessary to use some device for making the rod plumb. Spirit levels attached to brass "angles" which may be secured to a corner of the rod are very convenient, as shown in Fig. 4.17.1. In some rods, such as those used for precise leveling, the levels are set permanently into the rod itself.

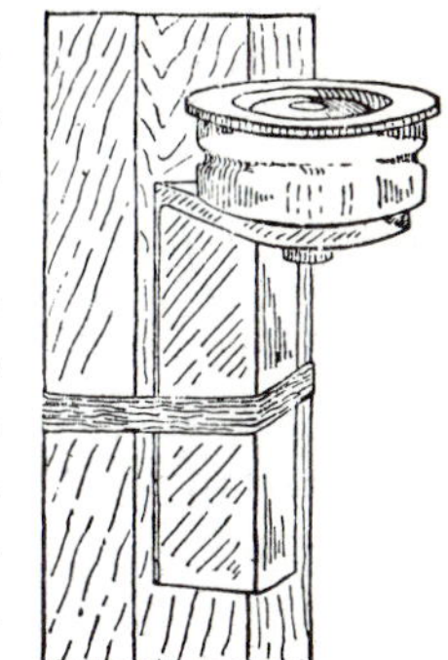

FIG. 4.17.1. ROD LEVEL.

4.18. Errors in Length of Rod.—Changes of temperature do not affect seriously the length of the rod since the coefficient of expansion of wood is small. The effect of moisture is greater but is indeterminate. It is well known that rods vary in length, and consequently if very accurate leveling is to be done the length of the rod should be tested frequently with a steel tape. Rods which are accurate when manufactured may show errors, after a time, as great as 0.01 ft. in 12 ft. In accurate work the error of the rod should be determined and recorded, and the heights should be corrected for this error just as horizontal distances are corrected for the error of the tape. In some makes of rod the graduations are painted on a metal strip which fits into grooves on each side of the face of the wooded rod. The strip may be attached so that is is not affected by changes in the length of the wood. It can also be replaced when it becomes worn or damaged. For high precision work, the metal of the strip is invar, which has a very small temperature expansion coefficient.

USE OF THE LEVEL AND ROD

4.19. PRINCIPLE OF LEVELING.—To obtain the difference in elevation between two points, hold the rod vertically at the first point and, while the instrument is level, take a rod-reading. This is the distance that the bottom of the rod is below the line of sight of the telescope. Then take a rod-reading on the second point; the difference between the two rod-readings is the difference in elevation of the two points.

4.20. To Level the Instrument.—Set up the instrument in such a position and at such a height that the horizontal cross-hair will strike somewhere on the rod when held on either point. Time will be saved if the habit is formed of doing nearly all of the leveling by moving the tripod legs, using the leveling screws only for slight motions of the bubble in bringing it to the middle of the tube. Turn the telescope so that it is directly over two opposite leveling screws. Bring the bubble to the center of the tube **approximately**; then turn the telescope until it is over the other pair of leveling screws and bring the bubble **exactly** to the center. Move the telescope back to the first position and level carefully, and again to the second position. Repeat until the bubble is exactly in the center in both positions. If the instrument is in adjustment and is properly leveled in both directions, then the bubble will remain in the center during an entire revolution of the telescope about the vertical axis. The instrument should seldom be clamped, but this may be necessary when a strong wind is blowing.

This type of repeated centering in both positions is necessary for the dumpy level, but for the tilting level or the automatic level a simpler procedure obtains. One need only center the bulls-eye or circular bubble of such an instrument and one is thereby assured of staying within the range of the tilting screw or the automatic compensator.

Both four screw and three screw leveling heads are in common use (see Arts. 2.34 and 2.35).

4.21. To Take a Rod-Reading.—The rodman holds the rod on the first point, taking pains to **keep it as nearly plumb as possible.** The levelman focuses the telescope on the rod, and brings the bubble to the center **while the telescope is pointing at the rod,** because leveling over both sets of screws will not make the bubble remain in the center in **all** positions unless the adjustment is perfect. If a target rod is used, the target should be set so that the horizontal cross-hair bisects it **while the bubble is in the center of the tube.** It is not sufficient to trust the bubble to remain in the center; it should be examined just before setting the target and immediately afterward, **at every reading.** It is helpful in making fine adjustments of the bubble if the instrument can be set up so that an opposite pair of leveling screws is in line with the telescope when taking a reading on the rod.

With the self-reading rod the levelman reads the feet and tenths and notes approximately the hundredths; then, **after verifying the position of the bubble,** he makes the final reading of hundredths and half-hundredths.

If, however, the target is being used the levelman signals the rodman to move the target up or down. When the center of the target coincides with the horizontal cross-hair the levelman signals or calls to the rodman "all right" and the rodman clamps the target. Before the rodman reads the target he should allow the levelman time to check the position of the target to be certain that it has not slipped in clamping. The reading is then recorded in the note-book. For readings to hundredths of a foot it is not necessary to clamp the target or the rod; the rodman can hold the target in position or the two parts of the rod firmly together while he reads the scale.

Most levelers prefer to have the rodman stand behind the rod, facing the instrument, so that he can watch the levelman. Other levelmen prefer that the rodman stand beside the rod and plumb it at right angles to the line of sight, while the levelman directs the rodman which way to plumb the rod so that it will be parallel to the vertical cross-hair. **It is extremely important that the rod be held plumb.** Vertical lines on buildings are a great aid to the rodman in judging when his rod is plumb. If there is no wind blowing he can often tell when the rod is plumb by **balancing** it on the point. If the rodman promptly and attentively holds the rod plumb it not only increases the accuracy of the results but also the rapidity with which the work may be carried on.

4.22. Signals.—While the rodman is seldom very far away from the levelman in this work still it is often convenient (in noisy city streets, for example) to use hand signals. The following are commonly used in leveling.

Raise or Lower Target.—The levelman motions to the rodman by raising his extended arm above his shoulder for an upward motion and dropping his arm below his waist for a downward motion. As the target approaches the desired setting the arm is brought toward the horizontal position.

All Right.—The levelman extends both hands horizontally and moves them up and down, or, if not far away, he merely shows the palms of both hands.

Plumb the Rod.—The hand is extended vertically above the head and moved slowly in the direction it is desired to have the rod plumbed.

Take a Turning Point.—The arm is swung slowly in a circle above the head.

Pick Up the Level.—When a new set-up of the level is desired the chief of party signals the levelman by extending both arms downward and outward and then raising them quickly.

Some surveyors use an improvised system of signals for communicating the rod-readings, but mistakes are liable to be made unless great care is used.

4.23. DIFFERENTIAL LEVELING.—Differential leveling is the name given to the process of finding the difference in elevation of any two points. In Art. 4.19 the simplest case of differential leveling is described. When the points are far apart, or when the difference in elevation is great, or when for any other reason the measurement cannot be completed from a single set-up of the instrument, the difference in elevation is found as follows: The instrument is set up and a rod-reading is taken on the first point; this is called a *backsight*, or *plus sight*, and is usually written *B. S.* or +*S.* in the notes. Next the rod is taken to some well-defined point which will not change in elevation (such as the top of a firm rock, the top of a hydrant, or a spike in the root of a tree) and held upon it and a reading taken; this is called a *foresight*, or *minus sight*, and is written *F. S.* or −*S.* The difference between the two readings gives the difference in elevation between this new point and the first point. This second point is called a *turning point* and is written *T. P.* The level is next set up in a new position and a backsight taken on the same turning point. A new turning point further ahead is then selected and a foresight taken upon it.

This process is continued until a foresight is taken on the final point. The elevation of the last point above the first is equal to the sum of all the backsights minus the sum of all the foresights. If the result is **negative,** i.e., if the sum of the foresights is the **greater,** then the last point is **below** the first. The field work is illustrated by Fig. 4.23.1.

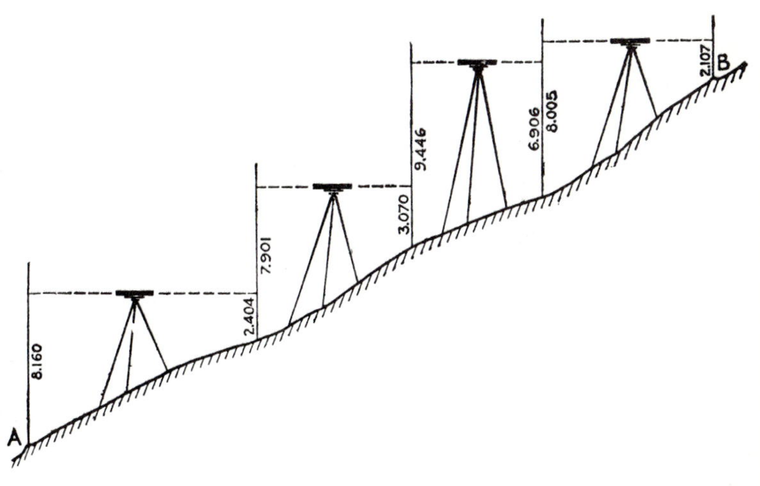

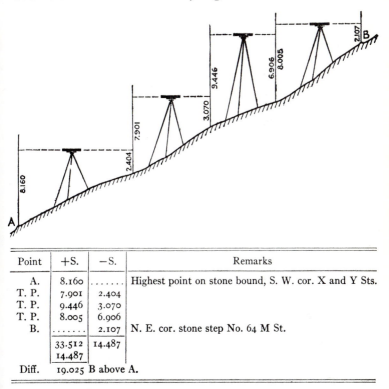

Point	+S.	−S.	Remarks
A.	8.160	....	Highest point on stone bound, S. W. cor. X and Y Sts.
T. P.	7.901	2.404	
T. P.	9.446	3.070	
T. P.	8.005	6.906	
B.	....	2.107	N. E. cor. stone step No. 64 M St.
	33.512	14.487	
	14.487		
Diff.	19.025 B above A.		

FIG. 4.23.1. DIAGRAM ILLUSTRATING DIFFERENTIAL LEVELING.

4.24. The Proper Length of Sight.—The proper length of sight will depend upon the distance at which the rod appears distinct and steady to the levelman, upon the variations in readings taken on the same point, and also upon the degree of precision required. Under ordinary conditions the length of sight should not exceed about 300 ft. where elevations to the nearest 0.01 ft. are desired. "Boiling" of the air due to irregular refraction is frequently so troublesome that long sights cannot be taken accurately.

If the level is out of adjustment the resulting error in the rod-reading is proportional to the distance from the instrument to the rod. If the level is at equal distances from the rod the errors are equal and since it is the **difference** of the rod-readings that gives the difference in elevation, the error is eliminated from the final result if the rodman makes **the distance to the point where the foresight is taken equal to the backsight** by counting his paces as he goes from one point to the other.

4.25. Effect of the Earth's Curvature and of Refraction on Leveling.—Since the surface of the earth is very nearly spherical, any line on it made by the intersection of a vertical plane with the earth's surface is virtually circular. In Fig. 4.25.1 the

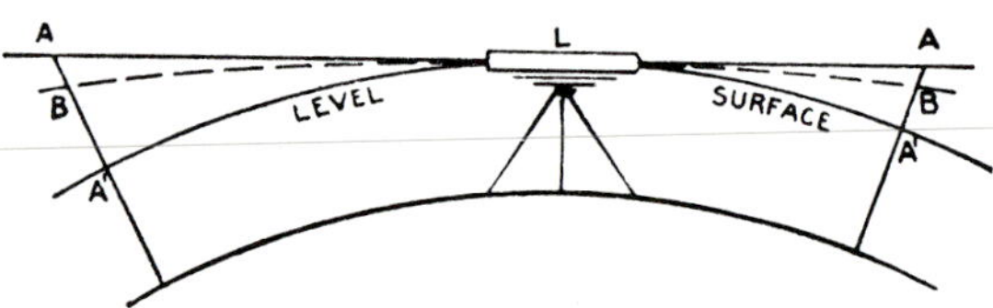

FIG. 4.25.1. DIAGRAM ILLUSTRATING EFFECT OF EARTH'S CURVATURE AND OF REFRACTION.

distance AA' from the horizontal line to the level line, varies nearly as $\overline{A'L}^2$. This is the amount by which the rod-reading is too large owing to the earth's curvature. The ray of light that enters the telescope horizontally at L, however, is that starting from B, not that from A because the ray from B is bent into a curve. The effect of the refraction of the atmosphere is to make this offset from the tangent appear to be $A'B$, which is about one-seventh part smaller than $A'A$. This offset, corrected for refraction, is about 0.57 ft. in 1 mile and varies as the square of the distance. Corrections (in feet) for curvature and refraction are given in Table 4.25.1 for every 100 ft. of distance up to 1000 ft. The table is based on this formula:

$$C_{cr} \text{ (in feet)} = 0.0205 \text{ D}^2$$

where D is in 1000's of feet. Another common version of the same formula is:

$$C_{cr} \text{ (in feet)} = 0.572 \text{ D}^2$$

where D is in miles.

TABLE 4.25.1

Distance in Feet	Correction (in Feet) for Earth's Curvature and Refraction	Distance in Feet	Correction (in Feet) for Earth's Curvature and Refraction
100	0.000	600	.007
200	.001	700	.010
300	.002	800	.013
400	.003	900	.017
500	.005	1000	.020

If the rod is equally distant from the instrument on the foresight and backsight the effect of curvature and refraction is eliminated from the result.

4.26. Precautions in Level Work.—Nearly all of the precautions mentioned in Art. 2.40, for the theodolite, are also applicable to the level. Care should be taken not to strike clamped extension rod on the ground before it has been read. This may cause the parts of the rod to slip, and it may also affect the elevation of the turning point.

4.27. Common Sources of ERROR in Leveling.—

1. Improper focusing (parallax).
2. Bubble not in middle of tube at instant of sighting.
3. Rod not held plumb.
4. Foresights and corresponding backsights on turning points not equally distant from the instrument.
5. Poor turning points selected. (See Art. 9.8)
6. Erroneous length of rod.

4.28. COMMON MISTAKES.—

1. Foresight and backsight not taken on exactly the same point. (T. P.'s should be marked before using.)
2. Neglecting to set target accurately when "log rod" is used.
3. In the use of the self-reading rod neglecting to clamp the rod at the proper place when "long rod" is used.
4. Reading the wrong foot-mark or tenth-mark.
5. In keeping notes—entering F. S. in B. S. column or *vice versa*.
6. In working up notes, adding F. S. or subtracting B. S.

ADJUSTMENT OF THE LEVEL

4.29. TESTS FOR ADJUSTMENT.—With careful use and handling level instruments should stay in adjustment. By equalizing backsights and foresights the effect of error in line of sight can be eliminated (Art. 4.24). It is advisable to check the adjustment of the instrument periodically. No adjustment should be made, however, before making sure that the fault is in the instrument and not in the test. Extensive repairs should be made by the manufacturer. The tests described in Arts. 4.30–4.31 apply particularly to American-type instruments. They should be made in the order given. European instruments have complicated optical systems which cannot be adjusted in the field. The "peg test" applies to these levels (Arts. 4.32, 4.33).

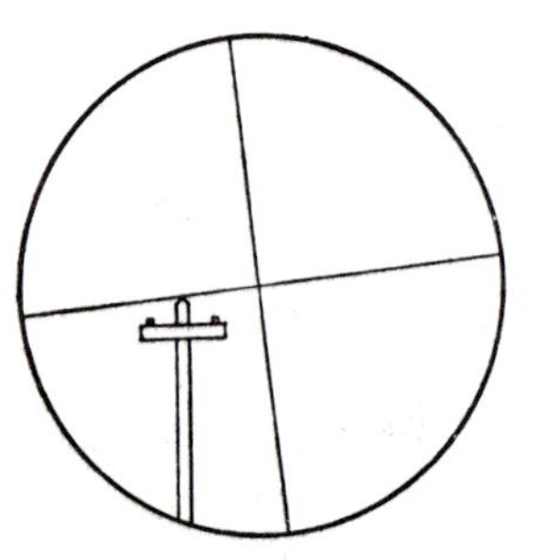
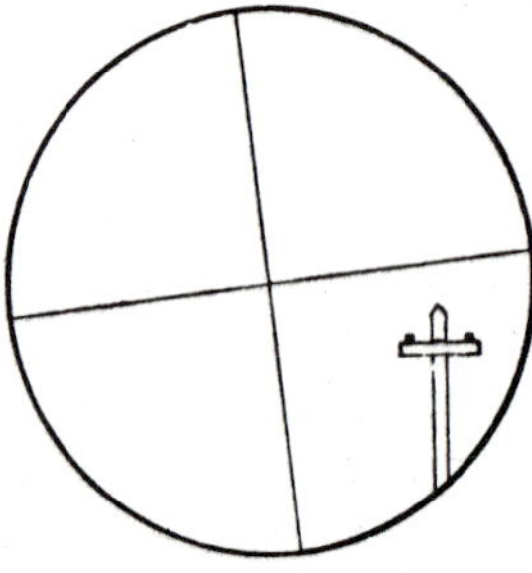

FIG. 4.30.1. TESTING THE HORIZONTAL CROSS-HAIRS.

ADJUSTMENTS OF THE DUMPY LEVEL

4.30. Adjustment of the Cross-Hair.—If the horizontal cross-hair is not truly horizontal when the instrument is level it should be made so by rotating the cross-hair ring as described in the adjustment of the transit, Art. 2.59 (see Fig. 4.30.1).

4.31. Adjustment of the Bubble Tube.—To Make the Axis of the Bubble Tube Perpendicular to the Vertical Axis. Owing to the construction of the dumpy level it is necessary to make this adjustment **before** making the line of sight parallel to the bubble tube. It is done by centering the bubble over one pair of leveling screws, and turning the instrument 180° about the vertical axis. If the bubble does not remain in the center of the tube, move it half-way back to the center by means of the adjusting screws on the level tube.

4.32. The Direct, or "Peg," Adjustment.—To Make the Line of Sight Parallel to the Axis of the Bubble.—(See Fig. 4.32.1.) Select two points *A* and *B*, say, 200 ft. or more apart. Set up the level close to *A* so that when a rod is held upon it the eye-piece will be about a quarter of an inch from the rod. Look through the telescope **wrong-end-to** at the rod and find the rod reading at the cross-hair, if visible, otherwise the reading opposite the center of the field. After a little experience it will be found that this can be done very accurately. Owing to the fact that only a small portion of the rod is visible it will be found convenient to set a pencil-point on the rod at the center of the small field of view. Turn the telescope toward *B* and take a rod-reading on it in the usual way, being certain that the bubble is in the middle of the tube. The difference between these two rod-readings is the difference of elevation of the two points **plus** or **minus** the error of adjustment. The level is next taken to *B* and the above operation is repeated. The result is the difference in elevation **minus** or **plus** the same error of adjustment. The mean of the two results is the true difference in elevation of points *A* and *B*. Knowing the difference in elevation between the two points and the height of the instrument above *B* the rod-reading at *A* which will bring the target on the same level as the instrument may be computed.

The bubble is brought to the center of the tube and the horizontal cross-hair raised or lowered by means of the adjusting screws on the cross-hair ring until the line of sight strikes the target. In this method the small error due to the curvature of the earth (nearly 0.001 ft. for a 200-ft. sight) has been neglected.

Example

(See Fig. 4.32.1)

Instrument at A.

Rod-reading on	$A = 4.062$
Rod-reading on	$B = 5.129$
Diff. in elev. of A and	$B = 1.067$

Instrument at B.

Rod-reading on	$B = 5.076$
Rod-reading on	$A = 4.127$
Diff. in elev. of B and	$A = 0.949$

$$\text{Mean of two diff. in elev.} = \frac{1.067 + 0.949}{2} = 1.008 \text{ true diff. in elev.}$$

Instrument is now 5.076 above B.
Rod-reading at A should be $5.076 - 1.008 = 4.068$ to give a level sight.

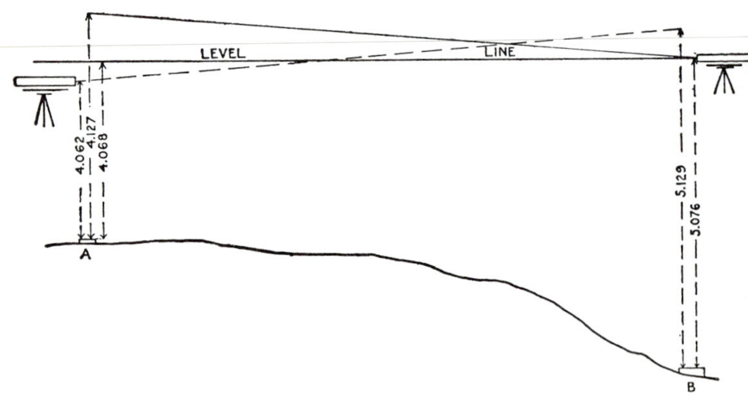

FIG. 4.32.1. PEG ADJUSTMENT.

The peg method may be used for adjusting the transit, the difference being that in the dumpy level the axis of the bubble tube is first made horizontal and then the line of sight is brought parallel to it, while in the transit the line of sight is first made horizontal and then the axis of the bubble tube is made parallel to it. Consequently, in adjusting the dumpy level the cross-hair ring is moved, whereas with the transit the adjustment is made in the bubble tube.

4.33. Alternate Procedure for Peg Adjustment.—Another way of making the peg adjustment is as follows. Set the level half-way between the two bench-marks and determine their difference in elevation. This difference will be correct, even though the level is not in adjustment, because the instrument is equally distant from the two bench-marks. Then set the level in line with the bench-marks so that both points will be in front of the level, but one close to the level. Take readings on the two points. If they differ by the true difference in elevation the instrument is in adjustment. If they do not, then move the horizontal cross-hair so it will read on the distant rod the same as the near reading plus or minus the difference in elevation. Then sight again on the near point. This time the reading will be changed slightly. Repeat the operation until the two readings differ by exactly the correct difference in elevation. When the near point is sighted merely note the change in reading. When the distant point is sighted move the cross-hair so that it sights the desired rod-reading.

Example.—Suppose that the two bench-marks are at the same elevation and that the instrument when sighting at A

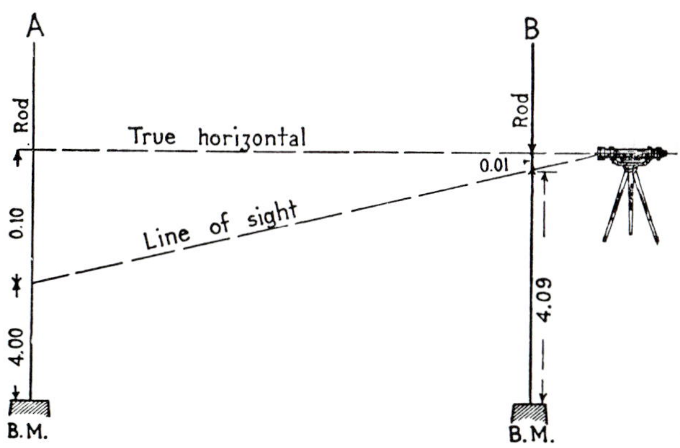

FIG. 4.33.1. PEG ADJUSTMENT.

reads 4.00 ft., and when sighting at *B* reads 4.09 (Fig. 4.33.1). If the cross-hair is now moved until it reads 4.09 on *A*, it will be found to read 4.099 ft. on *B*. If it is again raised until it reads 4.099 on *A*, it will read (nearly) 4.10 on *B*. The line of sight is now horizontal because the reading is 4.10 on each rod.

4.34. ADJUSTMENT OF TILTING LEVEL.—The object is to make the coincidence of the ends of the bubble occur when the line of sight is horizontal. This is done by making the peg test (Art. 4.33). After the target has been set on level line through the instrument, bring the line of sight on it by turning the micrometer screw. There are three capstan-headed screws at the back of the bubble housing. Loosen the two outside screws. Turn the middle screw until the ends of the bubble are in coincidence, then retighten the two outside screws.

4.35. ADJUSTMENT OF THE SELF-LEVELING LEVEL.—To Make the Circular Bubble Centered When the Axis Is Vertical.—This adjustment is important because, if the circular level is out of adjustment, there is no assurance that the compensator is free to swing when the bubble is centered. To perform the test, center the bubble with the leveling screws, turn the telescope through 180° and see if bubble remains in the center. If it does not, unscrew the circular level cover or the Observation Prism, if the instrument is so equipped. This will expose the three adjusting screws. The screws should be kept under firm tension but not tight. By manipulating the screws, bring the bubble half-way toward the center. Relevel and repeat until the bubble remains centered.

4.36. To Make Line of Sight Level.—Make peg test as described in Art. 4.33. After the target has been set on level line, unscrew the reticule cover. This is a circular cap 1$\frac{5}{16}$ inches in diameter at the end of the telescope just in front of the eyepiece. A small capstan-headed screw will be exposed just above the eyepiece. This raises and lowers the cross lines against a spring loading. Bring the cross line on the target by regulating this nut.

PART II

SURVEYING METHODS

CHAPTER V

LAND SURVEYING

5.1. Property Surveys.—Surveys of land are usually made for one or more of the following purposes:—(1) to furnish an exact description of the boundaries, (2) to determine the enclosed area, or (3) to furnish data for making a plan. The instruments and methods used will depend upon the land value and the use to be made of the results of the survey. A *cadastral* survey is the survey of legal boundaries of land.

Property surveys, or boundary surveys, or land surveys, or cadastral surveys—all virtually the same thing—are made only under the direction of a registered or licensed land surveyor (L.S.) in each of the states.

The usual practice is to begin at any convenient corner on the boundary and measure distances and angles (or distances and bearings) in the order in which they occur. This connected series of lines and angles is known as a *traverse*. Whenever it is practicable to do so surveys of this character should be made in closed loops, that is, the survey should return to the point of beginning so that it will be possible to calculate by trigonometry, or to plot, from the starting point continuously around the figure to the closing point; the computed position of this last point should, theoretically, be coincident with the starting point. Such a figure is spoken of as a "closed traverse." It furnishes one of the best means of checking the accuracy of all of the measurements. Since all the measurements contain errors, large or small, the position of the final point will never agree perfectly with that of the initial point. The distance between these two points is the actual error of closure.

This error alone is rather meaningless. It is generally considered together with the total distance covered by the survey and then presented as either "error per foot of distance" or as a relative accuracy such as "one part in 5000 parts, 1:5000, 1/5000 or one in 5000."

5.2. Surveying for Area with the Compass.—The compass was formerly used extensively as a direction instrument in land surveying. It has little application today except as an aid in running old lines or for a rough survey of low-priced land. If the compass is used, the direction of each boundary is determined by its magnetic bearing and the length measured with a tape. Where it is the custom of the locality to refer to distances as chains and links these will be used instead of feet because in that case previous surveys in this neighborhood will probably be given in these same units.* A heavy tape marked in chains and links is suitable for such work.

One important matter, which should not be overlooked, is to determine the magnetic declination *at the time and place* of the survey, and record it. When comparing with other surveys the difference in declination must be taken into account. It is desirable to have included in the description of the land the date at which the declination was determined. Much confusion arises from not knowing positively the declination at the time of the survey, as well as from doubt as to whether the date is that of the survey itself or merely the date of copying an old description. A definite statement, such as "Bearings are magnetic; magnetic declination, 1930, was 14° 30′ West" will clarify the record. To be of value this should be incorporated into the deed or recorded on the plan.

When making a survey enclosing an area it is customary to begin at some convenient corner, which can be easily identified and described, and to take bearings and distances in order around the field. As the measurements are made they are recorded immediately in the field note-book. It is not really necessary to measure the distances and bearings in the order in which they occur, but since they must be arranged in this order for the purpose of computation it will be conve-

*In many of the older states nearly all of the very old surveys are recorded in rods and links. Sometimes *rood*, *perch* or *pole* are used, all of which equal a rod.

nient to have them so arranged in the original notes. Mistakes are less likely if this is always done.

If the length and the bearing of one side are omitted the area is nevertheless completely determined; but since these omitted measurements would furnish a valuable check on the accuracy of all the measurements they should never be omitted if they can be taken. **It is of the utmost importance in every survey that check measurements should be taken.** Even a few rough checks in the field which require only a little extra time often prove to be of great value in detecting mistakes. Both a forward bearing and a reverse (or back) bearing should be taken at each corner; from these two bearings the angle at a corner can be obtained which is free from error of local attraction of the needle.

It is often impossible to set the compass up at the exact corners of the property; to overcome this difficulty lines running parallel to the property lines can be surveyed as described in Art. 5.4 and the area thereby determined. Sometimes the compass can be set on a property line at an intermediate point and the bearing of the line obtained; but the surveyor must be sure that there is no local attraction of the needle at this point. All points where the compass is set should be marked and described so that they can be found again. If any instrument point is not otherwise defined it may be marked temporarily by a small stake and several reference measurements made from this stake to permanent objects nearby which may be readily identified. From these mea-

(LEFT-HAND PAGE) (RIGHT-HAND PAGE)

Survey of Wood Lot of John Smith, Northboro, Mass.						N Brown, Surveyor J. Long, Chainman Oct. 7, 1957 — 2 rod chain – K & E Compass chain 0.1 link too long.
Sta.	Bearing	Reversed Bearing	Distance (chains)			Remarks
A	Due E	N 32$\frac{1}{4}$° W	77.75			Stake and stones cor. J. Smith, B. White and L. Richards
B	N 58$\frac{3}{4}$° E	N 89$\frac{1}{2}$° W	13.55			Pine Stump
C	N 1$\frac{1}{2}$° E	S 58$\frac{1}{4}$° W	32.36			Oak Stump
D	S 85$\frac{3}{4}$° W	S 1$\frac{1}{2}$° W	23.75			Cedar Stk. 5′ S.E. of large oak.
E	S 23$\frac{1}{2}$° W	N 85$\frac{3}{4}$° E	30.94			Stone bound, E. side Pine St.
F	S 32$\frac{1}{4}$° E	N 23$\frac{1}{2}$° E	11.16			Stone bound, E. side Pine St.

FIG. 5.2.1. NOTES OF CHAIN AND COMPASS SURVEY.

surements the position can be relocated definitely and positively, if the stake is lost. These measurements are called *ties*. (Art. 5.11)

Notes of the traverse are often recorded as shown in Fig. 5.2.1.

5.3. Survey of a Field with Transit and Tape.—The survey of a field for area can usually be made in one of the three following ways.

(1) By setting up the transit at the corners of the property and measuring the angles directly, the distances being measured directly along the property lines.

(2) When the property lines are so occupied by buildings or fences that the transit cannot be set up at the corners, but the distances can still be measured along the property lines, then the angles at the corners are obtained by measuring the angles between lines which are made exactly parallel to the property lines.

(3) If the boundaries of the property are such that it is not practicable to set the transit up at the corners or to measure the distance directly on the property lines, a traverse is run approximately parallel to the property lines and these lines connected with the traverse by means of angles and distances.

5.4. Transit and Tape Survey with Obstacles.—In case (2) the parallel lines are established in the following manner. Set the transit up at some point *E* (Fig. 5.4.2) preferably within 2 or 3

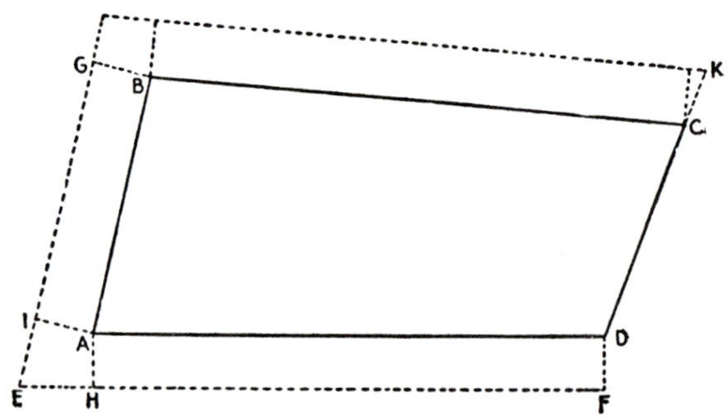

FIG. 5.4.2. TRANSIT LINES PARALLEL TO THE SIDES OF FIELD.

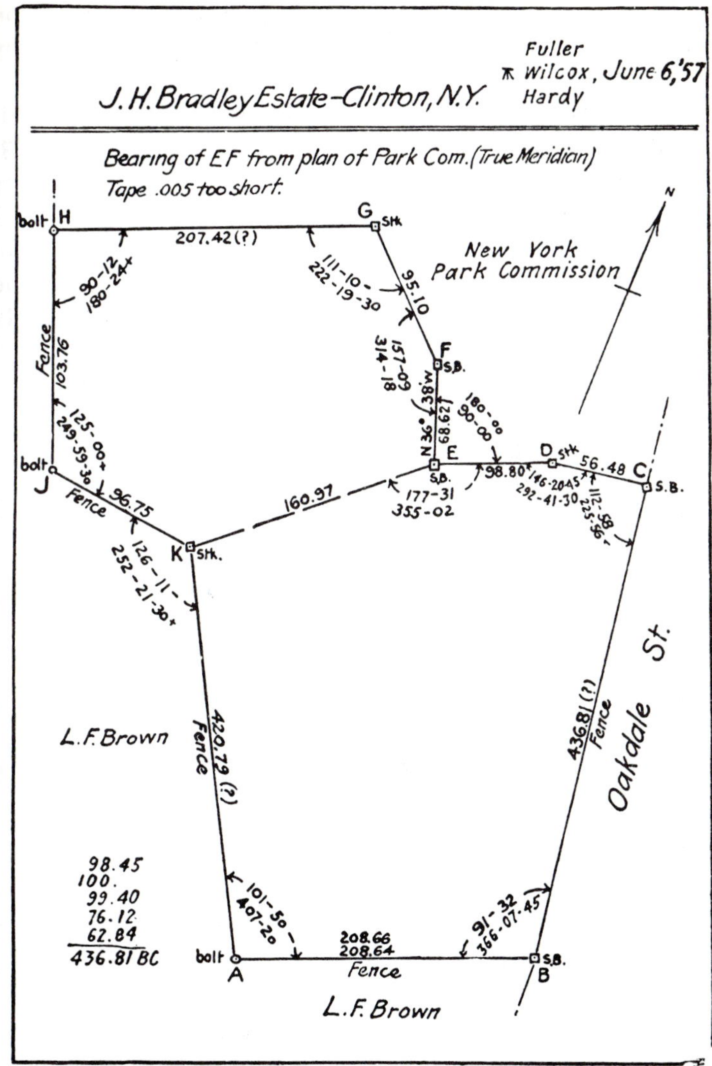

FIG. 5.4.1. NOTES OF SURVEY WITH TRANSIT AND TAPE.

ft. of the corner A. Establish the line EF parallel to AD by making $DF = AH$ by trial. Point H cannot be seen through the telescope, but it is so near the instrument that, by sighting by eye from a position behind the plumb-line at E toward point F, point H can then be accurately lined in; or, a swing offset AH, can be measured. Similarly EG is established parallel to AB. Then the angle FEG is measured; and this is the property angle at A. It is evident that the values of AH and DF and of AI and BG are of no permanent use, so they are not recorded in the notes. When it can be done it is advisable to place the transit point (K for example) exactly on one of the property lines, or on its prolongation. Fig. 5.4.1 is a set of notes illustrating either case (1) or (2).

5.5. Transit and Tape Survey with Arbitrary Points.—In case (3) the transit can be set up at an arbitrary point marked by a stake and placed far enough from one of the corners so that the telescope can be focused on this corner. In this way all the corners of the traverse are chosen so that the traverse lines will be approximately parallel to the sides of the field. The angles and distances of this traverse are then measured. To connect the property lines with this traverse, the angle and distance are measured at each transit station to the corresponding corner of the property before the instrument is moved to the next point. Fig. 5.5.1 is a set of notes illustrating this case. Time can be saved in the computations and a good check on the work may be obtained if the property lines are also measured directly. These distances should always be measured when possible. These lengths are not only useful as checks on the accuracy of the survey, but they will be needed in writing a description of the property.

These three methods which have been described may be combined in any survey according to circumstances.

In the three preceding surveys the distances have all been measured by holding the tape horizontal. Fig. 5.5.2 shows the notes of a traverse in which the measurements are taken by the slope method. Each distance was measured twice using a 200-ft. tape. Where the distance was over a tape-length long the slope of the tape was maintained by placing nails in the

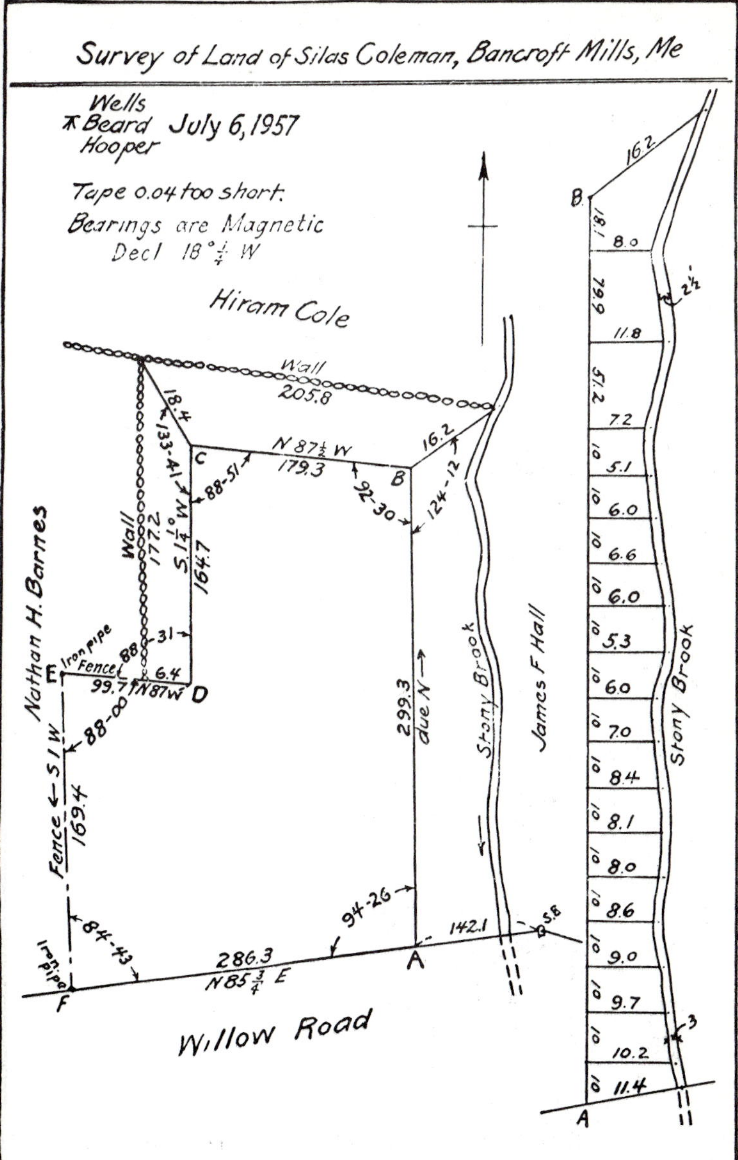

FIG. 5.5.1. NOTES OF SURVEY WITH TRANSIT AND TAPE.

Survey of center lines of roads bounding Block 5 - Cedar Brook Acres Land Development						Hunt - Chief, White - X, McKay - Chain, Wood - " April 21, 1957
Sta.	**Hor. Ang.**	**Vert. Ang.**	**Slope Dist.**	**Corr.**	**Hor. Dist.**	**Remarks & Sketch**
1						Stake
1-2		6°-55'	67.79	0.49	67.30	
2-1		10°-08'	68.38	1.07	67.31	
2	80°-36' L			Use →	67.30	Stake
	161°-12'					
	241°-48'					
Use →	80°-36'					
2-2A		0°-40'	217.35	0.02	217.33	
2A-2		2°-20'	217.52	0.18	217.34	
2A	180°-00'			Use →	217.33	Tk. in 8" Stump
2A-2B		5°-00'	177.30	0.68	176.62	
2B-2A		3°-33'	176.95	0.34	176.61	
2B	180°-00'			Use →	176.61	Stake.
2B-3		1°-25'	110.53	0.03	110.50	
3-2B		0°-37'	110.48	0.01	110.47	
3	171°-21' L			Use →	110.48	Drill hole in rock
	342°-41'-30"					
	514°-02'					
Use →	171°-20'-40"					
3-3A		1°-08'	198.14	0.04	198.10	
3A-3		0°-00'	198.09	0.00	198.09	
3A	180°-00'			Use →	198.09	Stake
3A-4		0°-00'	65.05	0.00	65.05	
4-3A		2°-15'	65.09	0.05	65.04	
4	74°-20'-30" L			Use →	65.04	Drill hole in ledge
	148°-41'					
	223°-00'-30"					
Use →	74°-20'-10"					
4-5		8°-25'	259.58	2.80	256.78	
5-4		7°-13'	258.79	2.05	256.74	
5	156°-14'-30" L			Use →	256.76	Stake
	312°-30'					
	468°-45'					
Use →	156°-15'					
5-6		7°-17'	105.50	0.85	104.65	
6-5		4°-19'	104.95	0.30	104.65	
6	112°-02'-30" L			Use →	104.65	Drill hole in rock
	224°-05'-30"					
	336°-08'-30"					
Use →	112°-02'-50"					
6-6A		0°-00'	286.99	0.00	286.99	
6A-6		1°-26'	287.14	0.09	287.05	
6A	180°-00'			Use →	287.02	Stake
6A-1		5°-38'	348.47	1.68	346.79	
1-6A		6°-30'	349.02	2.24	346.78	
1	125°-25'-30" L			Use →	346.78	
	250°-50'					
	376°-16'					
Use →	125°-25'-20"					
2						

FIG. 5.5.2. SHOWING DISTANCES TAKEN BY SLOPE METHOD.

sides of trees or in temporary stakes which were lined in on the proper slope by means of the transit, and measurements taken to the nails. Some of the long lines, where the slope of the ground changed considerably, have been broken into sections (2 to 2A, 2A to 2B, etc.). Under average conditions prevailing in the field this method is likely to give better results than horizontal taping.

5.6. Irregular Curved Boundaries.—When a tract of land is bounded by an irregular curved line, such as a brook, it is customary to run the traverse line near it, sometimes crossing it several times, and to take perpendicular offsets to the brook. If it is a winding brook with no distinct turns in it, offsets at regular intervals are measured from the transit line as in the portion near point *A* of Fig. 5.5.1. Near point *B* in this figure the portions of the brook between turns are almost straight, in which case the proper measurements to make are the offsets to those points where the course of the brook changes and the distances along the transit line between these offset lines. Since they are usually short the right-angle offset lines are laid off in the field by eye.

5.7. Survey of a Field by a Single Set-up of the Transit.—When it is necessary to economize time in the field at the expense of accuracy and of the time required to calculate the survey the following method may be used. If possible set up at a point within the field, preferably near the middle, from which all the corners can be seen, and measure the angles and distances to each corner. The field is thus divided into several oblique triangles in each of which two sides and the included angle have been measured; from these the area and the third side (property line) can be computed. As a check on the measured angles their sum should be 360°; there is no check on the property lines unless they are measured directly.

This method of surveying a field may be employed as a check on one of the other methods which have already been described, but it is not recommended as a method to be used by itself except in emergencies. One weak point in it is the low degree of precision with which the angles are usually measured. Here the effect of an error of, say, 30 seconds in an angle may often be much larger than the errors in the measured distances. The additional measurement of the property line gives the length of all three sides of the various triangles into which the field is divided. If the area is calculated from the three sides of the triangles, using the measured angles as checks only, an accurate result may be obtained, but at the expense of office work.

5.8. Survey of a Field with a Tape Only.—Sometimes in an emergency it may be necessary to survey a field when a transit is not at hand. This can be done by dividing the field into several triangles and measuring all their sides. To insure accuracy the triangles should be so chosen that there are no angles in them less than 30° or greater than 150°, because, in general, weak intersections cause large errors in the results. Sometimes, however, a small angle may enter in such a way as to do no harm. Lining in by eye is sufficiently accurate for obtaining the length of a line. But if a point is set on line by eye for the purpose of measuring perpendicularly to the line, such as the altitude of a triangle, only approximate results can be expected.

5.9. Selecting the Corners.—If a corner is marked by a stone bound, an iron rod or a pipe the exact point may be easily found; but where it is described as the intersection of stone walls or fences the surveyor will have to examine all evidence as to its position and use his judgment in deciding where the true corner is located (Art. 5.2.8.). When the property is bounded by a public way or a town boundary, data relating to the location of these lines must be obtained from the proper local authorities, usually from the town clerk. After determining the positions of the corner points, the surveyor should use precisely the same points in all distance or angle measurements. If stakes are used the exact point should be marked by a tack or nail driven into the top of the stake.

In deciding upon the location of the boundary lines from an examination of artificial features it should be borne in mind that fences or walls along highways should be built entirely on private property so that the face of the wall or fence is on the side line of the highway. In country districts, however, this is not always the custom.

The rerunning of the road lines of public ways in towns and villages so that they will agree with the recorded description is usually next to impossible. Their description is usually so indefinite and inaccurate that one soon finds that occupancy is the better evidence of the position of these public ways. In defining these ways, it is advisable to adhere to the width of road given in the description and to make the alignment of the road fit the positions of the walls, fences, and other objects which are evidence of occupancy. Then it is advisable to get the town authorities to establish that line by proper municipal ordinance.

In most cities the baseboard of a fence is usually built so that its face is on, or nearly on, the street line, but the location of the fences has no weight when the street line is defined by permanent marks or by city ordinance (Art. 10.2). In large cities street lines are defined by records in the City Engineer's office. These sometimes refer to stone bounds, but a better system, adopted by most city engineering offices, defines these lines by **a record of many swing-offset measurements** taken from **permanent** parts of existing buildings, such as the stone faces of party walls or the masonry underpinning of existing buildings.

Since the boundary line between adjacent owners has no width and since every fence occupies some width of land, a reasonable amount of land may be taken for the fence, one-half from each owner, because each owner must contribute his share of land for the division fence. It is the surveyor's duty to set the fence so that it occupies equal amounts of the abutting owners' land. For boundaries between private lands the legal line is the center of the stone wall or Virginia rail fence; in New England it is customary in the case of a tight board or picket fence to set the back side of the board (the face of the rangers) on the property line. This puts the boards entirely on one property and the rangers and posts which support the boards on the other. Since the posts are 6 to 8 ft. apart and the lower ranger is a foot or so above the ground this practice seems equitable. When the entire area is occupied by buildings as in the business portions of large cities the legal line often is the middle of the party-wall (Arts. 10.40–10.49). If the deed reads "to thread of stream" the boundary is the line midway between the shores when the stream has its average regimen of flow.

Not infrequently woodland is marked off by blazing the trees on one or both sides of the boundary line, the blazing being done on the side of the tree nearest the boundary line. If a tree comes directly on the line it is blazed on both sides at the points where the line strikes the tree. The blazes are sometimes painted red. A small pile of stones, sometimes with a stake in the center of the pile, is often used to mark the corners of such land. Farmland and woodland corners are sometimes marked by a substantial wooden stake with no stones piled around it. In the New England states round cedar posts from 4 to 6 inches in diameter are often used. In Canada and other parts of the continent corners are marked by heavy squared stakes. The early surveyor used to mark these stakes with his private mark. One surveyor named Morrow always cut a pair of crossed swords with initial "M" between them; another cut a pair of axes, and still another carved his two initials "AL" on every corner stake that he placed. The stakes usually extended 3 or 4 ft. into the ground and from 2 to 4 ft. out of the ground. This personal marking of stakes is of great value for it gives not only a clue to the age of the stakes but also to their reliability, for in the past, as in the present, certain surveyors were reputed for their care and accuracy, and where there is evidence of two different lines (and in woodlands this is a common condition) obviously more reliance will be placed on the line defined by a surveyor who had a reputation for accurate work than on the work of the haphazard surveyor. It is not uncommon to find, in the woods, two or more blazed lines near together. The oldest line takes precedence unless there has been acquiescence in a line of later date for a sufficiently long period to make the law of "adverse possession" determine the line (Art. 5.23). The age of blazed lines can be determined readily by cutting into the tree crosswise of the grain and counting the annular rings from the bark to the face of the old blaze. Usually the scar in the tree made by the old blazing has healed over so that only a faint vertical line on the bark is left to indicate where the tree was originally blazed. Consequently the surveyor is liable to overlook the old blazes unless he is resourceful.

5.10. Method of Procedure.—In deciding where the traverse shall be run the surveyor should keep in mind both convenience in fieldwork and economy in office work. Frequently a method of procedure which shortens the time spent in the field will increase greatly the amount of labor in the office. Circumstances will determine which method should be used. If there is no special reason why the time in the field should be shortened, the best arrangement of the traverse will be the one that will make the computation simple, and thereby diminish the liability of mistakes. If the lines of the traverse coincide with the boundary, as in cases (1) and (2) (Art. 5.3), the amount of office work will be the least. If in case (3) the traverse lines are approximately parallel to and near the boundaries of the property this simplifies the computation of the small areas to be added to or subtracted from the area traverse.

5.11. Ties.—All important points temporarily marked by stakes should be "tied in," i.e., measurements should be taken so that the point may be readily found or replaced in the future. There should be at least three *horizontal* ties which intersect at angles not less than 30°. They should be taken from points which are definite and easily recognized, such as nails driven into trees, drill holes in ledge, stone bounds, fence posts, or buildings. All such measurements should be recorded carefully, usually by means of a sketch. Fig. 5.11.1

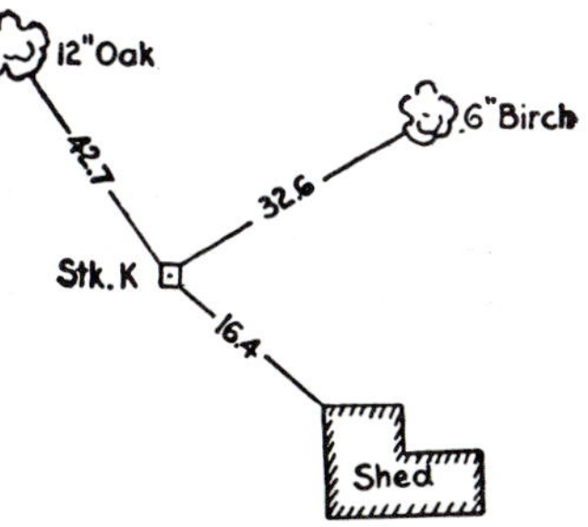

Fig. 5.11.1. Approximate Ties.

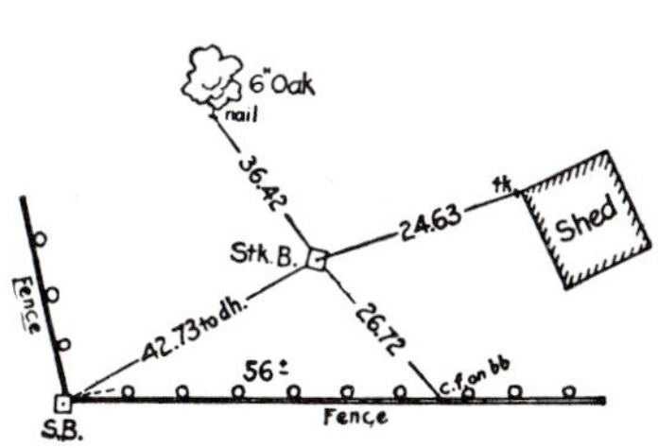

Fig. 5.11.2. Exact Ties.

shows a stake located by ties measured to tenths of a foot; these approximate ties are taken simply to aid in finding the stake. It is just as easy and is better practice to take the ties so that the exact point can be replaced. The surveyor should mark carefully by nail the exact points from which measurements (taken to 1/100 ft.) are made, and record in the notes as shown in Fig. 5.11.2. It will assist in finding a tie point if a circle is painted around it. In the woods, especially in winter, bright blue paint is the best color. For further details regarding ties see Art. 10.7.

5.12. Measurement of the Angles of the Traverse.—There are four common ways of measuring traverse angles; by measuring the *angle to the right*, by measuring the *interior angle*, by measuring the *deflection angle* (equal to the difference between the interior angle and 180°), or by measuring the *azimuth*.

5.13. MEASUREMENTS OF ANGLES TO THE RIGHT.—In this method all angles are measured to the right (clockwise). After backsighting along the previous line with the horizontal circle reading 0° the telescope is turned *to the right* (clockwise) and the angle measured to the next station (angle point). Then the telescope is inverted and the angle remeasured, "doubled." Half the final reading is the value used for the angle. This is good practice because it is simple to follow, applies to both linear and closed traverses and eliminates most instrumental errors.

5.14. MEASUREMENT OF INTERIOR ANGLES.—An interior angle is measured by sighting on the previous traverse station with the circle reading 0° and with the telescope direct. The angle is then measured to the traverse point ahead, being sure to measure the arc which represents the *interior* angle of the closed field, even if it is greater than 180°. Leaving the horizontal plate clamped, invert the telescope and repeat the process thereby "doubling" the angle. Half of the final reading is the required angle. Additional repetitions may be made if greater precision is required.

5.15. MEASUREMENT OF DEFLECTION ANGLES.—This is

he angle between the backsight produced by reversing plunging) the telescope and the forward traverse line; it is ecorded in the notes as R or L to indicate whether the telecope was turned to the right (clockwise) or to the left (counerclockwise) in measuring the deflection angle. Evidently a ingle measurement of this angle would be seriously affected y any error in the adjustment of the line of sight or of the tandards; such errors as may exist are doubled by plunging he telescope between the backsight and foresight which proess is required. Consequently, it is especially necessary to "double" the deflection angle to eliminate such errors of adustment. Best practice therefore requires making the first acksight with the instrument *direct* so that when the second latter) foresight is taken the instrument will again be in the *lirect* position and hence ready for conveniently carrying on he survey.

5.16. Measurement of Azimuths.—The common use of he term azimuth applies to the measurement of angles lockwise (to the right) from the *true meridian* or from the *nagnetic meridian.* In this case it is customary in surveying ractice to set the circle to read 0° when sighting toward the orth; consequently a sight toward the east would read 90° and ne toward the west would read 270°.

If a true meridian line has been established set up the ransit at the south end of it, set the "A" vernier at 0°, sight at he north end of the meridian, and then clamp the lower lamp. If the upper clamp is now loosened and the telescope ighted at any point, then vernier "A" will read the true zimuth to that point. The direction of this meridian may then e carried over to other traverse points as will be explained.

If the meridian is not actually marked out, but the true zimuth of some line is known, such as *AB* (Fig. 5.16.1), the ransit is set up at station *A*, the vernier set at the known zimuth of *AB* (say, 113° 16′) and the upper plate clamped. If he telescope is sighted at *B* and the lower clamp tightened he circle is oriented. If the upper clamp is now loosened and he telescope sighted at any other point the vernier which was originally set at the azimuth reading will read the true azimutl of that point from station A.

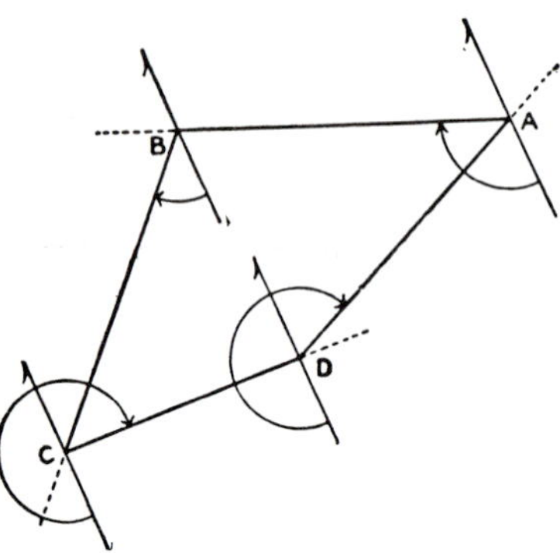

FIG. 5.16.1. AZIMUTHS.

Any line through A may b assumed arbitrarily as the 0 direction and azimuths ma be read from this direction Magnetic north is sometime used as the 0° azimuth.

Irrespective of the direc tion that is assumed as the 0 direction, the azimuth is car ried ahead to station B as ex plained in the next para graph.

The azimuth of BC may be obtained in one of two ways (1) Invert the telescope and backsight on A, the circles remain ing clamped together at the reading of B taken at A; ther clamp the lower plate, turn the telescope to its direct position loosen the upper clamp and sight on C. The circle then read the azimuth of BC referred to the same meridian as th azimuth of AB. The disadvantage of this method is that th error of collimation enters the azimuth each time. (2) Add 180 to the azimuth of AB, set this off on the circle and vernier, an sight on A. The telescope may then be turned directly to C (without inverting) and the azimuth of BC can be read directl on the circle. The disadvantages of this method as compare with the former are that the error in the angle between oppo site verniers (eccentricity and constant angle) enters the re sult, that time is consumed in setting the circle at each set-u of the instrument, and that there is an opportunity for mistake in calculating the angle and in making the setting on the ver nier. Therefore if the collimation adjustment is good the firs method may be preferred, but if there is doubt about this, the the second method may be safer.

In order to obtain a check on the angles of the travers occupy station A after the angle is measured at D, and deter mine the azimuth of AB by backsight on D. This new azimut of AB should agree with the original azimuth of AB.

5.17. Checking the Fieldwork.—Reading the horizontal angle to one minute will be suitable for usual rural land surveys. "Doubling the angle" is always advisable to prevent mistakes. Referring to Fig. 5.4.1, it will be seen that the angles were quadrupled where the sides were long, and doubled where they were short. In this instance the angles were repeated to increase the precision as well as to avoid mistakes.

Errors in the angles may be detected by observing the magnetic bearings and calculating the angles from these bearings; or by starting with one observed bearing (assumed to be correct), calculating the other bearings in succession by means of the measured angles, and noting whether the observed bearings agree approximately with the *calculated bearings*.

If a **very short line** enters into a traverse, the errors in angle at either end of this line caused by errors in centering the instrument over the point may be eliminated by first marking a foresight 200 ft. or more distant in the prolongation of this short line and then measuring all angles from this foresight. The angle between the lines on either side of the short line will then have an accuracy consistent with the other angles of the traverse.

The accuracy of the transit work may be tested by adding the measured angles. The sum of the interior angles of the field should equal $(n - 2) \times 180°$, where n is the number of sides in the field. If the deflection angles are used the sum of all the right deflections should differ from the sum of all the left deflections by 360°, or in other words, **the algebraic sum of the deflection angles should be 360°.**

If there is any doubt as regards the correctness of the measurement of a line it should be remeasured, preferably in the **opposite direction,** so that the same mistake will not be repeated. (See line *AB* in Fig. 5.4.1; also Art. 5.5 and Fig. 5.5.2.) If the traverse lines do not coincide with the boundaries, an independent check is obtained by measuring along the boundaries as well as on the traverse lines, as illustrated in Fig. 5.5.1. This furnishes at once a **rough** check on the distances in the field and a **close** check after the survey has been calculated. It is often advisable to run a line across the traverse, especially when there are many sides to the field, thus dividing the field into two parts, as in Fig. 5.4.1. If any mistake has been made it is then possible to tell in which portion of the traverse it occurred (Art. 15.13). If the transit is provided with stadia hairs the traverse distances can be checked by stadia readings.

5.18. PRECISION REQUIRED.—That the precision in the measurement of distances shall be consistent with that of the angles it is necessary that great care should be exercised in holding the tape horizontal, in plumbing, in aligning, and in securing proper tension. On sloping ground inclined distances can be measured. (See also Fig. 5.5.2.)

If the angles are measured to the nearest minute and the distances to the nearest tenth of a foot, it will be sufficiently accurate to use sighting-rods in "giving line." The error of closure of such a survey should be not greater than 1/5000, but would seldom be less than 1/10000.

If the property is very valuable, as in the case of city building lots, it is well to use a more accurate theodolite. The angles should be repeated, not only as a check against mistakes, but to increase the precision of the measurement. The tape measurements should be made with special care, and should be taken to the nearest hundredth of a foot. In the best work the temperature correction should be applied, a spring balance should be used to give the right pull on the tape, the correction to the standard distance should be determined (Art. 10.2), the alignment given with the transit, and great care taken in plumbing. Sights are given by holding a pencil vertically on top of the tack on the stake or by plumb-line. In this work it is important that the property line should be followed, when possible, to insure the most accurate results. In such work an error of closure of 1/20000 to 1/40000 can be expected. It is customary on city work to neglect the effect of temperature and to omit the use of the spring balance, the pull being carefully judged. Under these circumstances the closures obtained may be 1/20000 or higher, but the actual error due to neglecting temperature may be much more than this, since a variation of only 8° F. will introduce an error of about 1/20000, and a variation of tension of, say, 3 or 4 lb., may introduce a similar error. (See Precise City Surveys, Arts. 10.40-10.52.)

To illustrate the relation between the precision of angle and of distance measurements, suppose an angle of 42° 37′ is read with a one-minute transit. The true value of the angle can be anywhere from 42° 36′ 30″ to 42° 37′ 30″. The error in setting at 0° may be ±05″ to ±10″, thus either increasing or decreasing the angular error. The largest error in the angle may be from ±35″ to ±40″; this assumes that both the error of setting at 0° and the reading of the angle are the greatest they can be and are both in the same direction. The least error in the angle may be 0″; in this case the circle is assumed to be set exactly at zero and the true angle to be exactly a full number of minutes. Neither of these extreme cases is likely to occur. It seems reasonable, therefore, to assume that the average error of reading a single angle is probably about ±15″, or possibly ±20″.

If the sides of a 300-ft. square are measured precisely to the nearest 0.01 ft., this degree of linear precision, expressed as an offset of 0.01 ft. in 300 ft., corresponds to an angle precision of 07″. To obtain this degree of precision with a one-minute transit, the angles should be tripled, thus giving single readings to 20″, which, on the average, should be within ±05″ of the correct reading. Uncertainties in setting the plate exactly on zero might introduce an additional error of ±02″ or ±03″.

In order to secure the same precision in angles and distances, not only should the proper transit be selected, but the necessary number of repetitions of angles should be determined.

5.19. Organization of Transit Party.—Transit surveys can be carried on readily by a party of three men. The note keeper, who is in charge of the party, directs the entire work; the transit-man, who has the instrument always in his care, sets it up where directed by the note keeper, reads the angles and gives line when it is desired; the tape-man, generally acting as head-tape-man, and the note keeper, as rear-tape-man, measure all distances. There is often economy, however, in having another tape-man so as to leave the note keeper free to keep the notes and to direct the party. Where much clearing of underbrush is required an axeman is also a desirable or even a necessary addition to the party. The axemen should be trained in the duties of cutting and in marking and driving stakes, preparing for ties, and otherwise assisting in the surveying.

The speed of the party depends upon the way in which the chief of party directs his men. He should plan the work and train his men so that every man realizes his responsibility and anticipates the particular function he is to perform so that there is little necessity for calling off orders. Some surveying parties can perform in a day twice the work of others owing solely to the fact that the chief has trained his men well and insisted upon the kind of discipline that produces accurate results in a short time. The chief of party should check mentally, by estimate, all distances and angles measured, so as to catch a large error the moment the dimension is called off to him. He should be at the spot where important measurements are made, so that he can check the readings himself if he has any doubt of their accuracy.

5.20. NOTE KEEPING.—All measurements should be recorded in a special note-book **as soon as they are made** and never left to be filled in from memory. When a dimension is called off the note keeper should record it immediately in his note-book. If his sketch is not prepared for that dimension, or if he does not know just where he wants to record it, he should, nevertheless, **write it somewhere in the book at once** and then decide promptly where it should be placed permanently. Even then it is advisable not to erase the original entry. To depend upon carrying a dimension in one's mind even for a few moments is dangerous practice. **The notes should be neat and in clear form so that there will be no doubt as to their meaning.** Great care should be taken that they shall not be susceptible of any interpretation except the right one. They are usually recorded in pencil, but they should always be regarded as permanent records and not as temporary memoranda. As other persons, not familiar with the locality, will probably use the notes and will depend entirely on what is recorded, it is very important that the notes contain all necessary data without any superfluous information. If the note keeper will bear in mind constantly how the survey is to be calculated or plotted it will aid him greatly in judging which measurements must be taken and which ones are unnecessary. Clearness is of utmost importance in note keeping, and to attain it the usual custom is not to attempt to sketch closely to scale; and yet in special surveys where much detail is desired, such as the details of a doorway and steps, it is well to carry out the sketches in the note-book approximately to scale. Care should be taken not to crowd the notes—paper is cheap—and an extra page of the note-book devoted to a survey may save hours of time in the office consumed in trying to interpret a page of crowded notes, and may also avoid mistakes. Too much stress cannot be laid on the importance of being careful not to lose the note-book; not infrequently a note-book contains data which could not possibly be replaced. Such notes should be photostated, and the original book kept in the office vault.

Although sufficient fullness to make the notes clear is always desirable, it is customary to abbreviate the names of the artificial features most commonly met with by the surveyor. To interpret a set of notes properly one must be familiar with these symbols and abbreviations, some of the more common of which are given below.

angulation Station. △
ansit Traverse Point. ⊙
adia Station. ⊡
one bound. S.B.
onument. Mon.
ake. Stk.
ike. Spk.
ill-hole. dh.
nch-mark. B.M.
rning point. T.P.
ow-foot (a mark like this V or ↓). c.f.
t crow-foot (cut into wood or stone). c.c.f.
nce. —·—·—
nce, showing on which side the posts are. —o—o—o

Line of building; the outside line is the base-board, the cross-hatched part is the line of the stone or brick underpinning.
Center. C.
Center line. ℄
Nail. na.
Tack. tk.
Curb. cb.
Catch basin. C.B.
Manhole. M.H.
Telephone pole. Tel.
Foresight. F.S.
Backsight. B.S.
Base-board of fence. b.b.

Distances should always be recorded in such a way as to indicate the precision with which they were taken. For example, if they were read to hundredths of a foot and a measurement happened to be just 124 ft. it has to be recorded as 124.00, not as 124. The two zeros are of as much consequence as any other two digits which might have come in their places. Angles that have been read to the nearest half-minute, however, are recorded as follows: 6° 47′ 30″. It will be seen that this is not consistent with the foregoing. A more proper way of reading and recording this angle would be 6° 47½′, but this is not common practice.

In addition to the measurements, every set of notes should contain the following information: the name of the job, the location and character of the work, the date, and the names of the field party. It is well also to state the names or numbers of the instruments used and the error of the tape. Where the notes of a survey are continued through several pages of a bound book the date may be placed at the top of every page; other data need not be repeated, but if loose-leaf notes are used the sheets should be carefully "paged," and **every leaf headed with the name of the job.** The underlying fault of the loose-leaf system is that the note keeper may fail to put the title on each leaf, with the result that any separate pages of the notes that become misplaced may have no identifying mark on them to indicate the survey to which they belong. Fig. 5.2.1, Fig. 5.4.1, Fig. 5.5.1 and Fig. 5.5.2 are given as illustrations of fieldnotes.

If there are many pages of valuable notes, and especially if these are being sent back and forth between office and field, it is advisable to have photostats made as soon as possible. The originals may then be kept at the office and photostats sent out to the field party.

On the front cover of every note-book should be plainly lettered the name and address of the surveyor, the fact that the book is of great value and, if deemed advisable, a note to the effect that the finder of the book will be rewarded upon its return to the owner.

5.21. Survey of a Field for a Deed.—If a description for a deed is to be written, the lengths and bearings of all the boundaries are desired. The traverse lines should therefore follow the property lines, if possible. The bearings desired are not those observed, but are those calculated by means of the balanced transit angles as explained in Art. 15.4, and therefore are relatively as accurate as the angles themselves. If a true meridian is found by astronomical observation (Chapter VIII) the bearings should be referred to this and marked **true bearings** on the plan, and this information should also be contained in the deed.

A plan which is to accompany a deed should show such features as watercourses, highways, buildings, and adjoining property lines, as well as stone bounds, stakes, fences, walls, or other artificial objects which mark the boundaries of the property.

This plan should contain the following information.

1. Lengths of all property lines together with their calculated bearings or the angles at the corners.
2. Location and description of corner bounds.
3. Conventional sign or name on walls, fences, etc.
4. Names of highways, streams or ponds, and names of adjacent property owners.
5. Scale of drawings and direction of the meridian used (true or magnetic or assumed or grid meridian). It is better to refer all bearings to the state plane coordinate grid meridian if feasible, or alternatively to the true meridian, rather than to an assumed or the magnetic meridian.*

---

*As magnetic bearings are unreliable (Art. 2.1) true bearings should be used wherever their adoption does not entail too much additional expense. In those parts of the country which have been subdivided by the U. S. Bureau of Land Management, formerly General Land Office, true meridians can be obtained from the government surveys; in many of the older (Eastern) states, the meridians or grid meridians have been established by local authorities. If the survey can be connected with any plane coordinate system or geodetic system established by federal or local bodies, then, since the true bearings of all of the triangulation lines are known, the bearings of the traverse lines can be obtained. The true bearing of one of the traverse lines may be obtained by an observation on polaris (Art. 8.3) or on the sun (Art. 8.9).

6. The title should include a simple and complete statement giving the name of owner, place, date, and name and address of surveyor. An explanatory note, such as a statement whether bearings refer to true or to magnetic meridian, may also be necessary. (See Art. 18.1)

5.22. Deed Description.—A deed description should give the bearings (or angles) and distances along each property line and should state whether bounded by a highway, stream, or private property together with the name of the owner. The description should run *to* and *along* lands of adjoining owners to aid in discovering mistakes which may have been made through carelessness of the conveyancer or copyist. The following paragraph is a deed description of the property shown in Fig. 5.5.1. The magnetic bearing of the traverse line *AB* was assumed correct and from it the bearings of the walls and fences bounding the property were computed.

"Beginning at a point in the northerly line of Willow Road in the town of Bancroft Mills, Maine, at an iron pipe sunk in the ground at the S. E. corner of land now or formerly belonging to Nathan H. Barnes, and running along the said northerly line N 85° 34′ E a distance of two hundred ninety-seven and seven-tenths (297.7) feet to the thread of channel of Stony Brook at land now or formerly belonging to James F. Hall; thence turning and running in a northerly direction, by thread of channel of said Stony Brook and land of said Hall, a distance of about three hundred and eight (308±) feet to a stone wall at land now or formerly belonging to Hiram Cole; thence turning and running along the middle of said stone wall and by land of said Cole N 86° 45′ W a distance of two hundred and five and eight-tenths (205.8) feet to the middle of another stone wall at land of said Barnes; thence turning and running by said latter stone wall and land of said Barnes S 0° 53′ E a distance of one hundred and seventy-seven and two-tenths (177.2) feet to a fence; thence turning and running by said fence and land of said Barnes N 87° 09′ W a distance of ninety-three and three-tenths (93.3) feet to an iron pipe sunk in the ground; thence turning and running by a fence and land of said Barnes S 1° 51′ W a distance of one hundred and sixty-nine and four-tenths

(169.4) feet to the point of beginning; all the bearings being magnetic and the parcel containing a calculated area of 79,305 square feet more or less." Careful conveyancers prefer to write the bearings in words as well as in figures.

It is unfortunate that the description of the property in deeds in the vast majority of cases does not define the property in such a manner that it can be plotted from the description. Some descriptions in deeds are so loosely written as to contain nothing but the names of the owners of adjacent property, no bearings or distances being given.

5.23. Laws Relating to Boundaries.*—The surveyor really has no judicial functions; it is his function to *develop evidence* from his personal examination and measurements of the property and descriptions of it. But to do this intelligently he must obviously have some knowledge of the laws pertaining to property boundaries. All that will be attempted here is to state a few basic rulings to which the surveyor may expect that any court will adhere. It will be his further duty to inform himself as to the practice, professional and legal, prevailing in the particular region of his endeavor and to conform to it, for in this short text it will be impossible to discuss those many legal decisions which vary among the different states.

It is distinctly his duty to **find the position of the original boundaries** of the property and **not attempt to correct the orig-**

---

*For a more complete statement of the principles mentioned above, consult "The Surveyor and His Legal Equipment," Trans. Am. Soc. of Civil Engineers., Vol. 99, 1934, pp. 1155-1195, and "The Surveyor and the Law," Trans. Am. Soc. of Civil Engineers, Vol. 107, 1942, pp. 1379-1400, both by A. H. Holt.

"Boundary Control and Legal Principles," by Curtis M. Brown, published by John Wiley & Sons, Inc., New York.

A "Treatise on the Law of Surveying and Boundaries," by Frank Emerson Clark, Esq., published by The Bobbs-Merrill Co., Indianapolis, Ind., 2nd Ed., new printing 1948, gives especial emphasis to the U. S. Public Land System.

"Legal Elements of Boundaries and Adjacent Properties," by Ray H. Skelton, published by The Bobbs-Merrill Co., Indianapolis, Ind.

"Evidence and Procedures for Boundary Location" by C. M. Brown and W. H. Eldridge, John Wiley and Sons, New York, 1962.

"Clark on Survey and Boundaries" by J. S. Grimes, Bobbs-Merrill, Indianapolis, 1959.

inal survey even though he may be sure that an error exists in it. The surveyor must also fortify himself with the evidence that will convince the judge or jury that he has **actually retraced the old boundaries,** and that he has **followed the tracks of the original surveyor.**

Physical boundaries, provided they are mentioned in the deed, have the legal force of monuments and therefore take precedence over courses and distances. These boundaries (or monuments) may be either "natural monuments" such as trees, lakes, boulders, ledges, etc., or "artificial monuments" such as fences, stone walls, posts, stakes, pits, mounds, stone bounds or other physical objects placed by man for boundary marks. Natural boundaries take precedence over those that are artificial because they are more permanent.

Where no physical boundaries can be found after an exhaustive search,* then the testimony of old reliable witnesses may be resorted to. It is important that these witnesses shall have no personal interest in the property and shall have had a good opportunity to observe the lost marks, fences, trees, or corner posts which limited the property. If the witness is personally interested in the land, then to have his testimony valuable it is necessary either that he shall have made the statements depended upon before any controversy has arisen regarding the line or else that his statements are detrimental to his own interest. It is not uncommonly true, after all evidence of persons familiar with the location of the lost lines has been analyzed, that one or more of the physical corner marks can now be found even though careful search had previously been made. Sometimes only a spot of discolored earth at the place pointed out by a witness as the corner will indicate where a corner post or stake, which has long since decayed, was undoubtedly located.

---

*It must not be assumed that a boundary is missing because it is not at once visible. Stone bounds are often buried 2 or 3 ft. deep; the top of a stake soon rots off, but evidences of the existence of the stake are often found many years after the top has disappeared, and the supposed location should be carefully examined and then dug over to find traces of the old stake. Care should be taken, however, that the evidence is not destroyed in the process of digging Traces of the old wood or charcoal may often be found if great care is used.

It is to be assumed that the deed was drawn by the grantor with honest intent to convey the property to the grantee. It shall therefore be interpreted if possible so as to make it effectual rather than void. The deed should also be construed in the light of what was known at the time when the title was transferred. In its interpretation it is assumed that it was intended to convey property the boundaries of which will form a closed raverse. There it is within the jurisdiction of the surveyor to reject any evident mistake in the description when running out the property line, e.g., a bearing that has obviously been recorded in the opposite direction or an entire side or chainength omitted. When the area does not agree with the boundaries as described in the deed, then the boundaries control. All distances unless otherwise specified are to be taken as straight lines; but distances given as so many feet "along a wall" or "by a highway" are supposed to follow these lines even if they are not straight. When a deed refers to a recorded plan the dimensions on that plan become a part of the deed description of the property. (See Art. 5.25)

Regarding the boundary of ownership of property abutting on highways the state laws are not uniform. In some states the streets and roads are owned in fee by the state or municipality and when abandoned as highways do not revert to the former owners; in others the fee runs to the middle of the street subject to the public's use of the street; but when abandoned for such use it reverts to the respective private abutters.

The rights of owners of land bordering on streams, lakes, ponds, navigable and nonnavigable streams are different in different states. The subdivision of lands gained by accretion is also a question as to which there is considerable divergence of opinion and practice.

It not infrequently happens that a land owner *A* has occupied for many years not only the part within his deed boundaries but also abutting land that formerly belonged to another person *B*. If such an occupancy by *A* of *B*'s land has been open, adverse and continuous for 20 years *A* has become the owner of that part of *B*'s land which he has been so occupying. This is called obtaining title by "Adverse Possession." This period of

20 years is the time required in many states in the Union. In some states *A* can acquire part of the public highway in 40 years of continuous and notorious use.

It is a general rule that any excess or deficiency found upon subsequent measurement to exist in a line marked by permanent monuments at its ends only, must be distributed among the several subdivisions of the line in proportion to the original measurements. For example, if a city block originally surveyed as 600.00 ft. long and permanently monumented was subdivided into twelve lots of 50.00 ft. frontage, and the block is subsequently found to measure 600.24 ft. between monuments, then each lot should be given a frontage of 50.02 ft. This rule cannot be followed blindly. For example, if the frontage on one of the lots had been omitted from the plan or had been marked 50± ft., then the whole excess of 0.24 ft. should go into that unmarked or approximately marked frontage. This general rule applies also to dimensions of irregular shaped lots and to deficiency or excess in areas, as where the original dimensions appear to leave a strip or to overlap where two lots are intended to abut upon each other and where it is intended to convey the whole tract.

Whenever a plan of a single lot or of a plot in which this lot is contained is mentioned in a deed that plan or plot should also be recorded with the deed unless it has been recorded already.

5.24. RERUNNING OLD SURVEYS FROM A DEED.—Courses and distances recorded by the original survey are presumed to be correct. They seldom are, because of the change in declination of the needle, stretching of the chain or tape, errors due to poor plumbing, difference in pull of chainmen and even the customs of some of the earlier surveyors to give "full measure." Most distances given in old deeds will be found to be shorter than those obtained in recent surveys.

To follow the old survey is a relatively easy matter when its physical boundaries can be identified. If only one line of an old survey can be found marked by physical boundaries, its length and direction can then be measured and compared with the original survey record. They will doubtless be found to differ from the original, and this difference in course will show the **correction to apply to all of the original courses,** and similarly the difference in distance will give a **correction percentage to apply to all of the original dimensions.** Then with these new values for the courses and distances the old survey can be retraced.

If none of the old lines can be found but one old corner is identified, then the old survey must be rerun by courses and distances. As the old courses are usually magnetic bearings it is first necessary to find the declination of the needle at the date of the original survey as well as at the present time and to change all the bearings accordingly. The declination of the needle should appear on the original deed or plan; but unfortunately it seldom does, and the year the survey was made must then be obtained either from the original deed, the old plan, or from witnesses, and the declination of the needle at that time computed. Observations at different places and times have been compiled by the U. S. Coast and Geodetic Survey, now known as the National Oceanic and Atmospheric Administration, National Ocean Survey (NOAA/NOS) and these results may be found in convenient form for calculation in Reports and Special Publications of the Survey.* From these observations the approximate change in declination may be obtained probably as accurately as the original bearings were taken.

When taking magnetic bearings or declinations from a plan the surveyor should be on his guard. Not infrequently a plan is dated years after the bearings were observed, the plan being a copy, or partly a copy, of an older plan, old survey, or old deed description. He should discover additional supporting evidence before relying upon such bearings.

---

*See U. S. Coast and Geodetic Survey Serial 664, "Magnetic Declination in the United States 1945" and also Publication 40-1, "Magnetism of the Earth," 1962.

Additional information may be obtained by writing to the Director of the National Oceanic and Atmospheric Administration, National Ocean Survey, Rockville, Maryland.

5.25. How to Look Up a Recorded Deed.—In all the states of the Union the transfer of real property must be recorded in the respective county Registry of Deeds or in the office of the city or town clerk. At the Registry of Deeds is kept an exact copy of the deed, which can be examined by any one. It is frequently necessary for the surveyor to make use of these copies when it is not convenient to obtain the deed from the owner of the property or when it is necessary to look up the deed of adjacent property or previous transfers of any of them.

In every Registry of Deeds is kept an index of deeds, which is divided into two parts, the *grantor* index and the *grantee* index; the grantor being the party who sells the land and the grantee the one who buys it. The indexes are frequently divided by years and for this reason the surveyor should know not only the name of the party who bought or sold the property (both if it is convenient to get them), but also the approximate date of the transaction. With this information he can find in the proper index the name of the party, opposite which will appear the date of the transaction, the number of the deed book, and the page on which the copy of the deed is recorded. He then finds the deed book, from which he can copy whatever data he desires from the deed; usually the description of the property is all that concerns the surveyor. In the deed book there is usually a reference number in the margin, or in the text of the deed, which refers to the next preceding transfer of the same property or to any attachments, assignments, and the like which may have been made on it. This method of indexing and filing deeds is used in the New England States and in many of the other states; the general principles are the same throughout the country although the details may differ to some extent.

If a plan is referred to in a deed then the plan becomes a part of the description. In this case the filed plan should also be looked up at the Registry.

5.26. Land Court.—While many states* have a Land Regis-

*In California, Colorado, Illinois, Minnesota, New York, North Carolina, Ohio, Oregon, Washington, Nebraska, Virginia, North Dakota, South Dakota, Georgia, Hawaii and the Philippine Islands.

tration Act spelling out very exact procedures pertaining to deed registry and property boundary surveys, Massachusetts has established a Land Court for any needed adjudication of ownership and of boundaries. A decree issued by this Court registering a piece of land virtually makes the State the guarantor of both the validity of the title and the boundaries.

The usual procedure is for the land owner to petition the Court for registration, pay the required registration fees, and have a survey and plan made in accordance with the rules issued by the Court. The Court examines the chain of title to the land and notifies by mail and by public notice all persons that it has reason to believe may have any interest in the land, and a date for hearing is assigned. When the Court sits, the surveyor's plan of the land to be registered is produced and competent evidence as regards ownership and the boundaries shown by the plan is then heard. In controversial cases an attorney usually represents the petitioner, but in simpler cases the owner or his surveyor may present the evidence before the Court. The Court's decision, except upon matters of law, which may be reviewed by the Supreme Court, is final. Questions of fact may on petition be tried by jury, and if not, are determined by the Land Court.

The surveyor's plan is the basic evidence on which the Land Court starts. It is necessary, therefore, that this plan be prepared carefully and in accordance with the regulations of that Court, some of which regulations are stated in the next paragraph.

The surveyor is expected to indicate to the best of his knowledge the lines of ownership as well as boundary dimensions. His plan must be based upon a closed survey, accompanied by his field notes and computations. The errors of the survey must be balanced so that the boundary angles and distances mathematically and exactly define a polygon. (Art. 15.7.) The accuracy required in the fieldwork is a closure of 1 in 10 000 for valuable city or town land where the boundaries are the middle of partition walls, faces of buildings or fences, etc.; a closure of 1 in 8000 for residential and suburban districts; 1 in 5000 for country estates; and 1 in 3000 for farms and woodlands. Every survey must join two or more permanent bounds, preferably intervisible, from which to relocate the boundaries at any future time. If such bounds do not exist already, then they must be placed by the surveyor. They need not be artificial; they may be drill-holes in ledge or offsets from the masonry underpinning of permanent buildings. Where it is found practicable the survey should be connected with triangulation points of town, city or state survey systems and with municipal boundary lines. Most lands abut on a public or a private way; this necessitates first the rerunning of the side lines of such way. Records of these lines are usually found in the offices of the city or town engineers, county clerks, county commissioners, park, highway, or water commissioners. When re-establishing these lines it should be remembered that stone monuments which have been previously set at angle points or at the beginning or end of curves are sometimes displaced by frost action, or by other causes; these should not be assumed to be in their correct location until verified by accurate survey. Detailed statements of the requirements for land plans are usually issued by the engineer of the Land Court.

THE UNITED STATES SYSTEM OF SURVEYING THE PUBLIC LANDS

5.27. THE SYSTEM.—The United States System of Surveying the Public Lands, frequently referred to as the rectangular system, was planned in 1784 by the Continental Congress, and has been modified only slightly by subsequent acts of Congress. The law requires that the public lands "shall be divided by north and south lines run according to the true meridian, and by others crossing them at right angles so as to form townships six miles square," and that the corners of the townships thus surveyed "must be marked with progressive numbers from the beginning." It provides that the townships shall be subdivided into thirty-six sections, each of which shall contain six hundred and forty acres, as nearly as may be; the division to be by cardinal lines surveyed and marked at intervals of a mile.*

*Cardinal lines are those running either due north and south or due east and west.

In order to make the adjustments in the survey to provide for the convergency of meridians, a unique system has been developed which applies uniformly and effectively throughout the whole of the public domain where the jurisdiction over the land is vested in the Federal government.*

The work of the Public Lands Survey is carried on under the Director of the Bureau of Land Management. Where more detail is required than given in this text recourse should be had to the Manual† of Instructions prepared by the Bureau of Land Management, formerly the General Land Office.

The law requires that measurements be true horizontal distances expressed in miles, chains, and links.‡ The directions of all lines are expressed as bearing angles, which are intended to refer to the true meridian.

The general system will be described in Arts. 155 and 156, followed by a detailed discussion of the several processes. The practice is applicable within twenty-nine of the states and in the Territory of Alaska, and applies to the retracement of old surveys as well as to their extension into new territory.

5.28. Outline of the System.—It will be convenient to consider the process of subdivision as separated into several distinct operations, to be carried out in sequence. It must be understood, however, that one operation, for instance the division of the area into 24-mile tracts, is rarely or never completed over the entire area to be covered before the next operation in order is begun; a single surveying group may be

---

*The surveys and the granting of the title to the lands within the New England and Atlantic Coast States (excepting Florida), and in Pennsylvania, West Virginia, Kentucky, Tennessee, and Texas, are under the jurisdiction of the states.

†Manual of Instructions for the Survey of Public Lands of the United States, 1966, may be obtained from the Superintendent of Documents, Washington, D.C.

‡Chain = 66 ft.; link = .66 ft. These units were used in the early Colonial surveys of the English speaking settlers, and in the state grant surveys excepting in Texas, where the Spanish *vara* was employed; the early French settlers used the *arpent* unit. The arpent is a measure of area, one side of this square unit equals 191.994 ft. (Louisiana, Mississippi, Alabama and Northwestern Florida); 192.500 ft. (Arkansas and Missouri). The vara = 33.333333 in. (Texas); 33.372 in. (Florida); 32.99312 in. ("public domain" in Southwest).

carrying on two or three different operations, such as running township exteriors and immediately afterward subdividing the townships into sections.

Briefly stated, the subdivision work is carried on as follows:

FIRST.—The establishment of

(a) An *Initial Point* by astronomical observations.

(b) A *Principal Meridian* conforming to a true meridian of longitude through the Initial Point, and extending both north and south therefrom, and

(c) A *Base-Line* conforming to a true parallel of latitude through the Initial Point, and extending both east and west therefrom. This initial operation is indicated in Fig. 5.28.1.

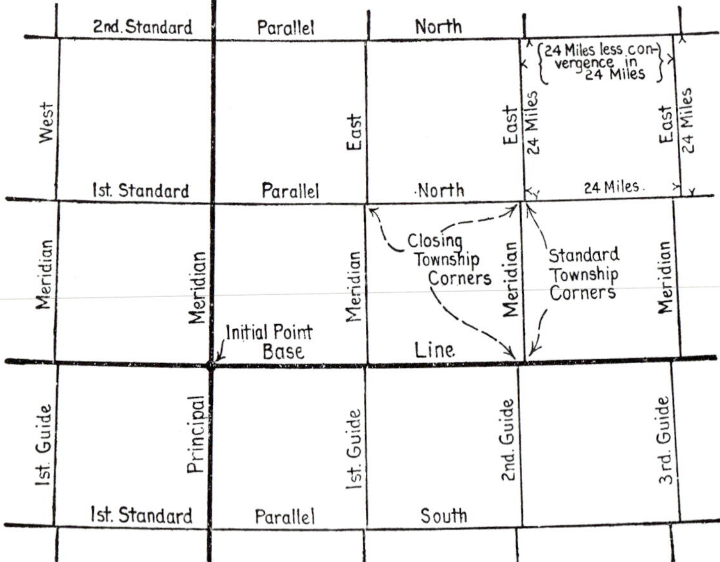

FIG. 5.28.1. SHOWING DIVISION INTO 24-MILE BLOCKS.

The principal meridian will be marked out on the ground as a straight line, and the base-line will follow the curve of a due east and west line, being at every point at right angles to the meridian through that point.

SECOND.—The division of the area to be surveyed into tracts

approximately 24 miles square (Fig. 5.28.1) by the establishment of

(a) *Standard Parallels* conforming to true parallels of latitude through the 24-mile points previously established on the principal meridian, and extending both east and west therefrom, and

(b) *Guide Meridians* conforming to true meridians of longitude through the 24-mile points previously established on the base-line and standard parallels, and extending north therefrom to an intersection with the next standard parallel or baseline.

Since the guide meridians converge, these 24-mile tracts will be 24 miles wide on their southern and less than this on their northern boundaries. Theoretically, both the east and the west boundaries should be just 24 miles in length, but, owing to discrepancies of field measurements, this is rarely or never the case.

THIRD.—The division of each 24-mile tract into *Townships*, each approximately 6 miles square, by the establishment of

(a) Meridional lines, usually called *Range Lines*, conforming to true meridians through the standard township corners previously established at intervals of 6 miles on the base-line and standard parallels, and extending north therefrom to an intersection with the next standard parallel, or to the base-line, and

(b) Latitudinal lines, sometimes called *Township Lines*, joining the township corners previously established at intervals of 6 miles on the principal meridian, guide meridians, and range lines. The division resulting from the first three operations is indicated in Fig. 5.28.2.

Neglecting the effect of discrepancies and irregularities in measurement, both the east and the west boundaries of all townships will be just 6 miles in length, but the north and south boundaries will vary in length from a maximum at the standard parallel or base-line forming the southern limit of the 24-mile tract to a minimum at that forming its northern limit.

FOURTH.—The subdivision of each township into *Sections*, each approximately 1 mile square and containing about 640 acres, by the establishment of *Section Lines*, both meridional and latitudinal, parallel to and at intervals of 1 mile from the eastern and southern boundaries of the township.

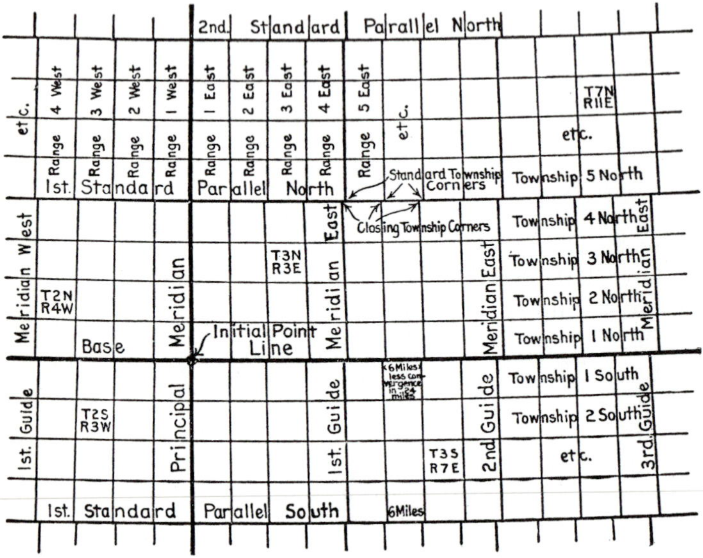

FIG. 5.28.2. SHOWING SUBDIVISION OF 24-MILE BLOCKS INTO TOWNSHIPS.

Assuming all the fieldwork to be done with mathematical exactness, this subdivision would result in sections exactly 80 chains (1 mile) on each of the four sides,* except the most westerly range of 6 sections in each township, which would be less than 80 chains in width by an amount varying with the distance from the southern boundary of the 24-mile tract. The extent to which this condition is realized in practice is indicated in Art. 5.40, wherein the usual field methods of subdividing a township are described in detail.

*These theoretical sections would not be exactly square, as may be readily perceived, but would be rhomboids.

5.29. Methods of Designating Lines and Areas.—The various principal meridians and base-lines of the Public Lands Surveys are designated by definite names or by number, as, for example, "The Fifth Principal Meridian and Base-Line," or "The Cimarron Meridian."*

The standard parallels are numbered in order both north and south from the base-line, and are so designated. The guide meridians are numbered in a similar manner east and west from the principal meridian. Fig. 5.28.1 illustrates the method.

Any series of contiguous townships or sections situated north and south of each other constitutes a *range*, and such a series situated in an east and west direction constitutes a *tier.*

The tiers of townships are numbered in order, to both the north and the south, beginning with number 1 at the base-line; and the ranges of townships are numbered to both the east and the west, beginning with number 1 at the principal meridian. A township is designated, therefore, by its serial number north or south of the base-line, followed by its number east or west of the principal meridian, as "Township 7 south, Range 19 east, of the Sixth Principal Meridian." This is usually shortened to "T. 7 S., R. 19 E., 6th P.M."

The sections of a township are numbered as in Fig. 5.29.1. In all surveys of fractional townships the sections will bear the same numbers they would have if the township were complete.

6	5	4	3	2	1
7	8	9	10	11	12
18	17	16	15	14	13
19	20	21	22	23	24
30	29	28	27	26	25
31	32	33	34	35	36

FIG. 5.29.1. DIAGRAM OF A TOWNSHIP ILLUSTRATING METHOD OF NUMBERING THE SECTIONS.

By the terms of the original law and by general practice section lines are surveyed from south to north and from east to west, in order uniformly to place excess or deficiency of measurement on north and west sides of the townships.

*These bases and meridians are shown on the large wall map of the United States published by the Bureau of Land Management, on the various official state maps, and on a special map entitled "United States, Showing Principal Meridians, Base Lines and Areas Governed Thereby," Superintendent of Documents, Washington, D. C.

5.30. Field Methods.—The work of subdivision of the Public Lands has already been largely completed, and the surveyor of today is usually concerned only with the retracing of old lines, the relocation of lost corners, or with the subdivision work that comes with increase in population. For all these, however, a thoroughgoing knowledge of at least the common field processes and methods that have been used in the original surveys is essential. Certain details of field practice have varied somewhat from time to time, but the leading features have remained fairly constant for all those areas that have been surveyed since the system became well established.

In the following pages is given a somewhat detailed description of the methods commonly employed in carrying out the operations briefly indicated in the preceding articles.

5.31. Initial Points.—No new initial points have been established in many years, excepting in Alaska. All of the thirty-four existing points were selected for the control of the surveys in agricultural areas to meet the demands for the settlement of the lands. After the selection of the point, the geographic position, in each location, was determined by field astronomical methods.

5.32. Base-Line—Establishing a Parallel of Latitude.—The base-line is a true parallel of latitude extending both east and west from the initial point.

There are three general methods for establishing a true parallel of latitude: (1) the solar method, (2) the tangent method, and (3) the secant method.

The solar method employs the solar attachment to the transit, by means of which the direction of the meridian may be quickly found at every set-up when the sun is visible. Once the direction of the meridian has been established, a line turned off at right angles to the meridian will determine the direction of the true parallel of latitude at that set-up. If sights between set-ups are not over 40 to 20 chains in length, the series of straight lines established by turning a right angle from the meridian at each set-up will not differ appreciably from the true curve of the parallel of latitude.

The solar transit is prescribed by the Bureau of Land Management for locating the meridian, particularly in forested regions, and in dense undergrowth where sights are short. This transit has an auxiliary telescope which may be pointed at the sun in such a way that the direction of the meridian is found directly in the field without calculations. (See Arts. 8.11, 8.12) The instrument that is used is the Smith solar transit which has been designed especially for the public-land surveying practice; it is equipped for making all necessary observations for time, latitude, and azimuth; the specifications require performance that ordinarily can be expected with a one-minute transit. The measurements are made with a long steel tape; slope distances are appropriately reduced to the true horizontal.

When the sun is not visible or when an instrument without a solar attachment is used, the true latitude curve may be determined by offsets from a straight transit line. For this purpose the tangent method or the secant method may be employed. The former is the one ordinarily employed in surveying practice. The azimuths are determined and tested at intervals by observations on the sun, or observations on Polaris at elongation or at any hour angle. (See Chap. VIII.)

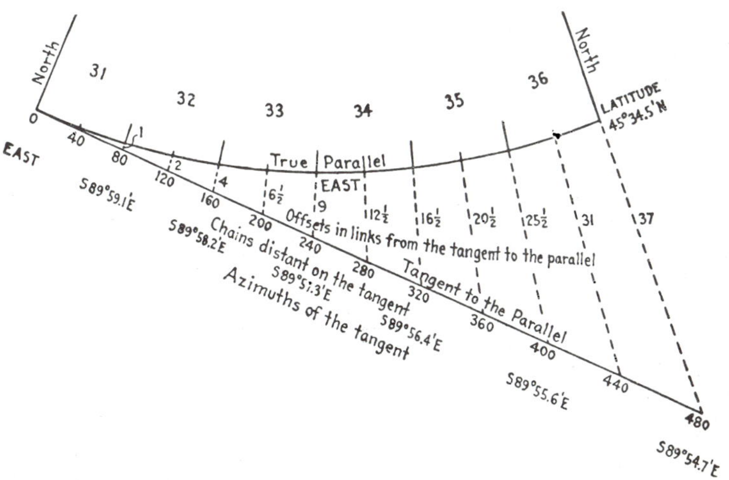

FIG. 5.33.1. LOCATING POINTS ON A PARALLEL BY TANGENT METHOD.

TABLE 5.33.1

AZIMUTHS OF THE TANGENT TO THE PARALLEL.

Lat.	1 mi.	2 mi.	3 mi.	4 mi.	5 mi.	6 mi.
°	° ′	° ′	° ′	° ′	° ′	° ′
25	89 59.6	89 59.2	89 58.8	89 58.4	89 58.0	89 57.6
26	59.6	59.2	58.7	58.3	57.9	57.5
27	59.6	59.1	58.7	58.2	57.8	57.4
28	59.5	59.1	58.6	58.2	57.7	57.2
29	59.5	59.0	58.6	58.1	57.6	57.1
30	59.5	59.0	58.5	58.0	57.5	57.0
31	59.5	59.0	58.4	57.9	57.4	56.9
32	59.5	58.9	58.4	57.8	57.3	56.8
33	59.4	58.9	58.3	57.7	57.2	56.6
34	59.4	58.8	58.2	57.7	57.1	56.5
35	59.4	58.8	58.2	57.6	57.0	56.4
36	59.4	58.7	58.1	57.5	56.9	56.2
37	59.3	58.7	58.0	57.4	56.7	56.1
38	59.3	58.6	58.0	57.3	56.6	55.9
39	59.3	58.6	57.9	57.2	56.5	55.8
40	59.3	58.5	57.8	57.1	56.4	55.6
41	59.2	58.5	57.7	57.0	56.2	55.5
42	59.2	58.4	57.7	56.9	56.1	55.3
43	59.2	58.4	57.6	56.8	56.0	55.2
44	59.2	58.3	57.5	56.7	55.8	55.0
45	59.1	58.3	57.4	56.5	55.7	54.8
46	59.1	58.2	57.3	56.4	55.5	54.6
47	59.1	58.1	57.2	56.3	55.4	54.4
48	59.0	58.1	57.1	56.2	55.2	54.2
49	59.0	58.0	57.0	56.0	55.0	54.0
50	59.0	57.9	56.9	55.9	54.8	53.8
51	58.9	57.9	56.8	55.7	54.6	53.6
52	58.9	57.8	56.7	55.6	54.5	53.4
53	58.8	57.7	56.6	55.4	54.2	53.1
54	58.8	57.6	56.4	55.2	54.0	52.8
55	58.8	57.5	56.3	55.1	53.8	52.6
56	58.7	57.4	56.2	54.9	53.6	52.3
57	58.7	57.3	56.0	54.7	53.3	52.0
58	58.6	57.2	55.8	54.5	53.1	51.7
59	58.6	57.1	55.7	54.2	52.8	51.4
60	58.5	57.0	55.5	54.0	52.5	51.0
61	58.4	56.9	55.3	53.8	52.2	50.6
62	58.4	56.7	55.1	53.5	51.9	50.2
63	58.3	56.6	54.9	53.2	51.5	49.8
64	58.2	56.5	54.7	52.9	51.1	49.4
65	58.1	56.3	54.4	52.6	50.7	48.9
66	58.1	56.1	54.2	52.2	50.3	48.4
67	58.0	55.9	53.9	51.8	49.8	47.8
68	57.9	55.7	53.6	51.4	49.3	47.2
69	57.8	55.5	53.2	51.0	48.8	46.5
70	89 57.6	89 55.3	89 52.9	89 50.5	89 48.1	89 45.8

TABLE 5.33.2.

OFFSETS, IN LINKS, FROM THE TANGENT TO THE PARALLEL.

Lat.	½ mi.	1 mi.	1½ mi.	2 mi.	2½ mi.	3 mi.	3½ mi.	4 mi.	4½ mi.	5 mi.	5½ mi.	6 mi.
°												
25	0	0	1	2	3	4	6	8	10	12	14	17
26	0	0	1	2	3	4	6	8	10	12	15	18
27	0	1	1	2	3	5	6	8	10	13	16	18
28	0	1	1	2	3	5	7	9	11	13	16	19
29	0	1	1	2	3	5	7	9	11	14	17	20
30	0	1	1	2	4	5	7	9	12	15	18	21
31	0	1	1	2	4	5	7	10	12	15	18	22
32	0	1	1	3	4	6	8	10	13	16	19	23
33	0	1	1	3	4	6	8	10	13	16	20	24
34	0	1	2	3	4	6	8	11	14	17	21	25
35	0	1	2	3	4	6	8	11	14	18	21	25
36	0	1	2	3	5	7	9	12	15	18	22	26
37	0	1	2	3	5	7	9	12	15	19	23	27
38	0	1	2	3	5	7	10	13	16	20	24	28
39	0	1	2	3	5	7	10	13	17	20	25	29
40	0	1	2	3	5	8	10	14	17	21	26	30
41	0	1	2	4	5	8	11	14	18	22	27	32
42	0	1	2	4	6	8	11	15	18	23	27	33
43	0	1	2	4	6	8	11	15	19	24	28	34
44	0	1	2	4	6	9	12	16	20	24	29	35
45	0	1	2	4	6	9	12	16	20	25	30	36
46	0	1	2	4	7	9	13	17	21	26	32	38
47	0	1	2	4	7	10	13	17	22	27	33	39
48	0	1	3	4	7	10	14	18	23	28	34	40
49	0	1	3	5	7	10	14	19	23	29	35	42
50	0	1	3	5	8	11	15	19	24	30	36	43
51	0	1	3	5	8	11	15	20	25	31	38	45
52	0	1	3	5	8	12	16	21	26	32	39	46
53	0	1	3	5	8	12	16	21	27	33	40	48
54	0	1	3	6	9	12	17	22	28	35	42	50
55	0	1	3	6	9	13	18	23	29	36	43	52
56	0	1	3	6	9	13	18	24	30	37	45	54
57	0	2	3	6	10	14	19	25	31	39	47	56
58	0	2	4	6	10	14	20	26	33	40	49	58
59	0	2	4	7	10	15	20	27	34	42	51	60
60	0	2	4	7	11	16	21	28	35	44	53	63
61	0	2	4	7	11	16	22	29	37	45	55	65
62	0	2	4	8	12	17	23	30	38	47	57	68
63	0	2	4	8	12	18	24	32	40	49	60	71
64	1	2	5	8	13	19	25	33	42	52	62	74
65	1	2	5	9	13	19	26	34	44	54	65	78
66	1	2	5	9	14	20	28	36	46	57	68	81
67	1	2	5	9	15	21	29	38	48	59	72	85
68	1	2	6	10	16	22	30	40	50	62	75	90
69	1	3	6	10	16	24	32	42	53	66	79	94
70	1	3	6	11	17	25	34	44	56	69	84	100

5.33. Tangent Method.—The tangent method of running the true latitude curve consists in establishing the true meridian at the point of beginning, from which a horizontal angle of 90° is turned to the east or west, as may be required; the projection of the line thus determined is called the tangent. The tangent is projected 6 miles as a straight line; at the distances for each corner the offsets are measured to the north (on meridional lines) from the tangent to the line of the true parallel, on which the monuments are established. The data for the azimuths of the line of the tangent, at mile intervals, and for the offsets from the tangent to the parallel, at one-half mile intervals, are shown in Tables 5.33.1 and 5.33.2. The offset method is illustrated in Fig. 5.33.1.

5.34. Secant Method.—This method consists in running a straight line 6 miles long on such a course that it intersects the true parallel of latitude at the first and fifth mile points. From the secant, points are set on the parallel by measuring north or south along meridian lines (Fig 5.34.1). The 0 and 6 mile

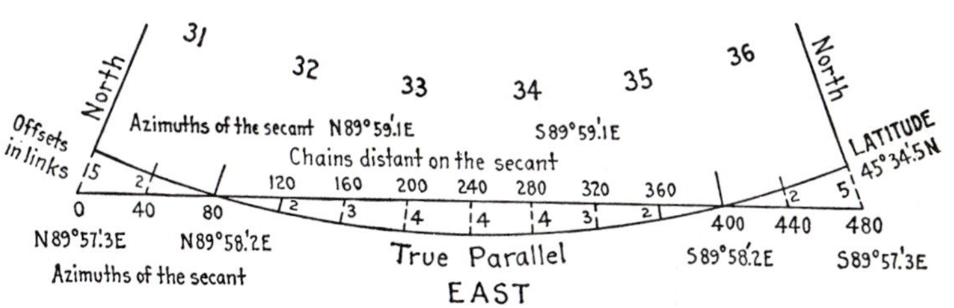

FIG. 5.34.1. LOCATING POINTS ON A PARALLEL BY SECANT METHOD.

points of the parallel will be north of the secant, and the 2, 3 and 4 mile points will be south of the secant. From the 0 to the 3 mile point the secant has northeasterly or northwesterly bearings (depending on the direction in which the parallel is being projected), and from the 3 to 6 mile points the secant has southeasterly or southwesterly bearings.

The instrument is set up south of the township corner where the survey is to begin, the distance from the corner being found in Table 5.34.2 in the column headed "0 mi." For exam-

TABLE 5.34.1.
AZIMUTHS OF THE SECANT.

Lat.	0 mi.	1 mi.	2 mi.	3 mi.	Deflection angle 6 mi.
°	° ′	° ′	° ′		′ ″
25	89 58.8	89 59.2	89 59.6	90° E or W.	2 25
26	58.7	59.2	59.6	" " "	2 32
27	58.7	59.1	59.6	" " "	2 39
28	58.6	59.1	59.5	" " "	2 46
29	58.6	59.0	59.5	" " "	2 53
30	58.5	59.0	59.5	" " "	3 0
31	58.4	59.0	59.5	" " "	3 7
32	58.4	58.9	59.5	" " "	3 15
33	58.3	58.9	59.4	" " "	3 23
34	58.2	58.8	59.4	" " "	3 30
35	58.2	58.8	59.4	" " "	3 38
36	58.1	58.7	59.4	" " "	3 46
37	58.0	58.7	59.3	" " "	3 55
38	58.0	58.6	59.3	" " "	4 4
39	57.9	58.6	59.3	" " "	4 13
40	57.8	58.5	59.3	" " "	4 22
41	57.7	58.5	59.2	" " "	4 31
42	57.7	58.4	59.2	" " "	4 41
43	57.6	58.4	59.2	" " "	4 51
44	57.5	58.3	59.2	" " "	5 1
45	57.4	58.3	59.1	" " "	5 12
46	57.3	58.2	59.1	" " "	5 23
47	57.2	58.1	59.1	" " "	5 34
48	57.1	58.1	59.0	" " "	5 46
49	57.0	58.0	59.0	" " "	5 59
50	56.9	57.9	59.0	" " "	6 12
51	56.8	57.9	58.9	" " "	6 25
52	56.7	57.8	58.9	" " "	6 39
53	56.6	57.7	58.8	" " "	6 54
54	56.4	57.6	58.8	" " "	7 9
55	56.3	57.5	58.8	" " "	7 25
56	56.2	57.4	58.7	" " "	7 42
57	56.0	57.3	58.7	" " "	8 0
58	55.8	57.2	58.6	" " "	8 19
59	55.7	57.1	58.6	" " "	8 38
60	55.5	57.0	58.5	" " "	8 59
61	55.3	56.9	58.4	" " "	9 22
62	55.1	56.7	58.4	" " "	9 46
63	54.9	56.6	58.3	" " "	10 11
64	54.7	56.5	58.2	" " "	10 38
65	54.4	56.3	58.1	" " "	11 8
66	54.2	56.1	58.1	" " "	11 39
67	53.9	55.9	58.0	" " "	12 13
68	53.6	55.7	57.9	" " "	12 51
69	53.2	55.5	57.8	" " "	13 31
70	89 52.9	89 55.3	89 57.6	" " "	14 15
	6 mi.	5 mi.	4 mi.	3 mi.	

TABLE 5.34.2.
OFFSETS, IN LINKS, FROM THE SECANT TO THE PARALLEL.

Lat.	0 mi.	½ mi.	1 mi.	1½ mi.	2 mi.	2½ mi.	3 mi.
°							
25	2 N.	1 N.	0	1 S.	1 S.	2 S.	2 S.
26	2	1	0	1	1	2	2
27	3	1	0	1	2	2	2
28	3	1	0	1	2	2	2
29	3	1	0	1	2	2	2
30	3	1	0	1	2	2	2
31	3	1	0	1	2	2	2
32	3	1	0	1	2	2	3
33	3	1	0	1	2	2	3
34	3	2	0	1	2	3	3
35	4	2	0	1	2	3	3
36	4	2	0	1	2	3	3
37	4	2	0	1	2	3	3
38	4	2	0	1	2	3	3
39	4	2	0	1	2	3	3
40	4	2	0	1	3	3	3
41	4	2	0	2	3	3	4
42	5	2	0	2	3	3	4
43	5	2	0	2	3	4	4
44	5	2	0	3	3	4	4
45	5	2	0	2	3	4	4
46	5	2	0	2	3	4	4
47	5	2	0	2	3	4	4
48	6	3	0	2	3	4	4
49	6	3	0	2	3	4	5
50	6	3	0	2	4	4	5
51	6	3	0	2	4	5	5
52	6	3	0	2	4	5	5
53	7	3	0	2	4	5	5
54	7	3	0	3	4	5	6
55	7	3	0	3	4	5	6
56	7	3	0	3	4	6	6
57	8	3	0	3	5	6	6
58	8	4	0	3	5	6	6
59	8	4	0	3	5	6	7
60	9	4	0	3	5	7	7
61	9	4	0	3	5	7	7
62	9	4	0	3	6	7	8
63	10	4	0	3	6	7	8
64	10	5	0	4	6	8	8
65	11	5	0	4	6	8	9
66	11	5	0	4	7	8	9
67	12	5	0	4	7	9	9
68	12	6	0	4	7	9	10
69	13	6	0	5	8	10	10
70	14 N.	6 N.	0	5 S.	8 S.	10 S.	11 S.
	6 mi.	5½ mi.	5 mi.	4½ mi.	4 mi.	3½ mi.	3 mi.

ple, in latitude 45° the transit would be set 5 links south of the corner. The direction of the first secant at its initial point may be found by observing on Polaris (Chapter VIII) to obtain the true meridian and then laying off the azimuth angle found in Table 5.34.1 under "0 mi." (See Fig. 5.34.1) This angle should be repeated several times to determine accurately the direction of the secant. This direction is then prolonged 6 miles. At each mile and half-mile point an offset is measured to establish a point on the curve, the offsets being shown in Table 5.34.2.

5.35. Running the Base-Line.—In running the base-line great care should be taken to secure instrumental accuracy. In prolonging the line at each set-up, two double reversals of the instrument are made and the mean of the observations is used.

The direction of the meridian obtained with the solar attachment should be checked, whenever practicable, by an observation on Polaris; and in all cases where error is discovered, the line should be corrected before proceeding with the survey. All observations and adjustments should be fully described in the field-notes.

To insure accuracy in the measurements two sets of chainmen are employed, one set checking the other. Each measures the 40-chain intervals, and if the difference in results is within the prescribed limits, the proper corner is placed midway between the end points of the two measurements. If the discrepancy exceeds 8 links on even ground, or 25 links on mountainous surface, the distance will have to be rechained.

Regular quarter-section and section corners are established at intervals of 40 chains, and regular township corners at intervals of 480 chains along the base-line. Meander corners are set wherever the line crosses meanderable streams, lakes, or bayous. (See Art. 5.48) When the corner falls in an inaccessible place, as on a cliff or in a lake, a witness corner is established. The monuments used to mark the positions of the township corners or section corners are marked with the letters *SC*, "standard corner," in addition to the other usual marks for the purpose of easily distinguishing them from the "closing corners," to be set later on the same line.

5.36. Principal Meridian.—The principal meridian is extended as a true meridian of longitude both north and south from the initial point. The methods used for the determination of directions, and the precautions observed to secure accuracy of measurement, are the same as those described in the preceding article, under the subject of "Running the Base-Line."

Also, as in the case of the base-line, all township, section, quarter-section, and other necessary corners are established in the proper places as the survey proceeds.

5.37. Standard Parallels.—Standard parallels, which are also sometimes referred to as *correction lines*, are extended both east and west from every fourth township corner previously established on the principal meridian. Sometimes, however, the distance between them is more or less than 24 miles, depending upon the requirements of the particular survey in question. For example, in Kansas, the correction lines occur at regular intervals of 30 instead of 24 miles. Where standard parallels are at intervals of 30 or 36 miles and where present conditions require additional parallels for initiating new, or closing the extension of an old survey, an intermediate correction line should be established and a local name given to it, such as "Cedar Creek Correction Line"; and the same will be run, in all respects, like the regular standard parallels.

Standard parallels are established as true parallels of latitude, and are run in the same manner and with the same precautions for accuracy as in the survey of the base-line.

Appropriate corners are established at the proper intervals, and the township and section corners are marked *SC*, the same as those on the base-line.

5.38. Guide Meridians.—Guide meridians are extended north from the base-line, or standard parallels, at intervals of 24 miles east and west from the principal meridian. They are run as true meridians of longitude, and are extended to an intersection with the next correction line north. At the point of intersection of the guide meridian with the correction line a *closing corner* is established, and the monument is marked with the letters *CC*, to distinguish it from the standard corners already in place. Also, the distance of the closing corner from the nearest standard corner is measured and recorded in the field-notes. This correction offset will vary with the latitude and with the distance of the corner from the principal meridian. At a distance of 15 or 20 ranges from the principal it may be so great that the closing corner will be nearer to the adjacent quarter-section corner than to the standard township corner. Furthermore, it is obvious that the closing corners will be west of the corresponding standard corners on the east side of the principal meridian, and east of them on the west side.

The mile and half-mile distances on the guide meridians are made full 80 and 40 chains in length until the last half-mile is reached, into which all excess or deficiency due to discrepancies of measurement is thrown.

The general method of running the guide meridians is the same as that used in running the principal meridian, and all the provisions for securing accuracy of alignment and measurement, and for establishing corners, prescribed for the latter apply to the former also.

Provision is made for running guide meridians from north to south where existing local conditions require this departure from the usual practice. In such a case the closing corner is first established on the correction line by calculating the proper correction distance and laying it off from the standard corner; and then the guide meridian is run due south from this point. This method may be used in case the standard corner from which the guide meridian would ordinarily originate is inaccessible, or for other adequate reasons.

The Manual also provides that where guide meridians have been placed at intervals exceeding 24 miles, and new governing lines are required in order to control new surveys, a new guide meridian will be established, and a local name may be assigned to the same, e.g., "Grass Valley Guide Meridian." These auxiliary guide meridians will be surveyed in all respects like regular guide meridians.

5.39. Township Exteriors.—The usual method of subdividing a 24-mile tract into townships is, as follows (see Fig. 5.28.2):

Beginning at the standard corner at the southeast corner of

the southwest township in the tract, the surveyor runs north on a true meridian of longitude a distance of 6 miles, setting all necessary corners by the way. From the township corner thus established he runs due west on a random line (Art. 6.25) to intersect the guide meridian (or the principal meridian, in case he is working in Range 1 East), setting temporary section and quarter-section corners as he goes. When he intersects the meridian, he notes the "falling"* of his random line, and, if this is within the limit prescribed, he then calculates the course of the true line joining the two township corners and runs back on it, setting permanent corners opposite the temporary ones previously set on the random line. In this way all the deficiency due to the convergence of the meridional boundaries of the township, and whatever excess or deficiency may arise from inaccuracies in measurement, is thrown into the most westerly half-mile of the latitudinal boundary.

The range line is now continued as a true meridian for another 6 miles, permanent corners being set as before. Then another random line is closed across to the western boundary of the range of townships, and is corrected back to the true line, in the same manner as that just described. This process is continued until the most northerly township in the 24-mile tract is reached, when the range line is merely continued as a true meridian to an intersection with the correction line, at which point a closing township corner is established. The half-mile intervals on the range line are made full 40 chains for the entire 24 miles, except the most northerly half-mile, into which all excess or deficiency due to irregularities of measurement is thrown.

The two other range lines of the 24-mile block are run in a similar manner, the latitudinal township lines being extended to the westward at the proper intervals and made to connect with the township corners previously established. At the township corners on the last range line, however, random lines are run also to the westward from the guide meridian,

*That is, the distance of the point at which the random line intersects the meridian from the objective corner.

and are then corrected back to the eastward on a true line between the township corners. This is done in such a way that the excess or deficiency of this line also is thrown into the most westerly half-mile.

In new surveys the township boundaries should be kept within 14 minutes of the cardinal direction in order to assure that the interior subdivision lines are kept within the prescribed limit of 21 minutes with the cardinal direction; but previously established township boundaries that come within the 21-minute limit will not be considered defective.

5.40. Subdivision of Townships.—The subdivision of a township proceeds in the order indicated in Fig. 5.40.1. Starting at the southwest corner of Section 36 a line is run toward the north parallel to the east boundary of the township; quarter-section and section corners are set at 40 and 80 chains respectively. From the section corner just set a random line is run eastward parallel to the south boundary of the township, a temporary quarter-section corner being set at 40 chains. The falling of this random line where it intersects the range line is noted, and also the distance by which it overruns or falls short of the length of the south boundary of the section. If the falling is not more than 50 links (33 feet, representing an angular deviation of 21 minutes), and if the distance overruns or falls short of the length of the southern boundary of Section 36 by not more than the same amount, a return course which will join the two section corners is calculated; this new line is then run toward the west, the permanent quarter-section corner being set at its middle point.

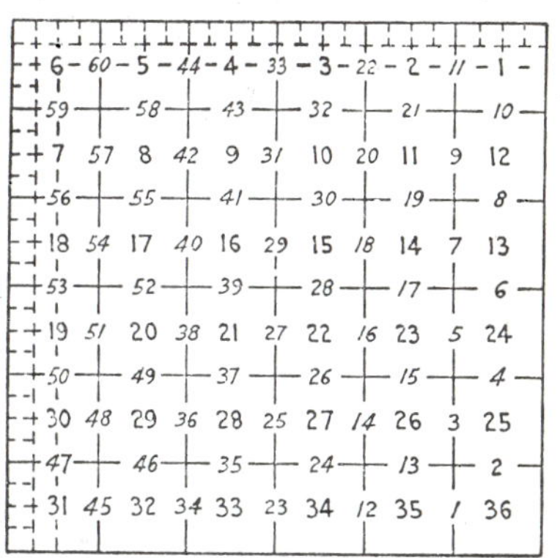

FIG. 5.40.1. ORDER OF RUNNING SECTION LINES. THE NUMBERS ON THE SECTION LINES INDICATE THE NORMAL ORDER IN WHICH LINES ARE RUN.

From the section corner just regained the survey is now continued north between Sections 25 and 26. At 40 and 80 chains on this line the quarter-section and section corners, respectively, are set, and from the section corner a random is run across to the range line, and a return course is calculated and run as before. This process is continued until five of the six sections in the series are enclosed. Then, if the north boundary of the township is not a base-line or standard parallel, from the section corner last established a random is run north to the township boundary, and from the data thus secured a true line is calculated and run from the section corner on the township line back to the initial corner. If the north boundary of the township is a baseline or standard parallel, however, the point at which the random intersects this boundary is established as a **closing corner** and its distance from the nearest **standard corner** is measured and recorded. In either case the permanent quarter-section corner is established at 40 chains north of the initial corner, the excess or deficiency being thrown into the most northerly half-mile. Here the meridional distance is checked by calculating the closure of the lines around the last section. The "limit of closure" is expressed by the fraction 1/452, provided that the limit of closure in neither latitude nor departure exceeds 1/640. An accumulative error of 12½ links per mile of perimeter in either latitude or departure cannot be exceeded in an accceptable survey.

The latitudes and departures of a normal section shall each close within 50 links; of a normal range or tier of sections, within 175 links; and of a normal township, within 300 links. The boundaries of each fractional section including irregular claim lines or meander traverses should close within the limit 1/640 for latitude and for departure considered separately.

In a similar manner the succeeding ranges of sections are enclosed, randoms being run across eastward to the section corners previously established and true lines corrected back. From the fifth series of section corners thus established, however, random lines are projected to the westward also, and are closed on the corresponding section corners in the range line forming the western boundary of the township. In correcting these lines back, however, the permanent quarter-section corners are established at points 40 chains from the initial corners of the randoms, thereby throwing all fractional measurements into the most westerly half-miles.

Meander corners are established wherever the section lines intersect meanderable bodies of water. The boundaries of the fractional sections formed by meander traverses (Art. 5.48 and Fig. 5.56.2) and section lines are checked for closure by the usual surveying methods. (See Chap. XV.)

Table 5.40.1, taken from the Manual, gives (to the nearest whole minute) the angular convergency of meridians from one to five miles apart. The meridional section lines, therefore, by reason of being (theoretically) parallel to the range line on the east boundary of the township, will depart from true meridians by the amounts indicated in the table.

From a consideration of the foregoing it will be apparent:

1. That interior meridional section lines are 80 chains in length, except those next to the north boundary of the township; and that the south half of these is 40 chains in length.

2. That interior latitudinal section lines are within 50 links of the length of the line forming the southern boundary of the range of sections, except those section lines next to the west boundary of the township; and that the east half of these is 40 chains in length.

3. That meridional section lines are run straight for five miles and adjustments made in sixth mile; latitudinal section lines are run straight between section corners on meridional lines.

4. That except in those section lines next to the north and west boundaries of the township, the quarter-section corners are placed equidistant from the two section corners on each side.

5. That meridional section lines are intended to be parallel to the range line forming the eastern boundary of the township; and similarly, that latitudinal section lines are intended to be parallel to the township line forming its southern boundary.

6. That the cumulative deficiency in latitudinal lines due to the convergence of the meridians is thrown into the most westerly half-mile of the township.

7. That in the older practice, quarter-section closing corners were not always established on correction lines for the use of the sections south of these lines but these are established in the current surveys.

5.41. CONVERGENCE OF BOUNDARIES.—The convergence of the meridians in a distance of 6 miles results in the north boundary of a township being shorter than the south boundary by an amount which must be taken into account in computing the closure of the township. The amount of the convergence varies in different latitudes. Table 5.41.1 shows the convergence of meridians 6 miles long, and 6 miles apart, in different latitudes. For shorter lengths the convergence is proportionately reduced. The theory of the convergence of meridians and its application to township boundaries and to other surveying problems are given in Art. 5.51.

5.42 MARKING CORNERS.—Corners are marked by various kinds of monuments. In the past, stone monuments were usually set where stone was plentiful. In timbered districts where suitable stones were difficult to find, wooden posts were used. In prairie regions, where neither stone nor timber was availa-

TABLE 5.40.1.

CORRECTIONS FOR CONVERGENCY WITHIN A TOWNSHIP.

Latitude	Correction to be applied to bearing of range lines at a distance of				
	1 mile	2 miles	3 miles	4 miles	5 miles
° °	′	′	′	′	′
30 to 35	1	1	2	2	3
35 to 40	1	1	2	3	3
40 to 45	1	2	2	3	4
45 to 50	1	2	3	4	5
50 to 55	1	2	3	5	6
55 to 60	1	3	4	5	7
60 to 65	2	3	5	7	8
65 to 70	2	4	6	8	10

TABLE 5.41.1.

Convergence of Meridians, Six Miles Long and Six Miles Apart, and Differences of Latitude and Longitude.

From "Standard Field Tables," U. S. General Land Office (Bureau of Land Management).

Lat.	Convergency			Difference of longitude per range			Difference of latitude for —	
	On the parallel	Angle		In arc		In time	1 mi.	1 Tp.
°	*Lks.*	′	″	′	″	*Seconds*		
25	33.9	2	25	5	44.34	22.96		
26	35.4	2	32	5	47.20	23.15		
27	37.0	2	39	5	50.22	23.35	0.871	5.229
28	38.6	2	46	5	53.40	23.56		
29	40.2	2	53	5	56.74	23.78		
30	41.9	3	0	6	0.26	24.02		
31	43.6	3	7	6	3.97	24.26		
32	45.4	3	15	6	7.87	24.52	0.871	5.225
33	47.2	3	23	6	11.96	24.80		
34	49.1	3	30	6	16.26	25.08		
35	50.9	3	38	6	20.78	25.39		
36	52.7	3	46	6	25.53	25.70		
37	54.7	3	55	6	30.52	26.03	0.870	5.221
38	56.8	4	4	6	35.76	26.38		
39	58.8	4	13	6	41.27	26.75		
40	60.9	4	22	6	47.06	27.14		
41	63.1	4	31	6	53.15	27.54		
42	65.4	4	41	6	59.56	27.97	0.869	5.216
43	67.7	4	51	7	6.29	28.42		
44	70.1	5	1	7	13.39	28.89		
45	72.6	5	12	7	20.86	29.39		
46	75.2	5	23	7	28.74	29.92		
47	77.8	5	34	7	37.04	30.47	0.869	5.211
48	80.6	5	46	7	45.80	31.05		
49	83.5	5	59	7	55.05	31.67		
50	86.4	6	12	8	4.83	32.32		
51	89.6	6	25	8	15.17	33.03		
52	92.8	6	39	8	26.13	33.74	0.868	5.207
53	96.2	6	54	8	37.75	34.52		
54	99.8	7	9	8	50.07	35.34		
55	103.5	7	25	9	3.18	36.22		
56	107.5	7	42	9	17.12	37.14		
57	111.6	8	0	9	31.97	38.13	0.867	5.202
58	116.0	8	19	9	47.83	39.19		
59	120.6	8	38	10	4.78	40.32		
60	125.5	8	59	10	22.94	41.52		
61	130.8	9	22	10	42.42	42.83		
62	136.3	9	46	11	3.38	44.22	0.866	5.198
63	142.2	10	11	11	25.97	45.73		
64	148.6	10	38	11	50.37	47.36		
65	155.0	11	8	12	16.82	49.12		
66	162.8	11	39	12	45.55	51.04		
67	170.7	12	13	13	16.88	53.12	0.866	5.195
68	179.3	12	51	13	51.15	55.41		
69	188.7	13	31	14	28.77	57.92		
70	199.1	14	15	15	10.26	60.68	0.866	5.193

ble, a mound of earth was commonly raised over the corner and a small marked stone, a charred stake, a quart of charcoal, broken glass, or other permanent distinguishable object was placed beneath it. Occasionally in timbered country when the corner fell on a spot occupied by a tree too large to be removed the tree itself was made the monument. The type of marking used on any corner should be thoroughly described in the notes.

If stones are set, they are marked with notches (or grooves) to indicate their respective positions. Section corners on range lines, including under this term principal and guide meridians, are marked with notches on their north and south edges, the number of notches being equal to the number of miles to the next adjacent township corner north or south. In a similar manner the section corners on the township lines are notched on their east and west edges. Township corners, being located on both range and township lines, are marked with the township and range numbers on each of the four sides. In addition to being *grooved* for the number of miles as just indicated, corners on correction lines are marked *SC* on their northern or *CC* on their southern faces, depending on whether they are standard or closing corners. Section corners in the interior of a township are given notches on their east and south edges corresponding to the number of miles to the east and south boundaries of the township. Thus, the corner common to sections 20, 21, 28 and 29 (Fig 5.29.1) would have two notches on the south and four on the east edge. Quarter-section corners are marked with the fraction "¼," those on meridional lines on their west, and those on latitudinal lines on their north faces.

In current practice an iron post is set, and the identifying marks inscribed on a bronze cap.

5.43. Witnessing Corners.—Wherever possible the monument set at a corner is witnessed by several nearby objects, called "accessories," which may be easily found by anyone looking for the corner itself, and which are comparatively permanent. In timbered country the stone or post is usually witnessed by "bearing trees" located near the corner. The process of establishing a witness tree is to take its bearing and distance from the corner, then to blaze off the bark from a short section of the trunk on the side facing the corner and to cut into the wood with scribing tools certain letters and numerals indicative of the section in which the tree is located. For example, the tree northeast from the corner mentioned above would be marked T 7 S R 15 E S 21 B T (written vertically), the marks being low on the trunk, the letters and figures being abbreviations of "Township 7 south, Range 15 east, Section 21, Bearing Tree." Usually one tree is marked in each of the sections to which the corner refers, provided suitable trees can be found within a reasonable distance of the corner.

In prarie regions small rectangular pits are dug near the corner, the earth taken from them being scattered so as not to refill the pits. Marks of this kind are of much greater value than might at first be supposed, for, although the sharp outlines are quickly worn away, the grass sod grows down into the pits and preserves them from entire obliteration.

5.44. Witness Corners.—In case a regular corner falls in a creek, pond, or in any other place where it is impracticable to set or maintain a monument, a *witness corner* is set on one line leading to this corner. This is marked with the letters *WC* in addition to the markings that would be appropriate to the corner of which it is a witness. Witness corners are, in turn, referenced by bearing trees, pits, and other objects, the same as true corners.

5.45. Meander Corners.—Where a surveyed line intersects a meanderable body of water (Art. 5.43), a *meander corner* is established. The distance from the nearest section or quarter-section corner is measured and recorded in the notes, and the stone or post set as a monument is marked *MC* on the side facing the water, and the point is referenced by bearing trees or by other accessories. If practicable, the line is then carried across the stream or other body of water by triangulation to another meander corner set in line on the further bank, and the survey is continued.

5.46. Index of Corner Markings.—The following is an index of the ordinary markings common to all classes of corners and accessories:

MARKS	TO INDICATE	MARKS	TO INDICATE
A M C	Auxiliary meander corner.	R	Range.
A P	Angle point.	S	Section.
B O	Bearing object.	S	South.
B T	Bearing tree.	S C	Standard corner.
C	Center.	SE	Southeast.
C C	Closing corner.	S M C	Special meander corner.
E	East.	SW	Southwest.
M	Mile.	T	Township.
M C	Meander corner.	TR	Tract.
N	North.	W	West.
NE	Northeast.	W C	Witness corner.
NW	Northwest.	W P	Witness point.
P L	Public land (unsurveyed).	¼	Quarter section.
		1/16	Sixteenth section.

5.47. Marking Lines Through Timber.—The act of May 18, 1796, which is still in force, requires the marking of trees along the surveyed lines as positively as the erecting of monuments. Those trees that are intersected by the line have two hacks or notches cut on each of the sides facing the line; these are called sight trees or line trees. Other trees that stand within 50 links of the line are blazed on two sides quartering toward the line.

5.48. MEANDERING.—All navigable bodies of water and important rivers and lakes are segregated from the public lands at mean high-water elevation. The traverse of the margin of a permanent natural body of water is termed a meander line.

Numerous decisions in the United States Supreme Court and in many of the state courts assert the principle that meander lines are not boundaries defining the area of ownership of tracts adjacent to waters. The general rule is set forth in the decision (10 Iowa, 549) that in a navigable stream, as the Des Moines River in Iowa, high-water mark is the boundary line. When by action of the water the river bed changes, and therefore high-water mark changes, the ownership of the land also changes.

Mean high-water elevation is found at the margin of the area occupied by the water for the greater portion of each average year; at this level a definite escarpment in the soil will generally be traceable, at the top of which is the true position for the meander line.

In the survey of lands bordering on tide waters, meander corners are set at the intersection of the surveyed lines with the margin of mean high tide. When the exact corner cannot be readily monumented, a witness corner is set at a secure point near the true point for the meander corner.

All meander courses are referred to the true meridian. For convenience the courses of meander lines are taken to the nearest quarter degree, with angle points at distances of whole chains, or at multiples of 10 links, and with odd links only in the final course.

Following a river downstream, the bank on the left hand is termed the left bank and that on the right hand, the right bank. Navigable rivers and bayous, and all rivers not embraced in the class denominated "navigable" the right-angle width of which is 3 chains and upward, are meandered on both banks at the ordinary mean high-water mark by taking the courses and distances of their sinuosities. In general, rivers not classed as navigable are meandered only to the point where the average right-angle width is less than 3 chains. However, when such streams are so deep, swift, and dangerous as to be impassable, they may be meandered where good agricultural lands along the banks require separation into fractional lots for the benefit of settlers.

Tidewater streams, whether more or less than 3 chains wide, are meandered at ordinary high tide and as far as tidewater extends.

All lakes of 25 acres and upward in area are meanderable. The section and quarter-section lines are projected theoretically across such lakes, and meander corners are located wherever the shore is intercepted. If a meanderable lake lies wholly within a quarter-section, an "auxiliary meander corner" is established at some suitable point on its margin, and a connecting line is run to a regular corner on the section boundary.

A connecting traverse line will be recorded, if one is run, but it will also be reduced to the equivalent direct connecting course and distance, all of which will be stated in the field-notes, and the course and length of the direct connecting line will be shown on the plat of the survey.

Islands above the mean high-water elevation of any meanderable body of water are located by triangulation or by any other available method, and the section and quarter-section lines projected thereon. If the island lies wholly within a quarter-section, an auxiliary meander corner is set on its margin and tied in to a regular corner on the mainland. In the case of both islands and lakes, the point set at the intersection of a projected quarter-section line and the high-water line is called a "special meander corner."

5.49. Topography.—Generally the most essential topographic data only are shown upon the plat. Distances to stream crossings, roads and trails, springs, notable changes in the ground slope or elevation, entering and leaving timber, and similar important items of topography, are recorded when running the true lines. The development of the map features within the sections varies greatly in accordance with the needs; this development may be carried out by sketching, or with the use of air-photos; sometimes it is done by traversing, or by a plane-table survey. The intersections are noted at the lines of mineral and other irregular surveys; connecting lines are run to the mineral-location monuments, and to all bench marks and geodetic stations.

5.50. Field-Notes.—The field-notes are the written record of the survey. All are permanently filed for reference purposes; the notes are accessible for examination, and copies may be secured. This technical record shows the appropriate identification of the lines established by prior surveys, from which the new surveys may be extended, and the older lines upon which they may close. This is the base record for all directions and lengths of lines, for description of monuments and the "calls" that refer to natural objects, line trees, and items of topography, such as distances to stream crossings. Fig. 5.50.1 shows a typical page of field-notes in proper form.

Subdivision of T. 15 N., R. 20 E.

Chains.

I commence the subdivisional survey at the cor. of secs. 1, 2, 35, and 36, on the S. bdy. of the Tp., which is a sandstone, 8 x 6 x 5 ins. above ground, firmly set, marked and witnessed as described in the official record.

N. 0° 01′ W., bet. secs. 35 and 36.

Over level bottom land.

20.00 Enter scattering ash and cottonwood.

29.30 SE. cor. of field; leave scattering timber.

31.50 A settler's cabin bears West, 6.00 chs. dist.

39.50 Set an iron post, 3 ft. long, 1 in. diam., 27 ins. in the ground, for witness ¼ sec. cor., with brass cap mkd.

W C
¼
S 35 | S 36
1925

dig pits 18 x 18 x 12 ins., N. and S. of post, 3 ft. dist.

Enter an ungraded road, bears N. along section line, and E. to Mound City.

40.00 True point for ¼ sec. cor. falls in road.

Deposit a sandstone, 14 x 8 x 5 ins., mkd. X, 24 ins. in the ground.

50.50 NE. cor. of field.

51.50 Road to Bozeman bears N. 70° W.

57.50 Enter heavy ash and cottonwood, and dense undergrowth, bears N. 54° E. and S. 54° W.

72.00 Leave undergrowth.

80.00 Set an iron post, 3 ft. long, 2 ins. diam., 27 ins. in the ground, for cor. of secs. 25, 26, 35, and 36, with brass cap mkd.

T 15 N S 26	R 20 E S 25
S 35	S 36

1925

from which

A green ash, 13 ins. diam., bears N. 22° E., 26 lks. dist., mkd. T 15 N R 20 E S 25 B T.

A green ash, 23 ins. diam., bears S. 71¼° E., 37 lks. dist. mkd. T 15 N R 20 E S 36 B T.

A green ash, 17 ins. diam., bears S. 64° W., 41 lks. dist., mkd. T 15 N R 20 E S 35 B T.

A cottonwood, 13 ins. diam., bears N. 21¼° W., 36 lks. dist., mkd. T 15 N R 20 E S 26 B T.

Land, level bottom; northern 20 chs. subject to overflow.

Soil, alluvial, silt and loam; 1st rate.

Timber, green ash and cottonwood; undergrowth, willow.

FIG. 5.50.1. SPECIMEN OF FIELD-NOTES.

5.51. TOWNSHIP PLATS.—Township plats are developed from the field-notes; copies may be secured from the Bureau of Land Management, formerly the General Land Office. Each plat carries a certificate that reads substantially as follows:

"The above plat of Township . . . . ., Range . . . . ., of the . . . . . .Meridian, is strictly conformable to the field-notes of the survey thereof which have been examined and approved."

There are many supplemental plats showing new or additional lottings within one or more sections; these supersede the original lottings. There are also many plats of the survey of islands, and other fragmentary parcels of public land where such areas were surveyed after the completion of the main parts of the township. These supplemental plats should be considered locally in tracing descriptions that are based upon the new lottings.

5.52. Computation of Areas.—The meridional section lines (except in the sections along the north boundary of the township) are normally surveyed as 80.00 chains in length; the latitudinal section lines (except in the sections along the west boundary) will vary from 79.50 to 80.50 chains (called "within limits") and are regarded as being 80.00 chains for the purpose of showing the official area; these sections are treated as containing an exact 640 acres. There are twenty-five of these regular sections in every normal township (the numbers, 8 to 17, 20 to 29, and 32 to 36). The *aliquot parts* are the regular quarter-sections of 160 acres and the quarter-quarter-sections of 40 acres each, including the regular portions of the eleven sections (1 to 7, inclusive, and 18, 19, 30 and 31) which have the *fractional lottings* along the north and west boundaries of the township.

The deficiency in area that results from the convergence of meridians is placed normally in the fractional lots adjoining the west boundary of the township. The surplus or deficiency in area that results from the discrepancy in the meridional measurements between the exterior boundaries and the subdivision lines is placed normally in the fractional lots adjoining the north boundary of the township.

Fig. 5.52.1 illustrates the adjustments and areas of lots next to the westerly boundary of the township. In computing the areas of lots 1 to 4, Sec. 30, the latitudinal dimensions are calculated by assuming these dimensions to vary proportionally from 18.35 on the south of lot 4 to 18.21 on the north of lot 1. Each meridional dimension is taken as 20 chains.

Fig. 5.52.2 illustrates the adjustments and areas of lots in the northwest section (Sec. 6) of a township. The meridional dimensions of lots 1 to 4 and the latitudinal dimensions of lots 4 to 7 (not shown in the plat) are found by proportion.

Fractional areas abutting on irregular boundaries, such as bodies of water, are computed by the usual methods for closed traverses (Chap. XV).

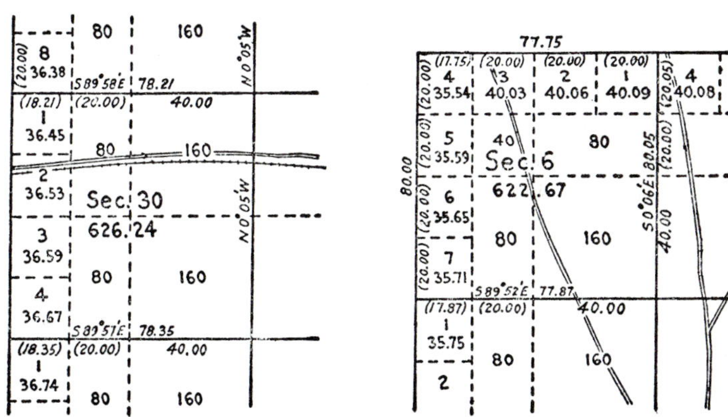

FIG. 5.52.1.

FIG. 5.52.2.

PORTIONS OF PLATS SHOWING AREAS OF LOTS.

5.53. DESCRIPTIONS AND RECORDS.—The description of parcels of land should conform to the accepted nomenclature in use by the Bureau of Land Management, formerly the General Land Office; it should include the name of the proper reference meridian, the appropriate township and range numbers, and the established subdivisions of the township.

Example:

> . . . township 9 north, range 12 east of the Black Hills meridian.

Each reference meridian has its own base-line, and therefore the words "and base-line" are usually omitted. For example, "T. 5 S., R. 11 E. of the Willamette meridian" means "township 5 south of the associated base-line, in range 11 east of the Willamette meridian."

Parts of the township are referred to by section number according to the official plat. Parts of the section are described by half-section or quarter-section (N½; E½; SW¼; etc.), and half-quarter or quarter-quarter-section (N½NE¼ sec. 18, meaning "the N½ of the NE¼ of section 18"). These terms give the so-called aliquot parts of 640 acres as contained in a regular section.

Lot numbers are employed to designate fractional units of the section—that is, other than quarter-quarter units of 40 acres each; but occasionally a subdivision smaller than a quarter-quarter-section is used, such as E½NW¼SW¼. Where terms for aliquot parts are spelled out, "half" and "quarter" are used, not "one-half" and "one-quarter."

The description of lots, other than aliquot parts where the numbers are omitted, is by fractional legal parts of the section, i.e., fractional W½NE¼ sec. 5, containing 49.52 acres; fractional NW¼SW¼ sec. 6, containing 44.73 acres.

In some of the very old surveys the township subdivisions and the units within the sections do not conform to the examples shown above, and there may be other complications that make it advisable to obtain a statement from the Bureau of Land Mangement, formerly the General Land Office, in reference to the appropriate terms to be employed.

In those states where the original surveys have been completed, application may be made to the proper offices for permission to examine the records and to make copies of the same. The list of these offices follows:

Alabama: Secretary of State, Montgomery
Arkansas: Commissioner of State Lands, Little Rock
Florida: Commissioner of Agriculture, Tallahassee
Illinois: Auditor of Public Accounts, Springfield
Indiana: Auditor of State, Indianapolis
Iowa: Secretary of State, Des Moines
Kansas: Auditor of State and Register of State Lands, Topeka
Louisiana: Register State Land Office, Baton Rouge
Michigan: Director, Department of Conservation, Lansing

Minnesota: Director, Division of Lands and Minerals, St. Paul
Mississippi: Land Commissioner, Jackson
Missouri: Secretary of State, Jefferson City
Montana: Public Survey Office, Helena
Nebraska: Commissioner of Public Lands and Buildings, Lincoln
Nevada: Public Survey Office, Reno
New Mexico: Public Survey Office, Santa Fe
North Dakota: State Engineer, Bismarck
Ohio: Auditor of State, Columbus
Oklahoma: Bureau of Land Management, Silver Springs, Md.
Oregon: Regional Land Office, Portland
South Dakota: Commissioner of Schools and Public Lands, Pierre
Utah: Regional Survey Office, Salt Lake City
Washington: Public Survey Office, Olympia
Wisconsin: Commissioner of Public Lands, Madison
Wyoming: Public Survey Office, Cheyenne

5.54. Retracement of Surveys.—It has been the common experience that many of the monuments and marks originally established on the lines of the Public Land Surveys become lost or obliterated with time. Witness and line trees are cut down. Pits and mounds marking the corners are quickly destroyed. The responsibility for the maintenance of the monuments and marks, rebuilding the old ones and setting new reference marks, in the interest of the landowners, rests primarily with the county surveyors, who are authorized by law to perform such work, and to file the record.

In retracing old surveys certain general rules* must be observed, which are controlling upon the location of all lands that have been granted or patented. These rules are summarized in the following paragraphs:

First. That the boundaries of the public lands when approved and accepted are unchangeable.

Second. That the original township, section, and quarter-section corners must stand as the true corners that they were intended to represent, whether in the places shown by the field-notes or not.

---

*Restoration of Lost or Obliterated Corners, and Subdivision of Sections, March 13, 1883, 1 L. D. 339; 2d Edition 1 L. D. 671; revised October 16, 1896, 23 L. D. 361; revised June 1, 1909, 38 L. D. 1; reprinted in 1916; revised April 5, 1939; revised 1952; revised 1963: Superintendent of Documents, Washington, D. C.

Third. That quarter-quarter-section corners not established in the original survey shall be placed on the line connecting the section and quarter-section corners, and midway between them, except on the last half mile of section lines closing on the north and west boundaries of the township, or on the lines between fractional or irregular sections.

Fourth. That the center lines of a section are to be straight, running from the quarter-section corner on one boundary to the corresponding corner on the opposite boundary.

Fifth. That in a fractional section where no opposite corresponding quarter-section corner has been or can be established, the center line must be run from the proper quarter-section corner as nearly in a cardinal direction to the meander line, reservation, or other boundary of such fractional section, as due parallelism with the section boundaries will permit.

Sixth. That lost or obliterated corners are to be restored to their original locations whenever it is possible to do so.

The surveyor has no right to alter the position of an established and recorded corner even though his resurvey may show its position to be faulty. The Bureau of Land Management, formerly the General Land Office, possesses no authority to change the boundaries of the lands that have been disposed of by the Federal government. The law requires that no resurvey or retracement shall be so executed as to impair the bona fide rights or claims of any claimant, entryman, or owner of lands affected by such resurvey or retracement.

Since the instructions for laying out public lands have been changed from time to time, the surveyor should familiarize himself with the instructions that were in effect at the time the original survey was made.

Whatever the purpose of the retracement may be, whether it calls for rerunning the true lines of the original survey, or subdividing a section, the rules as outlined require some or all of the definite steps which follow:

a. Secure a copy of the original plat and field-notes;
b. Secure all available data regarding subsequent surveys;
c. Secure the names and contact the owners of the property adjacent to the lines that are involved in the retracement;
d. Find the corners that may be required

First: By the remaining physical evidence;

Second: By collateral evidence, supplemental survey records, or testimony, if the original monument is to be regarded as obliterated, but not lost; or,

Third: If lost, by the application of the rules for proportionate measurement from other available corners;

e. Reconstruct the monuments as required, including the placing of reference markers where improvements of any kind might later interfere with the monuments or if the site is such as to suggest the need for supplemental monumentation;

f. Note the rules for the subdivision of sections where these lines are to be run; and

g. Prepare and file* a suitable record of what was found, the supplemental data that were employed, a description of the methods, the direction and length of lines, the new markers, and any other facts regarded as important.

5.55. Restoration of Lost or Obliterated Corners.—The Bureau of Land Management, formerly the General Land Office, defines the condition of previously established corners as follows:

An **existent corner** is one the position of which can be identified by verifying the evidence of the monument or its accessories by reference to the description that is contained in the field-notes or one at which the point can be located by an acceptable supplemental survey record, some physical evidence, or testimony.

Even though its physical evidence may have entirely disappeared, a corner will not be regarded as lost if its position can be recovered through the testimony of one or more witnesses who have a dependable knowledge of the original location.

An **obliterated corner** is a point at which there are no remaining traces of the monument or its accessories, but at which the location has been perpetuated or may be recovered beyond reasonable doubt by the acts and testimony of the interested landowners, competent surveyors, or other qualified local authorities, by witnesses, or by some acceptable recorded evidence.

---

*In many of the states there is a well-established practice of filing of field-notes and plats of surveys, usually in one of the county offices; otherwise the record is the property of whoever pays the cost of the survey, and ordinarily it would be filed as an exhibit with a deed, or agreement, or court decree, etc.

A position that depends upon the use of collateral evidence can be accepted only as duly supported, generally through its proper relation to known corners, and its agreement with the field-notes regarding distances to natural objects, stream crossings, line trees, and off-line tree blazes, etc., or through unquestionable testimony.

A **lost corner** is a point of a survey the position of which cannot be determined beyond reasonable doubt, either from traces of the original marks or from acceptable evidence or testimony that bears upon the original position, and the location of which can be restored only by rerunning of lines from one or more interdependent corners.

If there is some acceptable evidence of the original location, that position will be employed in preference to the rule that applies to a lost corner.

No decision should be made in regard to the restoration of a corner until every means has been exercised that might aid in identifying its true original position.

Existing original corners cannot be disturbed; consequently, discrepancies between the new and the recorded measurements will not in any manner affect the measurements beyond the identified corners, but the differences will be distributed proportionately within the several intervals along the line between the corners, i.e., the new measurement will be distributed among the several parts of the line in the same proportion as the originally recorded measurement was distributed.

The term "single proportionate measurement" is applied to a new measurement made on a single line to determine one or more positions on that line.

The term "double proportionate measurement" is applied to a new measurement made to determine the position of a point with respect to four corners: two on a meridional line and two on a latitudinal line.

In determining positions, those derived from points on standard parallels will be given precedence over other township exteriors, and ordinarily the latter will be given precedence over subdivisional lines; section corners will be relocated before the positions of lost quarter-section corners can be determined.

A lost corner of four townships (or sections) is restored by double proportional measurement, as in Fig. 5.55.1 where *A*, *B*, *C* and *D* are four original corners from which the lost corner *X* is to be restored. First *E* is located where *X* should be by proportional measurement from *A* and *B*; similarly *F* is located from *C* and *D*. The intersection of a latitudinal line through *E* and a meridional line through *F* is adopted as the restored position of *X*. The point *X* satisfies the proportional requirements, but does not lie exactly on either *AB* or *CD*.

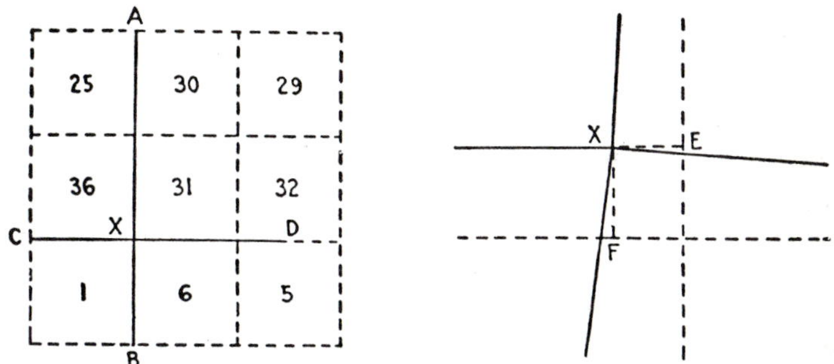

FIG. 5.55.1. REPLACING A LOST CORNER BY PROPORTIONAL MEASUREMENT.

In many of the surveys the field-notes and plats indicate two sets of corners along township boundaries, and frequently along section lines where parts of the township were subdivided at different dates. In these cases there are usually corners of two sections at regular intervals, and closing section corners that are placed upon the same line, but which were established at a later date. The quarter-section corners on such lines are usually controlling for one side only in the older practice.

In the more recent surveys where the record calls for two sets of corners, those that are regarded as the corners of the two sections first established and the quarter-section corners relating to the same sections will be employed for the retracement, and will govern both the alignment and the proportional measurement along that line. The closing section corners, set at the intersections, will be employed to govern the *direction* of the closing lines.

In order to restore a lost corner on a township line by single proportional measurement, a retracement will be made connecting the nearest identified regular corners on the line; a temporary stake (or stakes) will be set on the trial line at the original record distance (or distances); the total distance and the falling at the objective corner will be measured.

On meridional township lines an adjustment will be made at each temporary stake for the proportional distance along the line, and it will then be set over to the east or to the west for falling, counting its proportional part from the point of beginning.

On east-and-west township lines and on standard parallels the proper adjustment should be made at each temporary stake for the proportional distance along the line, for the falling, and in order to secure the latitudinal curve*; i.e., the temporary stake will be either advanced or set back for the proportional part of the difference between the record distance and the new measurement, then set over for the curvature of the line, and last corrected for the proportional part of the true falling.

Lost standard corners will be restored to their original positions on a base-line, standard parallel or correction line by single proportionate measurement on the true line connecting the nearest identified standard corners on opposite sides of the missing corner or corners, as the case may be.

Corners on base-lines are to be regarded the same as those on standard parallels. The corners that were set first in the running of a correction line will be treated as original standard corners; those that were set afterward at the intersection of a meridional line will be regarded as closing corners.

All lost section and quarter-section corners on the township boundary lines will be restored by single proportional measurement between the nearest identified corners on opposite

*The term latitudinal curve as here employed denotes an easterly and westerly line, properly adjusted to the same mean bearing from each monument to the next one in regular order, as distinguished from the long chord or great circle that would connect the initial and terminal points.

sides of the missing corner, north and south on a meridional line, or east and west on a latitudinal line, after the township corners have been identified or relocated.

All lost quarter-section corners on the section boundaries within the township will be restored by single proportional measurement between the adjoining section corners, after the section corners have been identified or relocated.

Lost meander corners, originally established on a line projected across the meanderable body of water and marked upon both sides, will be relocated by single proportional measurement after the section or quarter-section corners upon the opposite sides of the missing meander corner have been duly identified or relocated.

A lost closing corner will be reestablished on the true line that was closed upon, and at the proper proportional interval between the nearest regular corners, to the right and left.

In order to reestablish a lost closing corner on a standard parallel or other controlling boundary, the line that was closed upon will be retraced, beginning at the corner from which the connecting measurement was originally made; a temporary stake will be set at the record connecting distance, and the total distance and falling will be noted at the next regular corner on that line on the opposite side of the missing closing corner; the stake will then be adjusted as in single proportional measurement.

A closing corner not actually located on the line that was closed upon will determine the direction of the closing line, but not its legal terminus; the correct position is at the true point of intersection of the two lines.

5.56. Subdivision of Sections.—The sections are not usually subdivided in the field by the United States surveyors, but certain of the subdivision-of-section lines are "protracted" (i.e., drawn in) upon the township plats to indicate how the lottings are prescribed by law; the boundaries of the quarter-sections also are generally shown.

Preliminary to subdivision it is essential to identify the boundaries of the section, since it cannot be subdivided until the section corners and quarter-section corners have been found, or restored by proper methods, and until the resulting courses and distances have been determined by survey.

The order of procedure is: First, identify or reestablish the section boundary corners; next, fix the lines of quarter-sections; then, form the smaller tracts by proportional division.

In subdividing a section into quarter-sections straight lines are run between opposite quarter-section corners previously established or reestablished on the boundaries of the section. The point of intersection of these two lines is the common corner of the four quarter-sections and the legal center of the section.

Quarter-section corners on lines closing on the north and west boundaries of a regular township were established originally at 40 chains to the north or west of the last interior section corner, the excess or deficiency in measurement being thrown into the half-mile next to the township or range line. If such quarter-section corners are lost, they are reestablished by proportional measurement based upon the original record.

When the plat calls for double sets of section corners on township and range lines, the quarter-section corners for the sections south of the township line and east of the range line were not usually established in the original surveys. In subdividing such sections, new quarter-section corners should be so placed as to satisfy their calculated areas, as expressed on the official plat. Proportional measurements should be adopted for the sides of the quarter-sections where the new measurements of the north and west boundaries of the section differ from the recorded distances.

Sections that are invaded by meanderable bodies of water, or by approved claims not conforming to the regular legal subdivisions, are subdivided by protraction into regular and fractional parts, as shown by the dashed lines in Figs. 5.56.1 and 5.56.2.

Fig. 5.56.1 represents a regular section except for the parts that are set aside for mineral claims; two of the quarter-sections and two of the half-quarter-sections are normal. Fig. 5.56.2 shows a section that is made fractional by the lake; it is subdivided into aliquot parts in the usual manner as far as possible. The descriptions that follow illustrate the principles covered by the general rules in Art. 5.54.

To subdivide the section shown in Fig. 5.56.1, both center lines *ab* and *cd* are run to connect the opposite quarter-section corners; corner *b* on the north boundary, within the mineral claim, will be in place; the intersection at point *e* is the center. Points *f*, *g*, *h*, *i*, *j* and *k* for the sixteenth-section corners are set

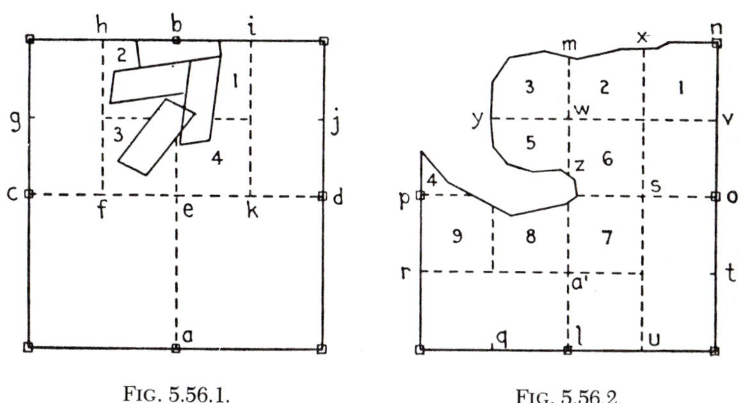

Fig. 5.56.1.

Fig. 5.56.2.

Examples of Subdivision of Fractional Sections.

at mid-points along the boundaries of the two northern quarter-sections, and a stake will be needed at mid-point between *b* and *e*. The center lines of these quarter-sections are then run, thus establishing the boundaries of lots 1, 2, 3 and 4. Intersection of these lines with the boundaries of the mineral claims are established and marked.

In Fig. 5.56.2 the boundaries of the section cannot be completed because of the lake. To subdivide this section, the north and south center line *lm* is run on the mean of the bearings of the east and west boundaries of the section, completing the line across the water by triangulation or by a meander traverse. The east and west center line *op* is run, connecting the opposite quarter-section corners, and completing the line across the water by triangulation or meander traverse. The theoretical center of the section will be fixed at the calculated intersection of the center lines (in the lake). The points q, r, s, t and u for the sixteenth-section corners along the boundaries of the two southern quarter-sections are set, all at mid-point. The point a' is set at mid-point on the line between l and the theoretical center of the section. The two other sixteenth-section corners v and w for the northeast quarter-section will be placed at the calculated proportional distances between o and n, and between the theoretical center of the section and m respectively, to agree with the intent of the representations on the official township plat. The north and south center line of the northeast quarter-section sx will be run on the mean of the bearings of its east and west boundaries. The line wy between lots 3 and 5 will be run parallel to op, the east and west center line of the section.

5.57. Convergence of the Meridians.—The angular convergence of the meridians, given in Tables 5.41.1 and 5.57.1, may be computed as follows. In Fig. 5.57.1 AB is an arc of a parallel of latitude and EQ the arc of the equator intercepted by the meridians through A and B. AT and BT are lines tangent to the meridians at A and B, meeting the earth's axis, prolonged, at T. It will be seen that the angle BTO equals the angle BOQ, which is the latitude of point A and B. The angle $AO'B$ is the difference in longitude of points A and B. The angle between the meridians at A and B is the angle ATB.

In the sector $AO'B$ (Fig. 5.57.1), $\dfrac{AB}{BO'} = \text{angle } AO'B$

In the sector ATB, $\dfrac{AB}{BT} = \text{angle } ATB$ (approximately)

But $$BT = \frac{BO'}{\sin BTO'} = \frac{BO'}{\sin BOQ}$$

$$\therefore \text{ angle } ATB = \frac{AB}{BO'} \sin BOQ$$

$$= \text{angle } AO'B \sin BOQ,$$

i.e., the angular convergence equals the **difference in longitude** times the **sine of the latitude.**

If the points A and B (Fig. 5.57.1) are not in the same latitude, tangents to the meridians through these points will not meet at a point. However, if the difference of latitude is reasonably small, the angular convergence may still be calculated very closely by using the **middle** latitude in the above formula; i.e., the angular convergence equals the **difference of longitude** times the **sine of the middle latitude.**

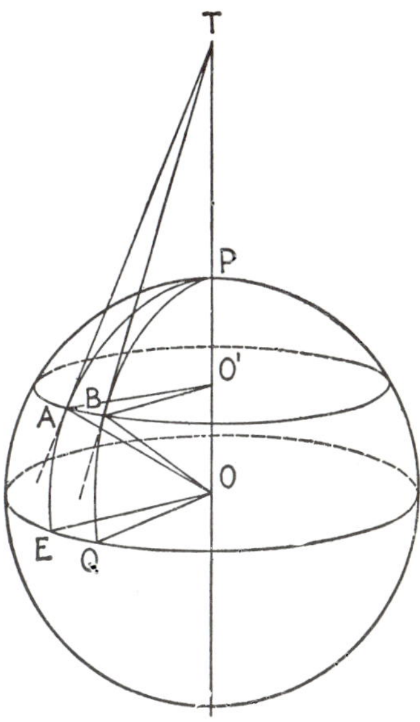

FIG. 5.57.1. CONVERGENCE OF MERIDIANS.

This principle may be applied to a township where 6 miles is the distance between the meridians through A and B along the southern boundary and 6 miles the distance between the parallels forming the northern and southern boundaries of the township. If the latitude of the southern boundary is 37° 00′ N, then the northern boundary is 6 × 5280 feet further north. There are approximately 6080 feet in one minute of latitude along a meridian on the earth's surface, so that 5280 × 6 ft. equals 5.2 minutes of latitude, and the latitude of the northern boundary is 37° 05′.2 N and that of the middle parallel of the township is 37° 02′.6 N. From Table 5.57.1 the length of 1° of longitude in the middle latitude (37° 02′.6 N) is 55.28 miles.

$$\text{Angular convergence} = \frac{6}{55.28} \times \sin 37° \, 02'.6 \times 60 = 3'.92.$$

The linear convergence or the amount by which the northern boundary is shorter than the southern boundary is shown by ab in Fig. 5.57.2. The full lines in Fig. 5.57.2 represent the meridians and parallels through A and B; the line Ab is a line

TABLE 5.57.1

LENGTH OF A DEGREE OF LONGITUDE.

Lat.	Degree of Longitude Statute Miles.	Lat.	Degree of Longitude Statute Miles.	Lat.	Degree of Longitude Statute Miles.
0	69.172	30	59.956	60	34.674
1	9.162	31	9.345	61	33.623
2	9.130	32	58.716	62	32.560
3	9.078	33	8.071	63	31.488
4	69.005	34	57.407	64	30.406
5	68.911	35	56.725	65	29.315
6	8.795	36	6.027	66	28.215
7	8.660	37	55.311	67	27.106
8	8.504	38	54.579	68	25.988
9	8.326	39	53.829	69	24.862
10	68.129	40	53.063	70	23.729
11	67.910	41	52.281	71	22.589
12	7.670	42	51.483	72	21.441
13	7.410	43	50.669	73	20.287
14	7.131	44	49.840	74	19.127
15	66.830	45	48.995	75	17.960
16	6.510	46	8.136	76	16.788
17	6.169	47	47.261	77	15.611
18	65.808	48	46.372	78	14.428
19	5.427	49	45.469	79	13.242
20	65.026	50	44.552	80	12.051
21	64.606	51	43.621	81	10.857
22	4.166	52	42.676	82	9.659
23	63.706	53	41.719	83	8.458
24	3.228	54	40.749	84	7.255
25	62.729	55	39.766	85	6.049
26	2.212	56	38.771	86	4.842
27	61.676	57	37.764	87	3.632
28	1.122	58	36.745	88	2.422
29	60.548	59	35.716	89	1.211

drawn through *A* parallel to the eastern meridian through *B*. The triangle *Aba* may be considered as a plane triangle with a right angle at *a*. The angle *aAb* is equal to the angular convergence just computed. Therefore in the triangle *Aba*

$$ab = 6 \times 5280 \times \sin 0° \, 3'.92 = 36.1 \text{ feet.}$$

Whenever traverses of considerable extent are being run, it is very desirable to check the bearings along the traverse by astronomical observations or by tying into triangulation stations (Art. 10.58) and checking the bearings against an azimuth from such stations. The convergence of meridians must be taken into account when the distance east or west of the starting point is large in order to obtain a check on the angular measurements. The same formulae employed above may be used, but it will probably be more convenient to make use of Table 5.57.2. The total east or west distance between the points *C* and *D* (Fig. 5.57.3) may be found from the difference

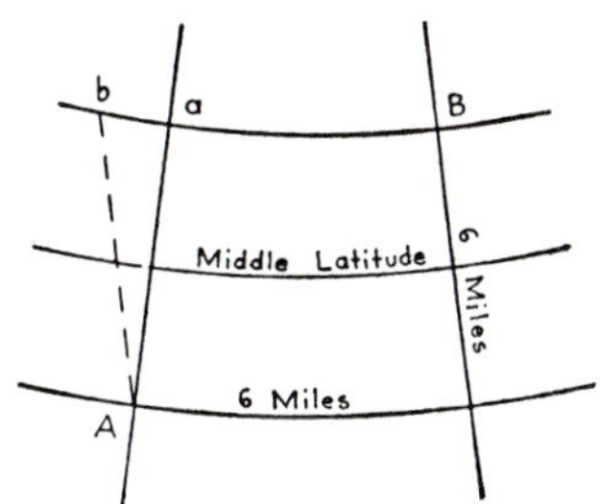

FIG. 5.57.2.

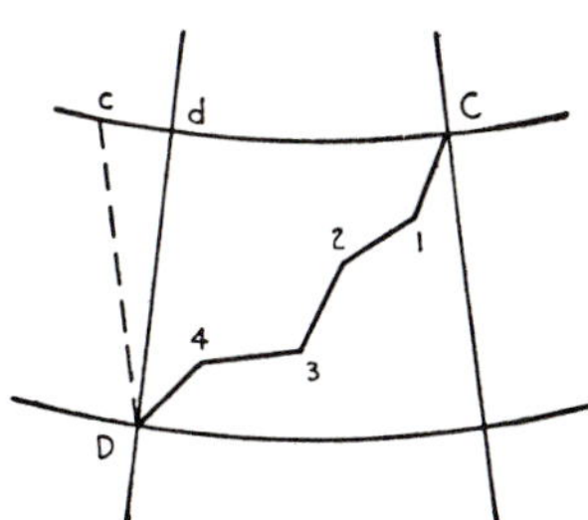

FIG. 5.57.3.

in the coördinates of these points as explained in Art. 15.9. In the column containing the number of thousands of feet of this east or west distance and on a line with the mean latitude will be found the angular convergence of the meridians. The convergence for numbers not in the table may be found by combining those that are given. For instance, that for 66,500 feet, in mean latitude 40°, may be found by adding together 10 time the value for 6000, and one-tenth the angle for 5000. The result is 550″, which is the correction to apply to the second observed azimuth to refer the line to the first meridian.

EXAMPLE.—In Fig. 5.57.3 assume that at station *C* the azimuth of the line *C* to 1 is 28° 42′ 40″ and that the survey proceeds in a southwesterly direction to point *D* where the azimuth of line *D* to 4 is found by observation to be 226° 17′ 20″. The calculation of the survey shows that *D* is 3100 feet south and 15,690 feet west of C. The mean latitude of the traverse is 40° N.

From Table 5.57.2 the convergence for 15,690 feet by parts is 83 + 41 + 5 + 1 = 130″ or 2′ 10″. This is the correction that must be added to the observed azimuth of D–4 to make it refer to the meridian through C. The azimuth at D referred to the meridian at C is, therefore, 226° 19′ 30″ and the difference in the directions of lines C–1 and 4–D is 226° 19′ 30″ − 180° − 28° 42′ 40″ equals 17° 36′ 50″ which should agree with the algebraic sum of the deflection angles of the traverse C to D, within the range of the precision expected.

Note that in the above computations 180° was subtracted from the azimuth of D–4 to change its direction from D–4 to 4–D so that the azimuth of C–1 and of 4–D would be taken in the same order.

TABLE 5.57.2

CONVERGENCE OF THE MERIDIAN IN SECONDS FOR EACH 1000 FEET ON THE PARALLEL

	Distance East or West (in feet)								
Latitude	1000	2000	3000	4000	5000	6000	7000	8000	9000
°	″	″	″	″	″	″	″	″	″
20	3.5	7.0	10.5	14.0	17.5	21.0	24.5	28.0	31.5
25	4.6	9.2	13.8	18.4	23.0	27.6	32.2	36.8	41.3
30	5.7	11.4	17.1	22.7	28.4	34.1	39.8	45.5	51.2
35	6.9	13.8	20.7	27.6	34.5	41.4	48.3	55.2	62.1
40	8.3	16.5	24.8	33.0	41.3	49.6	57.8	66.1	74.3
45	9.8	19.7	29.5	39.4	49.2	59.0	68.9	78.7	88.6
50	11.7	23.5	35.2	46.9	58.6	70.3	82.1	93.8	105.5

CHAPTER VI

TRAVERSE LINES—LOCATION OF BUILDINGS—MISCELLANEOUS SURVEYING PROBLEMS

TRAVERSE LINES

6.1. Traverses Which Do Not Close—Route Surveys.—A great many surveys, such as route surveys for highways, railroads, pipe lines, or canals are based on traverses which do not return to the starting point. In these surveys the traverse line is measured continuously and recorded in 100-foot "stations." The starting point is called "station 0," the next station 100 feet away is "station 1," the next "station 2," etc. Every 100-foot length is a *full station,* and any fractional distance is called a *plus.* The distance from station 0 to any point, **measured along the traverse line,** is the station of that point and is recorded always by the number of the last station with the plus station in addition, e.g., the station of a point at 872.4 ft. from station 0 is 8 + 72.4.

At the angle points it is customary to measure the **deflection angles** rather than the interior angles because the former are usually less than 90°. **These should be checked in the field by "doubling" the angles.** (See Arts. 5.12-5.17.)

The notes are kept so as to read **up the page.** The left-hand page is for the traverse notes and the right-hand page for the sketch, the stations in the sketch being opposite the same station in the notes. Fig. 6.1.1. is a set of notes illustrating this kind of traverse. Frequently no notes are kept in tabular form, all of the data being recorded on the sketch. Figs. 6.24.1 and 7.31.3 illustrate this style of notes.

In highway and railroad practice the preliminary line is usually run as a series of straight lines between angle points. The final location line, however, consists of traverse lines (*tangents*) connected by circular curves. Between tangents the final stationing follows the curves (Art. 10.11-10.24). When changes are made in alignment which affect the stationing, or where an error is discovered, a *chainage equation* is introduced to avoid changing all the subsequent stationing. This

(LEFT-HAND PAGE) (RIGHT-HAND PAGE)

Preliminary Survey for Pipe Line, Knowlton, N.D.

Redman Rolf r.c.
Lyons ⊼ Grover l.f.
Noyes h.c. Baird r.f.
Oct. 17, 1957

Sta.	Point	Defl. Angle	Observed Bearing	Calculated Bearing
10				
9			N 13°E	N. 13°06'E
8	⊙ +84.2	43°17'L		
+94.5				
+46.5				
7			N 30°W	N30°11'W
6	⊙ + 70.2	18°43'L		
+17.2				
5			N 11½°W	N 11°28'W
4	⊙ +62.1	16°17'R		
3				
+42				
2				
1			N 27¾°W	N 27°45'W
0	⊙			

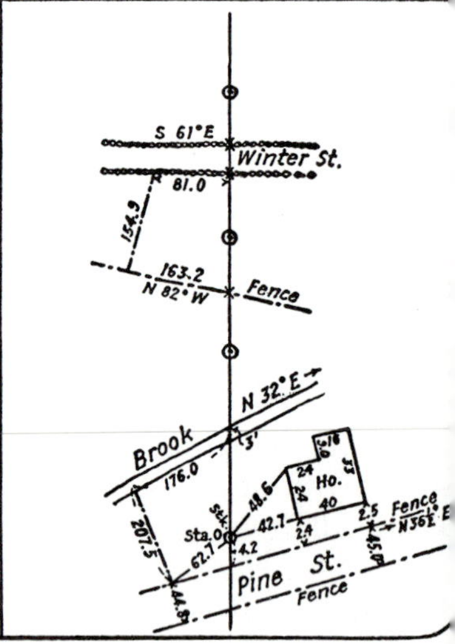

FIG. 6.1.1. TRAVERSE NOTES.

equation is written and recorded in the notes and on the plans where the new and old alignments come together, such as, for example, 34 + 15.62 (new stationing) = 35 + 07.80 (old stationing). When such equations are used the forward stationing obviously no longer indicates the distance from the beginning of the survey.

6.2. CHECKING BY ASTRONOMICAL METHODS.—The angles of any traverse may be checked by determining the azimuths of the first and last lines by astronomical methods. On long traverses requiring many days for their completion it is advisable to make such observations daily. The two most useful methods are (1) direct solar observations, which interrupt the regular work only a few minutes, and (2) observations of Polaris, which are simple and accurate but usually necessitate a special night trip to the place of observation. For details of these methods see Chapter VIII. Since the meridians converge toward the pole it is necessary to make proper allowance for this convergence when comparing the azimuths; the amount of the correction can be obtained from Table 5.41.1, or Table 5.57.2.

6.3. Checking by Cut-Off Lines.—The angles may be checked occasionally by cutting across from one point on the traverse to another at a considerable distance ahead, and measuring the angles from the traverse line at each end of this cut-off line, thereby obtaining all the angles of a closed traverse in which the length of one side only (the cut-off line) is missing. Sometimes the angle at only one end of the cut-off line can be measured, but the calculations for checking the angles are not so simple as when all the angles are known. When both angles have been measured the check consists in simply obtaining the algebraic sum of the deflection angles, whereas the traverse must be computed as explained in Arts. 15.15-15.16, if an angle and a side are missing.

6.4. Checking by Angles to a Distant Object.—A practical and useful method of checking the azimuth of any line of the traverse is as follows. At intervals along the line, measure carefully the angle from the traverse line to some well-defined distant object, such as a distinct tree on a hill or the steeple of a church. If the survey is plotted and it is found by laying off the angles taken to the distant object that these lines do not meet at one point on the plan there is a mistake in the angles, and a study of the plot will show the approximate location of the mistake.

6.5. Checking by Connecting with Established Points.—An accurate and desirable method of checking both the angles and distances of a traverse is to connect it with identifiable triangulation points or with established points on a State or local plane coördinate system (Arts. 10.53 and 10.54). The latitude and longitude of triangulation points and distances between them can be obtained from government sources (Federal, State or Municipal). By connecting the traverse lines with triangulation points by angles and distances, a closed traverse is obtained which can be checked for error of closure (Chap. XV). Checks may be made between coordinate points by comparing known differences in coördinates with those computed from the traverse distances and angles.

Information concerning triangulation and coördinate points can be obtained from the U. S. Coast and Geodetic Survey, the U. S. Geological Survey, State coördinate surveys, and frequently from City or Town surveys.

LOCATION AND COORDINATION OF BUILDINGS FROM TRANSIT LINE

6.6. METHODS OF LOCATING AND COORDINATING BUILDINGS.—When locating any objects, such as buildings, from the survey line the measurements should be such as will permit of accurate and rapid plotting. In city plans, for instance, accurate location is of great importance and should be made the first consideration whereas in some topographic maps rough locations will suffice, and accuracy may be sacrificed to speed. In the following pages the examples are limited to locations of buildings because these illustrate all the necessary principles involved in any location. Locations by (3) (angle and distance) are not discussed because the method is evident.

The principles involved are the same for any accuracy. In order to make clear the various methods used in the location of buildings it will be well, before giving particular cases occurring in practice, to enumerate the geometric principles to be used.

A **point** may be located and therefore coördinated from a traverse:

1. By rectangular coördinates, i.e., by its station and perpendicular offset.
2. By two ties from known points.

3. By an angle and a distance from a known point (polar coördinates).
4. By an angle at each of two known points (intersection).
5. By a (perpendicular) swing offset from a known line and a tie from a known point.
6. By (perpendicular) swing offsets from two known lines which are not nearly parallel.
7. By an angle from one point and a distance from another point.

A **line** may be located:

1. By locating two points on the line.
2. By locating one point on the line and the direction of the line. (Angle or Range.)

6.7. Running a Straight Line—One Point Visible from the Other.—There are several ways in which a straight line may be fixed on the ground, depending upon the existing conditions. If the line is fixed by two end points, one of which is visible from the other, the method of setting intermediate points would be to set the transit over one point, take a "foresight" on the other and place points in line. **For very exact work the instrument has to be used in both the direct and reversed positions** (Art. 2.54). This will eliminate errors of adjustment such as failure of the telescope to revolve in a true vertical plane, or failure of the objective tube to travel parallel to the line of sight.

6.8. Running a Straight Line.—Neither Point Visible from the Other.—If neither point can be seen from the other then it is necessary to find some point, by trial, from which the terminal points can be seen. The transit is set up at some intermediate point estimated to be on the line, a "backsight" is taken on one of the points and the instrument clamped. The telescope is then reversed on its horizontal axis. If the vertical cross-hair happens to cut the second point, then the estimated position happens to be on the line; if not, then the distance this trial line is to one side of the second point can be estimated (or measured) and from this the distance the transit is off line can be computed mentally by proportion and the instrument moved over (toward the line) by this amount. In this way, by successive trials, the true point is attained. The final tests should be made with the instrument in direct and reversed positions to eliminate errors of adjustment of the line of sight and the horizontal axis. To eliminate errors in the adjustment of the plate bubbles the plate level which is perpendicular to the line should be re-leveled just before making the second backsight and while the telescope is pointing in that direction. This can be more readily done if, when the transit is set up, one pair of opposite leveling screws is turned into the line; then the other pair will control the level which is perpendicular to the line of sight.

The method of running a line between two points when neither one is visible from an intermediate set-up is to run first a trial line, called a *random line*, as described in Art. 6.26.

6.9. Prolonging a Straight Line.—If a line is fixed by two points A and B, Fig. 6.9.1, not far apart, and it is desired to prolong this line in the direction AB, the instrument should be set up at A, a sight taken on B and other points C, D and E set in line beyond B. For the best results the transit should be

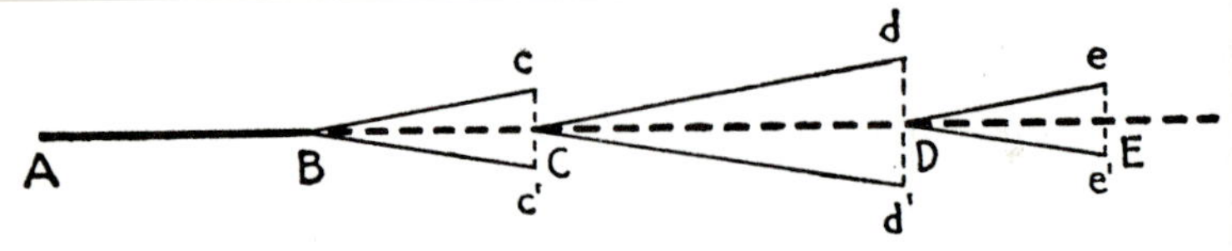

FIG. 6.9.1. PROLONGING A LINE BY BACKSIGHTING AND FORESIGHTING.

used in the direct and reversed positions and the mean result used; this is especially important if the distances are such that it is necessary to change the focus of the telescope. When running toward a foresight the points ahead tend to come nearer and nearer into line as the transit approaches the terminal point. If, however, there is any doubt about the objective slide's traveling parallel to the line of collimation it is advisable to use the instrument in the direct and reversed positions.

But when it is impossible to see beyond *B* from *A*, the transit should be set up at *B* and points ahead should be set by the method of backsighting and foresighting as follows. With the transit at *B* a backsight is taken on *A* with the telescope in its direct position and the plates clamped. The telescope is then inverted and a point *c* set ahead in line. The position of this point *c* is then verified by repeating the process, the backsight on *A* being taken with the telescope in the inverted position. If, when the telescope is again plunged point *c* is not on the cross-hair, set point *c′* beside point *c*. The true position of *C* is midway between *c* and *c′*. The transit is then moved ahead to the new point *C* and the whole process is repeated.

When prolonging a line by backsighting and foresighting, using the transit **in but one position** of the axis, the error in the sight line is always on the same side, the resulting line is curved, and the points lie farther and farther from the true line. **It is very important, therefore, to observe with the transit in the two positions and to use the mean result.**

6.10. Methods of Showing Sights.—If the point sighted is within a few hundred feet of the transit a pencil (held vertical) may be used to show a point for the transitman to sight on. Sighting-rods or range poles are used on long distances.* Where only the top of the pole is visible an error will be introduced if it is not held plumb. A plumb-line is much more accurate for such work but may be difficult or impossible to see on long sights. Under conditions where the plumb-line cannot be seen readily some surveyors use for a sight an ordinary white card held so that the center of the card is against the string, or the card may be folded down the middle and the

---

*It is desirable that the object used for the foresight should be of such color that the cross-hair is clearly seen, and of such width that the cross-hair can easily bisect it. A yellow or light colored pencil against a dark background, or a dark pencil against the sky usually give good results.

string held in the tilted fold so that the card forms a V directly on line over the point.*

Whenever the instrument is sighted along a line which is to be used frequently or along which the transit is to remain sighted for a long time the transitman should select if possible some well-defined point which he observes to be in the line of sight, called a "foresight." If no definite point can be found, a nail may be driven through a white card or paper or a "crow-foot" marked with crayon or "keel." By means of this "fore-sight" the transitman can detect if his instrument moves off the line, and can reset the telescope exactly at any time without requiring the aid of another man to show him the line.

6.11. Signals.—In surveying work the distances are frequently so great that it is necessary to use hand signals. The following are in common use.

"*Right*" or "*Left*" when setting points.—The arm is extended in the direction of the motion desired, the right arm being used for a motion to the right and the left arm for a motion to the left. A slow motion is used to indicate that the

---

*It is the practice among some surveyors to use a 2-ft. rule for a sight. The rule is opened so that it forms an inverted V (Λ). The plumb-string is jammed into the angle of the Λ by pressing the two arms of the rule together. The rule is then held so that the plumb-string as it hangs from the rule appears to bisect the angle of the Λ.

Another device involves attaching to the plumb-line an ordinary fish-line float (shaped like a plumb-bob). This may be fastened so that its axis coincides with the string and so that it can be raised and lowered on the string. It should be painted with such colors that it can be seen against any background, say, the upper half white and the lower half red, or black.

The man showing the sight for the transitman should always try to stand so that the sun will shine on the object he is holding; on long sights it is difficult (sometimes impossible) to see an object in a shadow.

An effective way of setting a point on line in a place where there is poor visibility such as in dense undergrowth or at twilight is by using an ordinary flashlight in the following manner. With the corner of a file make a V-shaped groove in the metal ring of the flashlight rim. In using the flashlight hold its barrel at about 30° to the horizontal with the groove at the top. Over this groove hang a plumb-line so that it will not touch the glass face of the flashlight. Then put on the light. The transitman can readily see the light and the vertical plumb-line in front of it. Another less accurate way is to loop the plumb-line over the flashlight near the lighted end so that the plumb-bob will hang under the flashlight over the point, and then the transitman will sight directly on the light. In this case the flashlight barrel is held horizontal.

point must be moved a long distance and a quick motion a short distance.

"Plumb the Pole."—The hand is extended vertically above the head and moved slowly in the direction it is desired to have the pole plumbed.

"All Right."—Both arms are extended horizontally and moved up and then down; or, both hands shown at same time (without raising and lowering).

"Give a Foresight."—The transitman, desiring a foresight, motions to the rodman, by holding one arm vertically above his head.

"Take a Foresight."—The rodman, desiring the transitman to sight on a point, motions the transitman by holding one arm vertically above his head and then he holds his lining-pole or pencil vertically on the point.

"Give Line."—When the rodman desires to be placed "on line" he holds his lining-pole horizontally with both hands over his head and then brings it down to the ground in a vertical position. If the point is to be set carefully, as a transit point, the rodman waves the top end of pole in a horizontal circle above his head before bringing it to the vertical position or, he may wave the arm in a similar manner if a pole is not being used.

"Pick Up the Transit."—When the chief of the party desires to have the instrument set at another point he signals to the transitman by extending both arms downward and outward and then raising them quickly.

All signals should be distinct so as to leave no doubt as to their meaning. Care should be taken to stand so that the background will not prevent the signals from being seen distinctly. The palms of the hands should be shown in making the signals; for distant signals, a white handkerchief is often used. Where much distant signaling is to be done, flags are attached to the range-poles. Special signals may easily be devised for different kinds of distant work and for various conditions.

6.12. Ties, Offsets, Swing Offsets, and Range Lines.—In the preceding list, the word *tie* is used as meaning a direct horizontal distance measurement between two points.

An *offset* is the **horizontal** distance from a line, usually at right angles.

A *swing offset* is the **perpendicular** distance to a line and is found by trial. The zero end of the tape is held at the point to be located and the tape is swung in a short arc about this point as a center, the tape being **pulled taut and kept horizontal.** The tape is read from the transit in various positions; the shortest reading obtainable is the perpendicular distance desired.

A *range line* is a line produced to intersect the transit line or some other line.

6.13. General Suggestions.—By whatever method the buildings are located the following suggestions should be carried out.

1. All the sides of the building should be measured and checked by comparing the lengths of opposite sides.
2. Other things being equal, a long side of a building should be located in preference to a short side.
3. Two ties or two directions should intersect at an angle as near 90° as practicable, and never less than 30°. If method (7) is used, however, the tie and the sight line should make as small an angle as possible.
4. One or more *check measurements* should be taken in the location of each building.
5. In order to secure the best location the surveyor should **keep constantly in mind how the building, or other object which is being located, is to be plotted.**

In most work of this character it is customary to record the measurements to tenths of a foot. How precisely the measurements should be taken, however, depends upon the scale to which they are to be plotted.

6.14. Typical Cases.—Although each case will have to be dealt with according to circumstances there are certain typical cases which will serve as guides. These are illustrated by the following examples.

6.15. Example I. Building Near Transit Line and Nearly Parallel to It.—As will be seen in Fig. 6.15.1 swing offsets are taken at the two front corners; these ties, together with the tie from *A* to station 1 and the length of the front of the building locate points *A* and *B*. The general dimensions of the building furnish sufficient data for plotting and checking the remaining sides. It is assumed that the corners of the building are square unless it is obvious that they are not. (In city buildings this assumption may not be a safe one to make.) The tie from *C* to station 2 is a check against an error in the other measurements.

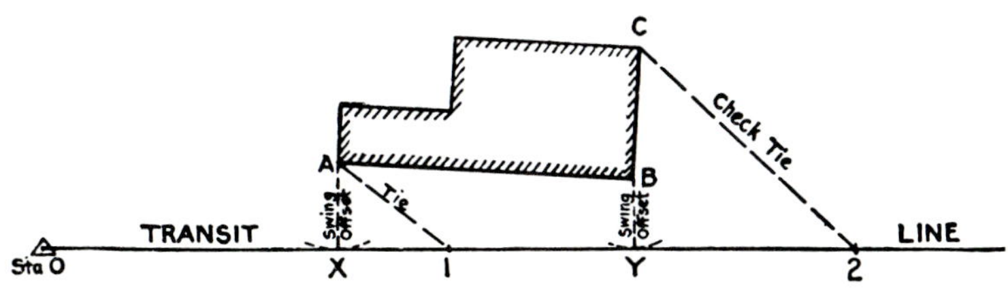

FIG. 6.15.1.

PLOTTING.—This building would be plotted thus:—lay off the distance *AX* perpendicular (estimated) to the transit line (say *A′X′*) and draw a line through *A′*, with triangles, parallel to the transit line; then scale *A*1 from station 1 to this parallel line. Point *A* is where the tie *A*1 cuts the parallel line. Point *B* is located in the same way, *AB* being used as the tie from *A*. Then by means of triangles and scale the building is completed and the distance *C*2 scaled and compared with the notes. Another way to plot point *A* would be to set on the compass the distance 1*A* and swing an arc about 1 as a center; then, keeping the scale perpendicular to the transit line, find where the distance *XA* will cut this arc, thus locating point *A*. Point *B* can be similarly located after *A* has been plotted. For the same degree of accuracy distances can be measured more rapidly with a scale than they can be laid off with a compass, therefore the former method is usually preferable.

This building might have been located by four ties *AO*, *A*1, *B*1, and *B*2. The plotting in this case would be slow because at least two of the ties must be swung by use of a compass, and inaccurate because the intersections would be bad.

6.16. Example II. Building Near Transit Line and Making a Slight Angle with It.—Fig. 6.16.1 illustrates two ways of locating a building in such a position that the intersection of the transit line by the long side (produced) can be readily obtained.

The left-hand building is located by the method of Example I. The tie $B1$ could have been taken instead of $B2$. It would have given a better intersection at B, but since it is a longer tie than $B2$ the fieldwork necessary is slightly greater. If $B2$ is taken $B1$ might be measured as a check tie although $A1$ would make a better check tie since it will also check the measurement of the side AB.

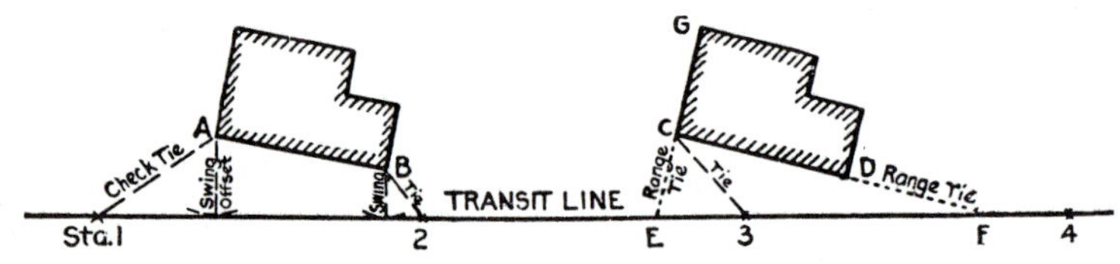

FIG. 6.16.1.

The right-hand figure illustrates another method of locating such a building. The front and side of the building are ranged out by eye, a method which is practicable and sufficiently precise for all ordinary purposes, and the plus stations of points E and F are measured. The range lines CE and DF are also measured and the check tie $C3$. $C2$ could have been taken as a check tie; it would have given a better intersection at C than the tie $C3$, but it is much longer.

PLOTTING.—The left-hand building is plotted as described in Example I. In plotting the right-hand building the plus stations on the transit line are first scaled. Then with the compass set at the distance EC an arc is swung from E as a center. From F the distance FC is scaled to intersect the arc, which locates point C and at the same time the direction of the side CD. The building is then plotted with triangles and scale. The check tie $C3$ should scale to agree with the notes and the line GC produced should strike point E.

There is little difference between these two methods in the amount of fieldwork, there being only one more measurement in the right-hand than in the left-hand figures, but one extra check is thereby obtained. In plotting, the method used in the right-hand figure is shorter.

6.17. Example III. Building Located Entirely by Direct Ties.—Any building can be located and checked by four ties as in Fig. 6.17.1 provided the pairs of ties intersect at favorable angles (say not less than 60° nor over 120°). This method has the advantage of being very simple and direct, especially in the field, but the plotting of the building calls for the use of the compass in two of the ties and hence is less rapid and accurate than where swing offsets or ranges can be used.

PLOTTING.—The plotting of this building is done by swinging the tie from one station to a corner of the building and scaling from the other station the tie to the same corner. Then the other corner is plotted in the same way or by using the side of the building as one of the ties if it gives a better intersection.

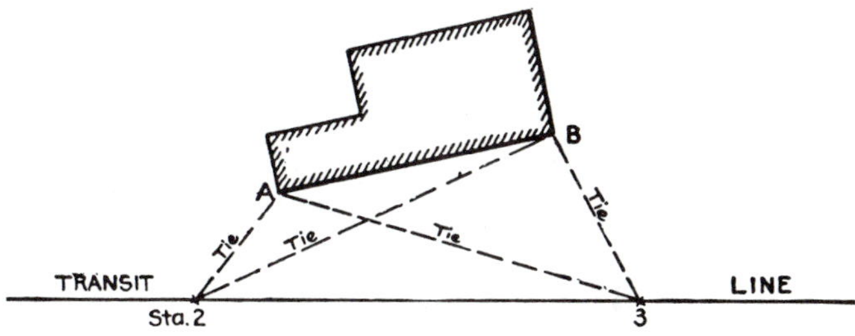

FIG. 6.17.1.

6.18. Example IV. Building Located at a Considerable Skew to the Transit Line.—A building which is at a considerable skew to the transit line can be located best by range ties as illustrated in Fig. 6.18.1. The range ties through *A* are sufficient to locate the building, provided *AE* and *AF* are not too short in comparison with the sides of the building. If these ranges are long enough, then *B*3 is a check tie; but if the ranges are short, *B*3 must be depended upon to determine the position of point *B* and in this event one of the range ties becomes a check. But if A is within two or three feet of the transit line it will be well to omit one of the ranges and take the additional tie $2C$ or the range tie DC produced.

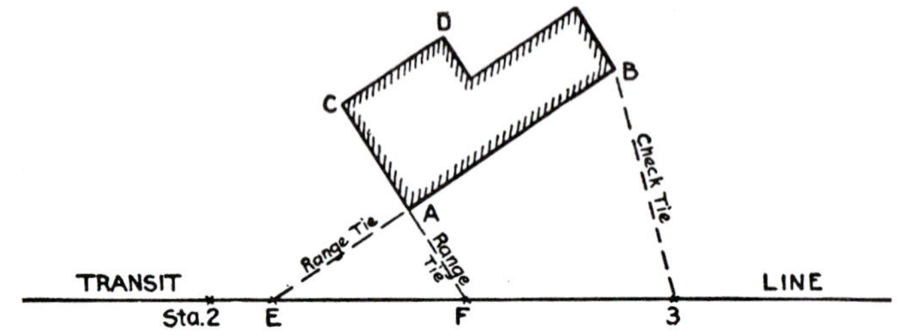

FIG. 6.18.1.

PLOTTING.—If the ranges are of fair length the building is plotted as explained for the right-hand building in Art. 6.16, but if the range ties are short point B is located either by swinging the arc with radius EB and scaling $B3$ or by arc $3B$ and scaling EB. The direction of AB is now determined and the building can be plotted. CA produced should strike at F, and AF should scale the measured distance.

6.19. Example V. Buildings at a Long Distance from the Transit Line.—It is evident that here (Fig. 6.19.1) the tape is not long enough to allow the use of swing offsets. Range ties may be used provided the building is not so far away that the

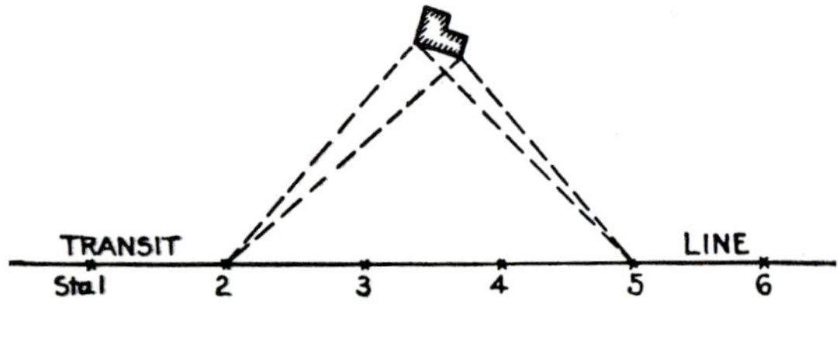

FIG. 6.19.1.

eye cannot judge the range line with reasonable accuracy. Sometimes the only methods available are long ties or angles or a combination of the two. In any specific case there may be some objections to any of these methods, and the surveyor will have to decide according to circumstances which method he will use. For example, where there are obstacles to the measurement of ties, the corners of the building may have to be located entirely by angles from two points on the transit line. Location by angles is objectionable because it is difficult to plot an angle quickly and at the same time accurately. It often happens, however, that when a building is at a long distance from the transit line its accurate position is not required, since as a rule the features near the transit line are the important ones. This method of "cutting in" the corners of the building by angle is often used in rough topographic surveying and is decidedly the quickest of all methods so far as the fieldwork is concerned.

PLOTTING.—The angles are laid off from the transit line with a protractor and the proper intersections determine the corners of the buildings. If the building is measured, the side between the corners located will serve as a check tie.

In some cases, e.g., in making a topographic map on a small scale, the buildings are not measured at all, their corners being simply "cut in" by several angles from different transit points, and the shape of the building sketched in the notes.

6.20. Example VI. Buildings Located from Other Buildings.—Buildings which cannot be located conveniently from the transit line, on account of intervening buildings, may be defined by ties from those already located. Fig. 6.20.1 shows several ways in which such buildings may be located. Any of the preceding methods are applicable, using the side of

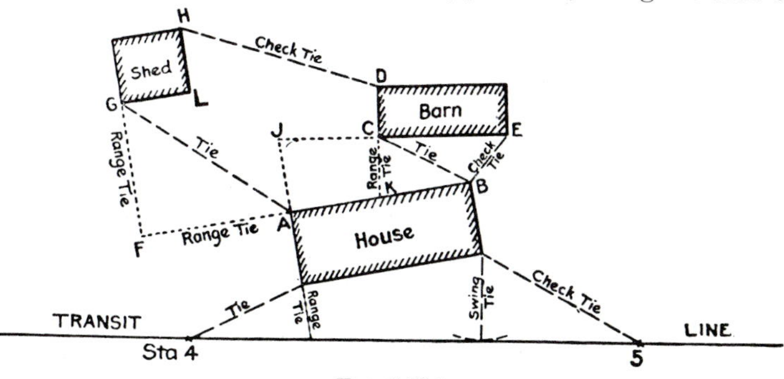

FIG. 6.20.1.

the house as a base-line, but it will be found that range ties are almost always preferable. For example, the barn is located by the distance *BK*, the range tie *KC* and the tie *BC*, and checked by the tie *BE*. Another location of the barn is the distance *AK* or *BK*, the range tie *KC*, and the two range ties *AJ* and *CJ*. By this latter method the directions of both sides of the barn are checked. Still another location of the point *C* would be to substitute in the place of the range tie *CK* a swing offset from *C* to the house. The shed is located by the range ties *AF* and *FG* and by the tie *AG*. The check tie *HD* checks in general the location of both the barn and the shed. If the side *HL* is ranged out instead of the opposite side it will be seen that the tie *AL* will give a poorer intersection at *L*. If convenient a tie from *L* to 4 or the range *GF* continued to the transit line may be measured as a check.

6.21. Example VII. Buildings of Irregular Shape.—Not infrequently a building of irregular shape has to be located. For example, the shop in Fig. 6.21.1 is located on the front by ties

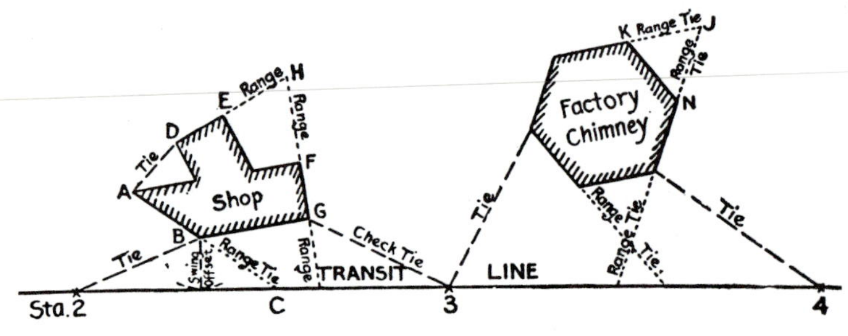

FIG. 6.21.1.

and swing offsets like Example I; then the direction of *AB* is determined by the range tie *BC*. The back corner *E* is determined by the ranges *FH* and *EH*, and by the dimensions of the building; *FA* is assumed parallel to GB. If the angle *F* is a right angle the tie *EF* may be taken instead of the range ties *FH* and *EH*, but even when *F* is a right angle it will be well if time will permit to take these range distances as they give valuable checks on the other measurements which the single tie *EF* does not furnish. *ED* is scaled along *HE* produced and the rest of the building plotted by its dimensions and checked by *AD*.

The ties shown on Fig. 6.21.1 to locate the factory chimney will locate its sides even if these do not form a regular polygon. If such a structure is situated at a considerable distance from the transit line probably the best way to locate it is by angles and distances to the corners, by the measurements of the sides, together with a few such ranges as *NJ* or *KJ*.

Any building in which the sides are short, interrupted by towers or curved walls, so that no one side can be located accurately, can always be located by angles and distances to the corners. (Method (3).)

6.22. Example VIII. Large City Buildings.—Fig. 6.22.1 illustrates the location of several buildings in a city block where the transit line runs around the block. The fronts of the build-

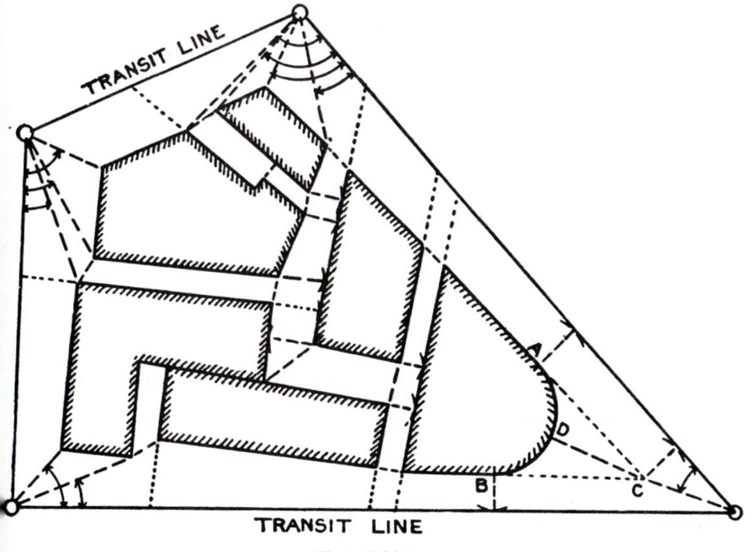

FIG. 6.22.1.

ings are located from the transit line and the rear corners are tied together. The range ties are shown by dotted lines and other ties by dashes. The angles measured are marked by arcs. At the curve *AB*, the side lines of the building are ranged out to point *C*, which is located from the transit line by an angle and distance, and checked by a swing offset; *CD* is also measured to locate point *D* on the curve.

Many large buildings have their walls reinforced by pilasters, and care should be taken when locating corners not to confuse the neat line of the wall with the line of the pilasters.

6.23. Example IX. Location of Buildings by Angles and Distances.—It will be seen from Figs. 6.22.1 and 6.24.1 that some of the buildings have been located by angles and distances from transit points. Any of the buildings in the preceding examples could be located by this method, and on account of the rapidity with which the work can be done in the **field** many surveyors prefer to use it almost exclusively.

For computing from "the angle and distance" obtained in the field, one now calculates coödinates using the transit line as a local coördinate axis. It is then merely necessary to compute perpendicular offsets and stations along the transit line. (See Art. 15.1.)

The use of ranges requires much more computations and is therefore not greatly desirable.

6.24. Location of Buildings and Fences from Transit Line.—Fig. 6.24.1 is a sample page from a note-book illustrating the above principles. It will be noticed that in the field-notes the letter *R* appears where the lines are ranges.

6.25. DEFECTIVE LOCATIONS.—In Fig. 6.25.1 are shown six locations which are weak or defective, given for the purpose of showing that some locations that appear at first glance to be complete may give trouble when an attempt is made to plot them. In No. 1 the arc defined by the tie nearly coincides with the line defined by the angle. The position of the corner of the building is therefore indefinite. A tie from 11 + 00 would have been better, but not so good as one from a point still nearer the transit. A range on the left side would fix the direction. In No. 2 the building is really located, but it cannot be plotted until calculations or geometric constructions have been made to determine another tie. That is, no one corner has been located directly from the transit line. For example, a tie from + 20 to the corner where the swing offset is taken would make it possible to plot that corner directly. A range on the

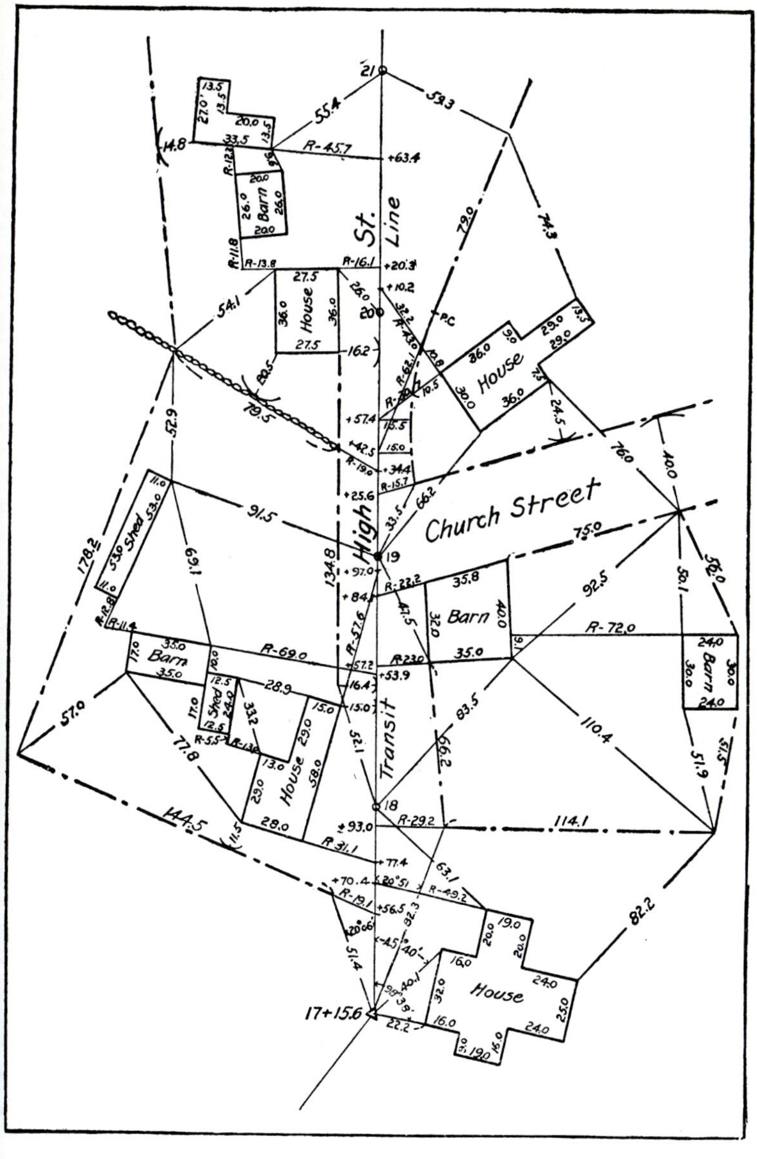
21
55.4
59.3
R-45.7
+63.4
14.8
Barn
High St. Line
R-16.1
+20.3
+10.2
20
House
54.1
27.5
36.0
16.2
P.C
House
36.0
29.0
79.0
74.3
79.5
52.9
+57.4
+42.5
+34.4
R-15.7
+25.6
Church Street
Shed
91.5
19
75.0
178.2
134.8
+97.0
R-22.2
+84.1
Barn
35.8
40.0
R-72.0
R-11.4
Barn
35.0
R-69.0
+57.2
R-23.0
+53.9
Shed
28.9
House
+16.4
+15.0
Transit
83.5
92.5
110.4
57.0
77.8
Barn
24.0
30.0
51.9
18
+93.0
R-29.2
114.1
144.5
28.0
R-31.1
+77.4
63.1
+70.4
R-49.2
R-19.1
+56.5
82.3
51.4
82.2
House
19.0
20.0
16.0
32.0
24.0
25.0
17+15.6
22.2

FIG. 6.24.1.

front side would be still better. In No. 3 the offsets were presumably estimated to be at right angles to the line. Upon this assumption the building can be plotted, but its location along the line is rather weak. If the offsets are not at right angles the

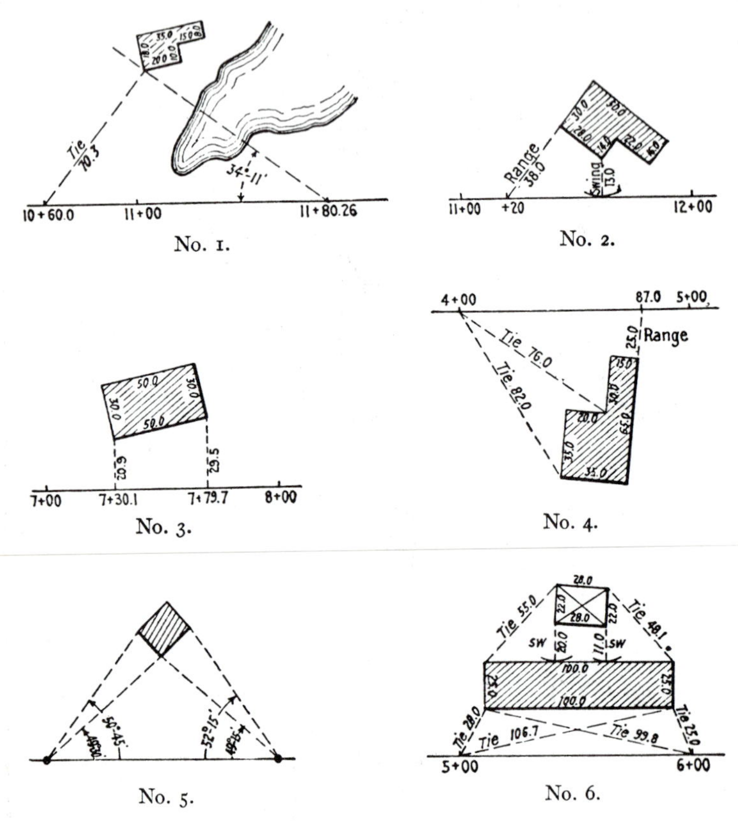

FIG. 6.25.1. SHOWING LOCATIONS THAT ARE WEAK OR DEFECTIVE.

location is indeterminate. One diagonal tie would make the location a strong one, although not checked. In No. 4 no one corner can be plotted directly until another distance has been calculated or found by construction. Additional measurements should be made. In No. 5 the building can be rotated a little about the front corner and yet the sight lines to the outer corners will pass through their respective corners. Only one corner of the building is really fixed. In the last figure (No. 6) the building in the rear cannot be plotted directly. A range on one side of the building instead of a swing, or, a tie to one of the nearer corners to intersect the swing, would remedy this defect.

PROBLEM

Indicate the best method of locating the fences and buildings shown below, by means of tape measurements.

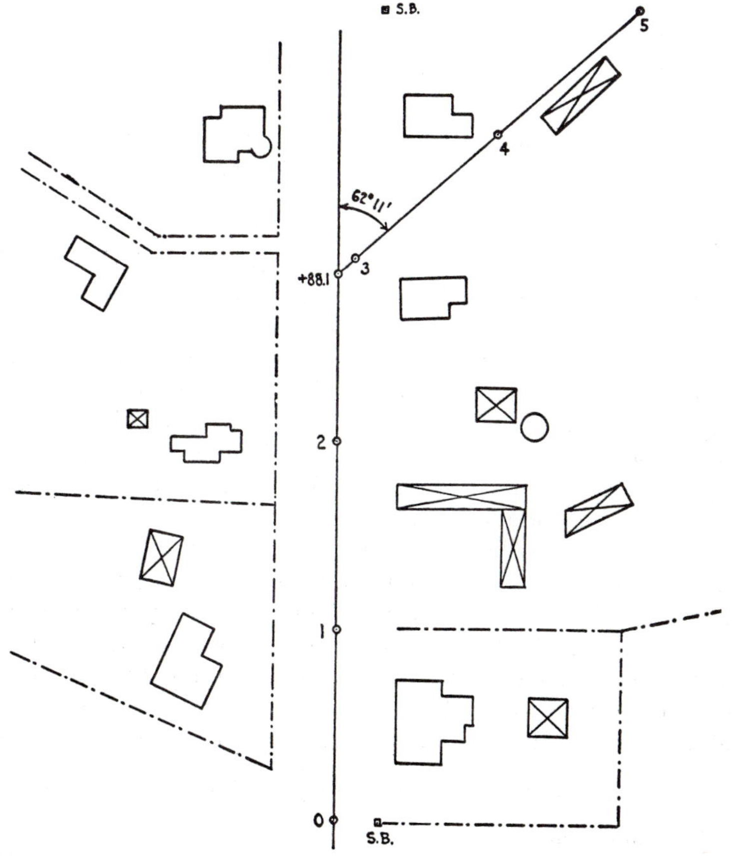

FIG. 6.25.2.

MISCELLANEOUS SURVEYING PROBLEMS

6.26. Random Line.—Not infrequently in attempting to run a straight line between two points A and B (Fig. 6.26.1) it is found impossible to see one point from the other or even to see both A and B from an intermediate set-up on a straight line between them. When this condition exists it is necessary to start at one point, e.g., A, and run what is called a trial, or *random*, line AC by the method explained in Art. 6.9, in the direction of the other end of the line as nearly as this can be judged.

Where the random line passes the point B the perpendicular offset YB is measured and also the distance to point Y along AC. Unless the random line is very close, say, within about two feet of the line AB, the point Y where a perpendicular to AC will pass through B cannot be accurately chosen by eye. The method resorted to is one which has general application in all kinds of surveying work, and is as follows.

With the transit at A point X is set carefully on the line AC and as nearly opposite point B as possible. Then the instrument is set up at X and 90° turned off in the direction XZ. If this line does not strike B (and it seldom will exactly) the distance BZ is carefully measured by a swing offset as described in Art. 6.12. The distance BZ is equal to the distance XY which, added to AX, gives the length of the long leg AY of the right triangle AYB. The distance YB is then measured, and AB and angle YAB are easily calculated.

Angle DAY has been measured from the previous course, AD; the addition of the angle YAB together with the known distance AB makes the traverse complete to the point B without any further fieldwork. If the transit is now moved to B with a view to carrying on the survey it will be found that, since A cannot be seen from B, there is no point on the line BA to use as a backsight. But any point such as E can be set on the line AB by making the offset $ME = BY\frac{AM}{AY}$. Another point can be set similarly on AB as a check on this backsight.

This random-line method is sometimes employed when AB is a boundary which is covered with shrubs. Although the view from *A* to *B* may not be obstructed, it may be so difficult to **measure** the line *AB* that its length can be more easily obtained by the use of the random line while the **angle** DAB may be measured directly at *A*. If it is desired to mark the line *AB* by several intermediate points these may be established by means of perpendicular offsets calculated as described above.

The random line is not necessarily a single straight line—it may be a traverse consisting of several lines, as shown in Fig. 6.28.1. If the last point (3) is in such a position that a good closing can be made on *B* (that is, accurate angles) then the distance and bearing of *AB* may be computed as described in Art. 4.34; but if the line 2–3 passes so close to *B* that it is not desirable to make 3–*B* a line of the traverse, then the connection may be made as shown in Fig. 6.26.1.

6.27. Obstacles on Line.—When an obstacle, such as a building or a small pond, lies on the transit line various methods are resorted to for prolonging the line through such obstructions; the most useful of these methods will be explained.

6.28. Offsetting Transit Line.—This method is illustrated by Fig. 6.28.1. It is desired to produce the line *AB* beyond the house. Point *B* is set on line and as near as is practicable to the house. The instrument is then set up at *B* and a right angle *ABF* laid off with the transit. *BF* is made any convenient distance which will bring the auxiliary line beyond the building. Similarly point *E* is set opposite point *A*, and sometimes a

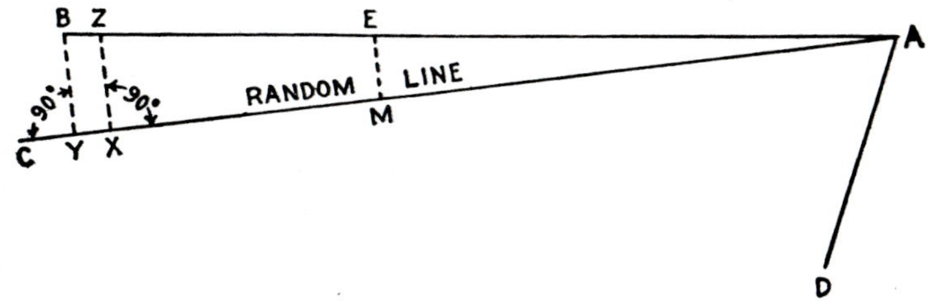

Fig. 6.26.1.

second point E' opposite A', points A and A' being **exactly** on the transit line. These points E and E' need not be set by means of a transit set up at A and at A' unless AE is quite long.

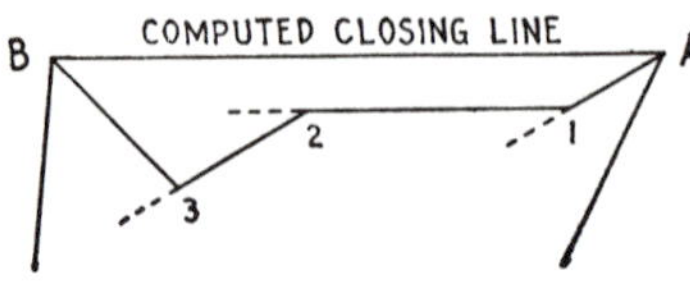

FIG. 6.28.1.

The instrument is then set up at F and backsighted on E, the sight is checked on E', the telescope inverted, and points G, H', and H set on line. Leaving the telescope inverted, another backsight is taken on E, and the process repeated as described in Art. 6.8. Then the transit is moved to point G, and a right angle turned off, and point C set on the right angle line, the distance GC being made equal to BF.

Then by setting up at C and sighting ahead on D ($DH = GC$), and checking on point D' ($D'H' = GC$), the transit line is again run forward in its original location. The distance FG is measured carefully and is used as the length for BC. Thus it is apparent why it is so necessary that the lines BF and GC shall be laid off at right angles by means of the transit. The other offsets AE, $A'E'$, DH, and $D'H'$ are not in any way connected with the measurement along the line; they simply define the direction of the line. So it is only necessary to show these distances as swing offsets for the transitman to sight on. It will be seen that offsets $A'E'$ and $D'H'$ are not absolutely necessary, but they serve as desirable checks on the work and in first-class surveying they should not be omitted. For obvious reasons the offsets AE and DH should be taken as far back from the obstacle as is practicable.

Should the house be in a hollow so that it is possible to see over it with the instrument at A, the point D, or a foresight of some sort (Art. 6.9) should be set on line beyond the house to be used as a foresight when the transit is set up again on the original line. The distance may be obtained by an offset line around the house or by slope measurements to the ridgepole. Sometimes it is possible to place exactly on line on the ridgepole of the house a nail or a larger wooden sight which gives an excellent backsight when extending the line on the other side of the building.

If the building has a flat roof it may not be out of the question to set a point on the roof exactly on line, move the instrument to this point on the roof, and prolong the line in the usual way. Under these conditions the transitman will have to be extremely careful in the use of his instrument as it will be set up on an insecure foundation. If he walks around the transit he will find that it affects the level bubbles and the position of the line of sight; it is therefore well for him to stand if possible in the same tracks while he backsights and foresights. Sometimes two men, one in front and one behind the transit, can carry on the work more accurately and conveniently. This method insures an accurate prolongation of the line, but the distance through the building must be measured by an offset method, unless it can be done by plumbing from the edge of the flat roof or by taking an inclined measurement and the vertical angle.

6.29. Short Transit Sights.—Sometimes the offset *BF* (Fig. 6.29.1) does not need to be more than 2 or 3 feet. The shorter this offset line can be made, and still clear the building, the better. But to lay off the short line *BF* will require a

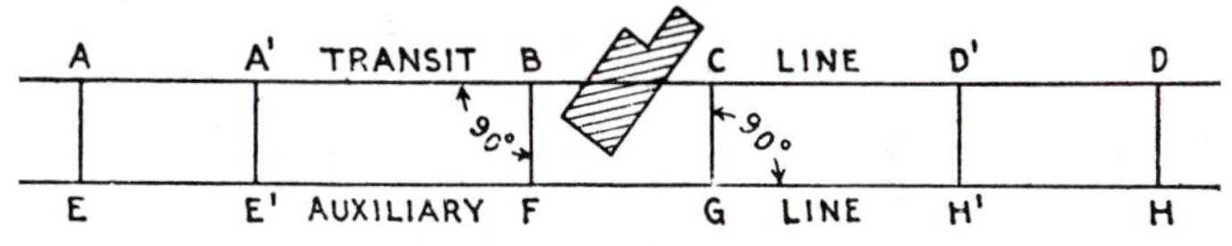

Fig. 6.29.1.

method somewhat different from any that has been explained heretofore. As the ordinary transit instrument cannot be focused on a point much less than about 5 ft. distant it is impossible to set point *F* directly. The method employed is to set a temporary point, by means of the transit, say 10 ft. distant, on which the transit can be focused, and on a line making 90° 00′ with the original transit line. From the transit point *B* to this auxiliary point a piece of string may be stretched and the point

F set at the required distance from B and directly under the string.

6.30. Bisection Method.—A method which is economical in fieldwork but not very accurate is the following. In Fig. 6.30.1 the instrument is set up at A, backsighted on the transit line, and equal angles turned off on each side of the transit line produced. Points B' and C' are carefully set on one of these lines and at convenient distances from A, and on the other line

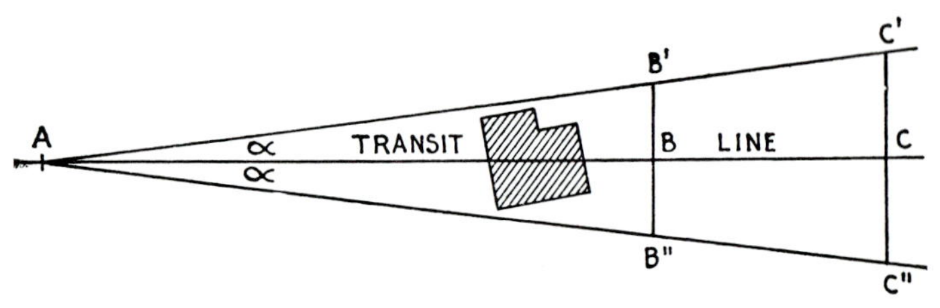

FIG. 6.30.1.

points B'' and C'' are set at the same distances from A. Then point B is placed midway between B' and B'', and similarly point C is set midway between C' and C''. The line BC is the prolongation of the transit line. Of course the distance $B'C'$ should be made as long as practicable. The two angles should be tested by the method of repetition and corrected if necessary.

The distance AB can be computed from the formula

$$AB' - AB = \frac{\overline{BB'}^2}{2AB} \text{ (approximately).}$$

6.31. Measuring Around a Small Obstacle.—In Fig. 6.31.1 the line AB runs through a tree, and points A, D, and B have been set on line. DE is made some convenient short distance judged by eye to be at right angles to the transit line. Then AE and EB are measured. The distance

$$AB = AE - \frac{\overline{DE}^2}{2AE} + EB - \frac{\overline{DE}^2}{2EB}.$$

When DE is taken as some whole number of feet the computation of the above is extremely simple.

This method of measuring around a small obstacle might be applied much more generally than it is at present if its accuracy and its simplicity were more fully realized by surveyors.

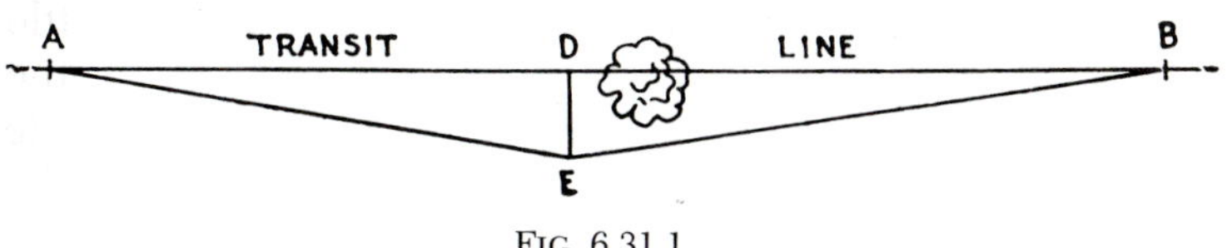

FIG. 6.31.1.

6.32. Equilateral Triangle Method.—While this method requires much less fieldwork than the offset method described above it is at the same time less accurate. Point *B* (Fig. 6.32.1) is set on the transit line as near the building as practicable but so that a line *BC* at 60° with the transit line can be run out. The instrument is set up at *B*, backsighted on *A*, and an angle of

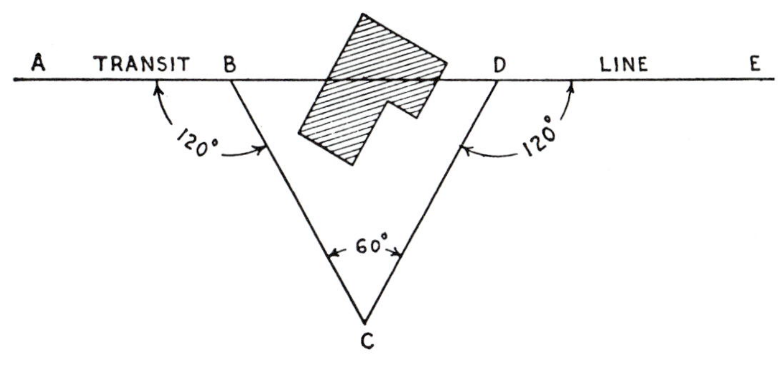

FIG. 6.32.1.

120° laid off; the line *BC* is made long enough so that when the instrument is set up at *C* and 60° is laid off from it, *CD* will fall outside the building. *BC* is measured and *CD* is made equal to *BC*. If the instrument is set up at *D* and angle *CDE* laid off equal to 120° the line *DE* is the continuation of the original transit line and the line *BD* = *BC*. This method is subject in three places to the errors incident to laying off angles and, when *BC* and *CD* are small, it has in two of its intermediate steps the disadvantages due to producing a short line. The results may be made more accurate by (1) repeating the angles, and (2) by running a crossline as explained at the end of Art. 5.17.

The triangle need not be exactly equilateral, although the latter is simpler to use. If the angle at *B* must be made a little greater or smaller than 120° on account of obstructions, the angles at *C* and *D* and the sides of the triangle may be made such that *D* is on *AB* prolonged and the length *BD* may be calculated. The triangle *BCD* must be solved by the formulas for oblique triangles, or as two right triangles by means of a perpendicular.

6.33. INACCESSIBLE DISTANCES.—If the obstruction is a pond, points on the farther side of it can be set and these should be used in producing the transit line. When the line can be produced across the obstacles the following methods may be used.

6.34. Inaccessible Distance by Right Triangle Method.—In Fig. 6.34.1 the line *AB* is made any convenient length and

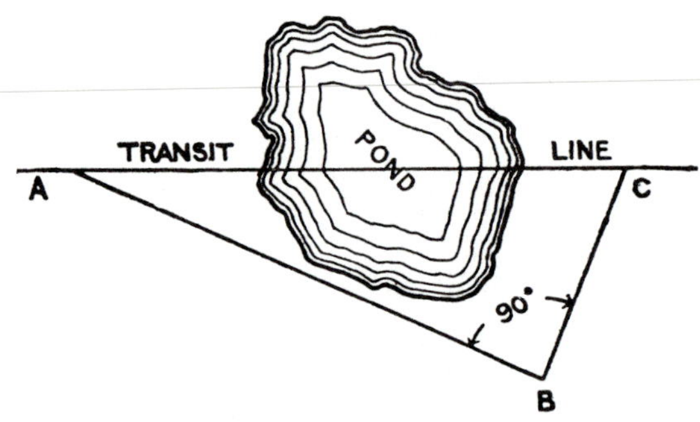

FIG. 6.34.1.

inclined at any convenient angle to the transit line. The line *BC* is laid off at 90° to *BA* (and verified by repetition); it is intersected with the transit line at *C* and the distance *BC* measured. *AC* is calculated from *AB* and cos *A* and checked by *BC* and sin A. Also the angle *ACB* can be measured; this will check the transit work.

6.35. Intersecting Two Transit Lines.—In many kinds of surveying work it is necessary to put in points at the intersection of two transit lines as at *C*, Fig. 6.34.1. It would be an easy matter to set the point if two transits could be used, one on each line, and the sight given simultaneously by the transitmen. As it is seldom practicable to use more than one transit in a surveying party the following method is much used.

An estimate is made by eye where the lines will cross each other and two temporary points not more than 10 ft. apart are set on one of the transit lines by means of the instrument, making sure that the second line will cross somewhere between the two temporary points. A string is then used to connect these two temporary points; the transit is set up on the other transit line and the point where the second line cuts the string is the intersection point. When the two transit lines cross at a very small angle, it is impossible to judge by eye within several feet where the lines will intersect, so a number of points must be set on the first line, because in practice the stretching line is seldom applicable for distances much over 15 ft. For short distances the plumb-line can be used as a stretching line.

6.36. Inaccessible Distance by Swing Offset Method.—If the distance across a pond or a river is not great the following method may be used. It has the advantage of requiring the minimum amount of fieldwork. With the instrument at *A* (Fig. 6.36.1) point *C* is set on the transit line on the farther side of the river. The instrument is then set up at *C* and the angle *ACB* measured (by repetition) between the transit line and a 100-ft. swing offset from point *A*.

A pencil is held vertically at the 100-ft. mark of the tape and while the zero-point is held firmly at *A* the tape, which is constantly kept **horizontal** and **taut,** is swung **slowly** in an arc *ab*. The transitman, using the tangent screw, can follow the pencil with the vertical cross-hair of the transit, stopping the cross-hair when the pencil is in its farthest position from *A*. Then as the tape is swung the second time he can check his setting; when this is determined the angle *ACB* is read. The distance *AC* then is easily calculated. It should be noted, however, that if AC is several times as long as AB the resulting error in AC may be so great as to prohibit the use of this method where very precise results are required. There is no reason why the swing offset could not be made at C while the instrument is set up at A if this is more convenient.

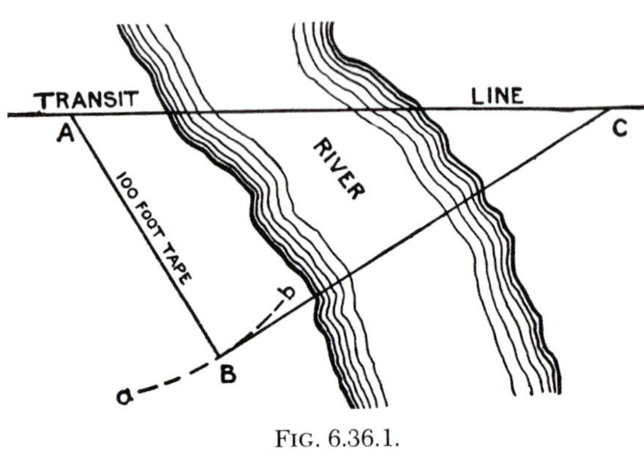

FIG. 6.36.1.

6.37. Inaccessible Distance by Tangent Offset Method.—In the method previously described the distance across the pond may be so great that 100 ft. will be too short a base to use, or point A may be situated on ground sloping upward towards B so that a swing offset cannot be made. In such cases the line AB (Fig. 6.37.1) can be laid off **exactly** at right angles to the transit line and of any convenient length. Then the angle ACB

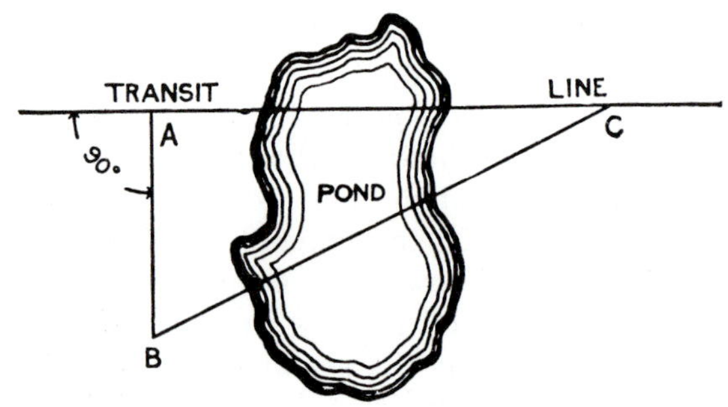

FIG. 6.37.1.

is measured (by repetition) and the line *AC* computed. By another set-up of the instrument the angle *B* can be measured as a check, and if the line *BC* does not cut across the pond its length can also be measured as a further check.

6.38. Inaccessible Distance by Oblique Triangle Method.—Often the shores of a stream are covered with trees so that none of the above methods is applicable. It may be possible, however, to measure a base-line *AB* (Fig. 6.38.1)

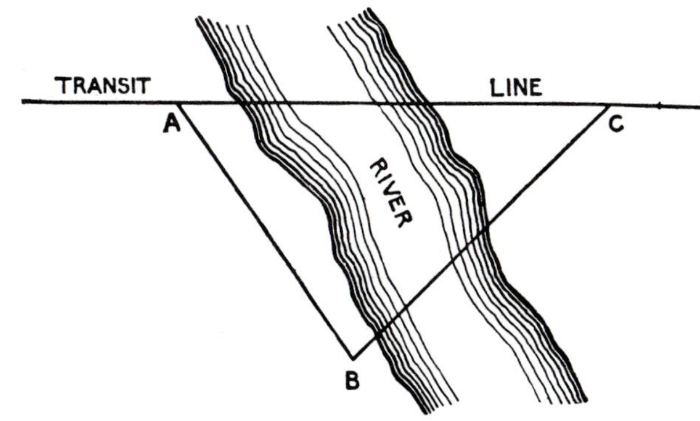

FIG. 6.38.1.

along the shore, to set the point *C* in line on the opposite side, and to measure the angles at *A* and *C*. The distance *AC* can then be computed. It will be well also to set up at *B* and measure the angle *B* as a check on the work. At the time when point *C* is set it is also good practice to set another point farther ahead on the line, to use as a foresight to check the transit line when the line is being prolonged from *C*.

6.39. To Obtain the Distance Between Two Inaccessible Points by Observation from Two Accessible Points.—In Fig. 6.39.1 the points *A* and *B* are inaccessible and it is desired to obtain the distance *AB* and the angle that *AB* makes with the transit line. From the point *D* the distance *DC* and the angles *BDA* and *ADC* are measured; similarly at *C* the angles *ACB* and *BCD* are measured. *AB* can then be calculated as follows:—in the triangle *CBD* compute *CB*; in triangle *ACD* compute *AC*; and in the triangle *ACB* calculate *AB*, the inaccessible distance. In the triangle *ACB*, angle *ABC* can be computed; this, together with the measured angle *BCD*, will give the difference in direction between *AB* and *CD*. It is not necessary that *DC* should have been measured as one straight line in the traverse; the traverse might have run as indicated by the dotted lines. If so, the distance *CD* and the necessary angles could be computed; the remainder of the computation would be the same as before.

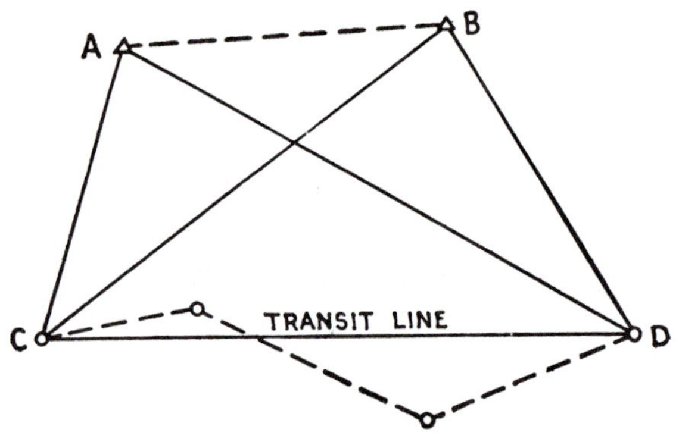

FIG. 6.39.1.

This problem occurs when the distance between two triangulation stations, *A* and *B*, and the azimuth of *AB* are desired and when it is inconvenient or impossible to measure the line *AB* or to occupy the points with the transit.

6.40. To Obtain the Inaccessible Distance Between Two Accessible Points by Observations on Two Inaccessible Points of Known Distance Apart.—In this problem (Fig. 6.39.1) *A* and *B* are the two accessible points and *C* and *D* are the two inaccessible points but the distance *DC* is known; the distance *AB* is required. With the transit at *A*, the angles *CAD* and *DAB* are measured; at *B* the angle *CBD* and *ABC* are measured. While it is easy to obtain *CD* in terms of *AB*, it is not easy to determine *AB* directly in terms of *CD*; it will be well therefore to use an indirect method. Assume *AB* as unity. Then by the same process as described in the preceding problem the length of *CD* can be found. This establishes a ratio between the lengths of the lines *AB* and *CD*, and since the actual length of *CD* is known the distance *AB* can be computed.

A problem of this sort would occur under the following circumstances. If the distance *CD* between two church spires were accurately known (from a triangulation system) and it is desired to use this line *CD* as a base-line for a survey, two points *A* and *B* could be assumed, and the distance between them and the azimuth of *AB* could be found by this method.

CHAPTER VII

TOPOGRAPHICAL SURVEYS

PRESENTATION OF TOPOGRAPHICAL DATA

7.1. Methods of Showing Relief.—The usual topographical map shows such physical features as water surfaces, limits of cultivation, fences, roads and buildings in their correct relative positions on a horizontal plane. These features of the map may be made as complete and accurate as desired by the use of horizontal distances and horizontal angles. When, however, it is desired to show on the map the vertical element, or relief of the ground, then some artificial system must be employed for this purpose. Writing the elevations on the plan does not convey to the eye a clear picture of the relief. The system used for showing the relief must fulfill two purposes. First, it should make the form of the surface apparent to the eye without great effort on the part of the user of the map; that is, the user of the map should be able to interpret the map as a model of the ground. Second, it should furnish also definite information regarding the elevations of points shown on the map.

7.2. Hill Shading.—Systems depending upon shadows cast by the elevated portions of the land have been used with striking effect. These make the general form of the surface apparent at once even to a person not experienced in the use of maps. Although useful for some purposes they tell nothing about actual elevations of the surface. Sometimes elevations are indicated by different tints or shades.

7.3. Hachures.—Systems depending upon the use of artificial shade lines, or *hachures*, have been used on old maps as illustrated in Fig. 7.3.1. These lines show surface form, but do not furnish information regarding heights. The short lines are drawn always in the direction of the steepest slope of the surface. The spacing of the lines, center to center, is always the same, but the weight (width) of the lines is increased with the steepness of the slope. The drawing of hachure lines is a slow and tedious process and does not provide the elevations required for engineering applications. The same type of symbol is used to some extent on modern maps to represent eroded banks, gravel pit sides and cuts and fills along a highway or railroad.

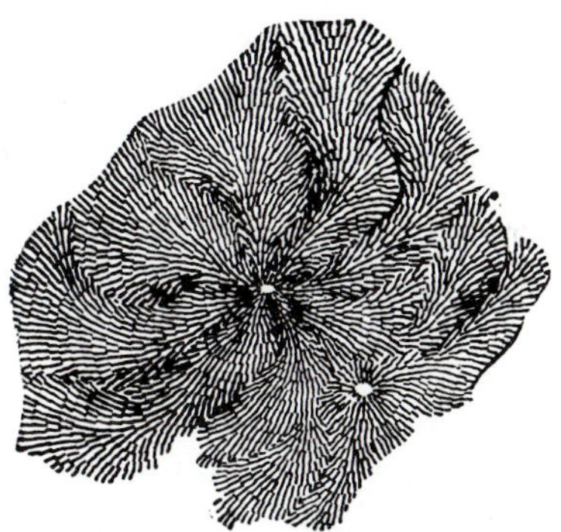

FIG. 7.3.1. HACHURE LINES.

The impression of relief may be had from color shading of various tints and intensities, different colors representing different ranges of elevation above sea level. A process has been developed for reproducing U. S. Geological Survey sheets on a raised base which actually conforms to the terrain. The vertical scale is sometimes exaggerated.

From stereo pairs of aerial photographs (Chap. XII), a three-dimensional image, or "model," can be created which will give an accurate reproduction of the terrain. By means of photogrammetric instruments distances and elevations can be obtained from this model and a map drawn showing planimetric and topographic features.

Air-brush shading to give a good pictorial rendition of the terrain is used, especially for mountainous areas, by the U. S. Geological surveys on their small-scale maps—along with contours. It is a modern improvement in the art of cartography. Another technique stemming from photogrammetry is the direct overprinting in light ink of an orthophotography developed from vertical stereophotos used for the mapping. Both these developments enhance the clarity of the presentation for the map-user and the latter especially adds greatly to the map's usefulness, especially in built-up areas.

7.4. Contour Lines.—The system now in general use for representing the form of the surface is that employing *contour lines*, or lines passing through points of equal elevation. The elevations of the contours are known definitely, so that the elevation of any point on the ground may be derived from the map. At the same time this system makes the form apparent to the eye, even to persons that have but little familiarity with maps. The location and relative heights of the hills is apparent at once; a close spacing of the contours indicates a steep slope, while a wide spacing means a flatter slope. With a little study one can learn to visualize the terrain from the contours, and to obtain accurate elevations from them. Fig. 7.4.1 shows the same mountain as is represented in Fig. 7.4.2, but represented by contour lines.

FIG. 7.4.1. CONTOUR LINES.

A contour line is the **intersection of a level surface with the surface of the ground.** A clearer conception of a contour line may be obtained from the following. Imagine a valley, or a depression in the surface of the ground, partly filled with water. The shore line of this body of water will be a contour line, because it is the intersection of a level surface with the surface of the ground. If the water stands at an elevation of 50 ft. the shore line is the 50-ft. contour. If the surface of the water

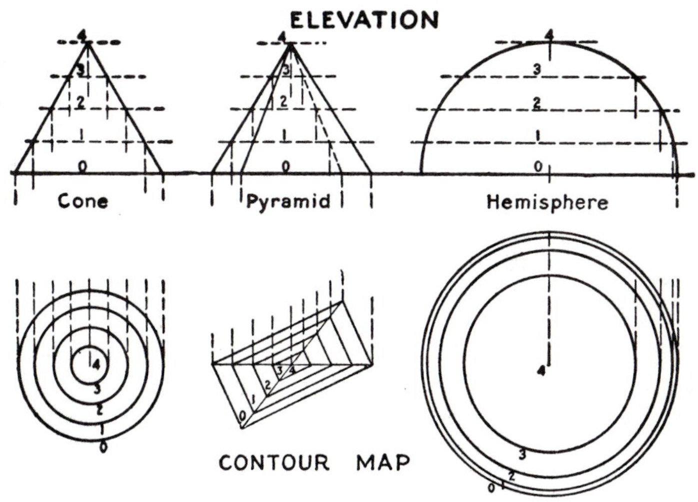

FIG. 7.4.2. CONTOUR MAPS OF SIMPLE SOLIDS.

were raised 5 ft. the new shore line would be the 55-ft. contour. Contour lines, if extended far enough, will therefore be closed curves, and all of the points on any one contour will have the same elevation above the datum. It is customary to take contours a whole number of feet or metres above the datum, spacing them in regard to height, so as to make the *contour intervals* equal, e.g., a contour may be taken at every 5 ft. or every 10 ft. of elevation. Since the contours are equidistant in a vertical direction their distance apart in a horizontal direction indicates the steepness of the slope.

Fig. 7.4.2 illustrates contour maps of simple geometric solids.

7.5. Characteristics of Contours.—The chief characteristics of contours are illustrated in Fig. 7.5.1, and may be summed up as follows.

FIG. 7.5.1. ILLUSTRATING CHARACTERISTICS OF CONTOURS.

1. All points on any one contour have the same elevation, as at *A*.

2. Every contour closes on itself, either within or beyond the limits of the map. If the contour does not close within the limits of the map it will run to the edge of the map, as at *B*.

3. A contour which closes within the limits of the map indicates either a summit or a depression. In depressions there will usually be found a pond or a lake; but where there is no

water the contours are usually marked in some way to indicate a depression, as at *C*.

4. Contours can never cross each other except where there is an overhanging cliff, and here there must be two intersections, as at *D*. Such cases occur seldom.

5. On a uniform slope contours are spaced equally, as at *E*.

6. On a plane surface they are straight and parallel to each other, as at *F*.

7. In crossing a valley the contours run up the valley on one side and, turning at the stream, run back on the other side, as at *G*. Since the contours are always at right angles to the lines of steepest slope they are at right angles to the thread of the stream at the point of crossing.

8. Contours cross the ridge lines (watersheds) at right angles, as at *H*.

9. In general the curve of the contour in a valley is convex toward the stream.

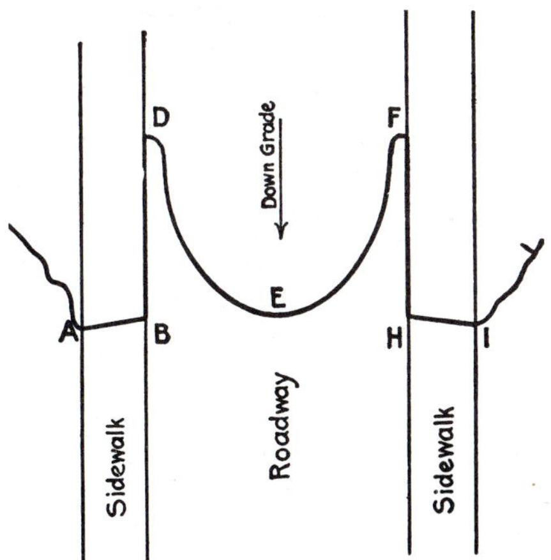

FIG. 7.5.2. CONTOUR CROSSING A STREET.

Fig. 7.5.2 shows a contour across an ordinary city street with sidewalks and curbstones, the street being located on a steep grade. In order to trace out the position of a contour it is necessary to keep in mind that it is a line all points on which are at the same elevation. It will be noticed that the contour from A to B crosses the sidewalk in a straight line but not perpendicular to the street line because the sidewalk is sloped toward the gutter. Turning at B it runs straight along the face of the curbstone until it strikes the gutter at D, and returns on the other side of the gutter along the surface of the road, the point E being where it swings around and travels back toward the other gutter. The other half of the street is similar. If the center of the road is at the same elevation as the top of the curb

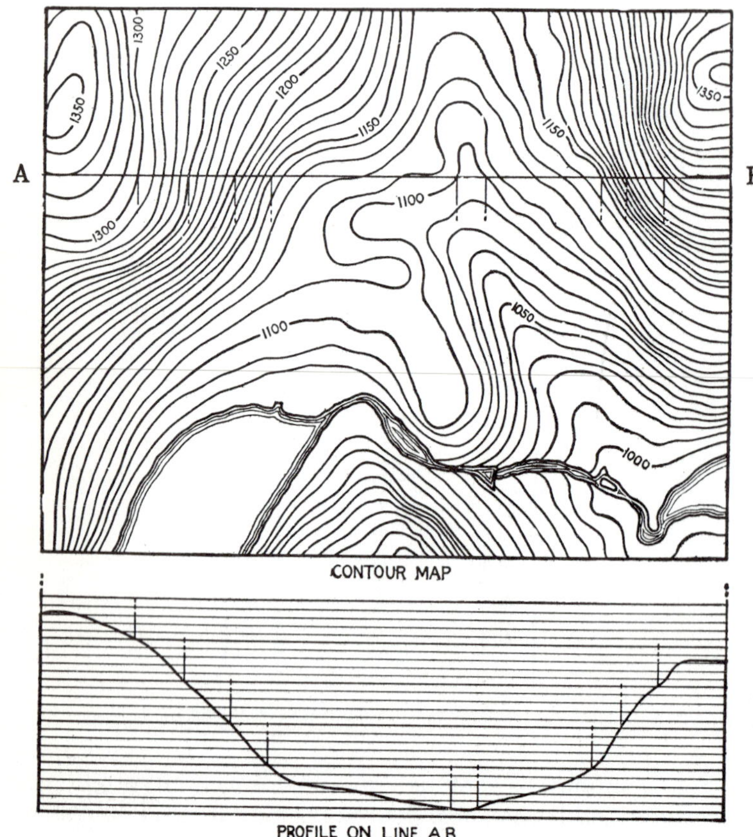

FIG. 7.6.1. PROFILE CONSTRUCTED FROM A CONTOUR MAP.

opposite, then *E* will be opposite *B*. This illustrates how contours run around valleys (gutters) and ridges (crown of street).

If the side of the street to the right (*HF*) were at a lower elevation than the left side then the contour at the point where it crosses the gutter, *F*, would be farther up the road from *E*, i.e., the contour would be unsymmetrical, *EF* being longer than *DE*.

7.6. Relation Between Contour Map and Profile.—If a line be drawn across a contour map the profile of the surface along that line may be constructed, since the points where the contours are cut by the line are points of known elevation and the horizontal distances between these points can be scaled or projected from the map. The profile shown in Fig. 7.6.1 is constructed by drawing first equidistant lines parallel to *AB*, their distance apart corresponding to the contour interval; from the points where *AB* cuts the contours, lines are projected to the lines of corresponding elevation on the profile. Conversely, if the profiles of a sufficient number of lines on the map are given it is possible to plot these lines on the map, mark the elevations of points on the lines, and from these points to sketch the contours as described in Art. 7.10.

7.7. Relation Between Contour Map and Side Elevation or Projection.—A photograph of a landscape taken horizontally is actually a perspective of the landscape, but if sufficiently distant it represents **approximately** a side elevation of the country. To construct such a projection from a contour map (Fig. 7.7.1), lines are drawn perpendicular to *AB*, the plane of projection, and tangent to the contours. These tangent points show the limits between the visible and invisible portions of the landscape, the observer being assumed to look in a direction perpendicular to the plane *AB*.

7.8. Drainage Areas.—The drainage area that supplies a stream or a pond is bounded by the *divide line*, which is a line drawn along the ridges surrounding a depression as indicated by the dotted line on Fig. 7.8.1. Since the perpendicular to the contour at any point is the direction of steepest slope the direction in which water will flow at any point can be determined at once by examining the contours. On the ridge there is

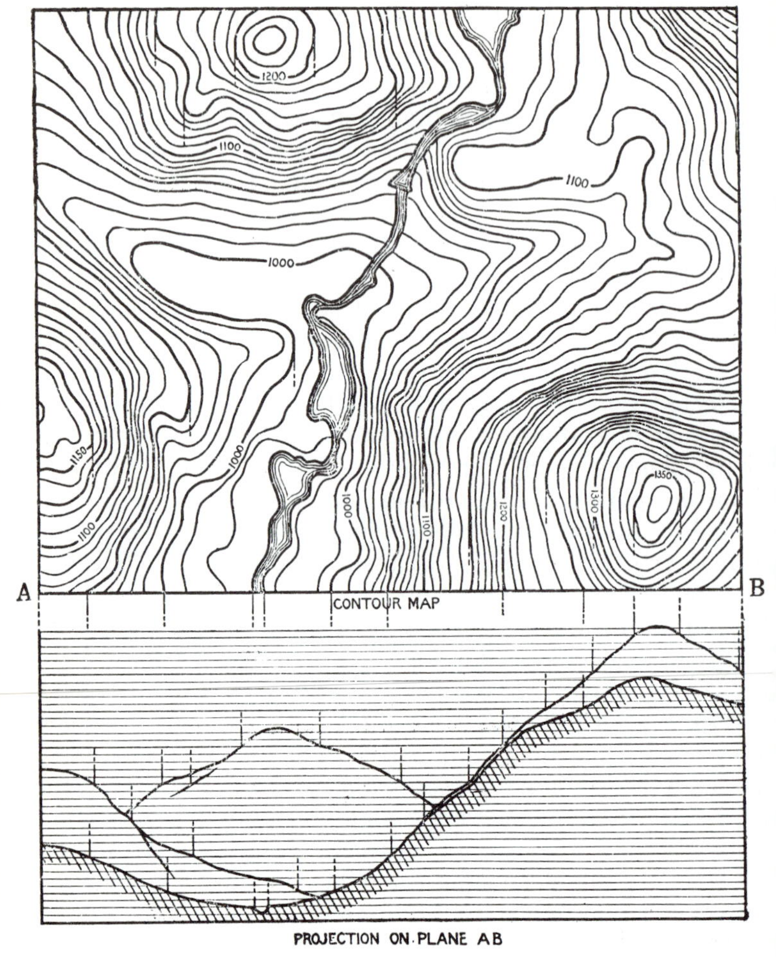

FIG. 7.7.1. SIDE ELEVATION CONSTRUCTED FROM A CONTOUR MAP.

a line (its summit) on one side of which water will flow down one of the slopes and on the other side of which it will flow down the other slope. This line is the divide line or *watershed line*.

If a dam were built as shown in Fig. 7.9, the elevation of the water back of the dam being 960 ft., the area actually flooded

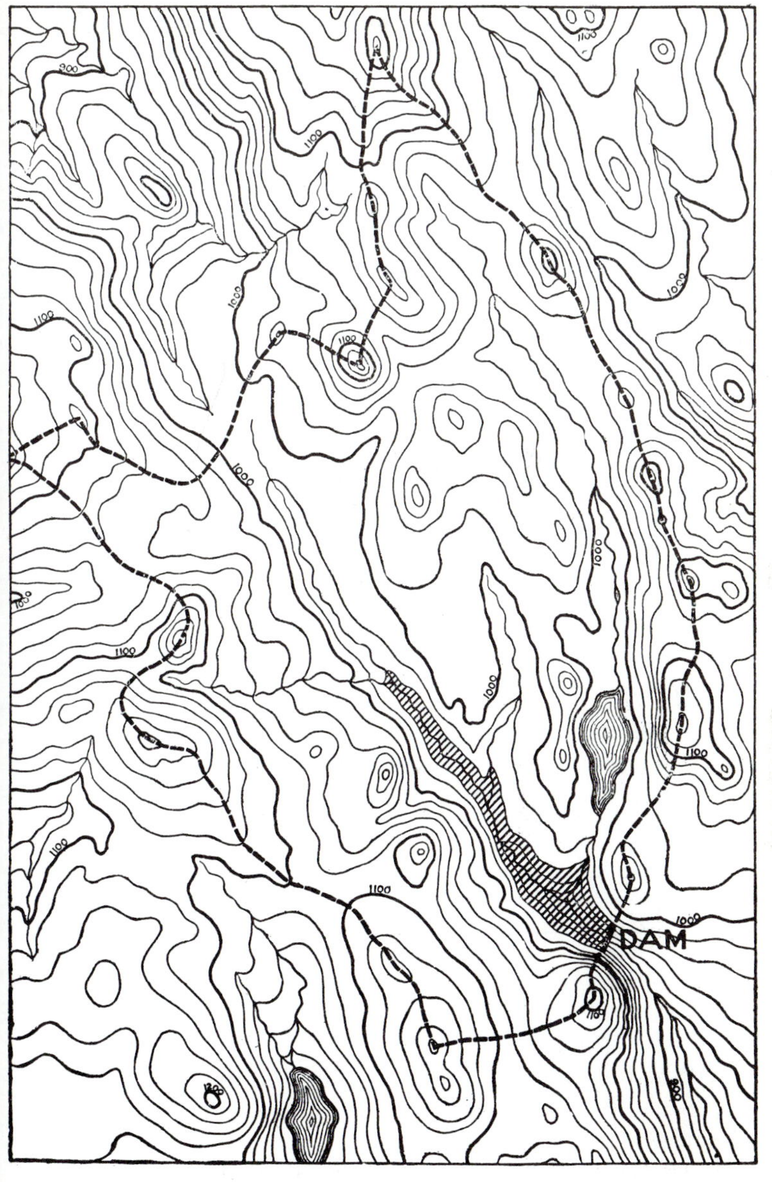

Fig. 7.8.1. Illustrating Flooded Area and Drainage Area.

by the water at this stage is the area included within the 960-ft. contour; this is indicated by the shaded section. The drainage area for the portion of the stream above the dam is the area included within the heavy dotted line, which follows the line of the divide.

7.9. Sketching Contours from Streams and Summits.—The present topography of some parts of the country is due almost entirely to erosion by streams. Consequently the position and fall of the streams give more information regarding the position of the contours than do any other topographic features. If a definite position of the contours is desired

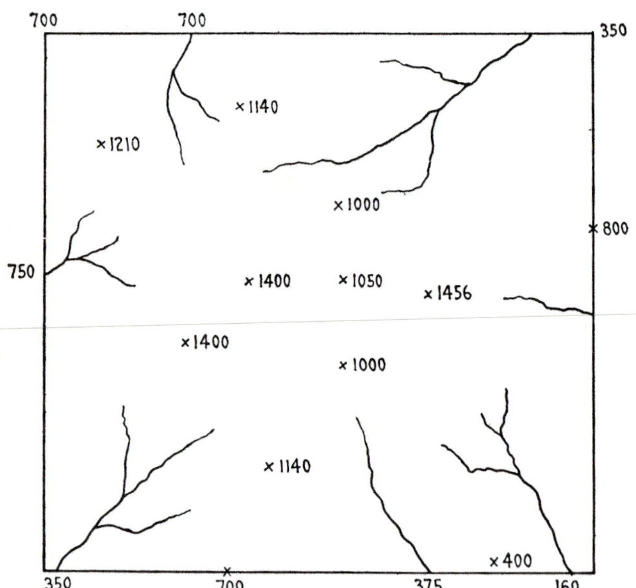

FIG. 7.9.1. MAP SHOWING THE LOCATION AND ELEVATION OF STREAMS AND SUMMITS.

it will be necessary to obtain the elevation of a few governing points on the ridges as well as the location and elevation of the streams, as shown in Fig. 7.9.1.

In sketching contours from these data it should be borne in mind that the contours cross the stream at right angles to its thread and that they curve around from the hill on either side so as to represent the valley of the stream. Since hills formed by erosion have rounded tops and relatively flat slopes near their bases the contours are farther apart at the top and bottom of the slope of an eroded hill than they are near the middle. A stream is usually steeper near its source than in the lower portion and therefore the contours are closer together near the source. This is true in general but the shape of the contours in any particular locality will depend upon the geological formation. Fig. 7.9.2 represents the same country as Fig. 7.9.1 but with the contours sketched on it, following out the general suggestions which have been mentioned.

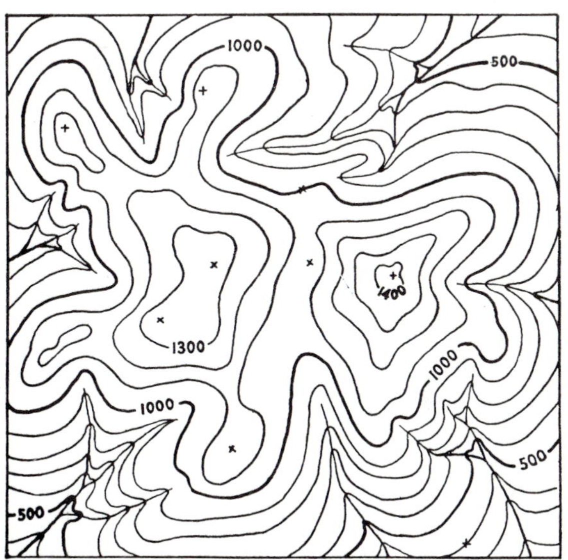

FIG. 7.9.2. CONTOURS SKETCHED FROM THE DATA GIVEN IN FIG. 7.9.1.

7.10. SKETCHING CONTOURS FROM KNOWN ELEVATIONS.— A portion of the country can be cross-sectioned as described in Art. 9.13, or profiles can be run on any desired lines as explained in Art. 9.9. From these known elevations contours can be sketched by interpolation. This is done usually by estimation and the principle involved is the same whether the elevations were obtained by cross-sectioning or by profiles.

Fig. 7.10.1 illustrates how contours can be sketched from cross-section notes. Elevations are taken at each grid corner and also wherever there is a change of slope, so that it may be

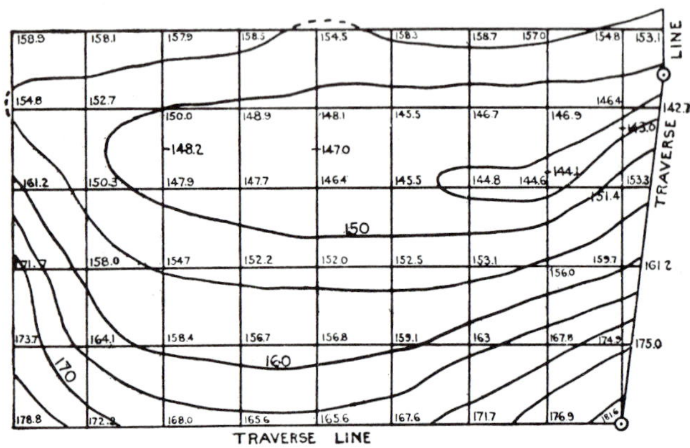

FIG. 7.10.1. CONTOURS SKETCHED FROM CROSS-SECTION NOTES.

assumed that the slope of the ground is uniform between any two adjacent points. Then by simple interpolation the contours may be sketched. This interpolation may be done by geometric construction, but for most topographic work it is accurate enough to interpolate by eye. In fact it is seldom that the ground is sufficiently regular to warrant interpolation by geometric construction.

7.11. MISTAKES IN SKETCHING CONTOURS.—Fig. 7.11.1 shows several examples of incorrect contour sketching. The locations of the streams on this map are assumed to be correct. The numbers on the illustrations refer to the tabulation of characteristics of contours given in Art. 7.5; these will assist in identifying the kind of error existing. The number 2, for instance, shows a contour line which stops short at the stream, whereas it evidently should cross the stream and continue on the other side of the valley. At 4 is a divided contour, which is an impossibility. The 7's all show places where the contours are evidently drawn without any regard to the position of the streams. For example, at the middle of the map the stream appears to be climbing up the valley, then over the saddle, and then flowing down the opposite valley.

FIG. 7.11.1. CONTOURS INCORRECTLY SKETCHED.

7.12. CONTOUR PROBLEMS.—There are many surveying problems involving earthwork which can be solved by use of a contour map. As a rule the smaller the contour interval, the more accurate will be the result of such computations. Contour studies occur in a variety of problems, so numerous that it would be difficult to attempt to cover the subject fully. Three typical problems, however, are illustrated and explained; and these contain the essential principles applicable to all contour studies.

7.13. EXAMPLE 1.—(Fig. 7.13.1). Given a contour map, the surface being represented by contours shown by full lines; a plane (extended indefinitely) is passed through the straight

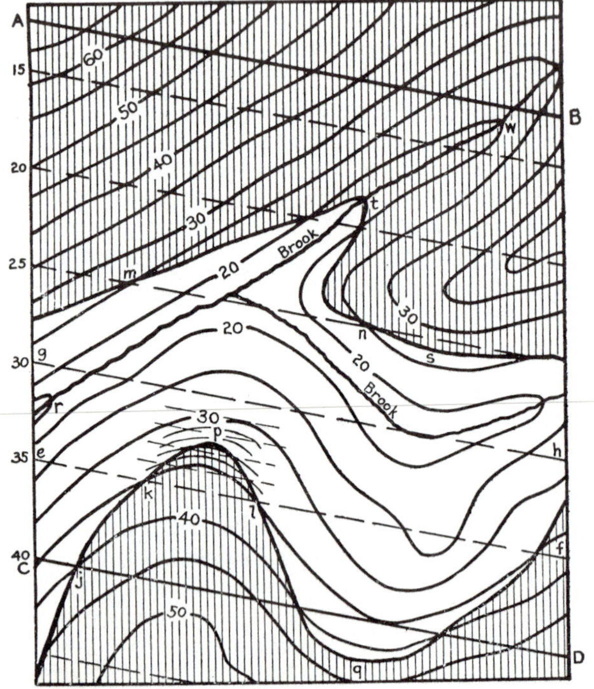

FIG. 7.13.1.

lines AB and CD, which are level and parallel, AB being at elevation 12.5 and CD being at elevation 40. It is required to find where this plane intersects the surface, and to shade the portion which is above the plane.

Since the proposed surface is a plane, contours on it will be parallel to AB and CD. The elevations of AB and CD being known, other contours, such as *ef* and *gh*, can be interpolated between *AB* and *CD*. Their interval is made 5 ft. the same as the contour interval for the original surface. Evidently the point where any of these parallel lines crosses a surface contour of the same elevation, as *j*, *k*, *l*, *m*, or *n*, is a point on the intersection of the plane with the surface. Joining these points gives the line of intersection of the plane with the original surface, which is indicated by the heavy full line on the figure. Such points as *q*, *s*, or *t* are determined by interpolation. Intermediate contours are drawn at one-foot intervals between the original surface contours; corresponding lines are interpolated between the straight contours that show the plane; in this way additional intersections are obtained, and the point *p* is determined. Again it will be seen that point *t*, with reference to the parallel straight contours, is at elevation 18.5; with reference to the original contours, it will be seen that *wt* is about three-tenths of *wr*, the distance between contours, and this makes the elevation of point *t* equal to 18.5.

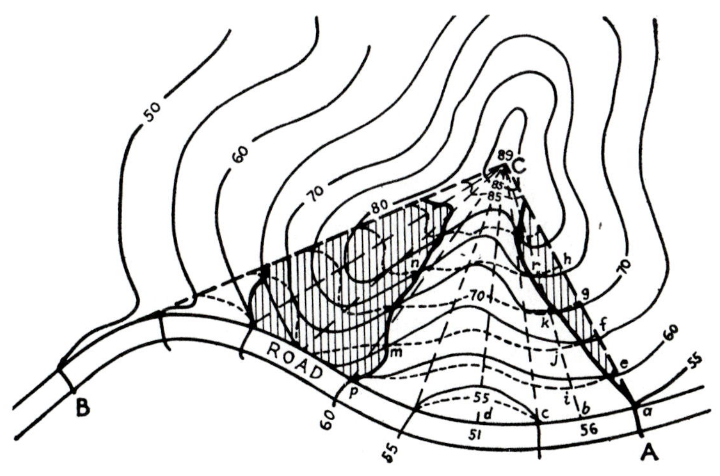

FIG. 7.13.2.

EXAMPLE 2.—Fig. 7.13.2 shows a road, and terrain on which the original contours are represented by full lines. It is desired that all of the road between *A* and *B* shall be visible from the ground at point C. Sketch on the map, and shade, the portions that will have to be cut down to fulfill this requirement.

The general method of solving this problem is to sketch a new set of contours on the map, which will represent a uniform slope from C to the nearer edge of the road. All above the surface represented by these new contours must be cut away.

First draw lines, such as Ca, Cb, and Cc, the points a, b, and c being points on the upper side of the road in such positions that it may be assumed that the slope is uniform from a to b and b to c. Along these lines interpolate points which will lie on the uniform slope from C to the road and also on the regular 5-ft. intervals which correspond to the contours. For example, along the line Ca from the summit which is at elevation 89 to the road at a which is at elevation 55, there is a drop of 34 ft., or a little less than 7 contour intervals. Points e, f, g, h, etc., are plotted so as to divide Ca as follows: ea, ef, etc., are each 5/34 of AC, and the upper division is 4/34 of AC. Similarly points i, j, k, etc., are plotted along the line Cb; but since the point b is at elevation 56, point i is plotted so that the distance ib is four-fifths of the other distances ij, jk, etc. When these points have been plotted on all of the necessary diagonal lines, the contours representing a uniform slope from C to the road are sketched on the map as shown by the dotted lines on the figure. The points, such as m, n, or r, where the new contours cut the old contours of equal elevation, are points of "no cut and no fill." A line connecting these points encloses portions of either cut or fill. The shaded portions of the figure, where the new contours are nearer C than the corresponding old ones, represent the portions where it will be necessary to excavate to the surface represented by the dotted contours. In the central portion of the figure, from point c to point p, the road can be seen without excavation. This problem embodies principles used in studies of military maps—to determine dead (not visible) areas.

EXAMPLE 3.—(Fig. 7.13.3). Given a contour map on which are shown the two side lines of a road, the contours being represented by full lines. The road is to be built on a 4% down grade starting at *A* at elevation 55. Scale 1 inch = 125 ft. Side slopes of road to be 1½ horizontal to 1 vertical. It is desired to sketch the new contours on the slopes of the road, to sketch on the map the top and foot of slopes, and to designate the portion in embankment and the portion in excavation.

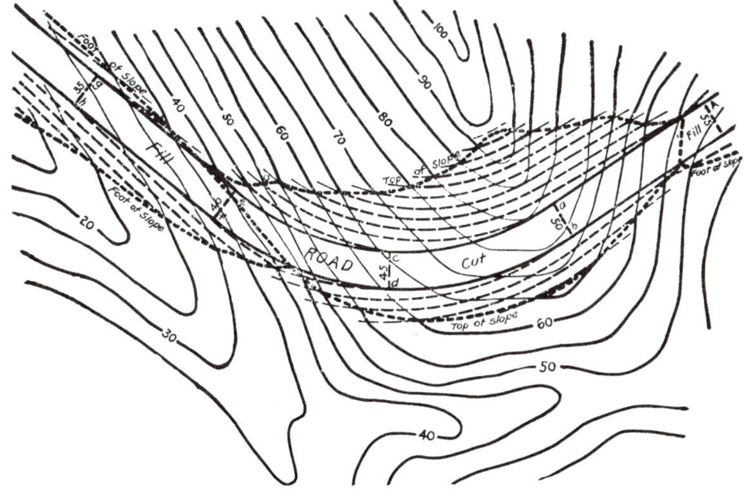

FIG. 7.13.3.

First, the new contours which are to cross the road are plotted at *ab*, *cd*, *ef*, *gh*. These will be 125 ft. apart, as a 4% grade falls 5 ft. in a distance of 125 ft. If for simplicity, the road is assumed to be level in cross-section, then these lines will cross the road at right angles as shown in the figure. From points *a* and *b*, on either edge of the road, the new contour lines will follow along the slope, e.g., the line *ao* represents the new 50-ft. contour. Where this contour *ao* passes point *c* it is just 5 ft. above the road. Since the slope of the cut is 1½ to 1, then the distance out from *c* must be 1½ × 5 = 7.5 ft.; opposite *e* it is 10 ft. above the road and similarly the distance out from *e* must be 15 ft. Where this new 50-ft. contour meets the old 50-ft. contour at *o*, is a point at the top of the slope. Similarly all of the new contour lines, which are represented on the figure by dash lines are plotted and their intersections with the corresponding contours of the original surface give points of "no cut" or "no fill," or top of slope (in excavated portions) and foot of slope (in embankment portions). These lines are shown in the figure by heavy dotted lines. Where this heavy dotted line crosses the road it marks a "no cut" and "no fill" line, i.e., the road bed cuts the surface of the ground.

7.14. Locating a Grade-Line.—In locating a highway, a railway or an aqueduct on a contour map a controlling factor of the location (apart from many other practical considerations) is the "grade line." In portions of the location the grade line ascends or descends at a uniform rate; and in some places may represent the maximum gradient permissible. A uniform grade-line may be drawn on the map by first computing the horizontal distance corresponding to the contour interval and the given gradient. If, for example, it is desired to lay out on the plan a line that will rise at a rate of grade of 5%, or 5 ft. in 100, the contour interval is 10 ft., and the scale of the map is 400 ft. to an inch, then the horizontal distance corresponding to a 10-ft. rise (one contour interval) is 200 ft., and this is represented by one-half inch on the map. If we set on the

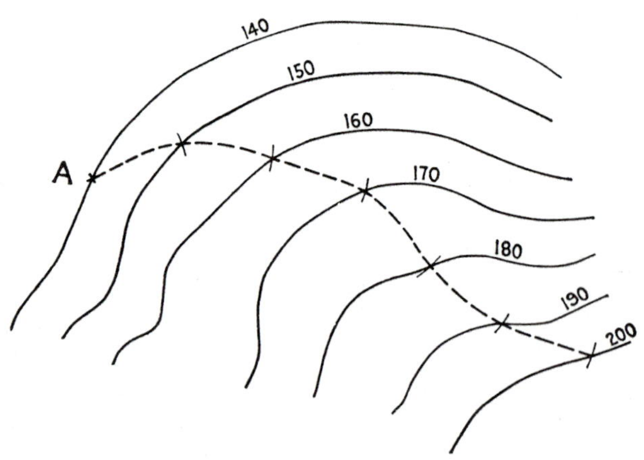

FIG. 7.14.1.

compass (or dividers) a radius of one-half inch and intersect the successive contours with short arcs beginning at some known point, as shown in Fig. 7.14.1, we obtain points on the contours which will lie on the 5% grade line. If the contours present irregularities it may be necessary to interpolate a contour half-way between and use a quarter-inch distance in finding the point on the interpolated contour.

7.15. Topographical Surveys.—This term may mean anything from a topographical survey for the purpose of locating a building, up to a survey for a topographic map of the whole country. If a large map is to be made showing contours of a large area, it is necessary first to execute a control survey, triangulation or traverse, according to circumstances, in order to obtain a coördinate frame for the whole survey. This also applies to vertical as well as to horizontal locations. In Chapter X is a brief description of how this is done in a city survey. The details of the topography may then be filled in by plane table, transit and stadia or by an aerial survey. The latter are commonly used to obtain topography for surveys of considerable extent or for large engineering projects. They are dependent upon an accurate ground survey (ground control) for establishing points of known position and elevation which can be identified in the photographs. Chapter XII gives a brief introduction to this field of surveying.

A topographic survey affords an opportunity to use a combination of several different survey methods. This is especially true of a large survey but may also be true of a smaller one. The subject of executing a control survey and filling in the topographic details by appropriate methods is taken up more fully in Vol. II.

For small, detached surveys, the methods used for obtaining contours will usually be simple and direct requiring only the transit, steel tape and level and stadia rods. The methods described in this chapter will usually be more practical and less expensive than an aerial survey producing the same degree of precision. Even with an aerial survey, some details not well defined on the photographs might have to be located or checked by ground surveying methods. This is particularly true of property-line corners, drainage structures and utility manholes.

7.16. LOCATION OF POINTS FROM THE TRANSIT LINE.—When the survey is being made by measuring angles with the transit and distances with the tape, such objects as fences, walls and buildings may be located as described in Chapter VI, but it will not be necessary to make the measurements with as great precision. Fig. 7.16.1 is a sample page of notes of a topographical survey where the transit and tape were used. On city plans, which are frequently drawn to a scale of 40 ft. to an inch, a fraction of a foot can be shown easily. On a topographic map the scale often is such that an error of a fraction of a foot becomes insignificant in the side measurements from the transit line, where such errors cannot accumulate. Sometimes it may be sufficient to obtain the distances by pacing, and the angles or directions by means of a pocket compass. Locations frequently may be checked by noting where range lines intersect the transit line. In making a series of measurements it is well to take each measurement with a little greater precision than is actually needed for plotting, in order to be sure that the accumulated error does not become too large.

In taking measurements the surveyor should constantly keep in mind **how the notes can be plotted**; often this will prevent the omission of necessary measurements. No matter whether an accurate or only a rough survey is desired **check measurements** should be taken on all important lines. An example of notes of a survey, using the stadia method, is shown in Fig. 7.31.3.

7.17. Locating Contours by Cross-Sections.—A very common as well as expensive method of locating contours is that of taking cross-sections. Elevations on the surface of the ground are usually taken at each cross-section corner to tenths of a foot. From these elevations the contours may be sketched by interpolating between these known elevations as explained in Art. 7.10. The accuracy may be increased materially by taking elevations at intermediate points when it appears that the ground does not slope uniformly between the grid points. The

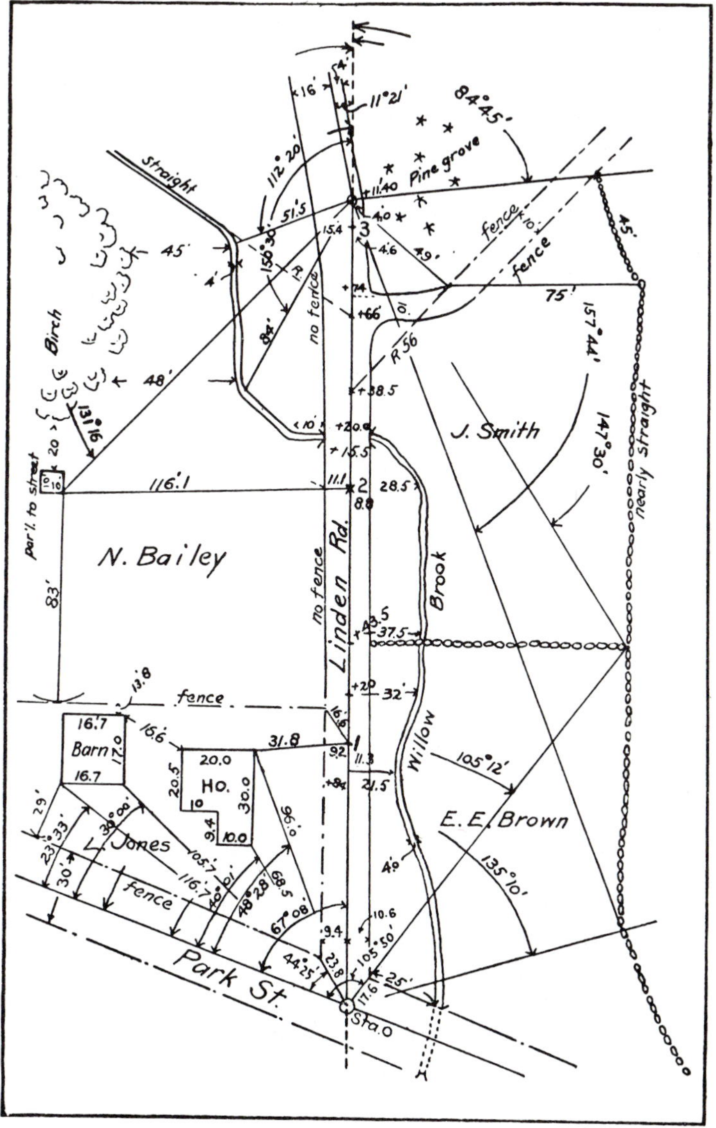

FIG. 7.16.1. FIELD-NOTES OF A PORTION OF A TOPOGRAPHICAL SURVEY WITH TRANSIT AND TAPE.

size of the squares used should depend upon the roughness of the surface and the contour interval employed.

7.18. Locating Contours by Profiles.—Where the ground is fairly smooth it is sufficient to take a few profiles on known lines, not necessarily at right angles to each other. These lines are stationed and elevations are taken at every full station and also at the points of marked change in slope. From these data the contours are sketched on the map by interpolation as described in Art. 7.10. This sketching should be done in the field while the terrain is before the eyes of the topographer.

7.19. Locating Points on the Contours.—If the contour interval is small, say one or two feet, and the topography is to be determined with considerable accuracy, it is advisable to find, in the field, points actually on the contours and thereby avoid the errors of interpolation. The rodman moves up or down the slope until the rod-reading indicates that the foot of the rod is on a contour. The position of the rod may then be located by an angle and a distance from some known line, the distance being taken with a tape or by stadia.

7.20. Locating Contours by the Hand Level.—Contours may be located more rapidly but less exactly by means of the hand level. The work is done by making profiles of lines whose positions on the map are known. A point on a contour is found in the following manner.

The first step is to measure to the nearest tenth of a foot the distance from the ground to the eye of the leveler, which may be, say, 5.4 ft. If the B.M. is at elevation 143.43 and it is desired to locate a point on the 140-ft. contour, the rodman holds the rod (or a tape) on the B.M. while the leveler attempts to place himself on the 140-ft. contour. When he is on the 140-ft. contour the elevation of his eye (H.I.) is 145.4 and the rod-reading at the B.M. must be $145.4 - 143.43 = 1.97$, or 2.0 to the nearest tenth of a foot. The leveler therefore travels along the line on which the point is to be located until he reads 1.97 on the rod. His feet are then on the 140-ft. contour, the position of which is located from some known point on the line. Sometimes this is done by measurement and sometimes by pacing.

A point on the 145-ft. contour could have been located first by applying the same principle, but if the 140-ft. contour is established it is very easy to locate a point on the 145-ft. contour as follows. The distance from the leveler's feet to his eye being 5.4 ft., if he stands on the 140-ft. contour and reads 0.4 ft. on the rod, the bottom of the rod must be on the 145-ft. contour. By trial then the point is found where the rod reads 0.4 ft.* Then the leveler walks up the hill and, standing on the point just found, places the rodman on the next higher contour by the same process.

In working down the hill to locate the 135-ft. contour, if the leveler is standing on the 140-ft. contour, the rod will be on the 135-ft. contour when it reads 10.4 ft. Or, when the 140-ft. contour has been found by the leveler the rodman comes forward and holds the rod on this spot and the leveler backs down the hill until he reads 0.4 ft. on the rod; he is then standing on the 135-ft. contour. Some prefer to cut a stick just 5 ft. long and hold the hand level on the top of it in taking sights.

The points thus found at regular contour elevations are then plotted on the corresponding line and the contours sketched by joining points of equal elevation by lines which represent as nearly as possible all the irregularities of the ground between such points. Where practicable it is well always to sketch the contours in the field rather than in the office.

7.21. Locating Contours by Cruising with Compass and Clinometer.—In dense woods or underbrush the cruising and sketching method herein described will often give results sufficiently close for the type of survey being made.

Traverse lines are first run through clearings in the woods at intervals of 1000 to 3000 ft., preferably parallel to each other, and profiles made of these lines by direct leveling or stadia. These traverse lines are plotted on field sheets and used as control for filling in topographic details between them, as follows.

---

*For very rough work sometimes the rod is not used, the leveler simply estimating where the rod-reading will come on the rodman's body and placing him so that his feet will be on the proper contour.

Starting on one of the control lines, sketch the contours by eye on both sides of the cleared lines for a short distance into the woods. The topographer then cruises the country between these control lines by running range lines by hand compass (usually at right angles to the control lines) at, say 100 to 300 ft. intervals, starting from one traverse line and with the elevation determined by leveling, and running at right angles through the woods to the next line, sketching contours for 100 ft. or more on either side of the range line, using the Locke level or clinometer (Fig. 4.11.1) for determining elevations. At the next traverse line he checks his location and ground elevation. A small sketching board is convenient for filling in the contours. Then he returns on the next range to the first control line, obtaining as he does so all of the topography in the belt which he has covered. These data are then transcribed directly onto a plan which has been left on the first control line. The topographer then proceeds to the next pair of ranges running, say south on one and back north on the other, checking out at each time on elevation and distance and on his position east and west on the control line, as he reaches the clearing where the traverse line was run. At the end of each cruised line he adjusts any errors in elevations and distance which may appear, provided they are not so great as to indicate a serious mistake. This method gives an independent control through the traverse on all of the topography. It provides a check on all of the cruising work and it gives a fairly accurately located contour even in heavily wooded districts. The transcription of the contours drawn on the sketching board onto the larger plan in the field discloses any apparent error; this should be corrected immediately in the field.

7.22. Location of Streams and Shore-Line.—A transit traverse run approximately parallel to a stream or a shore-line will form a base from which the shore-line can be located by right angle offset lines measured to bends and to other selected points along the shore-line. Sometimes a boulder or a tree can be identified at a bend of the shore-line and angles can be taken to it from two different transit points thus locating the bend by the intersection of the two lines. If these located points are plotted on a small sketching board the topographer can then sketch by eye the details of the shore-line between these located points and thereby produce a faithful map.

Shore-lines of ponds can be located economically from traverse lines run on the ice in winter. From the pond side of the shore its shape and characteristic changes can be seen more clearly.

The shore line required may be the high-water mark or the low-water mark, the average shore-line or the shore on the day the survey is made. If the shores are flat slopes the positions of these different lines may be several feet apart. The surveyor should ascertain before making the survey which shore-line is required; and his field-notes and the plan should show clearly which line is being located.

Where the high-water mark is required and the survey is made during a period of low water, it is often difficult to distinguish the high-water mark, particularly on marshy shores. Usually it is possible, however, to determine the elevation of high water from stains on some wharf or ledge or boulders lying in the water, and then to trace out the contour corresponding to that elevation on those portions of the shore-line where the high-water mark is indistinct, as described in Art. 7.19.

7.23. Use of Contours of Small Interval.—For portraying the shape of the ground surface contour intervals of 10 ft., 5 ft. or 2 ft. are commonly used depending upon the scale of the map and the roughness of the terrain. Smaller contour intervals of 1.0 ft., 0.5 ft. or 0.1 ft. are useful for representing more precisely the surface of proposed engineering works, such as airports, highway interchanges and complicated street intersections. By this means the overall shape of the layout can be visualized to an extent impossible from an inspection of profiles and cross-sections. Interpolation between contours will provide elevations with sufficient accuracy to set grades for landscaping. When an 0.1 ft. interval is used, pavement grades may be interpolated to 0.01 ft.

7.24. Topographic Surveys with Stadia.—Although the previously described methods are used for smaller projects, the main method—aside from photogrammetry—for topographical surveys is the stadia method. Its principle is described in Chapter IV.

7.25. Fieldwork.—In surveying by the stadia method points are located by means of (1) the azimuth, (2) the angle of elevation or depression, and (3) the distance. When the survey does not involve the determination of elevations the vertical angles are merely read close enough for computing the horizontal correction to the distances read. But when a topographical survey is made, the determination of elevations becomes necessary and all vertical angles must be read closer, usually to the nearest minute. The azimuths and distances are read with a precision that is consistent with the accuracy required in the final result. The azimuths are usually referred to 0° on any traverse line running through the set-up point, which must be recorded in the field notes.

Traverse lines which form the control of the survey are run out either by means of the transit and tape or by stadia, according to the accuracy demanded. Where a number of transit points are to be distributed over a large area, a system of triangulation may be employed in which a base and a few additional lines are measured with the tape, the horizontal angles are measured by repetition with the transit, and the lengths of all the lines are computed; the additional measured lines will serve as checks on the accuracy of the work. Where the survey is controlled by such a triangulation or by tape traverses the stadia work is confined to filling in details. Where no great precision is required the main traverses may be measured by stadia alone; but the distances should be read on both forward and backsights.

From the traverse line as a control, points taken for the purpose of locating details are determined by angles and stadia distances. These observations are commonly called "side shots." The precision with which these measurements are taken need not be so great as that of the traverse measurements, because any error in the measurements will affect but a single point, whereas an error in the traverse line will be carried through the remainder of the traverse. The side shots are usually numbered consecutively in the note-book. When taking side shots it is customary not to clamp the horizontal plates because the angles are read only to the nearest 5′, which is close enough for plotting.

In locating points as described above the readings are taken most conveniently in the following order. The vertical hair is set on the rod and the upper plate clamped; the distance is read by setting one stadia hair on a whole foot-mark and reading the position of the other stadia hair; finally the middle horizontal hair is set on that point on the rod to which the vertical angle is to be taken. After making this setting the transitman signals the rodman to proceed to the next point; in the meantime he reads the azimuth and the vertical angle.

7.26. Azimuths.—Azimuths are counted from the north point through the east up to 360° in accordance with geodetic practice. If the true azimuth of any line is known, all the azimuths of the survey may be referred to the true meridian. If the direction of the true meridian is not known an initial azimuth may be taken from the magnetic bearing of some line and all azimuths referred to the magnetic meridian. The latter method has the advantage that all azimuths may be checked directly by reading the magnetic bearings. If the magnetic meridian cannot be used, any direction may be arbitrarily assumed as a meridian and all azimuths referred to this direction. If a transit and tape traverse has been run previously it is convenient to assume one of the transit lines as the 0° line of azimuths, a new traverse line being taken as a reference line for the azimuths at each set-up if desired.

The stadia method often is used to supplement the data obtained by transit and tape, and for this purpose it is not considered important to orient the circle on the true meridian at each set-up of the instrument. At each set-up the horizontal circle is set at 0°, the telescope is sighted at the preceding station, and all azimuths are read (clockwise) from this backsight.

If azimuths are measured from the meridian the direction of

the meridian must be carried forward from one transit point to the next; there are two methods in common use, each of which has its advantages. Suppose that the work at point A has been completed, all of the azimuth angles about A having been referred to the meridian which was chosen as the 0° direction, and that the transit is to be moved to a new station B. Before leaving A the transit point B is located from A by its distance and azimuth. The transit is then set up at B and the azimuth of any line (BC) is determined in either of the following ways.

(1) Backsight on A with the telescope inverted, the horizontal vernier remaining at the same reading it had at A (the azimuth of line AB); clamp the lower plate, turn the telescope into its direct position, and, loosening the upper plate, turn toward C. The vernier will then read the azimuth of BC referred to the same meridian as the azimuth of AB. It is evident that this method does not eliminate any error that may exist in the line of collimation, so that the error in the azimuth will tend to accumulate on a long traverse. The advantage of this method is the rapidity with which the instrument can be oriented.

(2) Add 180 degrees to the azimuth of AB, set this off on the plate, and sight on A with the telescope in the direct position. Clamp the lower clamp, loosen the upper clamp and sight the telescope toward C; the angle read will be the azimuth of BC. The disadvantages of this method as compared with the former are that time is consumed in setting the circle at each new set-up of the instrument.

7.27. Vertical Angles.—When vertical angles are to be taken the middle horizontal cross-hair is sighted at a point on the rod whose distance above the foot of the rod is equal to the distance from the center of the transit to the ground (or the stake) beneath. This distance is known as the *height of instrument* (H.I.); it is not the same as the H.I. used in ordinary leveling; the latter is the height of the instrument above the datum plane. If the cross-hair is sighted at this H.I. point on the rod, it is evident that the line of sight is parallel to the line from the transit point to the foot of the rod, and also that the difference in elevation between the center of the instrument and the H.I. point on the rod is the same as the difference in elevation between the point under the transit and the foot of rod. It is common practice to fasten a wide, heavy rubber band or a strip of red cloth on the rod so that it can be set on the new H.I. point at each set-up.

7.28. Distances.—The distance is read by setting one of the stadia hairs on a whole foot-mark and counting the feet and tenths between the stadia hairs, the hundredths of a foot being estimated. If a Philadelphia leveling rod is used and the distances are short, the hundredths of a foot may be read directly. **Great care should be taken not to mistake the middle horizontal cross-hair for one of the stadia hairs.** This mistake is liable to occur when the telescope is of high power, because the stadia hairs appear to be far apart in the field of view and consequently the eye does not see all three hairs at once. In counting the number of feet in the rod interval between the stadia hairs great care should be taken to obtain this interval correctly. It can be checked by reading the interval between the middle hair and a stadia hair and observing if this is nearly one half of the whole interval. (Fig. 3.30.1)

In reading the distance it is customary to set the lower stadia hair on that foot-mark which will bring the middle cross-hair in the vicinity of the H.I. In finding the horizontal correction to apply to the distance read it is customary to use the vertical angle that was read when the middle cross-hair was on the H.I. Theoretically a slight error is thus introduced, because when the distance is read with a stadia hair on a whole foot-mark the inclination of the line of sight is not the same as the vertical angle read when the middle cross-hair is set on the H.I. point. The distance and the angle introduced into the formula do not correspond. The middle hair, however, need never be more than half a foot above or below the H.I. when the distance is read, and it is easy to show that the consequent error in horizontal distance is negligible.

Whenever a portion of the rod is obscured, by leaves for instance, or when the distance is so great that the two stadia hairs do not fall on the rod at the same time, an approximate value of the distance may be obtained by reading first the interval between the upper and middle hairs and then the interval between the middle and lower hairs, and taking the sum of the two readings. If the two spaces are found to be exactly equal it will be sufficient thereafter to take one reading and double it, but it should never be assumed that the two are equal until they are found to be so by test.

It is of great importance that the rod be held plumb when the distance is being read, as any inclination of the rod evidently will introduce an error into the observed distance. This error becomes greater as the inclination of the line of sight increases. For example, if a 10-ft. stadia rod is held so that the top is 0.5 ft. farther from the transit than when it is vertical (inclination of 2° 52′) then for a distance of 700 ft. and vertical angle of 5° the resulting error will be 0.04 ft. on the rod or 4 ft of distance. For 15° the error is 0.11 ft. on the rod or 11 ft. of distance. To hold a 10-ft. rod more nearly vertical than this requires great care. In some classes of stadia work it is desirable and even necessary to plumb the rod by means of a rod level whenever highly inclined sights are taken. (Art. 4.17.)

7.29. STADIA TRAVERSES.—In a stadia traverse the instrument is set at the first station and the telescope set on the meridian (or reference line) with the vernier reading 0°. The position of the second station is located by reading the distance, azimuth, and vertical angle. In determining the azimuth of the line it is well for the rodman to show the narrow edge of his rod as a foresight, so that the transitman can make a more exact setting of the vertical cross-hair. The transit is then moved to the second station and placed in position by backsighting on the first point as explained in Art. 7.26; at the same time the vertical angle and the distance are again read thus checking the distance between the two points and also the difference in elevation if stadia levels are also being taken. In reading the distances both ways there is also an opportunity to guard against an inaccurate reading due to poor illumination of the rod. By sighting both directions with the telescope erect the index error of the vertical circle is eliminated; th process is similar in principle to the peg method of testing a level. (See Art. 4.32.)

7.30. Checks on the Traverse.—In running a closed stadia traverse the azimuths may be checked by redetermining the azimuth of the first line from the last and noting whether this value checks the azimuth of the first line as determined at the beginning of the traverse. If these differ by less than 5 minutes of arc the result will be sufficiently accurate for most topographical purposes. The azimuths may be checked roughly at any point by reading the magnetic bearings of the lines. Where there are triangulation points connected with the survey the known azimuths of these triangle sides will furnish a complete check.

In running stadia traverses for several miles where no other check on the azimuths can be obtained it is advisable to make observations on the sun for true azimuth. (Art. 8.9.) Such observations can be taken quickly and if made in the morning or afternoon, when the sun is not very near the meridian, will give the azimuth within about one minute of arc, which is as accurate as is required for this purpose. The convergence of the meridians, if appreciable, should be allowed for. (See Art. 5.57 and Table 5.57.2.)

7.31. Stadia Notes.—To the beginner the taking of good stadia notes presents great difficulties. The general instructions with regard to transit notes apply equally well to notes for stadia surveys. (Art. 5.20.) A large amount of sketching or description of details is required in order to convey sufficient information to enable the draftsman, who may not be familiar with the locality, to plot the results. Furthermore it is necessary to locate a large number of points in order to be sure of sufficient data to sketch the details correctly. If the map can be plotted soon after the fieldwork is done and by the person who made the survey, and especially if the map can be taken afterward into the field and the sketching completed there, then much greater accuracy in regard to details can be secured; this field sketching on the map is important and should be done whenever it is practicable.

The notes for the survey of a lot of land are shown in Fig.

Stadia Survey of Pasture Land of L. K. Miller — Jan. 15, 1957 — King ⊼ Stone

⊼ at	Sta.	Dist.*	Az. ∠ 0° on Mag. S	Obs. Bearing	Vert. ∠	Remarks
A	E B	717 642	211° 21′	N. 31° 15′ E.	−2° 15′ ——	
B	A C	642 786	272° 47′	S 87° 15′ E	—— ——	
C	B D	784 971	359° 03′	S. 1° 00′ E	—— +4° 40′	Cedar Post
D	C E	973 897	76° 05′	S. 77° 00′ W	−4° 30′ ——	Stake and Stones
E	D A	895 718	157° 34′	N. 22° 15′ W	—— +2° 00′	Stake and Stones
A			211° 20′ "check			
A	1	467	199° 00′			
A	2	491	199° 50′			

* F + c = 1.2 ft.
(Not included)

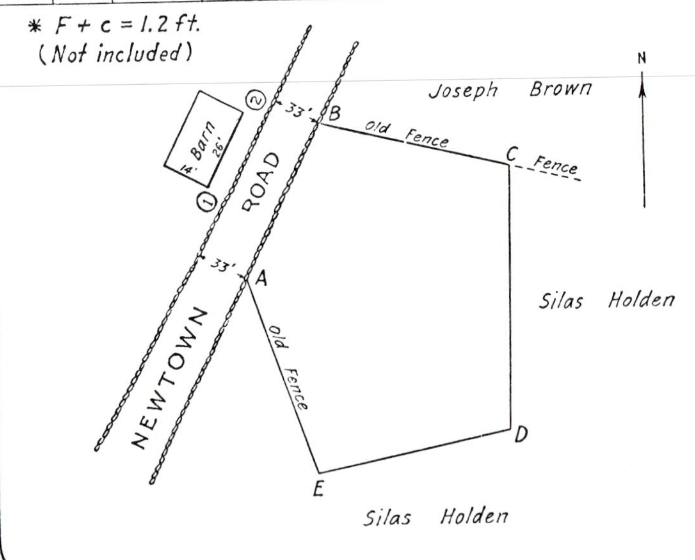

FIG. 7.31.1. STADIA SURVEY OF PASTURE LOT.

7.31.1. The notes given at the top in tabular form show in the first column the station at which the instrument was set up; the second column indicates the station to which it was sighted; the distances read are in the third column. The azimuths are read to the nearest minute because they are to be used for purposes of calculation; the observed bearings are taken as a check upon the azimuths. The vertical angles are taken only on those lines where the horizontal correction would change the distances as much as a foot, and they are read only to the nearest quarter degree or half degree, because for the distances and the vertical angles in these notes this will give the horizontal correction to the nearest half-foot. Evidently there is no necessity for indicating whether the angles were plus or minus. The traverse was measured in a right-hand direction around the field beginning at *A*, and finally the instrument was set up at *A* again, backsighted on *E*, and the azimuth of the line *AB* redetermined from the backsight on *E*. The notes indicate that an error of one minute was found in the azimuths. The notes also show that two shots, Nos. 1 and 2, were taken to the corners of a barn merely for the purpose of obtaining data for plotting it on the map. The figure in the lower part of the page is a sketch of the property, on which are lettered the names of the abutting owners, and the physical features which define the boundaries.

It will be observed that the distances on the traverse were measured both forward and backward, two independent readings being taken. Before computing the area of this field the average of these two readings should be taken as the correct distance and to this distance should be added the constant of the instrument (1 ft.). The horizontal correction should be applied to all inclined distances if the correction amounts to more than half a foot.

Fig. 7.31.2 & 7.31.3 is a double page of stadia notes of a survey in which elevations were required. The survey was made for the purpose of obtaining a preliminary map of a grade crossing. The traverse was run by transit and tape along the center line of the straight track and, at curves, running to the point of intersection of the tangents or else cutting across

Survey for Elimination of Grade Crossing at Westwood, A.&B.R.R.						
Sta.	Dist.	Az. Ang.	Bear.	Vert. Ang.	Diff. El.	Elev.
⊼ at ⊡ 6,		0° on 1+62.47,			H.I. = 4.6	58.3
B.M.	Top S.B. at P.T. 3+24.94			0° 00′ on 5.26		57.62
1	298 −1	221° 20′		+3° 27′	+18.0	76.3
2	238 −1	217° 15′		+2° 44′	+11.4	69.7
3	183	205° 10′		+1° 37′	+ 5.2	63.5
4	165	180° 00′		+0° 31′	+ 1.5	59.8
5	167	115° 40′		−0° 16′	− 0.8	57.5
6	177	142° 25′		−0° 42′	− 2.2	56.1
7	212	111° 00′		−1° 22′	− 5.1	53.2
8	323	80° 15′		−1° 57′	−11.0	47.3
9	426	70° 05′		−1° 47′	−13.3	45.0
10	409 −2	67° 30′		−3° 36′	−25.7	32.6
Etc.						
⊡ A	825 −2	221° 23′		+2° 34′	+37.0	95.3
1+62.47	437	0° 00′				
⊼ at ⊡ A,		0° on ⊡ 6,			H.I. = 4.7	95.1
⊡ 6	823 −2		S.7° 15′ E.	−2° 33′	−36.7	58.3
39	312	357° 10′		−1° 25′	− 7.7	87.4
40	133 −1	17° 35′		−3° 46′	− 8.8	86.3
41	156	22° 05′		−2° 46′ on 6.7	− 9.6	85.5
42	274 −1	3° 20′		−2° 24′	−11.5	83.6
Etc.						
⊡ 12+14.37	547 −2	48° 02′	S. 41° 00′ W.	−3° 07′	−29.8	65.3
⊡ 6		0° 01′				
⊼ at ⊡ 12+14.37,		0° on ⊡ 6,			H.I. = 4.8	65.2
⊡ 6		0° 00′		0° 00′ on 11.7	− 6.9	58.3
⊡ A	546 −2	269° 25′		+3° 08′	+29.9	95.1
52	640 −3	56° 25′		−3° 40′	−40.9	24.3

FIG. 7.31.2. STADIA SURVEY REQUIRING ELEVATIONS.

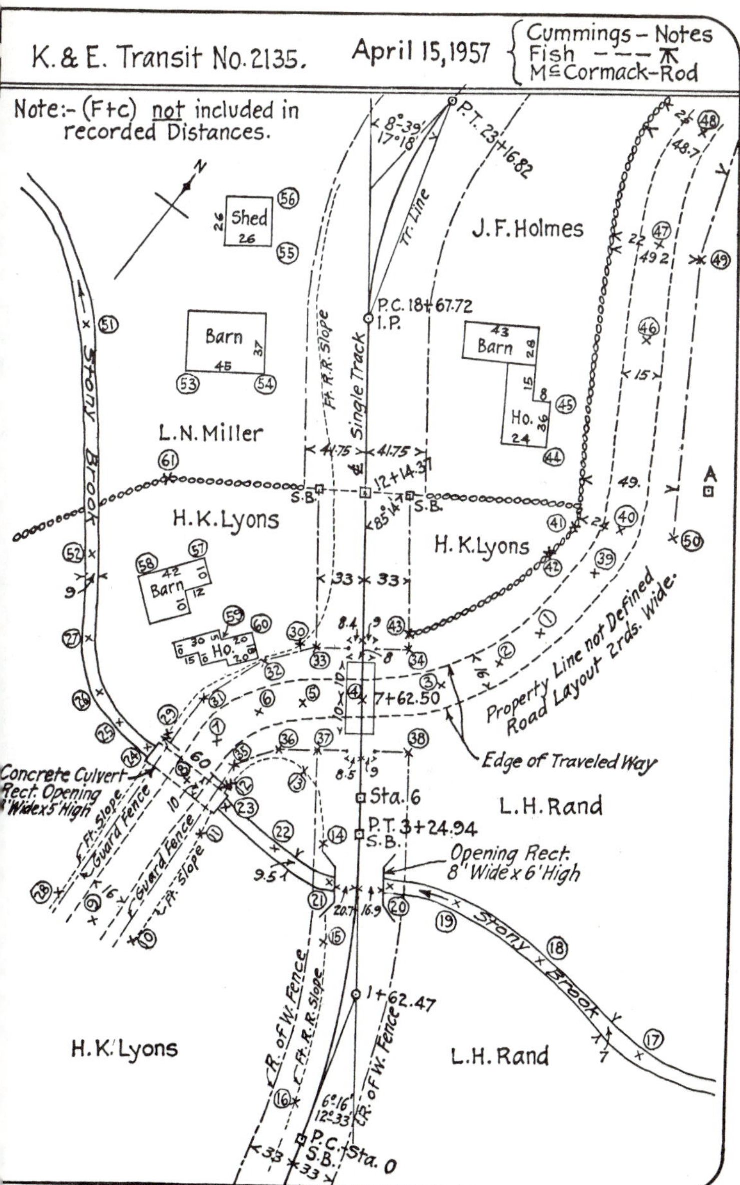

FIG. 7.31.3. STADIA SURVEY REQUIRING ELEVATIONS.

on long chords. The points occupied by the instrument are numbered and are designated by a ⊡, the usual symbol for a stadia station. The side shots are numbered consecutively either throughout the survey or through each day's work; the vertical angles are designated + or − to indicate whether they are above or below the horizontal line. If the middle cross-hair is sighted at any point other than the H.I. the point on which it is sighted should be noted directly under the vertical angle, as was done on side shot 41. The first five columns only are entered in the field; most of the values in the last two columns are computed in the office; those which are recorded with inclined figures were computed in the field for the purpose of checking elevations and for establishing station elevations to be used later in the same day's fieldwork. The erect figures were computed in the office.

Referring to the notes on Fig. 7.31.2 it will be seen that the elevation of ⊡ 6, where the transit was set up first, was obtained by means of a level reading of 5.26 taken on the B.M. The elevation of ⊡ 6 equals 57.62 + 5.26 − 4.6 = 58.3. A rectangle is drawn about this figure 58.3 because it is used for all computations of elevation of points that were sighted when the instrument was at ⊡ 6. After all necessary side shots were taken at ⊡ 6 the elevation of ⊡ A was determined by the stadia method as 95.3. The azimuths were checked by sighting again on sta. 1 + 62.47. The instrument was then moved to ⊡ A and a backsight taken on ⊡ 6. The elevation of ⊡ 6 was determined as 58.3 by direct leveling from the bench-mark. It is therefore recorded below opposite station ⊡ 6 as 58.3, and by means of the difference in elevation 36.7 the new elevation of ⊡ A is 58.3 + 36.7 = 95.0. Its elevation, however, was determined above from ⊡ 6 as 95.3 so there has been entered in these notes in the last column opposite "Instrument at ⊡ A" the figure 95.1, which is the mean (to the nearest tenth) of 95.0 and 95.3.

The side shots at ⊡ A are next taken and then sta. 12 + 14.3 has its elevation established by stadia as 65.3. Before picking up the instrument, ⊡ 6 was sighted to check the azimuth. If this azimuth is correct it indicates that the azimuths of all th side shots are probably correct, although it does not uncover blunders.

The instrument was then taken to sta. 12 + 14.37, and backsighted on ⊡ 6. Instead of using the elevation 65.3, previously obtained for the elevation of sta. 12 + 14.37, its elevation is obtained by direct leveling from ⊡ 6 using 58.3 as the elevation of ⊡ 6 which was obtained at the beginning of the work by a level reading on a bench-mark. The elevation of sta. 12 + 14.37 = 58.3 + 11.7 − 4.8 = 65.2; this elevation is more reliable than the one obtained by stadia.

It will be observed that in these notes the backsight reading was taken on 0° azimuth. The side-shot azimuths are all read to the nearest 5 minutes, this is close enough for plotting purposes. Occasional bearings have been taken to check the azimuths. In the second column the little minus quantities at the right are the horizontal corrections taken from a Table of Horizontal Corrections and entered in this column in the office. The notes, description and sketch are on the right-hand page. The numbers enclosed in circles are the numbers of the (rod) points which were located by stadia. These little circles are drawn so that these rod-point numbers will not be confused with the dimensions on the sketch.

7.32. Methods of Reducing the Notes.—The computation of the differences in elevation and of the horizontal distances is usually performed in practice by one of the following methods: (1) Stadia tables (2) stadia slide rule (3) stadia diagram. The use of these methods is described in **Arts. 7.34-7.35** for external focusing instruments. The same procedure applies to internal focusing systems except that $(F + c)$ is omitted.

7.33. Stadia Tables.—In spite of special stadia computation tools as mentioned in subsequent articles, tables provide the primary source for reduction of stadia readings. Figure 7.33.1 contains tables of horizontal corrections and vertical heights for vertical angle readings between 0° and 10°, which will cover most stadia readings. Rather than extending the tables, the reader is advised to use formulae and calculator if he happens to encounter a larger vertical angle. With the example provided, use of the table in Fig. 7.33.1 is self-explanatory.

FORMULAS TO USE WITH VERTICAL ANGLES GREATER THAN 10°

	Internal Focusing	External Focusing
Hor. Dist.	$100\,S\cos^2\alpha$	$100\,(S + 0.01)\cos^2\alpha$
Vert. Ht.	$100\,S\sin\alpha\cos\alpha$	$100\,(S + 0.01)\sin\alpha\cos\alpha$

Where: S = stadia intercept; α = vertical angle

TABLE A

HORIZONTAL CORRECTIONS FOR STADIA INTERCEPT 1.00 FT.

Vert. Angle	Hor. Cor. for 1.00 ft.	Vert. Angle	Hor. Cor. for 1.00 ft.	Vert. Angle	Hor. Cor. for 1.00 ft.
0°00′		5°36′		8°02′	
	0.0 ft.		1.0 ft.		2.0 ft.
1°17′		5°53′		8°14′	
	0.1 ft.		1.1 ft.		2.1 ft.
2°13′		6°09′		8°26′	
	0.2 ft.		1.2 ft.		2.2 ft.
2°52′		6°25′		8°38′	
	0.3 ft.		1.3 ft.		2.3 ft.
3°23′		6°40′		8°49′	
	0.4 ft.		1.4 ft.		2.4 ft.
3°51′		6°55′		9°00′	
	0.5 ft.		1.5 ft.		2.5 ft.
4°15′		7°09′		9°11′	
	0.6 ft.		1.6 ft.		2.6 ft.
4°37′		7°23′		9°22′	
	0.7 ft.		1.7 ft.		2.7 ft.
4°58′		7°36′		9°33′	
	0.8 ft.		1.8 ft.		2.8 ft.
5°17′		7°49′		9°43′	
	0.9 ft.		1.9 ft.		2.9 ft.
5°36′		8°02′		9°53′	
					3.0 ft.
				10°03′	

Results from Table A are correct to the nearest foot at 1000 feet and to the nearest 1/10 foot at 100 feet, etc.

With a slide rule, multiply the stadia intercept by the tabular value and subtract the product from the horizontal distance.

Example. Vertical angle, 4°22′; stadia intercept, 3.58 ft.

$$\text{Corrected Hor. Dist.} = 358 - (3.58 \times 0.6) = 356 \text{ ft.}$$

Table B gives the vertical heights for a stadia intercept of 1.00 ft. With a slide rule, multiply the stadia intercept by the tabular value.

Example. Vertical angle, 4°22′; stadia intercept, 3.58 ft.

$$\text{Vertical Height} = 3.58 \times 7.59 = 27.2 \text{ ft.}$$

FIG. 7.33.1. STADIA TABLES WITH EXAMPLE.

(Courtesy, Keuffel & Esser Co.)

TABLE B

VERTICAL HEIGHTS FOR STADIA INTERCEPT 1.00′

Min.	0°	1°	2°	3°	4°	5°	6°	7°	8°	9°
0	0.00	1.74	3.49	5.23	6.95	8.68	10.40	12.10	13.78	15.45
2	0.06	1.80	3.55	5.28	7.02	8.74	10.45	12.15	13.84	15.51
4	0.12	1.86	3.60	5.34	7.07	8.80	10.51	12.21	13.89	15.56
6	0.17	1.92	3.66	5.40	7.13	8.85	10.57	12.27	13.95	15.62
8	0.23	1.98	3.72	5.46	7.19	8.91	10.62	12.32	14.01	15.67
10	0.29	2.04	3.78	5.52	7.25	8.97	10.68	12.38	14.06	15.73
12	0.35	2.09	3.84	5.57	7.30	9.03	10.74	12.43	14.12	15.78
14	0.41	2.15	3.89	5.63	7.36	9.08	10.79	12.49	14.17	15.84
16	0.47	2.21	3.95	5.69	7.42	9.14	10.85	12.55	14.23	15.89
18	0.52	2.27	4.01	5.75	7.48	9.20	10.91	12.60	14.28	15.95
20	0.58	2.33	4.07	5.80	7.53	9.25	10.96	12.66	14.34	16.00
22	0.64	2.38	4.13	5.86	7.59	9.31	11.02	12.72	14.40	16.06
24	0.70	2.44	4.18	5.92	7.65	9.37	11.08	12.77	14.45	16.11
26	0.76	2.50	4.24	5.98	7.71	9.43	11.13	12.83	14.51	16.17
28	0.81	2.56	4.30	6.04	7.76	9.48	11.19	12.88	14.56	16.22
30	0.87	2.62	4.36	6.09	7.82	9.54	11.25	12.94	14.62	16.28
32	0.93	2.67	4.42	6.15	7.88	9.60	11.30	13.00	14.67	16.33
34	0.99	2.73	4.47	6.21	7.94	9.65	11.36	13.05	14.73	16.39
36	1.05	2.79	4.53	6.27	7.99	9.71	11.42	13.11	14.79	16.44
38	1.11	2.85	4.59	6.32	8.05	9.77	11.47	13.17	14.84	16.50
40	1.16	2.91	4.65	6.38	8.11	9.83	11.53	13.22	14.90	16.55
42	1.22	2.97	4.71	6.44	8.17	9.88	11.59	13.28	14.95	16.61
44	1.28	3.02	4.76	6.50	8.22	9.94	11.64	13.33	15.01	16.66
46	1.34	3.08	4.82	6.56	8.28	10.00	11.70	13.39	15.06	16.72
48	1.40	3.14	4.88	6.61	8.34	10.05	11.76	13.45	15.12	16.77
50	1.45	3.20	4.94	6.67	8.40	10.11	11.81	13.50	15.17	16.83
52	1.51	3.26	4.99	6.73	8.45	10.17	11.87	13.56	15.23	16.88
54	1.57	3.31	5.05	6.79	8.51	10.22	11.93	13.61	15.28	16.94
56	1.63	3.37	5.11	6.84	8.57	10.28	11.98	13.67	15.34	16.99
58	1.69	3.43	5.17	6.90	8.63	10.34	12.04	13.73	15.40	17.05
60	1.74	3.49	5.23	6.96	8.68	10.40	12.10	13.78	15.45	17.10

FIG. 7.33.1. (CONT.).

7.34. Stadia Slide Rule.—The most rapid means of reducing stadia readings is by the use of a slide rule which has, in addition to the ordinary scale of numbers (logarithms of the distances), two scales especially constructed for stadia work;

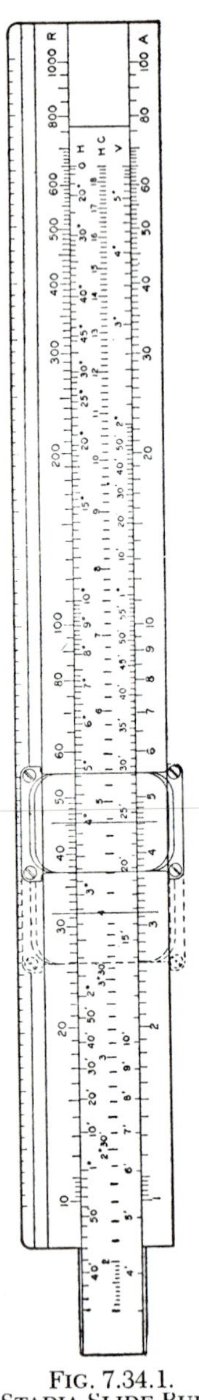

FIG. 7.34.1.
STADIA SLIDE RULE.

one consisting of values of $\log \cos^2 \alpha$, and the other of $\log \frac{1}{2} \sin 2\alpha$ for different values of α. On some rules the values of α range from 0° 34′ to 45°, on others from 0° 03′ to 45°. In some forms the horizontal distance is read directly; in others the horizontal correction $1 - \cos^2 \alpha$, or $\sin^2 \alpha$, is given. A 10-in. slide rule gives results sufficiently accurate for all ordinary purposes.

One type of stadia slide rule is shown in simplified form in Fig. 7.34.1; certain of the finer divisions that appear on the rule have been omitted. The stadia distance for which reductions are desired is set by moving the slide (middle portion of the rule) to the left until the zero on the right-hand end of the slide (H scale) coincides with the stadia distance reading on the upper scale of the rule (R scale); the setting in Fig. 7.34.1 is at 650 ft.

For a vertical angle of 4°, the correct horizontal distance is read on the upper scale of the rule opposite 4° on the upper right-hand scale of the slide. It will be observed that there are two angle scales on the upper edge of the slide; one for horizontal distances running from zero to 45° from right to left, and the other for difference of elevation running from 35′ to 45° from left to right. At the right-hand end of the H scale the first division mark close to zero is 5°, the second 10°, the third 15°, and the fourth 20°, as marked. The distance reading for 4° of vertical angle is taken a shade to the right of the first (5°) division and shows the horizontal distance to be 647 ft.

For small angles such as are commonly encountered in stadia work the horizontal distance is difficult to read to three significant figures on the stadia slide rule, since the divisions indicating horizontal distance are very close together especially near the zero end of the slide.

It is much **easier to read the horizontal correction** than the horizontal distance. In most cases the former is only required to one or two significant figures. In Fig. 96a a scale of horizontal corrections (H.C.) is engraved on the middle of the slide. To obtain the horizontal correction for a distance of 650 ft., the rider is moved to the left until the wire coincides with 4° on the middle scale. This reading is indicated in the figure by the dash-line position of the rider. The horizontal correction is read on the lower (A) scale; it is 3.15, or 3 ft. to the nearest foot, which agrees with the horizontal correction of 3 ft. indicated for the same set of conditions on the stadia diagram in Art. 230b.

The difference in elevation for a vertical angle of 4° and stadia reading of 650 ft. is found on the upper scale of the rule, opposite this angle on the upper left-hand scale of the slide; the reading is 45.3 ft. In Fig. 96a the rider (shown in full lines) is set at this reading.

If the vertical angle were small, such as 0° 45′, no discernible difference from 650 could be read for the horizontal distance. The angle of 0° 45′ on the upper scale of the slide falls off (outside of) the upper scale of the rule. Therefore the lower scale "A" of the slide must be used. Looking near the middle of the rule on the bottom pair of scales, it will be noted that 0° 45′ falls opposite 8.5 ft., which is the difference in elevation for a distance of 650 ft. and vertical angle of 0° 45′. In the stadia diagram, Fig. 96b, a difference in elevation of 8 ft. (to nearest foot) is indicated.

A condensed table of horizontal corrections can be easily prepared which may be carried in the field. The correction to apply to stadia distances to obtain the equivalent horizontal distances may be read from this table with ease and rapidity, the slide rule being used only to determine differences of elevation.

7.35. Stadia Diagrams.—Where stadia notes are reduced in the office it is sometimes convenient to determine the difference in elevation and the horizontal distance by means of a diagram, from which the results can be taken by inspection. There are several possible arrangements, but they all depend upon the same formulas. Fig. 7.35.1 shows a simple form of stadia diagram which can be made in the office on ordinary coördinate paper. The accuracy of printed coördinate paper is not always sufficient, however; for an accurate diagram it is better to construct the coördinate lines on heavy drawing paper. Only the degree lines are shown in Fig. 7.35.1, but the subdivision may be carried to 10′, or to 5′ of vertical angle, sometimes closer, by plotting additional points.

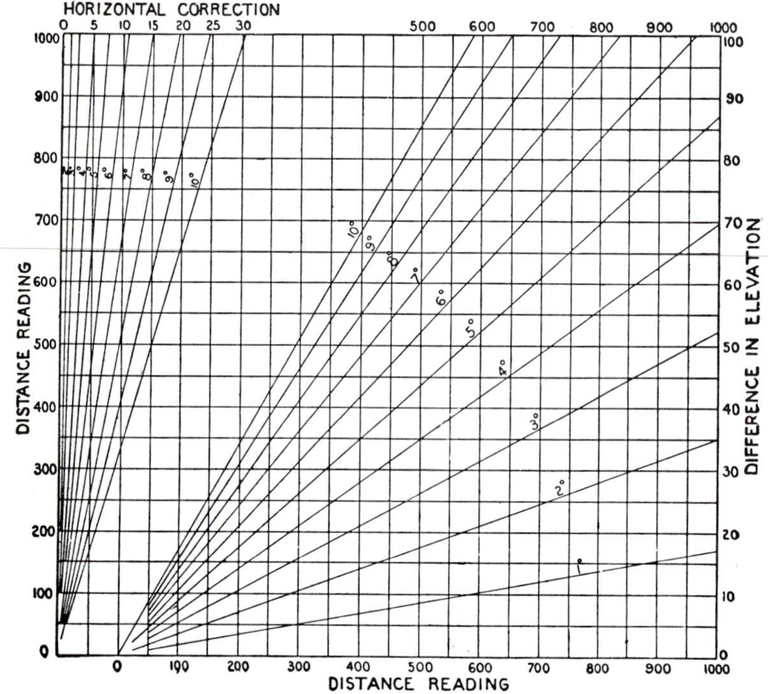

FIG. 7.35.1. DIAGRAM FOR REDUCING STADIA READINGS. (INCLUDE $(F + c)$ IN DISTANCE READING.)

In using this diagram the inclined line corresponding to the vertical angle is intersected with the vertical line corresponding to the distance reading; the difference in elevation for this point is read from the scale on the right, by interpolating between the horizontal lines. The horizontal correction to the distance is taken from the left-hand part of the diagram in exactly the same manner.

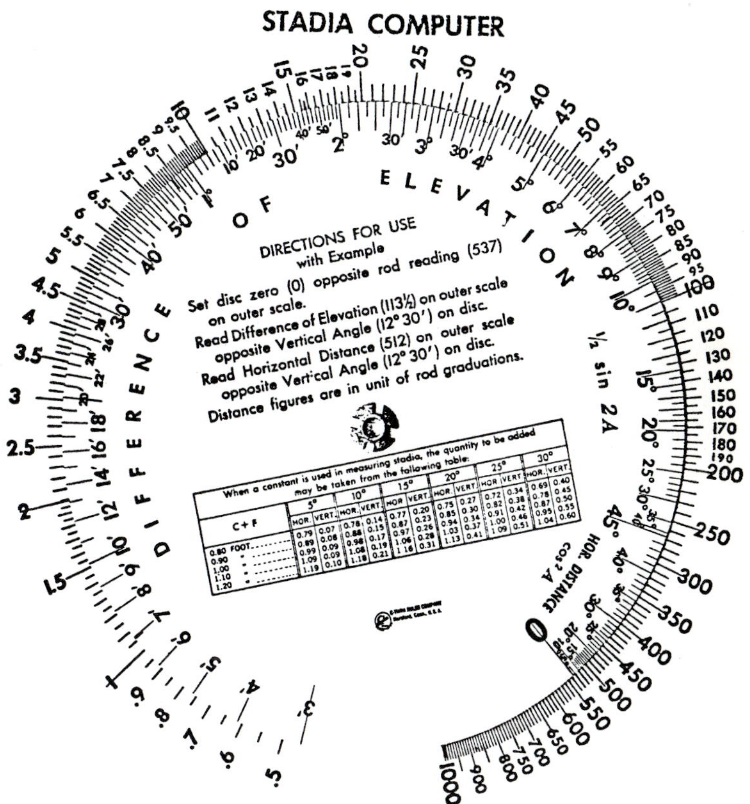

When a constant is used in measuring stadia, the quantity to be added may be taken from the following table:

C + F	5°		10°		15°		20°		25°		30°	
	HOR.	VERT.	HOR.	VERT.	HOR	VERT.	HOR.	VERT.	HOR	VERT	HOR.	VERT.
0.80 FOOT	0.79	0.07	0.78	0.14	0.77	0.20	0.75	0.27	0.72	0.34	0.69	0.40
0.90 "	0.89	0.08	0.88	0.15	0.87	0.23	0.85	0.30	0.82	0.38	0.78	0.45
1.00 "	0.99	0.09	0.98	0.17	0.97	0.26	0.94	0.34	0.91	0.42	0.87	0.50
1.10 "	1.09	0.09	1.08	0.19	1.06	0.28	1.03	0.37	1.00	0.46	0.95	0.55
1.20 "	1.19	0.10	1.18	0.21	1.16	0.31	1.13	0.41	1.09	0.51	1.04	0.60

FIG. 7.36.1. STADIA COMPUTER.

7.36. Stadia Computer.—Another fast device for reducing stadia observations is the Cox Stadia Computer, shown in Fig. 7.36.1. Again its use is described in the figure.

7.37. Stadia Arcs.—Some instruments have graduated on the vertical arc, or on an auxiliary arc, a series of lines corresponding to those vertical angles for which the difference in elevation is a simple multiple of the rod interval. Such angles would be 0° 34′ +, 1° 08′ +, 1° 43′ +, etc. If these angles are looked up in Table VIII they will be seen to correspond to the multiples 1.00, 2.00, 3.00, etc. To use such an arc for computing the difference in elevation the telescope must be inclined until the index mark is exactly opposite some line on the stadia arc. This changes the position of the horizontal hair on the rod, so the new reading of the middle hair must be taken. In order that the vertical angle shall be the true one, free from index

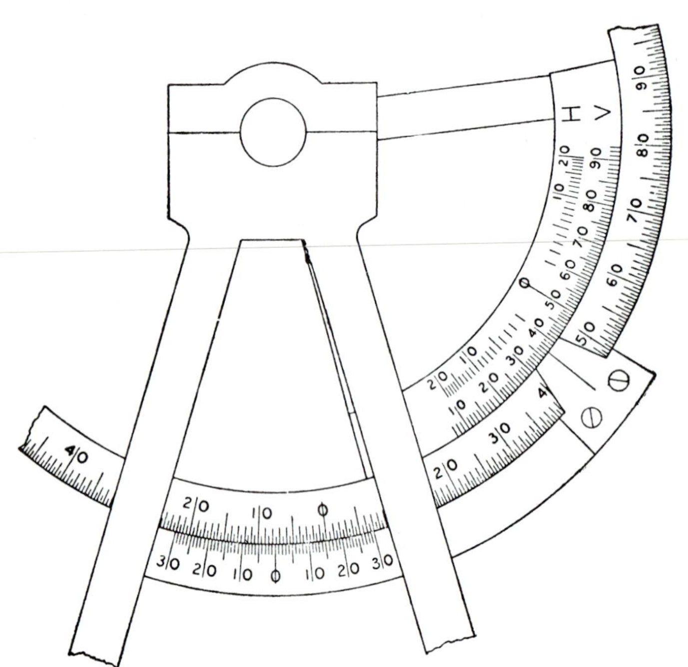

FIG. 7.37.1. THE BEAMAN STADIA ARC ATTACHMENT.

error, it is important that the telescope be leveled and the stadia index set at 0 before the sight is taken.

As an illustration suppose that the elevation of the transit point is 203.4 ft., the height of instrument is 4.0 ft., the rod interval (stadia) is 2.50 ft., and that the stadia index line is near the −13 line. When the −13 line is made to coincide exactly with the index line the middle hair is found to read 7.8 ft. on the rod. The elevation of the foot of the rod is therefore $203.4 + 4.0 - 13 \times 2.50 - 7.8 = 167.1$ ft. In this manner elevation may be computed rapidly in the field.

Various forms of stadia arc are manufactured. The original one was that designed by W. M. Beaman of the U.S. Geological Survey. On this arc the −13 would be marked as 37, the 50 mark being the zero, as shown in Fig. 7.37.1. From another set of graduations on the same piece of metal, the horizontal corrections (for a 1-ft. intercept) can be read off at the same time. In the setting shown in Fig. 7.37.1 this is between 1 and 2.

7.38. Methods of Plotting Stadia Notes.—Stadia notes are usually plotted by means of a circular protractor and a scale. If the main traverse is a transit and tape survey, or if the scale of the map is such that a protractor would not be sufficiently accurate, the traverse may be plotted by some more accurate method and the side shots put in afterward by the protractor and scale. In general, however, any measurement taken by stadia may be plotted with sufficient accuracy by means of a protractor. (See Chap. XVII for methods of plotting traverses.)

In setting the protractor in position for plotting it should be centered with care and turned to the proper azimuth as defined by a 0° line drawn through the point and extending each way beyond the circumference of the protractor. It is not safe to depend upon a line extending only one way, because the center of a protractor is usually marked in such a way that it is difficult to place it exactly over the transit point on the plan. Many protractors which are accurately graduated have the center point carelessly marked. The most accurate way to use a protractor is to draw two lines at right angles to each other, one of them being the meridian or reference line. These lines may be drawn at right angles as explained in Art. 17.8. The protractor is then **oriented,** i.e., turned in the proper direction by making the cardinal points on the circumference line up with lines drawn at right angles, without regard to the position of the center mark on the protractor.

The usual process is to place the protractor in position and plot all of the azimuths first, marking each by a light dot or a short radial line in the proper azimuth and writing opposite the mark the number of the shot. This work can be done conveniently by two persons, one reading the azimuths and the numbers while the other plots the angles. When all of the azimuths have been plotted the protractor is removed and the distances are scaled off, the proper elevation being written opposite each point. Sometimes the plotted position of the point is indicated by a dot enclosed by a small circle, the

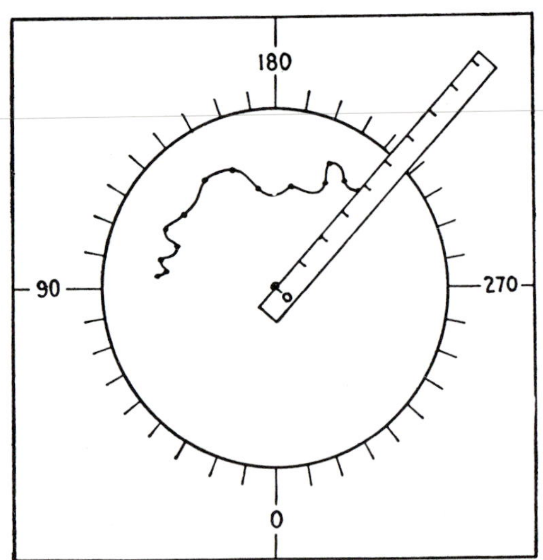

FIG. 7.38.1. PROTRACTOR FOR PLOTTING STADIA SIDE SHOTS.

height being written at one side. Another way, which is convenient when the plotted points are close together, is to write the whole number of feet of the elevation to the left of the point and the tenths to the right, the plotted point itself serving as the decimal point.

If much plotting is to be done it will be found convenient to use a large paper protractor from which the central portion has been removed, leaving the graduations close to the inner edge of the remaining paper. This enables the draftsman to set the zero of his scale on the station and the edge of the scale on the azimuth and to plot the points directly without the necessity for drawing lines or moving the protractor. Such a protractor can be made in the office. (Fig. 7.38.1.)

7.39. Locating Contours.—If contours are to be located by stadia there are two general methods which may be used. First, if the scale is comparatively large and the contours are to be located accurately it may be advisable to obtain points which are exactly on the contours. This process is slower but more accurate than the other. In obtaining these elevations the transit is used as a leveling instrument. The rod is first held on a benchmark and a backsight is taken to determine the H.I. The notes may be kept on a separate page in the form of ordinary level notes. The proper foresight to give a point on a contour is next found by subtracting the contour elevation from the H.I. Suppose, for instance, that the H.I. is 117.23 ft. and that the 110 contour is to be located. A rod reading of 7.23 will be obtained when the rod is held exactly on the 110 contour. It would be sufficient to read this as 7.2 when locating contours. The instrument man directs the rodman to move up the slope or down the slope until the foresight reading is 7.2. When this point is found it is located by azimuth and stadia distance in the usual manner. This, however, requires changing the inclination of the telescope in order to read the stadia distance, or else it requires making two readings and subtracting them. It is not therefore a rapid method.

Second, if the scale to be used is small, or if the required accuracy is not great, the rod may be held on such points as determine the various slopes of the ground, and these points may be located by distance, azimuth and vertical angle to the H.I. point on the rod. The points which will naturally be selected are those situated along the ridge lines and the valley lines, and especially those points where the slope changes. If all such points are located then it may be assumed that the slopes are uniform between rod points. There should be, however, enough points located so that errors may be detected. If the elevation of a single point is in error it will affect the contours over a wide area unless there are additional elevations to check it. It is better to have an excess of points and avoid such mistakes, than to have so few that the mistakes pass unnoticed. After these points are plotted the contours are found by interpolating between the elevations of the rod points.

Sketching Contours.—One of the weaknesses of the stadia method is that the plotting is usually done in the office. Very complete notes are necessary to give the draftsman a correct idea of the ground he is depicting, and complete notes are the exception, not the rule. Even if the draftsman did all the fieldwork himself he may not be able to remember all the details. It is not easy to foresee all the needs of the draftsman. The sketching done in the office, apart from that which is purely mechanical, is liable, therefore, to be in error.

An excellent way of supplementing the sketch is to take the map into the field, mounted on a drawing board, and to complete the details of the sketch while the ground is in sight Many little details of the contours may now be sketched which would probably be omitted in the notes, or which, if recorded would require that a large number of points be determined. By this procedure the stadia method can be made to give almost as good results as the plane table. It is possible also to supplement the fieldwork further by using the hand level and the pocket compass in connection with this sketching. It is perfectly feasible to have this plotting and sketching carried on simultaneously with the fieldwork if the size and importance of the job warrant it.

7.40. Stadia Leveling.—Lines of levels may be run by means of the transit and stadia, but it requires careful work to obtain good results. These elevations depend upon the measured vertical angles, and these angles are, as a rule, the least accurate measurements given by the transit. For rapid work when the closest results are not required the distances and vertical angles may be read but once, the transit being set up in between rod stations. For more careful work the transit stations and the rod stations should be the same points. The distances between stations should be read both forward and backward. The illumination of the rod often will be quite different on these two sights, so that one will give a good determination, while the other is at best only an approximate check on it. It is especially important to read the vertical angles both forward and backward. Careful attention should be given to leveling the plates and determine the index error. The horizontal hair, the long level on the telescope and the vernier of the vertical arc should be kept as closely as possible in adjustment. Large angles should be avoided if possible, because the effect of errors in the vertical angle is usually greater for steep grades. In leveling with external focusing transits, the $(F + c)$ cannot be neglected, because the resulting error tends to accumulate.

7.41. Traversing with Magnetic Needle.—Rough surveys can be made very rapidly by using the compass needle on the transit for obtaining directions and by measuring the distances by stadia. If the instrument and the rod are placed at alternate stations the amount of work is greatly reduced and ground can be covered rapidly. In this method there is no check on the accuracy. The transit is set up at station 2, and a distance and a compass bearing taken (backsight) to station 1. The rod is then taken to station 3 and a distance and bearing taken (ahead) to this point. The transit is then removed to station 4, and the process repeated. Side shots can be taken, when needed, by taking needle readings instead of azimuths. The notes must show clearly just what the different stations are and care must be exercised to read the needle correctly, and not to reverse a bearing. It must be remembered, however, that in using this method in a region where there is **local attraction** the results obtained may be **almost worthless.**

THE PLANE TABLE

7.42. The Plane Table.—The plane table consists of a drawing board mounted on a tripod in such a manner that it can be leveled and oriented, like a transit. This arrangement enables the operator to fasten his map to the drawing board and work directly on the map in the field.

The alidade is an instrument used on the plane table for taking sights and drawing lines when locating points on the map. It consists of a telescope, or else open sights, mounted on a metal straight-edge, with spirit levels for leveling the instrument.

Most alidades are equipped with stadia hairs and a vertical arc for measuring vertical angles. The vernier arc usually carries a level for eliminating the index correction. A striding level is also provided which may be set on top of the telescope when taking level sights. Some alidades have a prism on the eyepiece by means of which the observer may look vertically down into the telescope.

FIG. 7.42.1. PLANE TABLE AND ALIDADE.

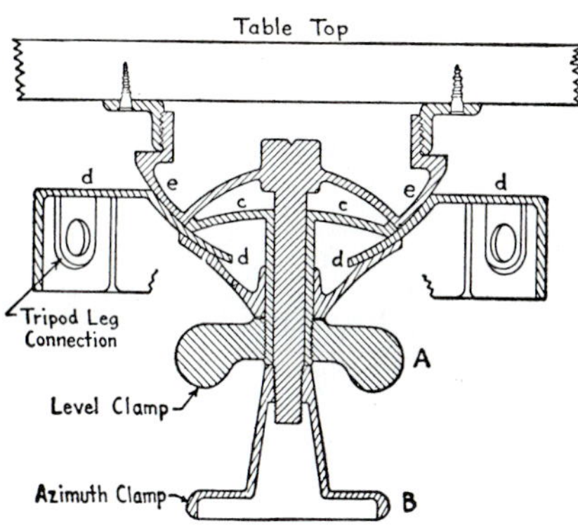

FIG. 7.42.2. JOHNSON PLANE-TABLE MOVEMENT.

Some types of plane tables have a leveling head and leveling screws similar to those on a transit; the orienting is controlled by means of a clamp and tangent screw. Another common type of plane-table head is that originally devised by W. D. Johnson of the U.S. Geological Survey, in which the leveling and orienting is done by means of a compact arrangement of spherical surfaces. Such a table is shown with alidade in Fig. 7.42.1. Referring to Fig. 7.42.2, which is a cross-section of the Johnson head, clamp *A* controls the leveling of the table, and clamp *B* controls the azimuth motion. After the board has been leveled by hand, using the bubble on the alidade as a guide, the upper clamp *A* is tightened. This forces the level-cup *c* against the tripod head *d*, the friction holding it in this position. The table is then turned to the desired azimuth, and the azimuth clamp *B* is then tightened, which forces the azimuth cup *e* against the tripod head, thus holding the table in the oriented position.

The alidade in Fig. 7.42.3 is equipped with a Beaman Stadia Arc (Art. 7.37). With this device differences in elevations and horizontal distances can be reduced in the field and recorded directly in the note book. The use of tables, slide rules and diagrams may be eliminated.

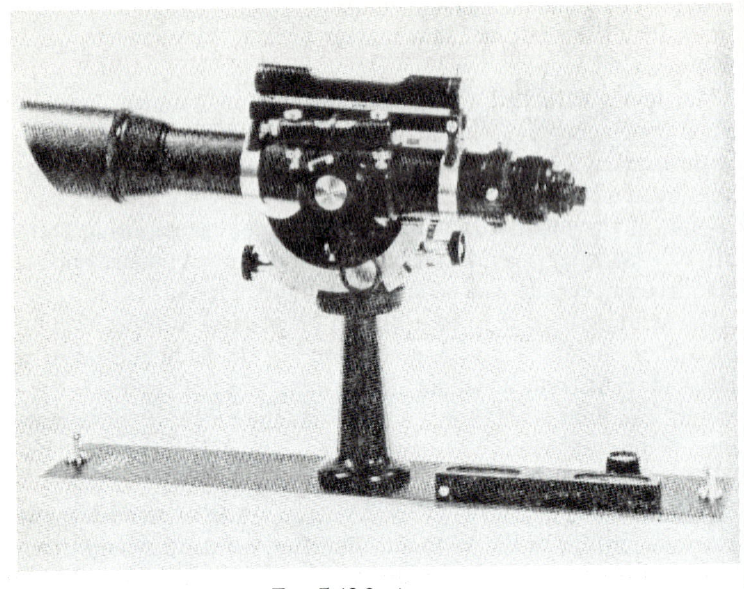

FIG. 7.42.3. ALIDADE.
(Courtesy, W. & L. E. Gurley)

7.43. Accessories.—The stadia rods, used when points are located by stadia distance and direction, are of the same patterns as those used in transit and stadia surveying. (See Fig. 3.30.1 & 3.30.2.) Plane tables intended for use on very large scale maps such as those required in landscape architects' work, are sometimes provided with an apparatus for plumbing a point on the map exactly over a station mark on the ground. This is needed only when using such scales as are capable of showing a fraction of a foot. Most plane tables are provided with a magnetic needle, or *declinatoire*. This may be fastened to the base of the alidade or it may be entirely separate.

7.44. Advantages and Disadvantages of the Plane Table.—The advantages of the plane table over the transit and stadia method are, 1. That all the sketching is done while the topographer can see the ground that he is representing, and he can therefore sketch more accurately. Also, he requires a smaller number of points to obtain the contours. He can decide what to include and what to omit. If any lines are sketched wrongly, or any points wrongly located, he is almost certain to observe this when in the field, but is not so likely to notice it if he is plotting it in the office. 2. The plane table can be set up at any assumed position and the corresponding plotted position of the station found quickly by resection; the position does not have to be found from the preceding station. This is not true of the transit, for the transit point must either be found from the preceding station or else from a trigonometric solution of the three-point problem. 3. It is much easier when using the plane table to determine at once whether the entire ground has been covered, than it is when making the survey with the transit.

The disadvantages of the plane table are, 1. That in rainy or in windy weather it may not be practicable to continue the fieldwork. 2. A longer time is usually required to complete the fieldwork than when a transit is used, although this is not always true. 3. Many surveyors' offices are not equipped with a plane table outfit, while all are equipped with transits. 4. The plane table outfit is bulkier and heavier to transport than the stadia outfit.

7.45. Adjustments of the Alidade.—The adjustments of the alidade are simple and easy to make. But the adjustments are not easily disturbed; so, as a matter of fact, they are not often required.

The levels attached to the base of the instrument may be tested by placing the alidade on the plane table, or on any horizontal surface, and bringing the bubbles to the centers of their tubes by the best means at hand, then reversing the alidade, end-for-end; if the bubbles move away from the centers, bring them half-way back by means of the adjusting screws. This applies to a circular level as well as to the ordinary levels.

The striding level may be adjusted by placing it in position on the collars on the telescope, and centering the bubble by means of the tangent screw. When the striding level is reversed, end-for-end, the bubble will move away from the center if the adjustment is imperfect. By means of the adjusting screw on the striding level bring the bubble half-way back to the center.

If the alidade carries a vernier level, first level the telescope by means of the striding level. Set the vernier to read zero. Then center the vernier bubble by means of the adjusting screw on this level. If there is no striding level the adjustment must be made by means of the "peg" method (see Art. 4.46). After the line of sight has been made horizontal the vernier is set at zero and the vernier bubble centered by means of its adjusting screw.

In some alidades the telescope is so mounted that it can be turned 180° in a sleeve. The point of intersection of the cross-hairs may be centered in the tube by sighting a point, turning 180°, and then bringing the cross-hairs half-way back by means of the capstan head screws on the diaphragm.

7.46. Locating Points.—In order to begin a survey with the plane table there must be at least two points on the ground whose positions are represented to some scale (known or unknown) on the map. If these points are the ends of a measured (or calculated) base-line and the table can be set at one or both of these stations, then the other points may be located at once by means of lines drawn along the straight edge of the alidade.

7.47. Intersection.—Suppose that *a* and *b* on the sheet represent *A* and *B* on the ground. (Fig. 7.47.1.) To find point *c* representing *C* on the ground, place the plane table so that *a* is vertically above *A* as nearly as the scale permits, the table being level. Then *orient* the table by pointing *ab* toward *B*. This is done by placing the straight edge along *ab*, turning the table until the vertical cross-hair is on the signal at *B*, and then clamping the table. If the alidade is centered on *a* and turned

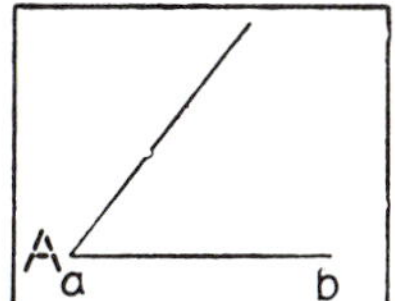

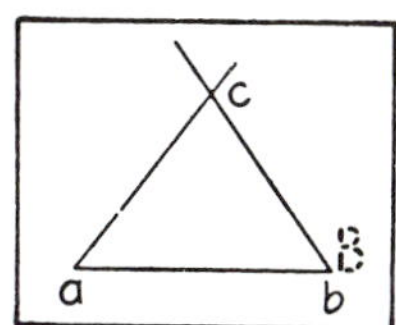

FIG. 7.47.1. LOCATING POINT BY INTERSECTION.

until C is sighted, then the line along the straight edge from a is directed to C. Therefore c is somewhere on this indefinite line. Next move the table to station B, level it and center point b over B. Direct ba toward A, and clamp the table. Then draw a line from b toward C by means of the alidade. These two indefinite lines intersect at the point c, which represents C to the scale of the map. This method of plotting is very accurate, and is the method chiefly used in locating the control points. As a check on this location the table may now be taken to station C, set up, leveled, and oriented by sighting ca toward A. If cb is found to be in the direction of B the work is checked. Locating points by this method is spoken of as "plane-table triangulation," or "graphical triangulation."

When centering the plane table over a point on the ground such as Sta. A, its plotted position a must be exactly above Sta. A when the table is oriented. To accomplish this, set the table in approximately an oriented position, then move it bodily so that a is vertically over A; then level the table and proceed with the work. This is sufficient for small-scale maps. For large-scale maps it may be found that, after the process just described, when the table is actually oriented by means of the alidade, it will have to be moved to one side or the other to get *a* exactly over *A*.

7.48. Direction and Distance.—Another method of locating points is that which is sometimes called "radiation." The alidade is centered on the point representing the plane-table station. Then any point, say *X*, Fig. 7.48.1, is located by sighting it with the alidade, drawing a line in this direction along the straight edge, and then determining the distance, either by stadia or by tape measurement. This distance is plotted to scale on the line at *x*. This method is used in locating details, after the principal stations have been located and checked. The radiating lines are not actually drawn, as a matter of fact, except in some special cases; for ordinary points the scale is laid down close to the straight edge and the point is plotted directly. This avoids covering the sheet with useless lines.

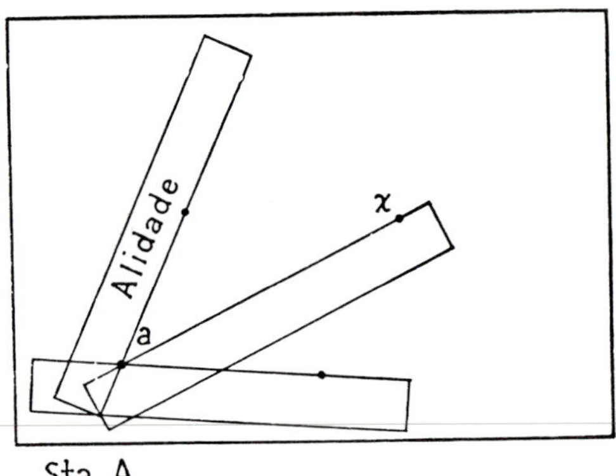

FIG. 7.48.1. LOCATING POINTS BY "RADIATION."

7.49. Sidesection.—If both ends of a base cannot be occupied with the table a new point can be located as readily as by intersection, provided one end of the base can be occupied. In Fig. 7.49.1, either point *B* cannot be occupied or else it is

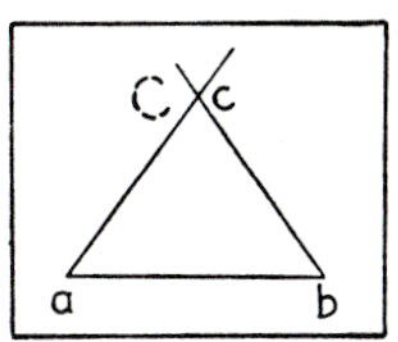

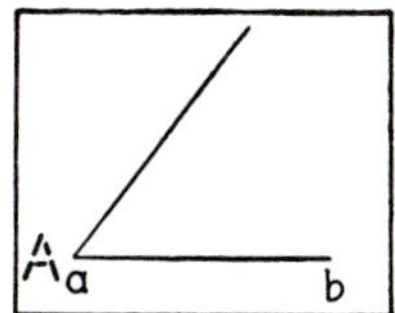

FIG. 7.49.1. LOCATING POINT BY SIDESECTION.

desired to avoid going to *B*. The table is set at *A* and oriented on *B* as before, and a line is drawn toward *C*. Then the table is taken to *C*, set up and oriented on *A* by means of the indefinite line *ac*. Then with the alidade on *b*, and pointing at *B*, a line is drawn backwards from *b* until it intersects the line from *a*. This locates *c* by *sidesection*. There is not the same opportunity to check in this case as there was when locating by intersection.

7.50. Three-point Problem (Resection).—If the table is set up at any unknown point D (Fig. 7.50.1) in the field its position d on the plane table sheet may be located provided three stations A, B and C whose positions are plotted on the sheet (at a, b and c), can be seen. Set up the table at D and roughly

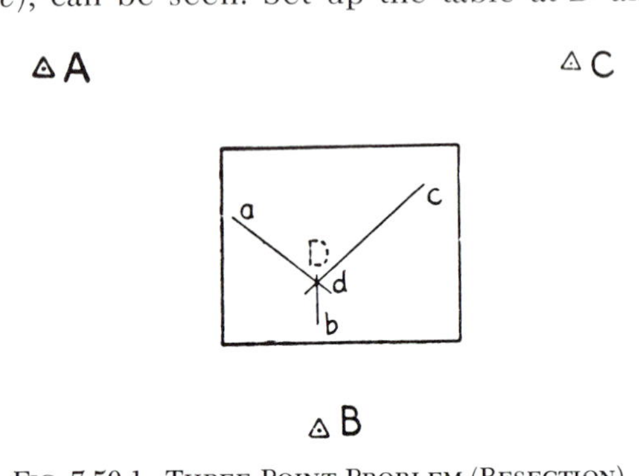

FIG. 7.50.1. THREE-POINT PROBLEM (RESECTION).

orient it by eye so that a, b and c appear to point toward A, B and C. Then mark a point on the plane table sheet to indicate where the planetabler thinks his position d is. If any topography has already been sketched on the sheet it will aid in locating the position of d. When d has been thus located the edge of the alidade is placed along da, and a short line drawn along the edge of the alidade through d. Leaving the alidade along da the table is then rotated about its vertical axis, until A is bisected by the vertical hair, and then clamped. Lines are then drawn by resecting from B and C; i.e., Bb and Cc are extended by drawing short lines along the edge of the alidade near point d. If these two resection lines intersect at d (representing D in Fig. 7.50.1) the point d is in the correct position and table is correctly oriented by the first guess. If the table had not been correctly oriented the three short lines would not have met at a common point but would have formed a triangle.

There is one case where the above method of orientation will not apply; this is when a circumference passing through ABC also passes through D.

In Fig. 7.50.2 is shown an example in which the first orientation was in error. This is indicated by the "triangle of error" formed by the resection lines from *a*, *b*, and *c*. As soon as the

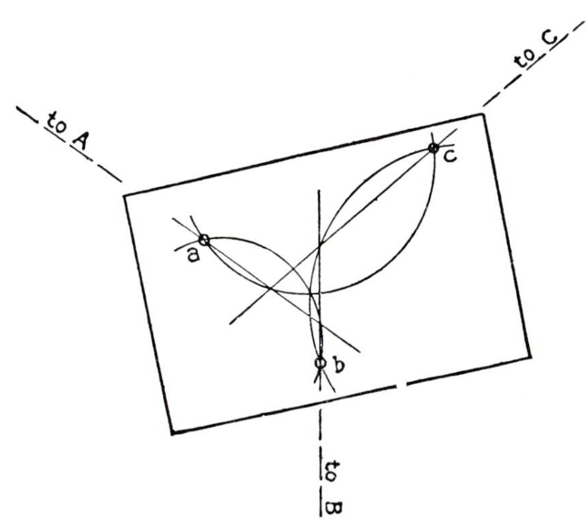

FIG. 7.50.2. TRIANGLE OF ERROR.

triangle of error is drawn we are able to determine at once the approximate position of *d*, the point representing *D*. A circle through *a*, *b*, and the vertex of the triangle of error which corresponds to the lines from *a* and *b* will also pass through the true point. Similarly a circle through *b* and *c* and the vertex formed by the lines from *b* and *c* will pass through the true point. The true point is therefore at their intersection. A circle through the points *c* and *a* and the corresponding vertex should also pass through the true point. (See also Figs. 7.50.3 & 7.50.4.)

Instead of constructing these circles carefully it is usually quicker to estimate a new position for d, correct the orientation by setting the straight edge on d and, say, c, and turning the table until C is sighted. A new triangle of error is then found as before. This time the size of the triangle of error should be much smaller. After two or three trials the triangle will be reduced to a point. This is the true point sought and the table is then correctly oriented.

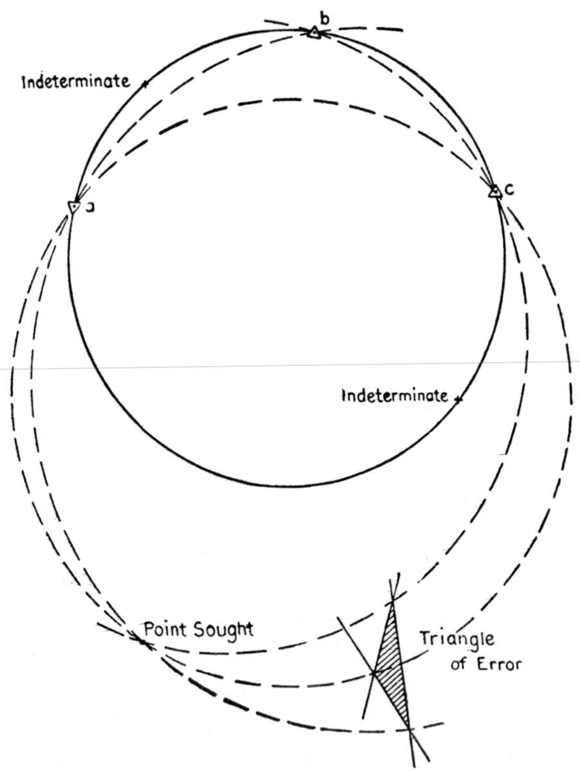

FIG. 7.50.3. UNFAVORABLE POSITION OF SIGNALS.

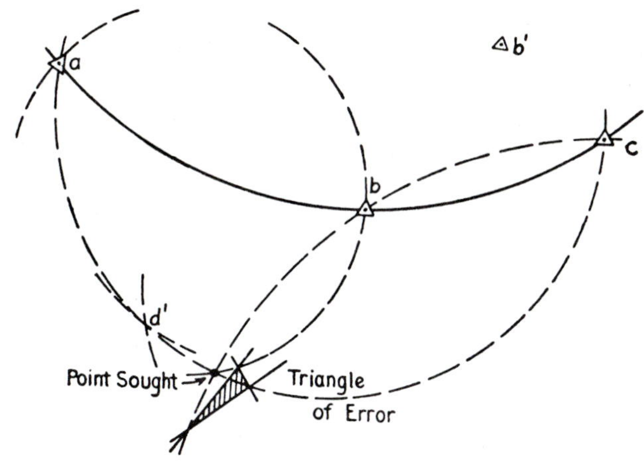

FIG. 7.50.4. FAVORABLE POSITION OF SIGNALS.

If the table happens to be on the circle through A, B, and C, the three resection lines will pass through a common point at the first trial, no matter whether the orientation is correct or not. Whenever the table is on this circle the location of the table is **indeterminate.** If, therefore, the triangle of error is a mere point at the first trial the relative positions of the signals and the table should be examined at once to see if the table is on the circle. If so this solution fails. The only way out of this difficulty is to choose another group of stations such that the circle through the three signals does not pass near to the table.

Whenever it is found that the three circles cut each other at small angles the position of the table will not be accurately determined, and it is better to abandon this solution and try another. When the plane-table station is inside the large triangle the true point is inside the triangle of error and the determination is usually accurate. If the plane table is on one side of the triangle, that is, "in range," the solution is accurate.

There are several other solutions of the three-point problem, some analytical, such as the trigonometric method which is demonstrated in Vol. II, Art. 1-57, and the geometric method presented and proved in Vol. II, Art. 7-42.

7.51. Two-point Problem.—If neither end of the base can be occupied with the table it is still possible to locate other points from this base, although the process is indirect, and the checks are insufficient. The solution consists essentially in assuming another base-line, unknown in length and position. The table is oriented by guess or by magnetic needle. From this second base-line the first base-line is located by triangulation. The difference in direction of the located position and the original plotted position of this base is the error in the original orientation of the table. The table is then turned through this angle so as to correct the orientation. The true position of the table can now be found by sighting to the ends of the original base.

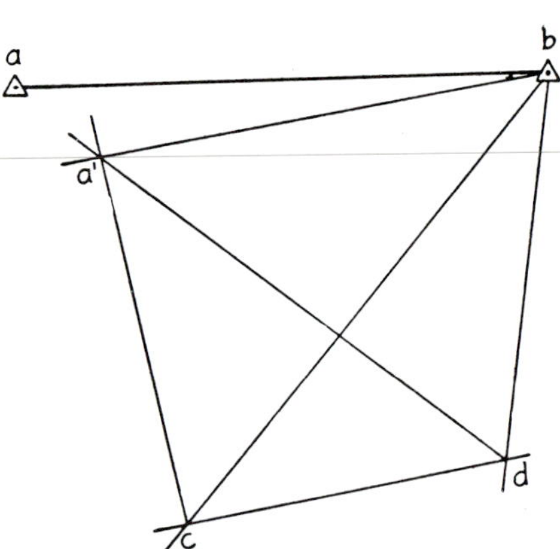

FIG. 7.51.1. TWO-POINT PROBLEM.

One method of carrying out this solution is indicated in Fig 7.51.1. Points *a* and *b* are the plotted positions of stations A and *B*. The table is set up at any point *D*; *d*, a preliminary point, is found by sighting to *A* and *B*. The alidade is centered on *d*, another point *C*, is sighted, and an indefinite line *dc* is drawn. The table is next carried to station *C*, leveled, and oriented by sighting *cd* toward *D*. By resecting from *B* with the alidade centered on *b*, the position of *c* is found. Line *dc* is the plotted base *DC*. Next draw from *c* a line toward *A*. This gives *a′* as the plotted position of *A* from the new (assumed) base. This shows that *ba′* is the true direction of *AB*. Since it was intended that *ab* should represent *AB*, the error in the orientation is evidently the angle *aba′*. To turn the table through this angle, set the alidade on line *ba′*, and find some point on the cross-hair, preferably a long distance away. Then set the alidade on line *ba*, unclamp, turn the table until the cross-hair is again on the same distant point; clamp the table. The table now is oriented correctly. Points *d*, *c*, *a′*, are no longer needed, and may be erased. New sightings to *A* and *B* will give the true position of *c*. If the position of *d* is desired it may be found by the method of sidesection. (See Art. 7.49.) There is no check on this result except such as may be obtained later from new points.

7.52. Traversing.—The plane table may be used for running traverses, although the results are not usually accurate, unless checked by triangulation. A rough check may be obtained by using the magnetic needle. The table is set at *A*, represented by *a* (Fig. 7.52.1), and a line drawn to *B*. This distance is determined by stadia, or other means, and *ab* is plotted. The table is next set at *B*; oriented by sighting *ba* toward *A*, and clamped. The alidade is then turned toward *C*, and *bc* is plotted in a similar manner. By this method the angles are plotted as accurately as the centering of the table permits, and the distances are as good as the scale permits. The method is inaccurate because the errors accumulate and there is not a suffi-

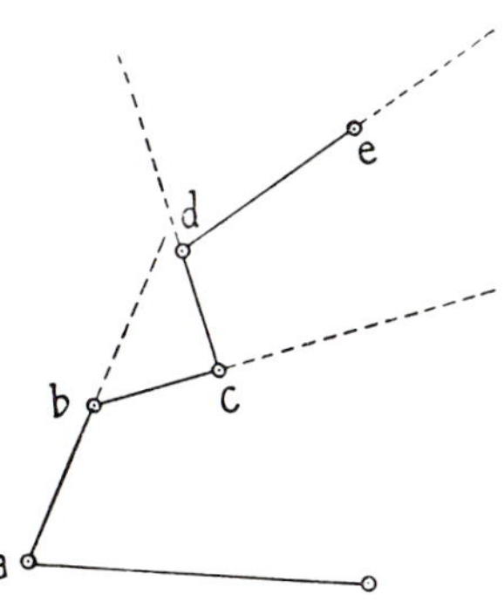

FIG. 7.52.1. PLANE-TABLE TRAVERSE.

cient check; nevertheless the method is often extremely useful. For rough work the table may be oriented by means of the magnetic needle, only alternate stations being occupied with the table, as described under the stadia method. (See Art. 7.40.) In Fig. 7.52.1, for instance, the table would be placed at *b* and *d*, and the rod held at *a*, and *c*, *e*. Distances may be paced, or even estimated, in some kinds of work.

7.53. Elevation of Table.—The elevation of the instrument or of the plane-table station may be found (1) by direct levels to the station (2) by stadia and vertical angle to a B.M. near the station, or (3) by a vertical angle to a triangulation point of known position and elevation.

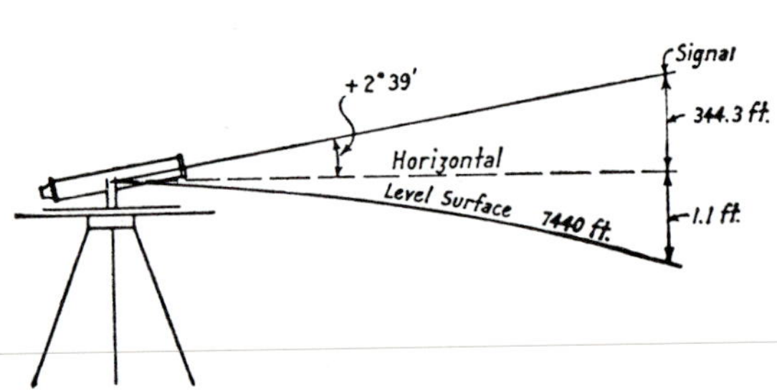

FIG. 7.53.1.

If the second method is employed the vertical angle will usually be taken to the H.I. point on the rod, so that the elevation of the plane-table station is the elevation of the B.M. plus or minus the calculated difference in elevation. If the third method is used the (horizontal) distance must be found by scaling it from the map. If the distance is greater than about half a mile it will be necessary also to allow for the effect of the curvature of the earth and of the refraction of the air. Suppose, for example, that the vertical angle to the top of a signal is + 2° 39′, the scaled distance to the point being 7440 ft., and the elevation of the signal being 1070.3 ft. The true difference in elevation = 7440 × tan 2° 39′ + correction for curvature and refraction.* The correction for a distance of 1000 ft. is nearly 0.02 ft. Therefore the correction for 7440 = $(7440)^2/(1000)^2 \times$ 0.02 ft. = 1.1 ft. The difference in elevation = 344.3 + 1.1 = 345.4 ft., and the elevation of the alidade is 10.70.3 − 345.4 = 724.9 ft. The alidade is 4.2 ft. above the station point; the elevation of the station point is therefore 720.7 ft. (Fig. 7.53.1.) If the vertical angle is negative the correction is subtracted from the difference in elevation.

7.54. Fieldwork.—When making a map on a comparatively large scale (as, 20 ft. to the inch, or 40 ft. to the inch) it is important that the plotted plane-table point be plumbed carefully over the station, that the straight edge which is vertically beneath the telescope be used for plotting all important points, and that the distances be determined with an accuracy consistent with the scale. This often will mean that the distances to corners of buildings, large trees, boundary posts, or plane-table points, must be measured with the tape. The stadia may be sufficient for locating contour points. When the plane table is set carefully over a station it must be nearly oriented before centering the point, otherwise the point will be thrown out of position when the table is turned in orienting.

If the scale is smaller, say 100 ft. or 200 ft. to the inch, the plane-table point need not be centered over the station with the same accuracy, the side straight edge of the alidade may be used for all sights, and stadia measurements may be used much more freely. The errors of all of these approximations will fall within the limits of a pencil dot.

The control of such a plane-table map should consist of a transit and tape traverse, or triangulation, to locate the main plane-table stations accurately, and direct levels to furnish benchmarks wherever needed. Only instrument stations of minor importance should be located with the table itself. The

*See Art. 4.27.

traverse points should be plotted by rectangular coödinates and the plotting carefully checked. The elevations of the B.M.'s should also be checked.

In locating contours on a large-scale map it may be advisable to place the rod exactly on the contour for each point located, thus tracing out the contour in a direct manner, as was described for stadia method in Art. 7.39. If less accuracy is required in any particular survey, the contours may be obtained by interpolating between elevations of points located by stadia and vertical angles, as is usual on small-scale maps.

While the plane table is preëminently an instrument for contour mapping, it may be used profitably for filling in details of other kinds of surveys. The control lines of the survey may be measured with a tape and plotted on the plane-table sheet by means of the alidade and scale. This gives a traverse which is sufficiently accurate from which to locate all details and does not require a long time to prepare. The topographic details are put in by setting the plane table at the established stations and locating points by the usual methods of intersection and radiation. (See Arts. 7.47. and 7.48.) These details are plotted as the survey progresses and are checked by sights taken from different stations. As much or as little tape work may be introduced as is required. All sorts of checks such as ties, ranges, etc., may be applied. By the use of this method the survey takes little longer, perhaps no longer, in the field than it would if made by the transit and tape method and the party returns to the office with an accurate pencil plan of the site, which then merely requires to be inked in by the draftsman.

Preliminary plans for the study of a highway location, or the elimination of a grade crossing, the survey of a cemetary, a map of the site of an accident, or a survey for the reconstruction of an existing bridge, are examples of plans which may be made economically by the plane-table method. It is also useful for filling in detail on maps prepared from aerial surveys. An enlarged reproduction of the aerial photograph is often used as a plane-table sheet.

CHAPTER VIII

OBSERVATIONS FOR MERIDIAN AND LATITUDE*

OBSERVATIONS FOR MERIDIAN

8.1. The Celestial Sphere.—When making observations on the sun, stars or other celestial objects with an engineer's transit and when calculating the results from such observations, it is convenient to assume that these celestial objects are situated on the surface of an imaginary sphere, called the celestial sphere, whose center is at the center of the earth and whose radius is infinite. This is perfectly consistent with what is actually seen by an observer on the earth because all celestial objects are so far away that they appear to the eye to be at the same distance and on the inside surface of a great sphere. The reason for adopting the celestial sphere is that all problems involving angular distances between points and angles between planes at the center of the sphere may readily be solved by spherical trigonometry.

Fig. 8.1.1 shows a celestial sphere with a transit set up at point *C* on the earth's surface. The earth is so small compared with the celestial sphere that, in astronomical calculations, vertical angles measured to celestial objects are practically the same whether measured at the surface or at the center of the earth, therefore the point *C* may be assumed as at the center of the sphere. Where this assumption would cause an appreciable error, the angle measured on the surface can be corrected to give the corresponding angle at the center. (See Art. 8.2)

---

*See also Chapter 2, Volume II.

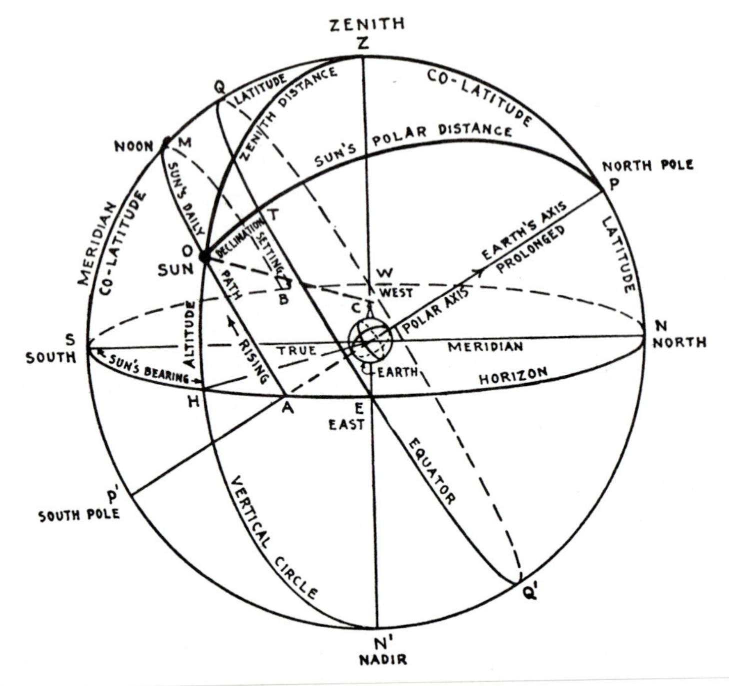

FIG. 8.1.1. CELESTIAL SPHERE.

The points where the axis of the earth intersects the sphere are the *celestial poles*. The great circle $EQWQ'$ is the *celestial equator*. It is the trace on the celestial sphere where the plane of the earth's equator if extended would intersect the sphere. It is everywhere 90° from the celestial poles. The point Z where the plumb-line of the transit prolonged upward intersects the celestial sphere is called the *zenith*. If prolonged downward it pierces the sphere at N', the *nadir*. The great circle $NESW$ midway between and everywhere 90° from the zenith and nadir is called the *horizon*. It is obvious that every observer has a different zenith and hence a different horizon. No matter what horizontal direction a telescope of a transit is pointing, if it is rotated about the horizontal axis until it points vertically, it will point to the zenith and will have traced on the sphere a great circle called a *vertical circle.* The vertical angle to any celestial object, called the *altitude,* is the angular distance the object is above the horizon, as *HO*. The great circle *SZPN* is the observer's *meridian*; it is also a vertical circle. It should be noted that the observer's meridian passes through the pole and his zenith. The small circle *AMB*, parallel to the equator, is a *parallel of declination;* the apparent daily path of a star or the sun (approximately) is a parallel of declination. The *declination* of the sun or a star is its angular distance north or south of the equator, as *OT.* The declination is considered positive when north, negative when south. Values of the declination of celestial bodies are given in the American Ephemeris and Nautical Almanac. The polar distance of the sun or a star is 90° minus the declination, as *OT.*

The *terrestrial latitude* of an observer is his angular distance measured along a meridian north or south from the equator. It will be noted that this corresponds on the celestial sphere to the arc of the meridian between the equator and the zenith, as *QZ*. Hence it is seen that the latitude is the declination of the observer's zenith. By geometry it is also seen to be equal to the altitude of the pole, as *NP.* The *azimuth,* or *bearing*, of the sun or a star is an arc of the horizon measured from the north or the south point to the vertical circle through the object. The bearing of the sun in Fig. 8.1.1 is *SH* (in the southeast quadrant). The azimuth of the sun measured clockwise from the south is *SWNEH*. The solution of the spherical triangle *PZO* would give the angle *PZO* = *NCH*, which corresponds to the bearing *NEH*. Such a bearing, being over 90°, would be changed to the arc *SH* (southeast quadrant) or to the arc *SWNEH*, the azimuth.

8.2. Corrections to Observed Altitudes.—In the solution of many astronomical problems it is necessary to know the true altitude of the center of the celestial body; i.e., the angular distance the center of the object is above the horizon, measured on a vertical circle. An observed altitude (approx. true altitude) is first obtained by measuring with the transit the vertical angle to the object. But, before the true altitude can be found it is necessary to apply certain corrections to the observed altitude.

INDEX CORRECTION.—The first correction to the recorded vertical angle is the *index correction.* It is determined and applied as explained in Art. 2.40.

REFRACTION CORRECTION.—Light rays passing through the air are refracted, and cause celestial objects to appear always at a greater altitude than they really are; for this reason the refraction correction is **always subtracted.** Values of this correction are given in Table 8.9.1.

SEMI-DIAMETER.—When the celestial object has an appreciable diameter such as the sun and moon it is customary to obtain the altitude by measuring the vertical angle to either the upper or lower limb (edge). As the required altitude is to the center of the object, the semi-diameter (angular distance from the limb to the center) is added to or subtracted from the observed altitude corresponding to the limb that has been observed. Values of the semi-diameter of the sun and moon are given in the American Ephemeris and Nautical Almanac and the smaller publication, the Nautical Almanac (see footnote, Art. 8.9). Stars are so far away that they appear in the telescope as small points and are bisected when the altitude is being observed.

PARALLAX.—Parallax is the difference in direction of a celestial object as seen from the surface and from the center of the earth. Altitudes of celestial objects are necessarily measured from the earth's surface. The coördinates given in the Ephemeris and the Nautical Almanac are referred to the earth's center. Hence the observed altitude must be reduced to the earth's center. In order to do this a correction called the parallax correction is applied to the altitude. It is **always added.** No parallax correction is necessary when stars are observed. They are so far away that the parallax correction is inappreciable. In the case of the sun the maximum correction is only 9 seconds and for the moon about one degree. For observations made with ordinary surveying instruments, values of the parallax correction are found in various publications. Table 8.18.2 is a table in which the refraction and parallax corrections are combined. This table can be used only for observations made on the sun. Corrections should be applied in the order given.

8.3. To Establish a True Meridian Line by Observation on Polaris (the Pole-Star).—Since the celestial pole is not visible it is necessary to make use of the celestial objects for establishing meridians. One of the simplest and at the same time the most accurate method for this purpose is to observe the Pole-star *(Polaris)*. On account of the earth's daily rotation on its axis all heavenly bodies appear to revolve once a day around the earth. Stars which are near the equator appear to revolve in large circles parallel to the equator. As we look farther north the apparent size of the circles grows smaller.

The Pole-star *(Polaris)* revolves about the celestial pole in a small circle whose radius (polar-distance) is about one degree. (Fig. 8.3.1.)

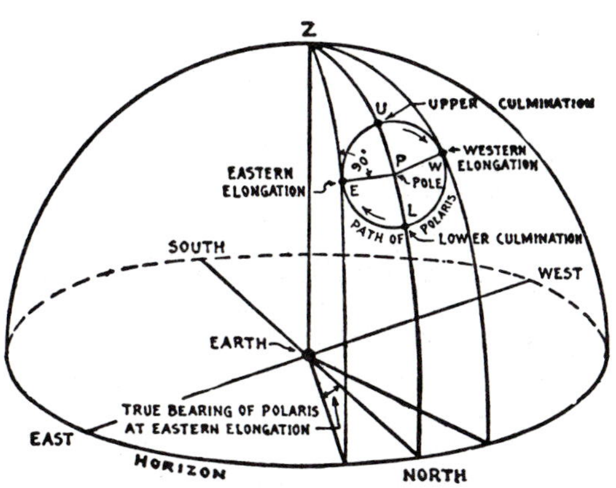

Fig. 8.3.1. Apparent Path of Polaris as Seen from Outside the Celestial Sphere.

When the star is directly above the pole it is in the plane of the meridian (bearing true north) and is said to be at *upper culmination.* About 12 hours later it will be directly below the pole (bearing true north) at *lower culmination.* About half-way between these two positions the star reaches its greatest east or west bearing, and at such times is said to be at its *greatest elongation.* At either eastern or western elongation the star's bearing is not changing because the star is moving vertically at that instant. The instant of elongation is therefore the most favorable time for an accurate observation for meridian. At culmination the star is changing its bearing at the maximum rate, so this is not so good a time to make an accurate observation. Polaris moves so slowly, however, that even at culmination its bearing can be obtained with fair accuracy.

This star can be found easily by means of two conspicuous constellations near it, *Cassiopeia* and *Ursa Major.* The seven bright stars in the latter constellation form what is commonly known as the "Great Dipper" (Fig. 8.3.2). The two stars forming the part of the bowl of the Dipper farthest from the handle are called the "pointers," because a line through them points almost directly at the Pole-star. On the opposite side of Polaris is *Cassiopeia,* shaped like a letter W. A line drawn from δ *Cassiopeiæ,** the lower left-hand star of the W, to ζ *Ursæ Majoris,* the middle star of the dipper handle, passes very close to Polaris and also to the pole itself.

8.4. OBSERVATION FOR MERIDIAN ON POLARIS AT ELONGATION.—When the Dipper is on the right and Cassiopeia on the **left,** Polaris is near its **western elongation;** when the Dipper is on the **left** Polaris is near **eastern elongation.** When the constellations are approaching one of these positions the transit should be set over a stake and leveled, and the telescope focused upon the star.† Unless the observation

*The Greek Alphabet will be found in the Appendix.

†It is difficult to find a star in the field of view unless the telescope is focused for a very distant object. The surveyor will find it a convenience if he marks on the telescope tube the position of the objective tube when it is focused for a distant object.

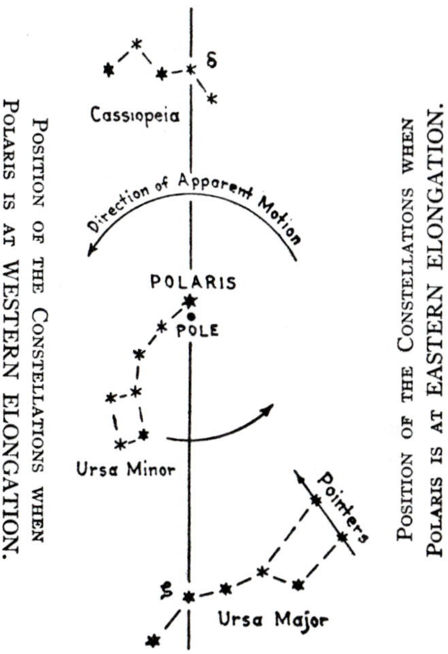

Fig. 8.3.2. Relative Position of the Constellations near the North Pole.

occurs at about sunrise or sunset it will be necessary to use an artificial light to make the cross-hairs visible. If the transit is not provided with a special reflector for throwing light down the tube a good substitute may be made by cutting a small hole in a piece of tracing cloth or oiled paper and then fastening it over the end of the telescope tube with a rubber band. If a light is then held in front and a little to one side of the telescope the cross-hairs can be plainly seen. The star should be bisected by the vertical hair and followed by means of the tangent screw in its horizontal motion until it no longer changes its bearing but moves vertically. (It will be seen from Fig. 8.3.1 that when the star is approaching eastern elongation it is moving **eastward** and **upward;** when approaching western elongation it is moving **westward** and **downward.**) As soon as this position is reached the telescope should be lowered and a point set in line with the vertical cross-hair at a distance of several hundred feet from the transit. Everything should be arranged beforehand so that this can be done quickly. Immediately after setting this point the instrument should be reversed, **re-leveled** if necessary, and again pointed on the star. A second point is then set at one side of the first. The mean of these two points is free from the errors of adjustment of the transit. On account of the great difference in altitude between the star and the mark the elimination of instrumental errors is of unusual importance (Art. 2.54). For 10 minutes of time on either side of elongation the bearing of the star does not change more than 5 seconds of arc and therefore there is sufficient time to make these two pointings accurately. The times of the elongations may be obtained from Tables 8.6.1 and 8.6.2.

TABLE 8.4.1

MEAN POLAR DISTANCE OF POLARIS*

Year	Mean Polar Dist.	Year	Mean Polar Dist.
1964	0° 54′ 09″	1971	0° 52′ 08″
1965	0° 53′ 52″	1972	0° 51′ 51″
1966	0° 53′ 35″	1973	0° 51′ 34″
1967	0° 53′ 17″	1974	0° 51′ 17″
1968	0° 53′ 00″	1975	0° 51′ 00″
1969	0° 52′ 43″	1976	0° 50′ 43″
1970	0° 52′ 26″	1977	0° 50′ 26″

*The above table was derived from data furnished by the Director of Nautical Almanac, Washington, D. C. The Mean Polar Distance is the polar distance the star would have at the beginning of the year, if unaffected by small periodic variations.

In taking the polar distance from the table for the purpose of looking up its sine the student should keep in mind the degree of precision desired in the computed result.

After the direction of the star at elongation has been found, the meridian may be established by laying off an angle equal to the azimuth, or true bearing of the star. Since this angle to be laid off is the **horizontal** angle between the star and the pole, it is not equal to the polar distance itself, but may be found from the equation:—

$$\text{Sin Star's True Bearing} = \frac{\text{Sin Polar Distance of Star}^{*}}{\text{Cos Latitude}}$$

The mean polar distances for each of several years may be found in Table 8.4.1. The latitude may be obtained from a reliable map or by observation (Arts. 8.14-8.15).

When the transit is set up at the south end of the line the angle thus computed must be laid off to the **right** if the elongation is **west,** to the **left** if the elongation is **east.** A convenient and accurate way of laying off the angle is by measuring the distance between the two stakes *A* and *B* (Fig. 8.4.1), and calculating the perpendicular distance *BC* which must be laid off at the north stake *B* to give a meridian *AC*.

Instead of placing a mark on the ground in line with Polaris at the instant of elongation, the direction to the star can be read, assuming a direction has been read to another fixed (and illuminated) ground point. Then by rapidly inverting the telescope and re-sighting Polaris to read a direction and then again reading the direction to the fixed ground point the angle will be established between the fixed point and the W. or E.

---

*This equation may be derived as follows: In Fig. 8.3.1, compute triangle *PZE* by spherical trigonometry,

$$\frac{\sin PZE}{\sin ZEP} = \frac{\sin PE}{\sin ZP}$$

But *PZE* is the angle between two vertical circles and equals the bearing. $ZEP = 90°$; *ZE* is tangent to the circle *UWLE*, which represents the path of Polaris. *PE* is the polar distance and *ZP* is $90° -$ latitude.

Hence,

$$\sin PZE = \frac{\sin PE}{\cos \text{lat}}$$

Elongation of Polaris (Fig. 8.4.2). This procedure in effect serves to obtain the true bearing of the line to the fixed point.

8.5. Observation for Meridian on Polaris at Culmination.—To observe Polaris at the instant of Upper Culmination or Lower Culmination would, of course, eliminate the need for more computation, for at that instant the star is on the true meridian. Two difficulties arise: first, the star is moving laterally quite rapidly and it is difficult to follow it with the cross-wire of the telescope and stop the telescope motion at the exact instant: second, it is difficult to have one's watch exact to the nearest second or better.

Reference to Fig. 8.3.2 will show that one can find an approximately correct meridian by observing with the transit the instant when Polaris is vertically aligned with δ Cassiopeiae or ζ Ursæ Majoris. When Polaris is above the pole (upper culmination), ζ Ursæ Majoris will be almost exactly underneath Polaris. When Polaris is below the pole (lower culmination) δ Cassiopeiæ will be almost directly below Polaris. Tables do exist (in the current American Ephemeris and Nautical Almanac) that give the exact time interval from the instant of alignment of these stars until the instant that Polaris is at its upper culmination (usually only a few minutes), so a fairly exact pointing could be made to obtain the true meridian direction.

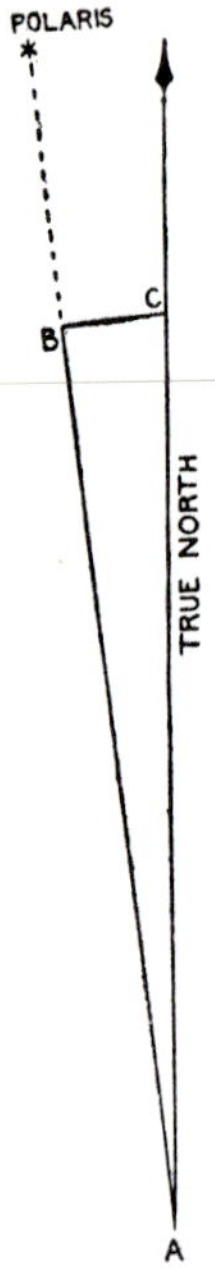

FIG. 8.4.1.

The observation to determine when the two stars are in the same vertical plane is at best only approximate, since the instrument must be pointed first at one star and then at the other; but since Polaris changes its azimuth only about one minute of angle in two minutes of time, there is no difficulty in getting fair results by this method. The vertical hair should be set first on Polaris, then the telescope is lowered to the approximate altitude of the other star to be used. As soon as this star comes into the field the vertical hair is set again carefully on Polaris. As it will take the other star about two minutes to reach the center of the field there will be ample time for this pointing.

Then the telescope is lowered and at the instant when the star passes the vertical hair the time is noted on a watch. This will be the time desired, with an error of only a few seconds. The time of culmination should then be computed as described above and the vertical hair set on Polaris when this computed time arrives. The telescope is then in the meridian, and this may be marked on the ground.

In this method the actual error of the watch has no effect on the result since it is used only for measuring the **interval** of a few minutes. The error in the meridian obtained by this method will seldom exceed one minute of angle.

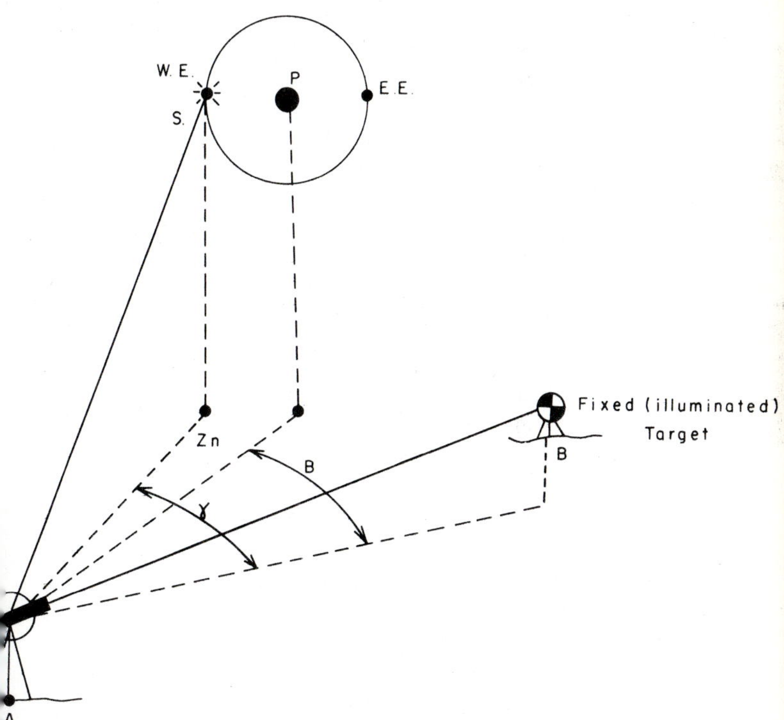

FIG. 8.4.2. OBSERVING POLARIS AT ELONGATION TO OBTAIN AZIMUTH OR BEARING OF LINE AB.

TABLE 8.6.1

POLARIS FOR THE MERIDIAN OF GREENWICH
LATITUDE 40° N, 1976

Universal Time or Greenwich Civil Time

Date	Upper Culmination	Previous East Elongation	Next West Elongation	Next Lower Culmination	Var. Per Day
1976	h m	h m	h m	h m	m
Jan. 1	19 27.2	13 31.0	1 23.4	7 25.2	
11	18 47.7	12 51.5	0 43.9	6 45.7	3.96
21	18 08.1	12 11.9	0 04.3	6 06.1	3.96
31	17 28.6	11 32.4	23 24.8	5 26.6	3.96
Feb. 10	16 49.0	10 52.8	22 45.2	4 47.0	3.96
20	16 09.5	10 13.3	22 05.7	4 07.5	3.96
Mar. 1	15 29.9	9 33.7	21 26.1	3 27.9	3.95
11	14 50.4	8 54.2	20 46.6	2 48.4	3.95
21	14 11.0	8 14.8	20 07.2	2 09.0	3.95
31	13 31.6	7 35.4	19 27.8	1 29.6	3.94
Apr. 10	12 52.2	6 56.0	18 48.4	0 50.2	3.94
20	12 12.9	6 16.7	18.09.1	0 10.9	3.93
30	11 33.6	5 37.4	17 29.8	23 31.6	3.93
May 10	10 54.4	4 58.2	16 50.6	22 52.4	3.92
20	10 15.2	4 19.0	16 11.4	22 13.2	3.92
30	9 36.0	3 39.8	15 32.2	21 34.0	3.92
June 9	8 56.9	3 00.7	14 53.1	20 54.9	3.91
19	8 17.8	2 21.6	14 14.0	20 15.8	3.91
29	7 38.7	1 42.5	13 34.9	19 36.7	3.91
July 9	6 59.6	1 03.4	12 55.8	18 57.6	3.91
19	6 20.5	0 24.3	12 16.7	18 18.5	3.91
29	5 41.5	23 45.3	11 37.7	17 39.5	3.91
Aug. 8	5 02.4	23 06.2	10 58.6	17 00.4	3.91
18	4 23.4	22 27.2	10 19.6	16 21.4	3.91
28	3 44.3	21 48.1	9 40.5	15 42.3	3.91
Sept. 7	3 05.2	21 09.0	9 01.4	15 03.2	3.91
17	2 26.0	20 29.8	8 22.2	14 24.0	3.91
27	1 46.9	19 50.7	7 43.1	13 44.9	3.92
Oct. 7	1 07.7	19 11.5	7 03.9	13 05.7	3.92
17	0 28.5	18 32.3	6 24.7	12 26.5	3.92
26	23 49.2	17 53.0	5 45.4	11 47.2	3.93
Nov. 5	23 09.9	17 13.7	5 06.1	11 07.9	3.93
15	22 30.6	16 34.4	4 26.8	10 28.6	3.94
25	21 51.2	15 55.0	3 47.4	9 49.2	3.94
Dec. 5	21 11.8	15 15.6	3 08.0	9 09.8	3.94
15	20 32.3	14 36.1	2 28.5	8 30.3	3.95
25	19 52.8	13 56.6	1 49.0	7 50.8	3.95
1977					
Jan. 4	19 13.3	13 17.1	1 09.5	7 11.3	3.95

From July 26 to Oct. 24 the East Elongation is for the previous day. From Jan. 1 to Jan. 22 and from Oct. 25 to Jan. 5 the West Elongation is for the next day.
From Jan. 1 to Apr. 22 and from Oct. 24 to Jan. 5 the lower culmination is for the next day.

8.6. To Find the Standard Time of Culmination and Elongation.—The approximate times of upper culmination (or transit) of Polaris may be found in Table 8.6.1. The times of elongation may be found by means of Table 8.6.2.

Eastern elongation precedes and western elongation follows upper culmination by the time interval given in Table 8.6.2. Lower culmination precedes or follows upper culmination by $11^h\ 58.0^m$.

(*a*) *To refer to any other than the tabular longitude* (90°).—Add 0.11^m for each 10° east of the nintieth meridian or subtract 0.11^m for each 10° west of the Greenwich meridian.

(*b*) *To refer to standard time.*—Add to the quantities in Table 9 four minutes for every degree of longitude the place of observation is west of the standard meridian (60°, 75°, 90°, etc.). Subtract when the place is east of the standard meridian.

8.7. Meridian Observation on Polaris at a Known Instant of Time.—This observation consists in measuring the horizontal angle, by repetition, from a reference mark to Polaris, the watch time of each pointing and the altitude of the star both being noted.

It is assumed that the average of all the azimuths is the same as the azimuth corresponding to the average of the watch readings. The average watch reading is therefore corrected for any known error in the timepiece and the resulting time taken as the Standard time of the observation. This Standard time is then converted into Local Civil Time by means of the longitude of the place. Next it is necessary to find the "hour angle" of Polaris at this instant. This is obtained by taking the difference between the local time just found and the time of the nearest Upper Culmination as given in Table 8.6.1. This difference, increased by 10^s for each hour in the interval, is the hour angle of Polaris, which should now be converted into degrees and minutes. (1° = 4 min.)

The azimuth of Polaris is computed by the formula

$$Z = p \sin P \sec h$$

in which p is the polar distance in seconds (or minutes), P is

TABLE 8.6.2

CORRECTIONS TO TIMES OF ELONGATION FOR DIFFERENT LATITUDES 1976

Latitude	10°	15°	20°	25°	30°	35°	40°	45°	50°
	m	m	m	m	m	m	m	m	m
West Elongation	+2.2	+1.9	+1.6	+1.3	+0.9	+0.5	0.0	−0.5	−1.2
East Elongation	−2.2	−1.9	−1.6	−1.3	−0.9	−0.5	0.0	+0.5	+1.2

Courtesy, Keuffel & Esser Co.

To obtain the precise Local Civil Time (*LCT*) of elongation or culmination at the meridian where the observation is to be made (the local meridian), add .011 m. for each degree east of Greenwich. Subtract if west.

To convert *LCT* to Standard Time, add 4 m. for each degree that the local meridian is west of the Standard Time Meridian. Subtract if east.

Example 1. Find Eastern Standard Time (75th mer. time) of upper culmination May 10, 1976 at 78° W. Long.

GCT of U.C. at G. (Table 8.6.1)	$10^h\ 54.4^m$
.011 × 78 = .86	− 0.9
LCT of U.C. at 78° W	$10^h\ 53.5^m$
4 × 3 = 12	+ 12.0
EST of U.C. at 78° W	$11^h\ 05.5^m$

Example 2. Find Pacific Time (120th Mer. time) of upper culmination June 17, 1976 at 122° W. Long.

From Table 8.6.1, U.C. at G., 1976 June 09 *GCT*	$8^h\ 56.9^m$
Var. per day = 3.91^m June 09 to June 17 = 8 days $8 \times 3.91^m = 31.3^m$	$-\ 31.3^m$
June 17 *GCT*	$8^h\ 25.6^m$
$.011^m \times 122 = 1.34^m$	$-\ 1.3^m$
LCT at 122° W	$8^h\ 24.3^m$
$4^m \times 2 = 8^m$	$+\ 8.0^m$
PST of U.C. at 122° W	$8^h\ 32.3^m$

In order to make ordinary latitude observation, it is sufficient to know the time of culmination of Polaris with an accuracy of five minutes.

the hour angle, and h is the altitude; Z is the azimuth in the same units as p.* If the observed time is earlier than the time of upper culmination the star is east of north (unless the hour angle is greater than 12 hours); if the observed time is later than the time of upper culmination, the star is west of north (unless the hour angle is over 12^h). This azimuth combined with the measured horizontal angle gives the azimuth of the reference mark.

EXAMPLE

Observations on Polaris for Azimuth, Oct. 20, 1976, by three repetitions between mark and star. Latitude, 42° 35′.5; Longitude, 71° 07′.5 W.

	Horizontal Angle (to the right)	Altitude		Watch (10ˢ fast of E.S.T.)
Mark	0° 00′			
Polaris 1st rep.	261 37	42° 55′		6ʰ 59ᵐ 30ˢ P.M.
" 2nd "		42 56		7 03 10
" 3rd "	64 52	42 57		7 05 30
Mean	261° 37′.3	42° 56′		7 02 43
		refr. −1	Watch fast	− 10
		h = 42° 55′		7 02 33
			Diff. Long.	+15 30 (3° 52′.5)
			Loc. Time	7 18 03 P.M.
				12
			Loc. Civ. Time	19ʰ 18ᵐ 03ˢ

$p = 0° 50'.53$

$Z' = p' \sin P \sec h$

$Z' = 50'.53 \dfrac{0.95983}{0.73234}$

$= 66'.2$

	N 1° 06′.2 E
Hor. Angle	261° 37′.3
Mark to N	260° 31′.1
Azimuth	99° 28′.9
Bearing	S 80° 31′.1

U.C. Oct. 17, 1976	0ʰ 27ᵐ 34ˢ
Corr. for 4 days	− 15ᵐ 41ˢ
U.C. Oct. 21, 1976	0ʰ 11ᵐ 50ˢ
Corr. for Long.	+ 13ˢ
Local Civil Time, U.C., Oct. 21	0ʰ 12ᵐ 03ˢ
" " " U.C., Oct. 20	24ʰ 12ᵐ 03ˢ
" " " Obs'n., Oct. 20	19ʰ 18ᵐ 03ˢ
Diff. in Time	4ʰ 54ᵐ 00ˢ
4.9 × 10ˢ	49ˢ
Hour Angle = P	4ʰ 54ᵐ 49ˢ
P	73° 42′ 15″ East

*Azimuths of Polaris for all hour angles and for latitudes from 10° to 70° are given in "Astronomical Phenomena."

8.8. SOLAR OBSERVATIONS.—Where great accuracy is not required many surveyors prefer solar observations because they can be made without much additional work, while star observations have to be made at night and require special arrangements for illuminating the field of view and the mark. If it is sufficient for the purpose in view to obtain the azimuth with ½ minute of angle solar observations will answer. In making these observations with the ordinary transit it is necessary to have some means of cutting down the sun's light so that it will not be too bright for the eye while making pointings. This is usually effected by placing a dark glass over the eyepiece. A dark glass in front of the objective will introduce error into the pointings unless the faces of this glass have been made plane and exactly parallel. If the instrument is not provided with a dark glass the observation may be made by holding a white card back of the eyepiece while the telescope is pointing at the sun. If the eyepiece tube is drawn out the sun's disc and the cross-hairs can both be focused sharply on the card. By this means pointings can be made almost as well as by direct observation.

8.9. OBSERVATION FOR MERIDIAN BY A SINGLE ALTITUDE OF THE SUN.—The most convenient method of obtaining the azimuth of a line is by measuring an altitude of the sun and computing the sun's azimuth by spherical trigonometry. This observation may be made in a few minutes time while the survey is in progress and is therefore preferred by many surveyors to the observation on Polaris, which consumes more time and usually requires a special trip to the point of observation.

To make this observation set up the transit at one end of the line whose azimuth is to be determined and set the plate vernier to read 0°. Sight the vertical cross-hair on the point marking the other end of the line, using the lower clamp and tangent screw. Place the colored shade glass over the eyepiece, loosen the upper clamp and point the telescope toward the sun. Before attempting to make the pointings focus carefully so that the edge of the sun is distinct, and then examine the

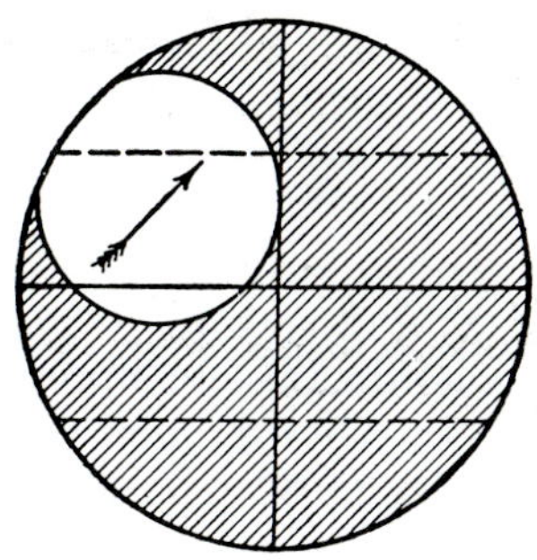

FIG. 8.9.1. SUN'S DISC BEFORE OBSERVATION. (A.M. Observations.) Northern Hemisphere.

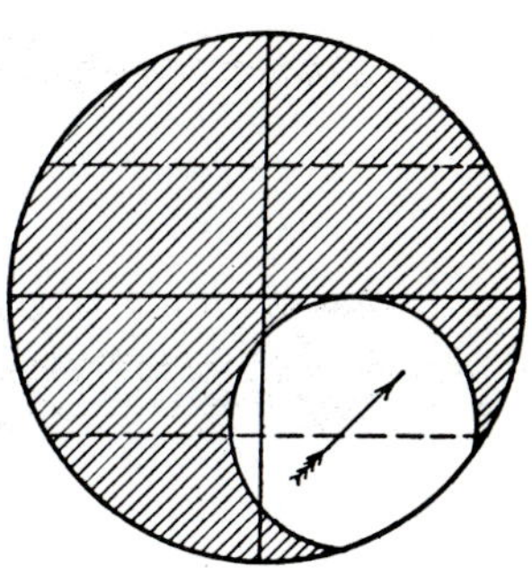

FIG. 8.9.2. SUN'S DISC BEFORE OBSERVATION. (A.M. Observations.) Northern Hemisphere.

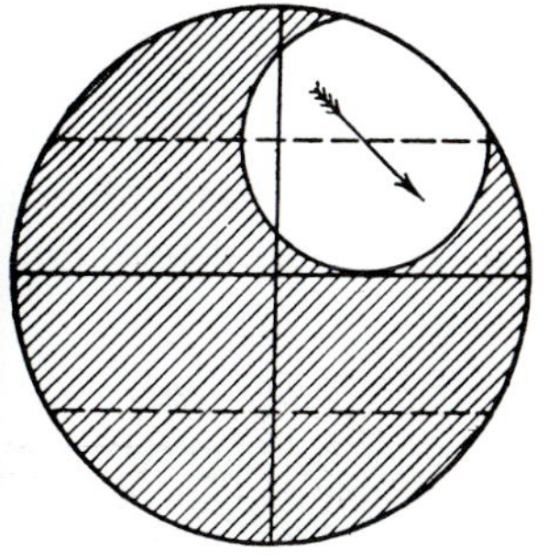

FIG. 8.9.3. SUN'S DISC BEFORE OBSERVATION. (P.M. Observations.) Northern Hemisphere.

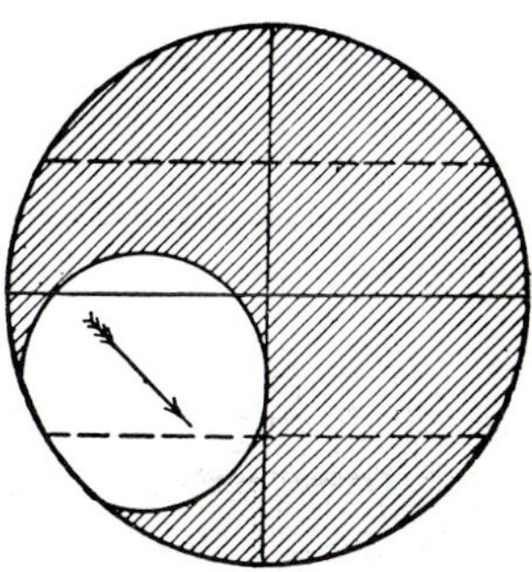

FIG. 8.9.4. SUN'S DISC BEFORE OBSERVATION. (P.M. Observations.) Northern Hemisphere.

field to be certain which is the middle horizontal hair. The three horizontal hairs cannot all be seen at once through the colored screen and there is danger of using one of the stadia hairs by mistake.

When making observations in the forenoon in the northern hemisphere it is best to observe first the righthand and lower edges of the sun's disc, and then the left-hand, and upper edges as shown in Figs. 8.9.1 & 8.9.2. In the afternoon the positions would be as shown in Figs. 8.9.3 & 8.9.4.* In order to avoid the necessity of setting on both edges at the same instant the forenoon observations may be made as follows. Set the horizontal hair so that it cuts a small segment from the lower edge of the disc, and the vertical hair tangent to the right edge.† Since the sun is now rising and moving to the right it is only necessary to keep the vertical hair tangent to the right edge with the upper plate tangent screw; the apparent motion of the sun itself will set the horizontal hair after a lapse of a few seconds. At the instant both hairs are tangent to the disc stop following the sun's motion and note the time by the watch. Read the vertical and horizontal circles, and record all three readings. The second observation is made in a similar manner except that the vertical hair is set a little way in from the left edge of the sun and the horizontal hair is kept tangent to the upper edge by means of the tangent screw on the standard. If it is desired to increase the precision several such observations may be made in each position, being careful to take the same number of pointings in each. If the transit has a full vertical circle the telescope should be inverted between

*In the diagram only a portion of the sun's disc is visible on account of the small angular diameter of the field of the telescope. In a telescope having a very large field the whole disc may be seen.

†It should be kept in mind that if the telescope has an inverting eyepiece the direction of the sun's apparent motion is reversed. If a reflecting prism is attached to the eyepiece, the upper and lower edges of the sun are apparently interchanged, but the right and left edges are not affected.

the two observations, thus eliminating errors of adjustment of the cross-hairs and the horizontal axis. Before the plate is unclamped the index correction should be determined and recorded. After the pointings on the sun are completed the telescope should be sighted again along the line whose azimuth is being determined and the vernier read to be certain that the plate has not moved during the observations.

Before attempting to calculate the azimuth examine the intervals between the successive watch times, vertical angles, and horizontal angles. These should be very nearly proportional. If there is any sudden change in the relation between these intervals it indicates a mistake and should be investigated. In the example on p. 344 the intervals are:

Horizontal	*Vertical*	*Time*
32′	19′	157^s
25′	13′	120^s

indicating that no large mistake has been made, since the lower interval is in all three cases about three-quarters of the upper one.

The mean of all the vertical circle readings, corrected for index error, is the apparent altitude. This altitude decreased by the correction for atmospheric refraction (Table 8.4.1 or where greater precision in solar observations is desired, Table 8.18.2) is the true altitude, h, in the formula given later in this article.

In order to compute the azimuth it is necessary to know the latitude of the place. The latitude may be obtained from a reliable map; or, if this is likely to be in error a half minute or more, it may be observed as described in Arts. 8.6-8.17. It is also necessary to know the declination of the sun at the instant of observation.

TABLE 8.9.1

REFRACTION CORRECTION

Altitude	Refraction	Altitude	Refraction
10°	5′ 19″	20°	2′ 39″
11	4 51	25	2 04
12	4 27	30	1 41
13	4 07	35	1 23
14	3 49	40	1 09
15	3 34	45	0 58
16	3 20	50	0 49
17	3 08	60	0 34
18	2 57	70	0 21
19	2 48	80	0 10

The sun's declination required in this observation may be taken from the "American Ephemeris and Nautical Almanac,"* Ephemeris of the sun, under the heading "apparent declination." The value of the declination given is that for the instant of 0^h (or midnight) each day at Greenwich. The tabulated difference for 24^h is given in the next column to the right. In order to find the declination for any instant of local or of Standard Time it is necessary first to find the corresponding instant of Greenwich Civil Time. In the United States, where Standard Time is in general use, the relation to Greenwich Time is simple. In the *Eastern* time belt the Standard Time is exactly 5^h earlier than at Greenwich; in the *Central* belt it is 6^h earlier; in the *Mountain* belt it is 7^h earlier; and in the *Pacific* belt it is 8^h earlier.† If the declination is desired for 7^h A.M. Eastern Time we must compute it for 12^h of Greenwich Time; if we wish the declination for 3^h P.M. Central Time we

*Published annually by the Nautical Almanac Office, U. S. Naval Observatory, and sold by the Superintendent of Documents. There is also the small "Nautical Almanac" which gives the declination of the sun. Various instrument makers publish a "Solar Ephemeris" containing this information.

†Daylight Saving Time is one hour later than Standard Time. If this kind of time is in use first subtract one hour and then proceed as described above.

must compute it for 9^h P.M. or 21^h of Greenwich Civil Time. If any instant of *local* time is to be converted into Greenwich Time this may be done by changing the west longitude of the place into hours, minutes and seconds, and adding it to the given local time.

For converting degrees into hours or hours into degrees the following table is convenient.

$15° = 1$ hour
$15' = 1$ minute
$15'' = 1$ second
also
$1° = 4$ minutes
$1' = 4$ seconds

It is important that the watch used should be nearly correct, at any rate within 2^m or 3^m. The correct Standard Time may be obtained by radio from U.S. Naval Observatory.

The correct standard time may be obtained by short-wave radio. The National Bureau of Standards beams Naval Observatory time from Fort Collins, Colorado (and also from Hawaii) every minute. In all, about 50 nations are transmitting the time constantly, primarily for purposes of navigation.

The declination for any hour of Greenwich Time is the value at 0^h increased (algebraically) by the "tabulated difference for 24^h" times the hours of Greenwich Civil Time divided by 24. That is, if the declination at 0^h is $-4° 44' 26''.8$, the tabulated difference for 24^h is $+1407''.5$, and the declination is desired for 10^h Greenwich Time, then the corrected declination equals $-4° 44' 26''.8 + (+1407''.5) \times 10 \div 24 = -4° 34' 40''.3$. North declinations are marked + and South declinations are marked −. An examination of the values for successive days will also show which way to apply the correction.

The azimuth of the sun's center may be computed by either of the formulas:

$$\cos Z_n = \frac{\sin D - \sin L \sin h}{\cos L \cos h}$$

or,
$$\cos Z_n = \frac{\sin D}{\cos L \cos h} - \tan L \tan h$$

in which L is the latitude of the place, h the true altitude of the sun's center, and D the sun's declination, + if North, − if South; Z_n is the azimuth of the sun east or west of the North point.

In making this computation five digits will be found sufficiently precise. Five-place tables define the angle within 5″ to 10″, which is a greater degree of precision than can be expected from such an observation. After the angle Z has been computed it is combined with the readings of the horizontal circle to obtain the azimuth of the line.

In order to determine the azimuth as accurately as possible by this method the observation should be made when the sun is nearly due east or due west, because the trigonometric conditions are then most favorable. Observations made near to noon are not reliable because small errors in the observed altitude and latitude cause large errors in the computed azimuth. Observations should also be avoided when the sun is less than 10° above the horizon, even though the sun is nearly east or west, because the correction for atmospheric refraction is uncertain. Table 8.9.1 gives values of the correction for an average temperature and pressure of the air; any deviation from these average conditions will produce changes in the correction, especially near the horizon. Table 8.18.2, contains the corrections for refraction and parallax for different air temperatures. (The parallax is never greater than 9″.)

8.10. Southern Hemisphere.—If the point of observation is in the southern hemisphere the following changes will be necessary: first, the pointings will be those indicated in Fig. 8.10.1 (provided the latitude is greater than the sun's declination); second, the formula must be modified suitably. This may be accomplished in either of two ways. The first would consist in treating the latitude L as negative and using the formula as in Example 1. The second method would be to consider L as positive and to reverse the sign of *D*. If this is done the azimuth *Z* will be that counted from the elevated (S) pole, either to the east or to the west. But if the sign of the right-hand side is changed then *Z* is counted from the north. In Example 2 the formula has the signs changed on the right-hand side. In using this formula the sign of the declination is reversed from that in the Ephemeris.

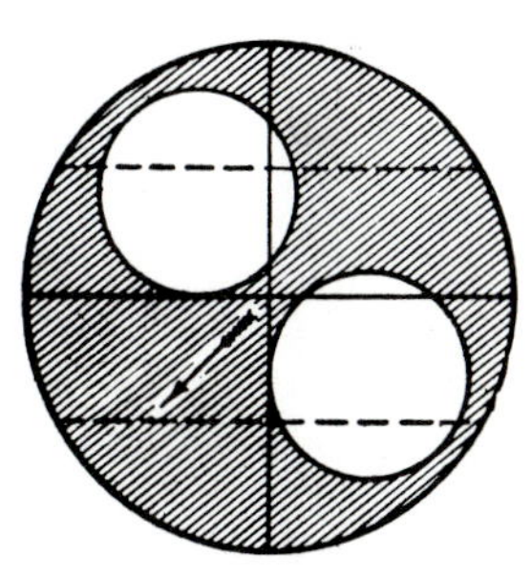

P.M. Observation

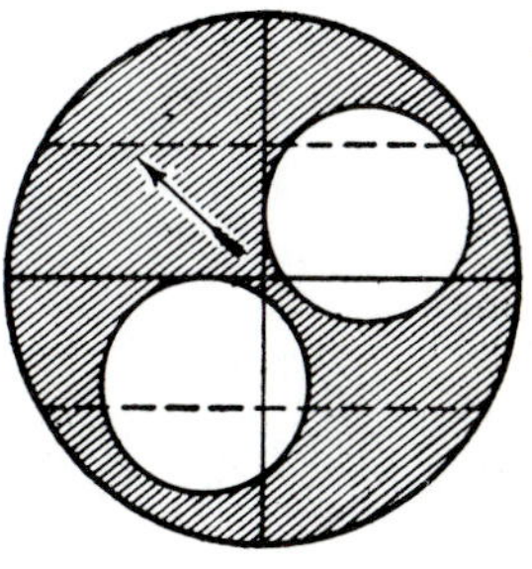

A.M. Observation

POSITION OF SUN'S DISC A FEW SECONDS BEFORE OBSERVATION. (Southern Hemisphere—Looking North)

FIG. 8.10.1.

If desired the time may be computed from the same data, the formula for the sun's hour angle being

$$\cos t = \frac{\sin h - \sin L \sin D}{\cos L \cos D}$$

Using the data of Example 1, p. 344, we find $t = 42° 14'.0$. This is the angle east of the meridian. Its equivalent in hours is $2^h 48^m 58^s$, and it represents the time that will elapse before noon. The local apparent time is therefore $9^h 11^m 04^s$. Since the apparent angular motion of the sun is not uniform, watch time is kept by what is known as the *fictitious* or *mean sun*, which is an imaginary point conceived to move at a uniform rate along the equator at such a speed as to make one revolution in the same time as the actual sun. The time kept by this sun is called *mean solar time*. The time kept by the real sun is called *apparent time*. The difference between these times is

EXAMPLE 1

OBSERVATION ON SUN FOR AZIMUTH

(Northern Hemisphere)

Latitude 42° 21′ N Longitude $4^h\ 44^m\ 18^s$ W
Date, Nov. 28, 1976, A.M.

	Horizontal Circle Vernier A	B	Mean	Vertical Circle		Watch (E. S. T.)
Mark	0° 00′	00′	00′			
Sun, R. and L. limbs	73 34	34.5	34.2	14° 41′		$8^h\ 38^m\ 42^s$ A.M.
Sun, R. and L. limbs	74 06	06	06	15 00		8 41 19
Instrument reversed						
Sun, L. and U. limbs	74 13	12.5	12.8	15 55		8 44 34
Sun, L. and U. limbs	74 38	37.5	37.8	16 08		8^h 46 34
Mark	0 00	00	00		Mean =	$8^h\ 42^m\ 47^s$ A.M.
						5^h
Mean horizontal angle (to the right)	74° 07′.7			Mean alt. 15° 26′.0 refr. (Table 11) −3 .5 h = true alt. 15° 22′.5		$13^h\ 42^m\ 47^s$ = Gr. Civ. T.

Using the formula:

$$\cos Z_n = \frac{\sin D - \sin L \sin h}{\cos L \cos h}$$

Declination at 0^h, Gr. Civ. T.	= −21° 17′ 24″
Tabular Diff.	= −0′.43/hr V.T.
− 0′.43 × 13^h.71	= − 05′ 54″
Corrected declination	= −21° 23′ 18″

$$\cos Z_n = \frac{-0.36469 - 0.67366 \times 0.26514}{0.73904 \times 0.96421}$$

Z_n = 139° 39′.7 (N tow. E)

Hour angle = 74° 07′.7

Bearing of mark N 65° 32′.0 E

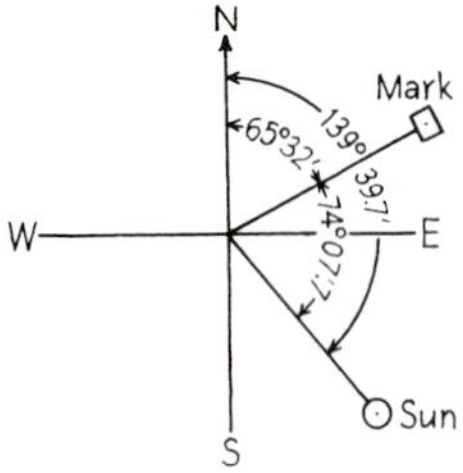

the *equation of time*. (Eq. T = App. T − Mean T.) It is given in Astronomical Phenomena, U. S. Naval Observatory, for 0^h Greenwich (midnight) for each day in the year. The equation of time for 0^h Greenwich on November 28, 1966, is $12^m\ 17^s.4$. Corrected for 9.18 hours after midnight, it is $12^m\ 17^s.4 - 20^s.1$

EXAMPLE 2

Observation on Sun for Azimuth

(Southern Hemisphere)

At a place in Latitude, 33° 01′ S; Longitude, 71° 39′ W, an observation is made on the sun Aug. 5, 1976, to determine the azimuth of A–B, as follows:

Instrument at Station A

(Air temperature 60° F.)

	Vernier A	Vertical arc	Watch (local time) P.M.
Sta. B.	0° 00′		
(sun symbol, limbs)	75° 15′	24° 22′	$3^h\ 00^m\ 10^s$
(sun symbol, limbs)	74 52	23 03	3 05 12
Mean	75° 03′.5	23° 42′.5	$3^h\ 02^m\ 41^s$
Refr. and Par., (Table 12)		− 2′.0	12
		h = 23° 40′.5	$15^h\ 02^m\ 41^s$

Longitude 71° 39′	$= 4^h\ 46^m\ 36^s$
Local Civil Time	= 15 02 41
Greenwich Civil Time	$= 19^h\ 49^m\ 17^s$
	$= 24^h - 4^h\ 10^m\ 43^s$
	$= 24^h - 4^h.18$
Decl. at 0^h Aug. 6	= +16° 43′ 36″
$0.68 \times 4^h.18$	= + 02′ 50″
	D = +16° 46′ 26″

Using the formula:

$$\cos Z_n = \tan h \tan L - \frac{\sin D}{\cos h \cos L}$$

Note: Sign of D is to be reversed!

$$\cos Z_n = 0.43845 \times 0.64982 - \frac{-0.28859}{0.91584 \times 0.83851}$$

Z_n= 48° 38′.7 northwest

Hor. angle= 75° 07 .5 (to right)

Bearing N 123° 46′.2 W

or S 56° 13′.8 W

$\times \dfrac{9.18}{24} = 12^m\ 17^s.4 - 07^s.7 = 12^m\ 09^s.7$. The local mean time or Civil Time is $9^h\ 11^m\ 04^s - 0^h\ 12^m\ 10^s = 8^h\ 58^m\ 54^s$. To obtain the corresponding Eastern Standard Time, the difference between the longitude of the place ($4^h\ 44^m\ 18^s$) and the Eastern Standard Time meridian (75° W. or 5^h) must be subtracted; i.e., $8^h\ 58^m\ 54^s - 0^h\ 15^m\ 42^s = 8^h\ 43^m\ 12^s$ E.S.T. Since the watch read $8^h\ 42^m\ 47^s$, it is 25^s slow.

FIG. 8.11.1. SOLAR ATTACHMENT TO TRANSIT.

(Courtesy, C. L. Berger & Sons, Inc.)

8.11. OBSERVATION FOR MERIDIAN BY MEANS OF THE SOLAR ATTACHMENT.—One of the auxiliaries to the engineer's transit is the *solar attachment*, illustrated in Fig. 8.11.1. It consists of a small instrument having motions about two axes at right angles to each other, like the transit itself. By means of this attachment a true meridian line can be found by an observation on the sun. This instrument seldom gives as great precision as direct solar observations but has the advantage that less calculation is required than with direct observations. In the form shown in Fig. 8.11.1, which is a modification of the Saegmuller pattern, the principal parts are the *polar axis*, which is attached to the telescope tube and is adjusted perpendicular to the line of sight and also the horizontal axis, and the *solar telescope*, which is mounted on the polar axis. The solar telescope is provided with clamps and tangent screws and can be revolved about the polar axis and can also be inclined to it at any desired angle. Attached to the solar telescope is a level tube, the axis of which is supposed to be parallel to the line of sight of the solar telescope. This is used when setting off the sun's declination by means of the vertical circle of the transit. At the base of the polar axis are four adjusting screws for making this axis perpendicular to the line of sight and to the horizontal axis of the main telescope. The solar telescope is provided with four additional cross-hairs which form a square whose angular diameter is equal to that of the sun. The eyepiece is covered with a colored glass to protect the eye while observing. A prismatic eyepiece attachment permits easy reading when the telescope is inclined. A counterpoise weight, shown near base of transit in Fig. 8.11.1, is attached to the bottom of the main telescope to balance the weight of the auxiliary telescope.

In Burt's attachment, the telescope is replaced by a small bi-convex lens and a metallic screen carrying ruled lines in place of cross-hairs. The sun's image is thrown on the screen and viewed with a magnifying glass; this device is really the equivalent of a telescope of low power. The sun's image is centered in the square formed by the ruled lines. The instrument is provided with a *declination arc* for setting off the declination of the sun when determining the direction of the meridian. To avoid the necessity of having a declination arc extending both ways from 0° the instrument has two sets of solar lenses and screens pointing in opposite directions; one is used for north and one for south declinations.

The Smith solar attachment is mounted on the side of one of the standards of the transit. The solar telescope serves as the polar axis. In front of the objective of the solar telescope is a mirror attached to a movable arm to which is attached the vernier of the declination arc. On the solar telescope is a special *latitude arc*, the vernier for which is at the top of the standard. The solar telescope may be rotated about its own axis. This instrument has the advantage that all of the settings may be made and allowed to remain without interfering with the use of the transit telescope for other purposes.

8.12. Observation on the Sun with Solar Attachment—Fieldwork.—If the polar axis is pointed to the celestial pole, made parallel to the earth's axis, then the small telescope can be made to follow the sun in its daily path by giving it an inclination to the polar axis equal to the sun's polar distance and revolving it about the polar axis.

1. To find the true meridian by an observation on the sun first make the angle between the polar axis and the solar telescope equal to the sun's polar distance at the time of the observation. This is done by turning the solar telescope into the same plane as the main telescope by sighting both on some distant object, and then making the angle between the two telescopes equal to the sun's declination. Some instruments are provided with a *declination arc* upon which the declination angle can be laid off directly. Others have a small spirit level attached to the small telescope, the vertical circle of the transit being used for laying off the declination angle. Incline the main telescope until the vertical circle reading equals the declination, and clamp; then level the solar telescope by means of the attached level. The angle between the polar axis and the solar telescope is then 90° plus or minus the reading of the vertical circle.

2. By means of the vertical circle of the transit incline the polar axis to the vertical by an angle equal to the **co-latitude of the place,** that is, 90° minus the latitude. The polar axis now has the same angle of elevation as the celestial pole. (For a method of finding the latitude by observation see Art. 8.17.)

3. If the observation is in the forenoon, place the solar tele

scope on the left of the main telescope (on the right if in the afternoon); then, by moving the whole instrument about the vertical axis and the solar telescope about the polar axis, point the solar telescope at the sun. The sun's image is brought to the center of the square formed by four cross-hairs, or ruled lines, in the solar telescope. The final setting is made

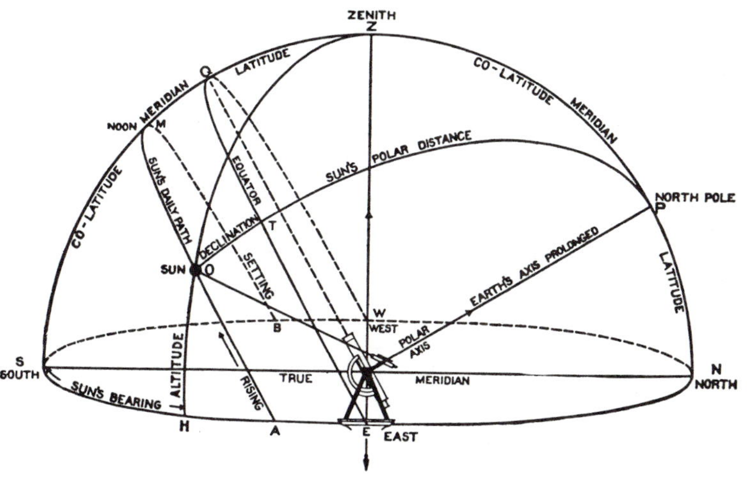

FIG. 8.12.1.

by the tangent screw controlling the horizontal motion of the transit and that controlling the motion of the solar telescope about the polar axis. **Only one position can be found where the solar telescope will point to the sun.** In this position the vertical axis points to the zenith, the polar axis to the pole, and the solar telescope to the sun. The instrument has thus solved mechanically the spherical triangle *ZPO* (Fig. 8.17.1). The horizontal angle between the two telescopes is equal to the sun's true bearing. Since the solar telescope is pointing to the sun the main telescope must be in the plane of the meridian. If all of the work has been done correctly the sun's image will remain between the cross-hairs which are parallel to the equator, and therefore the sun can be followed in its apparent path by a motion of the solar telescope alone. If it is necessary

TABLE 8.12.1

MEAN REFRACTIONS IN DECLINATION.*
TO BE USED WITH THE SOLAR ATTACHMENT

(Computed by Edward W. Arms, C. E., for W. & L. E. Gurley, Troy, N. Y.)

Hour Angle.†	DECLINATIONS.								
	For Latitude 2° 30′.								
	+20°	+15°	+10°	+5°	0°	−5°	−10°	−15°	−20°
0 h.	−18″	−12″	−07″	−02″	+02″	07″	12″	18″	23″
2	−18	−12	−07	−02	+02	07	12	18	23
3	−17	−11	−06	−01	+03	08	13	19	25
4	−15	−10	−05	0	+05	10	15	21	27
5	−10	−05	0	+05	10	15	20	26	32
	For Latitude 5°.								
0 h.	−15″	−10″	−05″	0″	+05″	10″	15″	20″	27″
2	−15	−10	−05	0	+05	10	15	20	27
3	−13	−08	−03	+02	07	12	17	23	29
4	−10	−05	0	+05	10	15	20	27	32
5	−05	0	+05	10	15	20	27	32	40
	For Latitude 7° 30′.								
0 h.	−13″	−08″	−02″	+02″	08″	13″	18″	24″	29″
2	−12	−07	−01	+03	09	14	19	25	31
3	−10	−05	0	+05	10	15	20	26	32
4	−05	0	+05	10	15	20	26	32	39
5	+07	12	17	23	29	36	43	51	1′01
	For Latitude 10°.								
0 h.	−10″	−05″	0″	+05″	10″	15″	20″	26″	32″
2	−07	−03	+02	07	12	17	22	28	34
3	−05	0	+03	08	13	19	25	31	38
4	0	05	10	15	20	26	32	39	46
5	+15	20	26	32	39	46	55	1′06	1′19
	For Latitude 12° 30′.								
0 h.	−08″	−02″	+02″	8″	13″	18″	24″	30″	36″
2	−06	00	+05	10	15	20	26	32	39
3	+02	07	12	17	23	29	36	43	51
4	04	09	14	20	25	31	40	48	55
5	21	27	33	40	48	57	1′08	1′23	1′41
	For Latitude 15°.								
0 h.	−05″	0″	+05″	10″	15″	21″	27″	33″	40″
2	−03	+02	07	12	18	23	29	36	43
3	+01	05	11	16	22	28	34	41	49
4	08	12	19	24	30	37	44	53	1′04
5	29	34	41	49	59	1′10	1′24	1′43	2 08

*Printed by permission of W. & L. E. Gurley.

†Hour angles are reckoned either way from local noon.

TABLE 8.12.1 (Cont.).

Hour Angle.†	DECLINATIONS.								
	For Latitude 17° 30′.								
	+20°	+15°	+10°	+5°	0°	−5°	−10°	−15°	−20°
0 h.	−02″	+02″	08″	13″	18″	24″	30″	36″	44″
2	0	05	10	15	21	27	33	40	48
3	+02	10	15	21	27	33	40	48	57
4	13	18	23	29	35	43	51	1′01	1′13
5	34	41	49	58	1′10	1′23	1′41	2 06	2 42
	For Latitude 20°.								
0 h.	0″	05″	10″	15″	21″	27″	33″	40″	48″
2	03	07	13	18	24	30	36	44	52
3	06	13	18	24	30	36	44	52	1′02
4	17	22	28	35	42	50	1′00	1′11	1 26
5	39	47	57	1′07	1′20	1′37	2 00	2 32	3 25
	For Latitude 22° 30′.								
0 h.	02″	08″	13″	18″	24″	30″	36″	44″	52″
2	06	11	15	21	27	33	40	48	57
3	11	15	21	27	33	40	48	57	1′08
4	20	26	32	39	46	56	1′07	1′19	1 37
5	45	53	1′03	1′16	1′31	1′52	2 21	3 07	4 28
	For Latitude 25°.								
0 h.	05″	10″	15″	21″	27″	33″	40″	48″	57″
2	08	14	19	25	31	38	46	54	1′05
3	12	18	24	30	37	44	53	1′04	1 18
4	23	29	35	45	53	1′03	1′16	1 31	1 52
5	49	59	1′10	1′24	1′52	2 07	2 44	3 46	5 43
	For Latitude 27° 30′.								
0 h.	08″	13″	18″	24″	30″	36″	44″	52″	1′02″
2	11	16	22	28	34	41	49	1′00	1 10
3	17	22	28	35	42	50	1′00	1 11	1 26
4	28	35	42	50	1′00	1′11	1 26	1 43	2 09
5	54	1′05	1′18	1′34	1 54	2 24	3 11	4 38	8 15
	For Latitude 30°.								
0 h.	10″	15″	21″	27″	33″	40″	48″	57″	1′08″
2	14	19	25	31	38	46	54	1′05	1 18
3	20	26	32	39	47	55	1′06	1 19	1 36
4	32	39	46	52	1′06	1′19	1 35	1 57	2 29
5	1′00	1′10	1′24	1′42	2 07	2 44	3 46	5 43	13 06
	For Latitude 32° 30′.								
0 h.	13″	18″	24″	30″	36″	44″	52″	1′02″	1′14″
2	17	22	28	35	42	50	1′00	1 11	1 26
3	23	29	35	43	51	1′01	1 13	1 28	1 47
4	35	43	51	1′01	1′13	1 27	1 46	2 13	2 54
5	1′03	1′15	1′31	1 53	2 20	3 05	4 25	7 36	

TABLE 8.12.1 (Cont.).

HOUR ANGLE.†	DECLINATIONS.								
	+20°	+15°	+10°	+5°	0°	−5°	−10°	−15°	−20°
	FOR LATITUDE 35°.								
0 h.	15″	21″	27″	33″	40″	48″	57″	1′08″	1′21″
2	20	25	32	38	46	55	1′05	1 18	1 35
3	26	33	39	47	56	1′07	1 21	1 38	2 00
4	39	47	56	1′07	1′20	1 30	1 59	2 32	3 25
5	1′07	1′20	1′38	2 00	2 34	3 29	5 14	10 16	
	FOR LATITUDE 37° 30′.								
0 h.	18″	24″	30″	36″	44″	52″	1′02″	1′14″	1′29″
2	22	28	35	42	50	1′00	1 12	1 26	1 45
3	29	36	43	52	1′02	1 14	1 29	1 49	2 16
4	43	51	1′01	1′13	1 27	1 49	2 14	2 54	4 05
5	1′11	1′26	1 54	2 10	2 49	3 55	6 15	14 58	
	FOR LATITUDE 40°.								
0 h.	21″	27″	33″	40″	48″	57″	1′08″	1′21″	1′39″
2	25	32	39	46	52	1′06	1 19	1 35	1 57
3	33	40	48	57	1′08	1 21	1 38	2 02	2 36
4	47	55	1′06	1′19	1 36	1 58	2 30	3 21	4 59
5	1′15	1′31	1 51	2 20	3 05	4 25	7 34	25 18	
	FOR LATITUDE 42° 30′.								
0 h.	24″	30″	36″	44″	52″	1′02″	1′14″	1′29″	1′49″
2	28	35	39	50	1′00	1 12	1 26	1 45	2 11
3	36	43	52	1′02	1 13	1 29	1 49	2 17	2 59
4	50	1′00	1′11	1 26	1 44	2 10	2 49	3 55	6 16
5	1′19	1 36	1 58	2 30	3 22	5 00	9 24		
	FOR LATITUDE 45°.								
0 h.	27″	33″	40″	48″	57″	1′08″	1′21″	1′39″	2′02″
2	32	39	46	52	1′06	1 19	1 35	1 57	2 29
3	40	47	56	1′07	1 21	1 38	2 00	2 34	3 29
4	54	1′04″	1′16″	1 33″	1 54″	2 24″	3 11″	4 38″	8 15″
5	1′23″	1 41″	2 05″	2 41	3 40	5 40	12 02		
	FOR LATITUDE 47° 30′.								
0 h.	30″	36″	44″	52″	1′02″	1′14″	1′29″	1′49″	2′18″
2	35	42	50	1′00	1 12	1 26	1 45	2 01	2 51
3	43	51	1′01	1 13	1 28	1 47	2 15	2 56	4 08
4	56	1′09	1 23	1 40	2 05	2 40	3 39	5 37	11 18
5	1′27	1 46	2 12	2 52	4 01	6 30	16 19		
	FOR LATITUDE 50°.								
0 h.	33″	40″	48″	57″	1′08″	1′21″	1′39″	2′02″	2′36″
2	38	46	55	1′06	1 18	1 35	1 57	2 28	3 19
3	47	56	1′06	1 19	1 36	2 29	2 31	3 23	5 02
4	1′02	1′14	1 29	1 48	2 16	2 58	4 18	6 59	19 47
5	1 30	1 51	2 19	3 04	4 22	7 28	24 10		

TABLE 8.12.1 (Cont.).

Hour Angle.†	DECLINATIONS.								
	+20°	+15°	+10°	+5°	0°	−5°	−10°	−15°	−20°
	For Latitude 52° 30′.								
0 h.	36″	44″	52″	1′02″	1′14″	1′29″	1′49″	2′18″	3′05″
2	43	50	59	1 11	1 26	1 42	2 23	2 49	3 55
3	50	1′00	1′11	1 26	1 45	2 11	2 51	2 58	6 22
4	1′05	1 18	1 35	2 10	2 28	3 19	4 53	8 42	
5	1 34	1 56	2 27	3 16	4 47	8 52			
	For Latitude 55°.								
0 h.	40″	48″	57″	1′08″	1′21″	1′39″	2′02″	2′36″	3′33″
2	46	55	1′05	1 18	1 34	1 56	2 30	3 15	4 47
3	55	1′06	1 19	1 35	1 58	2 30	3 21	4 58	9 19
4	1′10	1 23	1 42	2 06	2 43	3 44	5 49	12 41	
5	1 37	2 01	2 34	3 28	5 15	10 18			
	For Latitude 57° 30′.								
0 h.	44″	52″	1′02″	1′14″	1′29″	1′49″	2′18″	3′05″	4′37″
2	50	59	1 11	1 25	1 43	2 09	2 47	3 51	6 04
3	58	1′10	1 24	1 42	2 07	2 43	3 45	5 50	12 47
4	1′11	1 25	1 43	2 10	2 50	3 55	6 14	14 49	
5	1 41	2 06	2 42	3 42	5 46	12 26			
	For Latitude 60°.								
0 h.	48″	57″	1′08″	1′21″	1′39″	2′02″	2′36″	3′33″	5′23″
2	54	1′04	1 17	1 33	1 54	2 24	3 12	4 38	8 15
3	1′03	1 15	1 30	1 51	2 20	3 04	4 24	7 31	24 44
4	1 18	1 34	1 56	2 28	3 18	4 50	8 53		
5	1 45	2 11	2 50	3 57	6 21	15 32			
	For Latitude 62° 30′.								
0 h.	52″	1′02″	1′14″	1′29″	1′50″	2′18″	3′00″	4′17″	7′13″
2	58	1′09	1 23	1 41	2 06	2 43	3 44	5 50	12 44
3	1′07	1 23	1 38	2 01	2 35	3 30	5 16	10 24	
4	1 23	1 40	2 05	2 40	3 40	5 37	11 50		
5	1 48	2 17	2 59	4 14	7 03				
	For Latitude 65°.								
0 h.	57″	1′08″	1′21″	1′39″	2′02″	2′36″	3′33″	5′23″	10′51″
2	1′03	1 16	1 31	1 52	2 21	3 07	4 28	7 44	
3	1′12	1 27	1 46	2 12	2 52	4 02	6 33		
4	1 27	1 47	2 13	2 54	4 05	6 40			
5	1 52	2 22	3 08	4 30	7 52				
	For Latitude 67° 30′.								
0 h.	1′02″	1′14″	1′29″	1′50″	2′18″	3′00″	4′17″	7′13″	
2	1 08	1 22	1 40	2 03	2 39	3 37	5 32	11 28	
3	1 17	1 34	1 55	2 26	3 14	4 44	8 34		
4	1 32	1 53	2 23	3 14	4 35	8 05			
5	1 56	2 28	3 17	4 40	8 51				
	For Latitude 70°.								
0 h.	1′08″	1′21″	1′39″	2′02″	2′36″	3′33″	5′23″	10′51″	
2	1 14	1 29	1 50	2 18	3 00	4 17	7 13		
3	1 23	1 43	2 05	2 41	3 41	5 59	12 15		
4	1 37	2 00	2 34	3 28	5 20	10 12			
5	2 02	2 33	3 27	5 11	10 05				

to move the instrument about the vertical axis in order to point the solar telescope again at the sun this shows that the main telescope was not truly in the meridian.

After the meridian has been determined the main telescope may then be lowered and a point set which will be due north or due south of the instrument.

The declination of the sun for the time of the observation is computed as explained in Art. 8.9. After the correct declination is found it has still to be corrected for refraction of the atmosphere. The effect of refraction is to make the sun appear higher in the sky than it actually is. In the northern hemisphere, when the declination is north, the correction must be added numerically; when south, subtracted. Algebraically it is always added. The correction may be taken from Table 8.12.1 (not, in this case, from Table 8.9.1 or 8.18.2), the declination being given at the top of the column and the number of hours from local noon either way at the left of the page.

EXAMPLE

Latitude 40° N. Longitude 71° 15′ W.
Declination at Greenwich 0^h = Date: April 10, 1976
+7° 54′ 54″
Tab. diff. = + 0′.92/hr
Required declination settings for 2^h, 3^h and 4^h P.M., Eastern Standard Time.
2^h P.M., E. S. T. = 7 P.M. Gr. Time = 19^h Gr. Civ. Time
0′.92 × 19^h = 17′ 28″.8
Decl. at 2 P.M., E. S. T. = 7° 54′ 54″ + 17′ 28″.8
= 8° 12′ 22″.8

The values for 3^h and 4^h are obtained by adding 0′.92 = 55″.2 for each hour. The declination, refraction corrections and final settings (to nearest 15″) are as follows:

TIME E. S. T.	HOURS FROM LOCAL NOON	DECLINATION	REFR. CORR.	SETTING
2^h P.M.	2^h 15^m	8° 12′ 22″.8	0′ 45″	8° 13′ 08
3^h	3 15	8° 13′ 18″.0	0 57	8° 14′ 15
4^h	4 15	8° 14′ 13″.2	1 25	8° 15′ 38

Observation for meridian should not be made when the sun's altitude is less than about 10°, because the refraction correction will be unreliable. Observations near noon are to be avoided because a slight error in altitude produces a large error in the resulting meridian. For good results therefore the observation should be made neither within an hour of noon nor near sunrise or sunset.

Mistakes in Using the Solar Attachment.—

1. Solar on wrong side of main telescope.
2. Refraction correction applied wrong way.
3. Declination set off in wrong direction.

ADJUSTMENTS OF THE SOLAR ATTACHMENT

8.13. Adjustment of Polar Axis.—To make the Polar Axis Perpendicular to the Plane of the Line of Sight and the Horizontal Axis. Level the transit and the main telescope. Bring the bubble of the solar telescope to the center of its tube while it is parallel to a pair of opposite adjusting screws which are at the foot of the polar axis. Reverse the solar telescope 180° about the polar axis. If the bubble moves from the central position, bring it half-way back by means of the adjusting screws just mentioned and the other half by means of the tangent screw controlling the vertical motion of the solar. This should be done over each pair of opposite adjusting screws and repeated until the bubble remains central in all positions.

8.14. Adjustment of the Cross-Hairs.—To make the Vertical Cross-Hair truly Vertical. Sight on some distant point with all the clamps tightened and, by means of the tangent screw controlling the vertical motion of the solar, revolve the solar telescope about its horizontal axis to see if the vertical cross-hair remains on the point. If not, adjust it by rotating the cross-hair ring, as described in Art. 8.48.

8.15. Adjustment of Telescope Bubble.—To make the Axis of the Bubble Parallel to the Line of Sight. Level the main telescope and mark a point about 200 ft. from the instrument in line with the horizontal cross-hair. Measure the distance between the two telescopes and lay this off above the first point; this will give a point on a level with the center of the solar telescope. Sight the solar at this point and clamp. Bring the bubble to the center by means of the adjusting screws on the bubble tube.

OBSERVATIONS FOR LATITUDE

8.16. By the Altitude of Polaris at Upper or Lower Culmination.—When Polaris is approaching either upper or lower culmination (see Art. 8.3, and Fig. 8.3.1) set up the transit and point the horizontal hair on the star. Keep the cross-hair pointed on the star until the culmination is reached. Read the vertical arc and determine the index correction. The observed altitude is to be corrected for refraction by Table 8.9.1. This gives the true altitude. If Polaris is at upper culmination subtract from the true altitude the polar distance of the star at the date of the observation (Table 8.4.1). If the star is at lower culmination the polar distance is to be added. The result is the latitude of the place of observation.

8.17. By the Altitude of the Sun at Noon.—The observation consists in finding the maximum altitude of the sun's lower limb. This will occur (very nearly) when the sun is on the meridian. Begin the observation a little before *apparent* noon, remembering that this differs sometimes more than 16^m from *mean* noon. Furthermore it should be remembered that standard time may differ as much as half an hour from *mean* time. When the maximum altitude is found the following corrections are to be made: first, the refraction correction is to be subtracted (Table 8.9.1); second, the sun's semi-diameter (found in the American Ephemeris and Nautical Almanac)* is to be added; third, the sun's declination is to be subtracted if plus or added if minus. The result, subtracted from 90°, is the latitude of the place.

8.18. Meridian by Equal Altitudes of the Sun.—This observation consists in measuring in the forenoon the horizon-

---

*Or in "Ephemeris of Sun, Polaris and Other Selected Stars," by Bureau of Land Management, Dept. of Interior. Available from Superintendent of Documents.

EXAMPLE

Observed maximum altitude of the sun's lower limb on Jan. 15, 1976 = 26° 15′ (sun south of zenith); index correction = +1′; longitude = 71° 06′ W; sun's declination Jan. 15 at 0^h Greenwich Civil Time = −21° 14′; Tab. diff. for one hour = +0.45 .4; corrected equation of time = $-9^m\ 7^s$; semi-diameter = 16′ 17″.

Observed altitude	=	26° 15′
Index Correction	=	+1′
		26° 16′
Refraction	=	1′ 58″
		26° 14′ 02″
Semi-diameter	=	+16′ 17″
Altitude, sun's center	=	26° 30′ 19″
Declination	=	−21° 11′ 24″
Co-latitude	=	47° 41′ 43″
Latitude	=	42° 18′ 17″ N
or		42° 18′.3 N

Local App. Time	=	12^h
Longitude	=	4 44 24
Gr. App. Time	=	$16^h\ 44^m\ 24^s$
Equa. of Time*	=	−(−) 9 7
Gr. Civil Time	=	$16^h\ 53^m\ 41^s$

Decl. at 0^h	=	−21° 19′ 00″
0.45 × 16^h.90	=	7′ 36″
Declination	=	−21° 11′ 24″

tal angle between the sun and some reference mark at the instant when the sun has a certain altitude, and again measuring the angle when the sun has an **equal** altitude in the afternoon. If the distance of the sun from the equator were the same for the two observations the horizontal angles between the sun and the meridian would be the same in both observations, hence the mean of the two readings of the horizontal circle would be the reading for the meridian. But since the sun is changing its distance from the equator the measured angles must be corrected accordingly. The correction to the mean is computed by the equation

$$X = \frac{d}{\cos L \sin t}$$

in which X = the correction to the mean vernier reading, d = the hourly change in declination of the sun taken from Table 8.18.1 and multiplied by **half** the number of hours between the two observations, L = the latitude, and t = half the elapsed time converted into degrees, minutes, and seconds. Since the hourly change for any given day is nearly the same, year after year, an almanac is not necessary but the following table is sufficient.

---

*Corrected for hours since Greenwich 0^h. See p. 344.

TABLE 8.18.1

HOURLY CHANGE IN THE SUN'S DECLINATION

	1st.	10th.	20th.	30th.
January	+12″	+22″	+32″	+41″
February	+43	+49	+54	....
March	+57	+59	+59	+58
April	+58	+55	+51	+46
May	+45	+39	+31	+23
June	+21	+12	+01	−09
July	−10	−19	−28	−36
August	−38	−44	−49	−54
September	−54	−57	−58	−59
October	−58	−57	−54	−49
November	−48	−42	−34	−25
December	−23	−11	−02	+10

The observation is made as follows: *At some time in the forenoon, preferably not later than 9 o'clock, the instrument is set up at one end of the line the azimuth of which is to be found, and one vernier is set at 0°. The vertical cross-hair is then sighted at the other end of the line and the lower plate clamped. The upper clamp is loosened and the telescope turned until the sun can be seen in the field of view. The

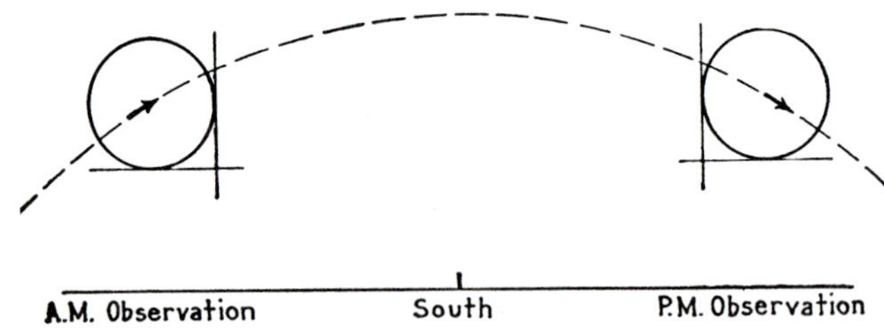

FIG. 8.18.1. MERIDIAN BY EQUAL ALTITUDES OF THE SUN.

horizontal cross-hair is to be set on the **lower** edge of the sun and the vertical cross-hair on the **right** edge (northern hemisphere). Since the sun is rising and also changing its bearing it

*The nearer the sun is due East or due West, the better the result.

is difficult to set both of the cross-hairs at once and it will be found easier to set the horizontal hair so that it will cut across the sun's disc, leaving it clamped in this position while the vertical hair is kept tangent to the right edge of the sun by means of the upper tangent screw. When the sun has risen until the lower edge is on the horizontal hair the instrument is in the desired position and after this position is reached the upper tangent screw should not be moved. As soon as this position is reached the time is noted. Both the vertical and the horizontal circles should now be read and the angles recorded.

In the afternoon, when the sun is found to be nearly at the same altitude as at the forenoon observation, the instrument should be set up at the same point and again sighted on the mark. The observation described above is repeated, the pointings now being made on the **lower** and **left** edges of the disc. The telescope is inclined until the vernier of the vertical circle reads the same as it did at the forenoon observation. When the sun comes into the field the vertical hair is set on the left edge and kept there until the lower edge is in contact with the horizontal hair. The time is again noted and the verniers are read. If desired, the accuracy may be increased by taking several pairs of observations. The mean of the two circle readings (supposing the graduations to be numbered from 0° to 360° in a clockwise direction) is now to be corrected for the sun's change in declination. The correction as obtained by the formula is to be added to the mean vernier reading if d is minus, and subtracted if d is plus, i.e., if the sun is going south the mean vernier reading is east of the south point, and *vice versa*. When the circle reading of the south point is known the true bearing of the mark becomes known and the bearings of other points may be found (see the following example).

The vertical cross-hair might have been set on the left edge in the forenoon and the right edge in the afternoon. The upper edge might have been sighted instead of the lower. The two observed positions must, however, be symmetrical with respect to the meridian.

The disadvantage of this method is that it is necessary to be at the same place both in the forenoon and afternoon, whereas

TABLE 8.18.2*

CORRECTION TO OBSERVED ALTITUDE OF THE SUN FOR REFRACTION AND PARALLAX

APP'T ALT.	TEMPERATURE.										APP'T ALT.
	−10° C. +14° F.	− 5° C. +23° F.	0° C. +32° F.	+ 5° C. +41° F.	+10° C. +50° F.	+15° C. +59° F.	+20° C. +68° F.	+25° C. +77° F.	+30° C. +86° F.	+35° C. +95° F.	
°	′	′	′	′	′	′	′	′	′	′	°
10	5.52	5.42	5.30	5.20	5.10	5.00	4.92	4.83	4.75	4.67	10
11	5.02	4.92	4.82	4.73	4.63	4.55	4.47	4.38	4.32	4.23	11
12	4.60	4.50	4.42	4.33	4.25	4.17	4.10	4.03	3.97	3.88	12
13	4.23	4.15	4.07	4.00	3.92	3.85	3.78	3.72	3.65	3.58	13
14	3.92	3.83	3.77	3.70	3.62	3.55	3.50	3.45	3.37	3.32	14
15	3.65	3.58	3.50	3.43	3.37	3.32	3.25	3.20	3.13	3.08	15
16	3.43	3.35	3.30	3.23	3.17	3.12	3.07	3.00	2.95	2.90	16
17	3.22	3.15	3.10	3.03	2.98	2.92	2.88	2.82	2.77	2.72	17
18	3.02	2.95	2.90	2.85	2.80	2.75	2.70	2.65	2.60	2.55	18
19	2.83	2.78	2.73	2.68	2.63	2.58	2.53	2.48	2.43	2.40	19
20	2.68	2.63	2.58	2.53	2.48	2.43	2.38	2.33	2.30	2.27	20
21	2.53	2.48	2.43	2.38	2.35	2.30	2.27	2.22	2.17	2.13	21
22	2.38	2.35	2.30	2.25	2.22	2.18	2.13	2.08	2.05	2.02	22
23	2.28	2.25	2.20	2.15	2.12	2.08	2.03	1.98	1.95	1.93	23
24	2.17	2.13	2.08	2.05	2.02	1.98	1.93	1.88	1.87	1.83	24
25	2.07	2.03	1.98	1.95	1.92	1.88	1.83	1.80	1.77	1.75	25
26	1.99	1.95	1.90	1.87	1.83	1.80	1.75	1.72	1.70	1.67	26
27	1.88	1.85	1.82	1.78	1.75	1.72	1.68	1.63	1.62	1.60	27
28	1.80	1.77	1.72	1.70	1.67	1.63	1.60	1.57	1.53	1.52	28
29	1.72	1.68	1.65	1.63	1.60	1.57	1.53	1.50	1.47	1.46	29
30	1.65	1.62	1.58	1.57	1.53	1.50	1.47	1.45	1.42	1.40	30
32	1.53	1.50	1.47	1.45	1.42	1.38	1.35	1.33	1.30	1.28	32
34	1.41	1.37	1.35	1.32	1.30	1.27	1.25	1.23	1.20	1.18	34
36	1.30	1.27	1.25	1.22	1.20	1.18	1.15	1.13	1.10	1.08	36
38	1.20	1.18	1.15	1.13	1.12	1.10	1.07	1.05	1.02	1.02	38
40	1.11	1.10	1.07	1.05	1.03	1.02	0.98	0.97	0.95	0.93	40
42	1.03	1.00	0.98	0.97	0.95	0.93	0.90	0.88	0.87	0.87	42
44	0.96	0.93	0.92	0.90	0.88	0.87	0.85	0.83	0.82	0.80	44
46	0.89	0.88	0.87	0.85	0.83	0.82	0.80	0.78	0.77	0.75	46
48	0.83	0.82	0.80	0.78	0.77	0.75	0.73	0.72	0.70	0.68	48
50	0.77	0.75	0.73	0.72	0.70	0.68	0.67	0.67	0.65	0.63	50
55	0.63	0.62	0.60	0.60	0.58	0.57	0.57	0.55	0.53	0.52	55
60	0.52	0.52	0.50	0.50	0.48	0.47	0.47	0.45	0.45	0.43	60
65	0.42	0.40	0.40	0.40	0.38	0.38	0.37	0.37	0.35	0.33	65
70	0.32	0.32	0.32	0.30	0.30	0.30	0.28	0.28	0.28	0.27	70
75	0.23	0.23	0.23	0.22	0.22	0.22	0.20	0.20	0.20	0.18	75
80	0.15	0.15	0.13	0.13	0.13	0.13	0.13	0.12	0.12	0.12	80
85	0.07	0.07	0.07	0.07	0.07	0.07	0.07	0.05	0.05	0.05	85
90	0.00	0.00	0.00	0.00	0.00	0.00	0.00	0.00	0.00	0.00	90

*From *Principal Facts of Earth's Magnetism, Coast and Geodetic Survey.*

in many cases the surveyor might in the afternoon be a long distance from where he was working in the forenoon.

EXAMPLE

Latitude 42° 18′ N. April 19, 1976

A.M. Observation

Reading on Mark, 0° 00′ 00″
Pointings on Upper and Left Limbs
Vertical Arc, 24° 58′
Horizontal Circle, 357° 14′ 15″ (to right)
Time $7^h\ 19^m\ 30^s$ A.M., E.S.T.
½ elapsed time = $4^h\ 26^m\ 21^s$ = 66° 35′ 15″

$$X = \frac{d}{\cos L \sin t}$$

$$X = \frac{228.7''}{0.73963 \times 0.91767}$$

$$X = 336''.9 = 5'\ 36''.9$$

P.M. Observation

Reading on Mark, 0° 00′ 00″
Pointings on Upper and Right Limbs
Vertical Arc, 24° 58′
Horizontal Circle, 162° 35′ 15″ (to right)
Time $4^h\ 12^m\ 12^s$ P.M., E.S.T.
Increase in declination in $4^h\ 26^m\ 21^s = 51''.6 \times 4^h.44 = 228.7''$

Mean circle reading	=	259° 54′ 45″ N
Mean circle reading	=	79° 54′ 45″ S
Correction	=	−5′ 37″
Bearing of Mark	=	S 79° 49′ 08″ E
Azimuth of Mark	=	280° 10′ 52″

8.19. Satellite Position Location.—Considerably newer but promising great changes in surveying for the future is the location of the position of any point on the surface of the earth from satellites passing overhead. A passive satellite (i.e., emitting no light or radio signal) must be used by photographing it against a star background when it is illuminated by the sun, from which the camera's ground position can be calculated. An active satellite (i.e., principally one emitting a radio signal) can be used to ascertain the geocentric coordinates of a ground station equipped with a proper receiver using the known orbit of the satellite.

Positional accuracy resulting from a fix by radio satellites can be ± 25 metres in each coördinate from a few satellite passes to less than ± 1 metre from many passes. The elapsed time is negligible, for the receiver runs unattended: an excellent fix can be obtained in less than 48 hours. Great expectations are held out for this method of control surveying—to find the X, Y, and Z geodetic coördinates of unknown points—soon to fix location within ± 0.5 metre or better. Such results will be easily comparable to the better order of control surveying, with the added benefit of great savings in time and expense.

CHAPTER IX

LEVELING

9.1. Definitions.—Leveling consists in ascertaining differences in elevation; there are two methods in common use, *Differential Leveling*, and *Trigonometric Leveling*. The principles of differential leveling and leveling instruments are covered in Chap. IV. Trigonometric leveling is based on vertical angle and distance measurements. (See Vol. II, Art. 3-22.)

Leveling is referred to a system of reference points called *bench marks* (B.Ms.), the relative heights of which are known accurately. These heights are usually referred to some definite surface such, for instance, as *mean sea-level* or *mean low water*, and the height of a point above this surface is called its *elevation*. This surface is known as the *datum*, or "datum plane." (See Art. 9.21, and Art. 10.4) Strictly speaking it is not a plane but a level surface, i.e., it is at every point perpendicular to the direction of gravity. (Art. 4.1.) If mean sea-level has not been established a datum can be assumed arbitrarily.

9.2. Leveling to Establish Bench Marks.—When it is necessary to run a line of levels to establish new bench marks the rod is first held on some bench mark the elevation of which is accurately known, and a *backsight* taken. Then differential leveling as described in Art. 4.23 is carried out.

When a bench mark is being established it should be **used as a turning point.** The elevation of this bench mark could be obtained by taking a foresight upon it and not using it as a turning point, but by making the bench mark also a turning point it becomes a part of the line of levels and if the level circuit checks, the elevation of the bench mark is also checked. Each bench mark should be recorded by a description or a sketch, or both. The elevations of the turning points are as accurate as those of the bench marks themselves. Consequently it is advisable to describe those turning points that can be identified readily so that they may be used later when an established bench mark is not available.

In leveling uphill the backsights can be long and the foresights short; in leveling downhill the reverse is true. To obtain the minimum number of set-ups of the level the long rod-readings should be near the top of the extended rod and the short rod-readings should be near the foot of the rod. The levelman must use good judgment in selecting the positions for his level and the rodman must apply good judgment in selecting the turning points, else much time will be lost. For example, in leveling downhill set up the level roughly (without pressing the tripod legs into the ground), turn the telescope toward the rod and then level it, approximately, in that direction. By sighting along the outside of the telescope, the approximate place where the line of sight will strike the rod can be noted and the distance the instrument should be moved up or down the slope can be estimated. Then move to the new position, level up carefully, and proceed to take the backsight. The rodman then proceeds downhill and selects a suitable point for a new turning point and one which he estimates will give a reading near the top of the rod. The levelman should be alert and, if the line of sight cuts above the top of the rod, he should immediately indicate this fact to the rodman and tell him (or signal to him) the approximate amount the top of the rod is below the line of sight so that the next point selected will be at about the right height. In carrying a line of levels uphill the same general procedure should be followed, but now the levelman stops the rodman when his ankles are on the line of sight, and the rodman should then select a turning point at the level of the ground on which he is standing.

9.3. Equal Back and Foresight Distances.—In this work it is very important to eliminate as far as possible errors of adjustment in the instrument. If at every set-up of the level the foresight and the backsight are taken at points which are equally distant from the instrument such errors will be eliminated. If the level is not in perfect adjustment the resulting error in any reading is proportional to the distance. At equal distances from the instrument the errors are equal, and, since it is the difference of the rod-readings that gives the difference in elevation, the error is eliminated from the final result by this method. By making the length of foresights and backsights equal on turning points it is possible to eliminate not only the error due to non-adjustment of the bubble but also any error due to non-adjustment of the objective tube, since this will occupy the same position in the telescope in each sight. The distance to the backsight is determined by the place where the instrument is set up, and the rodman, as he passes from one turning point to the next, can by pacing make the foresight distance approximately equal to that of the backsight. Where the shape of the ground may be so irregular that backsight distances cannot be made about equal to foresight distances and at the same time adhere to the suggestion above of making the long rod-readings near the top of the rod and the short rod-readings near the foot of the rod, the high and short rod-readings should be sacrificed to make the distance from the level to the rod approximately equal on backsights and foresights. This will, of course, entail more set-ups of the instrument. In running a line of levels from one B.M. to another some distance away, if the sum of the backsight distances is approximately equal to the sum of the foresight distances, even though each pair of backsights and foresights may not be taken on points equally distant from the level, the instrumental errors referred to above will be reduced. The line of levels should be "closed" by continuing the leveling until the original bench mark, or some other bench mark whose elevation is well established, is reached.

9.4. Level Notes.—The notes for this work may consist of five columns, as shown in Fig. 9.4.1. The height of instrument is obtained by adding the backsight rod-reading to the elevation of the point on which it is taken. The elevation of any point is found by subtracting the foresight rod-reading for that point from the corresponding height of the instrument. Notice that the calculations may be checked by adding the column of foresights and the column of backsights. The difference between these sums should be the same as the difference in elevation between the first and last points. This **page-check is important** and **should not be omitted.** (See Fig. 9.4.1.) The notes should be computed in the field to catch errors when they occur and then and there eliminate them. Also the note-keeper should compute mentally and not on a scratch pad or in the back of the book.

(LEFT-HAND PAGE) (RIGHT-HAND PAGE)

B.M. Leveling for Eastern Intercepting Sewer.					B. Jones ⊼ M. Brown Oct. 30, 1957
Point	B.S.	H.I.	F.S.	Elev.	Remarks.
$B.M._1$	4.122	93.139		89.017	Top S.E. cor. granite foundation S.E. cor. City Hall.
$T.P._1$	3.661	90.611	6.189	86.950	Curb
$T.P._2$	4.029	89.630	5.010	85.601	N.E. bolt top hydrant opp. #42 Main St.
$B.M._2$	3.901	86.161	7.370	82.260	S.W. cor. S.B. on N.W. cor. Main and Broad Sts.
$B.M._3$	3.512	83.056	6.617	79.544	N.W. cor. lower stone step #62 Broad St.
$T.P._3$	6.007	80.348	8.715	74.341	Cobble stone
$B.M._4$			9.070	71.278	Chisel cut N.W. cor. C.B. curb S.W. cor. Broad and State Sts.
	25.232		42.971		
			25.232	89.017	True elev. B.M. = 71.274 Book 27, P. 36.
			17.739	17.739	"Check"

FIG. 9.4.1. BENCH MARK LEVEL NOTES.

In this instance it is found that the difference of elevation is −17.739 ft, whereas the established (official) elevations of BM_1 and BM_4 show their difference to be −17.743 ft. If, as is usual, one must hold to these two established elevations and if the present work has been done to within prescribed limits of accuracy, then one distributes the discrepancy in proportion to the distance traveled, as explained later in Art. 9.25. Here, for example, the new-found elevation of BM_2 would be decreased by 3/6 (0.004) = 0.002, and down to 82.258 ft; the new-found elevation of BM_3 would be decreased by 4/6 (0.004) = 0.0027 or 0.003, down to 79.54 ft; and the new-found elevation of BM_4 would be decreased by 6/6 (0.004) = 0.004, down to the official elevation of 71.274 ft.

9.5. Three-Wire Leveling.—The three-wire differential leveling method is becoming common with the advent of levels with three horizontal wires, one central and two "stadia" wires. It is required in second-order control and metropolitan control leveling and can be advantageously used in miscellaneous local control work and on small engineering projects, topographic mapping, drainage studies, etc.

Simply enough, three readings are taken and recorded for each sighting of the level rod. This gives three different values, virtually three executions of the same job, which are averaged together for greater precision. There are inherent checks available during the operation to prevent blunders; the three readings are accomplished rapidly; and foresight distances can easily be kept equal to backsight distances.

The rod is read at each of the three "thread" or wire intercepts, usually directly to thousandths of feet (or yards, or metres, dependent on which level rod is used). The bubble is maintained level during the three readings. A quick comparison of the half-intervals is made to prevent blunders; if they differ by more than three or four units in the last (estimated) place the readings are immediately repeated. The following is an example:

Top wire	6.585	1.234 Upper Intercept
Middle wire	5.351	1.231 Lower Intercept
Bottom wire	4.120	

Since the intercepts are nearly alike, the triple reading is acceptable and the leveling is permitted to proceed. The total of the two intercepts (2.465) is itself an indication of the distance to the rod, for comparison with its paired foresight distance. If the stadia intercept ratio is 1:100, then the distance to the rod for this reading is 246.5 ft. Other stadia intercept ratios are used, a common one being 0.3:100 in some tilting levels and automatic levels.

9.6. Notes for Three-Wire Leveling.—Various forms of notes can be set up for three-wire leveling; the form of Fig. 9.6.1 will be explained to some extent to clarify the field procedure. The notes are set up to emphasize the pairing of readings taken from each instrument set-up, especially the distance relationships as seen in the interval (Int.) columns. The first half-intervals sum to 373 and 339, not a very good matching, so in the next set-up the effort is made to lengthen the

POINT	+ ROD	INT.	MEAN	POINT	− ROD	INT.	MEAN
BM E	2.819			TP 1	3.491		
		188				171	
	2.631		2.6300		3.320		3.3210
		185				168	
	2.446	373			3.152	339	
TP 1	8.266			TP 2	4.623		
		161				204	
	8.105		8.1037		4.419		4.4203
		165				200	
	7.940	699	10.7337		4.219	743	7.7413
TP 2	6.574			BM F	6.213		
		216				192	
	6.358		6.3577		6.021		6.0223
		217				188	
	6.141	1132	+17.0914		5.833	1123	13.7636
			−13.7636				
			+3.3278				

Given Elev. of BM E
= 802.621

Diff. Elev. = + 3.3278
Elev. BM F = 805.9488
or 805.949

FIG. 9.6.1. THREE-WIRE LEVELING NOTES (FOOT ROD).

distance to the forward rod. The result shows an overcompensation, the totals being 699 versus 743. In the final set-up, however, the balance for the whole run is quite well achieved, 1132 versus 1123.

Cumulative summations of the several columns, easily checked at every step, is done by the notekeeper as the work progresses. As noted before, the notekeeper requires the instrumentman to have the half-intervals check within three or four units (188 vs. 185, 171 vs. 168, etc.) or else reread the rod. He also directs the rodman and the instrumentman in setting the instrument and the turning points so as to achieve balanced distances to backsights and foresights.

In the use of the yard rod, the form of notes of Fig. 9.6.2 is used, and the procedure is very similar. One difference is in the case that a strip of tape calibrated in feet is glued to the back of the rod and is read for checking purposes at each sighting. Adding the three thread readings (in yards) gives, by coincidence, the mean value of the reading in feet: the close agreement between this sum in each case with the back-of-rod reading assures that no blunders occurred.

Distance comparisons can be made by use of the interval (Int.) figures, although these may not be recognized at once as foot distances. Since in this instance an automatic level was used having an intercept factor of 0.3/100, the sum of the backsight distance is $(0.396 \times 100) \div 0.3 = 132$ yd. = 396 ft. and the foresight distances total 378 ft. This use of a yard rod with an 0.3/100 instrument is seen to have a somewhat simplifying advantage.

In Fig. 9.6.1, were one to have used merely the middle wire readings in the calculation, the result would have been a difference of elevation of +3.334, giving an elevation of 805.955 ft. for BM F. The job as actually accomplished effectively gives (the mean of three measurements) a difference of elevation of +3.3278, for an elevation of 805.949 ft. for BM F. Had three successive levelings of the course been made with single-wire readings it is reasonable to expect that the mean of the three could approach the result of the three-wire leveling.

In the Mean column of both Fig. 9.6.1 and Fig. 9.6.2 the value can be obtained mentally by noting that if the two half-intervals are equal, the mean value is that of the middle wire; if the upper half-interval is one unit larger, the mean is 0.3 units higher than the middle wire; if two units larger, it is 0.7 units higher than the middle wire; etc. If the lower half interval is larger than the upper, the middle-wire value is decreased by 0.3, 0.7, etc. In practice, therefore, the mean will always round out to 0, 3, or 7 in the last place.

Point	+Rod	Int.	Mean	Ft. Rdg.	Point	−Rod	Int.	Mean	Ft. Rdg.
BM G	2.648				TP 1	3.174			
		60					55		
	2.588		2.5880	7.76		3.119		3.1200	9.35
		60					52		
	2.528	120				3.067	107		
	(7.764)					(9.360)			
TP 1	0.734				TP 2	3.897			
		76					72		
	0.658		0.6577	1.98		3.825		3.8243	11.48
		77					74		
	0.581	273	3.2457			3.751	253	6.9443	
	(1.973)					(11.473)			
	9.737					20.833			
TP 2	1.248				BM H	2.694			
		62					63		
	1.186		1.1863	3.56		2.631		2.6313	7.89
		61					62		
	1.125	396	4.4320			2.569	378	−9.5756	
	(3.559)					(7.894)		+4.4320	
	+13.296					−28.727		−5.1436	Yds.
	+13.296					+13.296		× 3	
					Diff. of Elev. =	=15.431	ft.	−15.4308	ft.

Fig. 9.6.2. Three-Wire Leveling Notes (Yard Rod).

Point	+Rod	Int.	Mean	Back of Rod (Feet)	Point	−Rod	Int.	Mean	Back of Rod (Feet)
BM J	2.063				TP 1	1.073			
		52					45		
	2.011		2.0117	6.59		1.028		1.0280	3.38
		50					45		
	1.961	102				0.983	90		
	6.035					3.084			
TP1	2.116				TP 2	1.011			
		81					83		
	2.035		2.0357	6.69		0.928		0.9273	3.05
		79	4.0474	13.28			85	1.9553	6.43
	1.956	262				0.843	258		
	12.142					5.866			
TP 2	2.318				BM K	0.097			
		34					40		
	2.282		2.2813	7.48		0.057		0.0567	0.17
		36	+6.3287	+20.76			41	−2.0120	−6.60
	2.248	332		−6.60		0.016	339		
	+18.990			+14.16 (Check Value, feet)		−6.036			
	−6.036								
	+12.954		+6.3287						
	÷ 3		−2.0120						
	+ 4.318 (Check Value, metres)		+4.3167	metres					
			×3.280833		Difference in elevation				
			+14.1624	feet					

FIG. 9.6.3 THREE-WIRE LEVELING NOTES (METRIC ROD).

9.7. Rod with Two Scales.—A similar checking procedure is apparent when using rods with two scales. One scale is the usual one, with zero at the base of the rod; the other is a strip of metallic tape affixed to the rod such that it begins with some arbitrary value, say 0.967, at the base of the rod. This will give a pair of readings for each point where the rod is held. Since differential leveling always utilizes the differences between backsight and foresight, such a constant shift has no influence on the resulting elevations.

For each sighting both scales are read. The difference between these readings, though a random amount, remains theoretically the same and is solely affected by reading errors. The note keeper can check this difference and detect reading errors immediately. This approach also checks the first B.S. and the last F.S.

9.8. Bench Marks and Turning Points.—Both the bench marks and the turning points should be such that their elevations will not change during the time they are needed. The only difference between the two is that turning points may be of use for only a few minutes while bench marks may be needed for many years. Bench marks should be described, very carefully and accurately, and their heights should be checked before being accepted as correct. They are frequently taken on such points as these: stone bounds, tops of boulders, spikes in trees, and on sills, stone steps, or underpinning of buildings. Curb stones or tops of hydrants are also used but are not so permanent. As it is often impossible in a new country to find existing points where bench marks can be established, it is usual to set stone monuments or iron rods and to determine their elevation carefully. The U.S. Geological Survey, and the U.S. Coast and Geodetic Survey use metal tablets set into concrete piers, into rock, or into the stone foundations of buildings. Some of the older bench marks of the Coast Survey and of the Missouri River Commission consist of stones the tops of which are buried 3 or 4 ft. under ground. The exact bench is the top of a spherical headed bolt set in the top of the stone. This is reached by lowering the rod through an iron pipe which extends from the surface of the ground to the top of the stone.

Bench marks should be established at frequent intervals for convenience in subsequent work. Some surveyors consider it advisable to have two bench marks in the same locality to serve as checks on each other. In choosing a bench or a turning point it is best to select a point which is slightly raised so that the rod will always rest on exactly the same point. A rounded surface is better than a sharp point, especially when it is on a rock, as the rod may chip off a small piece and alter the elevation. If a turning point is taken on a flat surface it is difficult to get the rod at exactly the same height each time. Bench marks are, however, sometimes established on flat level surfaces such as the coping stone of a masonry structure, because permanence is of more importance than great precision. Bench marks are not only described in the notes, but are themselves marked by paint, chisel marks, or drill-holes.

9.9. Leveling for Profile.—Profile leveling is for the purpose of determining the changes in elevation of the surface of the ground along some definite line like the center line of a highway. The line is first "stationed," i.e., marked at every hundred feet or such other interval as is desired. The level is set up and a backsight taken on a bench mark to determine the height of the instrument. Foresights are then read on as many full station points on the line as can be taken conveniently from the position of the instrument. Intermediate sights are taken at any points where there are marked changes of slope, and the plus stations of these intermediate points are recorded with the rod-readings. It will be remembered that the terms foresight and backsight do not refer to the **forward** and **backward** direction. **A backsight is a reading taken on a point of known elevation for the purpose of obtaining the height of the instrument. A foresight is a reading taken on a new point to determine its elevation.** Backsights are frequently called *plus sights* (+S) because they are added to an elevation to obtain the H.I., and foresights are called *minus sights* (−S) because they are subtracted from the H.I. to obtain elevation. When it is necessary to move the level to a new position in order to take readings on stations ahead, a turning point is selected and its elevation is determined. The level is then taken forward and the new height of instrument determined by taking a backsight on the turning point. This general process is continued until the end of the line is reached.

A line of levels should be checked by connecting with some reliable bench mark if possible. If there are any bench marks along the line of levels they should be used as turning points if convenient, or at least check readings should be taken on them in order to detect mistakes. When reading on such points it is evident that the reading taken on the bench mark is really a foresight since its elevation is being found anew from the height of instrument. Readings on bench marks and turning points should be taken to thousandths or to hundredths of a foot, according to the accuracy desired. If the elevations of the profile are desired to the nearest hundredth of a foot, as, for example, on a paved highway, the turning points should be taken to thousandths of a foot. Elevations on the surface of the ground will not usually be needed closer than to tenths, so the T.Ps. are taken only to hundredths. In calculating the elevations the results **should not be carried to more decimal places than the rod-readings themselves,** otherwise the results will appear to be more precisely determined than they really are.

9.10. Profile Notes.—Are often kept as shown in Fig. 9.10.1. In these notes also the computation of the heights of instrument and of the elevations of turning points may be checked by means of the sums of the foresights and backsights, provided that only the sights on turning points and the initial and final benches are included. If it seems desirable the computation of the elevations of stations may be checked by means of differences in foresights. The difference between the elevations of any two points that are obtained at the same set-up of the instrument is equal to the difference between the foresights taken on these points. In these notes the elevations of B.Ms. and T.Ps. are put in a different column from the surface elevations simply for the sake of clearness, but many prefer to put all the elevations in the same column. Another arrangement of columns which, if adopted, will be found convenient when plotting the notes, is to place the station column immediately to the right of the elevation column. If two foresight (−) columns are carried, one for T.Ps. and one for surface points, the verification of the B.M. elevations is more easily made.

(LEFT-HAND PAGE) (RIGHT-HAND PAGE)

Profile of Meadow Park Road.						Sept. 16, 1957 {Rowe ⊼, Harkins, Jacobs}
Sta.	+S	H.I.	−S	Elev.	B.M.& T.P. Elev.	Description
$B.M._3$	12.23	34.98			22.748	d.h. in wall near Sta. 0
0			9.8	25.2		
1			6.6	28.4		
2			3.0	32.0		
$T.P._1$	11.18	44.73	1.43		33.55	Stump
3			6.1	38.6		
+65			2.7	42.0		
4			3.7	41.0		
+20.7			5.2	39.5		
5			6.7	38.0		
6			11.2	33.5		
$T.P._2$	3.48	42.59	5.62		39.11	Nail in stump 80' W. Sta. 6+80.
7			10.2	32.4		
8			8.6	34.0		
9			7.6	35.0		
+62.4			4.0	38.6		
10			2.4	40.2		
+43			1.1	41.5		
11			2.6	40.0		
12			8.0	34.6		
$T.P._3$	0.42	31.89	11.12		31.47	Boulder
13			2.8	29.1		
14			8.7	23.2		
+23.8			11.2	20.7		
$B.M._4$	0.63	27.79	4.73		27.16	Elev. = 27.14 (Book 12. p26) Highest point large isolated boulder 200' E. Sta 16.
15			6.8	21.0		
16			7.2	20.6		
17			8.1	19.7		
18			9.0	18.8		
+54			9.2	18.6		

FIG. 9.10.1. PROFILE LEVEL NOTES.

Fig. 9.10.2 represents a rough plan and profile of the line of levels shown by the notes in Fig. 9.10.1. Angle points in the transit line are shown in the plan, but they do not appear in the profile of the line. It will be noticed that the T.Ps. and B.Ms. are not on the transit line in plan, and that they consequently do not appear on the profile. It is not customary to introduce any sketches into the profile notes except those used in describing bench marks or turning points.

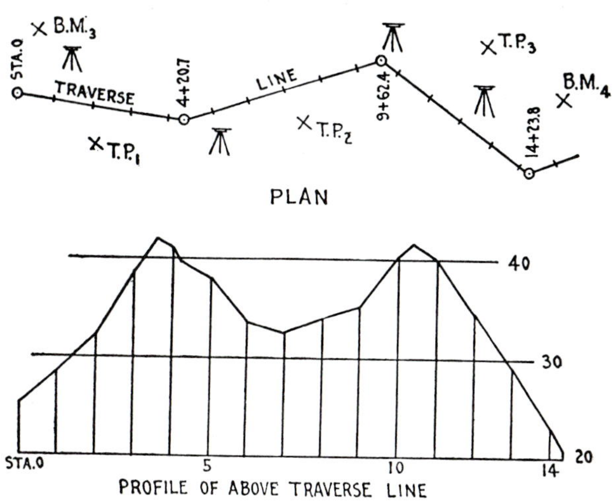

FIG. 9.10.2.

9.11. Checking of Ground Elevations.—As mentioned in the previous chapter, the ground elevations (−S) readings are virtually unchecked, since always a number of them are subtracted from the same H.I. In Fig. 9.11.1 a better way of recording is shown for part of the same profile.

In this case BS and FS refer to B.Ms. and T.Ps. only, while the other (−S) readings are listed as "Ground Readings" (GR). Rather than finding H.I., the differences (BS−FS) is formed and entered into a column "Rise" if positive, and "Fall" if negative. The elevation differences are determined for all points, using GR as FS and subsequently as B.S.

This method checks all the computations involving ground readings, which are unchecked when working with H.I. and the usual page check.

The recording in the field includes the actual readings only, while the rise and fall values can be derived in the office.

By calculating the elevations in this manner all steps are checked, because $\Delta H = (\sum BS - \sum FS) = (\sum \text{Rise} - \sum \text{Fall})$.

PROFILE ON MEADOW PARK ROAD

Sta.	BS	GR	FS	RISE	FALL	ELEV.
BM 3	12.23					22.748
0		9.8		2.43		25.18 → 25.2
1		6.6		3.2		28.38 → 28.4
2		3.0		3.6		31.98 → 32.0
TP1	11.18		1.43	1.57		33.55
3		6.1		5.08		38.63 → 38.6
+65		2.7		3.4		42.03 → 42.0
4		3.7			1.0	41.03 → 41.0
+20.7		5.2			1.5	39.53 → 39.5
5		6.7			1.5	38.03 → 38.0
6		11.2			4.5	33.53 → 33.5
TP2	3.48		5.62	5.58		39.10
7		10.2			6.72	32.39 → 32.4
8		8.6		1.6		33.99 → 34.0
9		7.6		1.0		34.99 → 35.0
+62.4		4.0		3.6		38.59 → 38.6
10		2.4		1.6		40.19 → 40.2
+43		1.1		1.3		41.49 → 41.5
11		2.6			1.5	39.99 → 40.0
12		8.0			5.4	34.59 → 34.6
TP3	0.42		11.12		3.12	31.47
13		2.8			2.38	29.09 → 29.1
14		8.7			5.9	23.19 → 23.2
+23.8		11.2			2.5	20.69 → 20.7
BM4			4.73	6.47		27.16
	27.31		22.90	40.43	36.02	(−22.75)
						4.41
		+4.41			+4.41	

FIG. 9.11.1. PROFILE LEVEL NOTES WITH COMPLETE PAGE CHECK.

9.12. CROSS-SECTIONING.—The term "cross-section" has several different but related meanings in surveying work. In one sense the term refers to short profiles of the ground taken at right angles to the center line of a linear structure, such as a railway, highway, dam, or canal. In another sense the term refers to the figure (in a vertical plane) at right angles to the center line which is formed by the intersection of the outlines of the proposed structure with the ground line (as in Figs. 9.7 and 14.14). In some types of surveys (such as those made for borrow-pits, or for the location of contours), a rectangular grid is laid out on the ground, and elevations are taken at the grid corners. This likewise is sometimes called cross-sectioning. (See Art. 9.16.)

9.13. Ground Cross-Sections for a Highway, Dam, or Canal. —The shape of the ground on the site of a linear structure, such as a highway, dam, or canal, is usually determined by running short profiles, called cross-sections, at **right angles to the center line** of the project. Cross-sections are usually taken at every 100-ft. station, and at intermediate points (pulses) **wherever the ground changes abruptly.** For highways, they are taken also at the starting points and at the ends of horizontal and vertical curves and at 25- or 50-ft. intervals on these curves; the cross-sections are extended to the right-of-way lines on both sides, and even farther if there is likelihood of the embankment or excavation extending beyond these limits.

The process of cross-sectioning is similar to that of profile leveling. The levelman obtains the height of instrument (H.I.) by taking a backsight on a bench mark (B.M.) of known elevation. Rod-readings (foresights) are then taken at a sufficient number of points on the cross-sections to determine the shape of the ground. The points at which rod-readings are taken are located by taping the distance from the center-line station to the place where the rod is held. The rod is read to the nearest 0.1 ft. when held on the ground to the nearest 0.01 ft. on T.Ps. and B.Ms. Distances are usually measured to 0.1 ft. and with a metallic tape. To obtain checks on the level work a rod-reading should be taken on every available B.M. along the line, and the levels must always be extended to, and checked on, a B.M. after the last cross-section has been taken.

A typical form of cross-section notes for a highway is shown in Fig. 9.13.1. Readings on B.Ms. and T.Ps. and the computed values of H.Is. and elevations are recorded in the usual form for level notes and are entered on the left-hand page of the note-book. On the right-hand page all cross-section readings are recorded opposite the stations where they were taken. A heavy line is drawn down the middle of the right-hand page to represent the center line of the highway. Rod-readings taken to the left of the center line are recorded on the left of this line, and those taken to the right of the center line are recorded on the right side of this line. In the field-notes the distance from the center line to the point where the rod-reading was taken is recorded above the rod-reading, and the computed ground

CROSS-SECTION NOTES
County Road, Ferndale, Mass.

Proj. S-52

Winslow
Rankin
Ludlam

Right of Way 80 ft. Oct. 30, 1957

Sta.	B.S.	H.I.	F.S.	Elev.		Left			¢	Right		
B.M.3	4.21	76.70	–	72.49								
10+00						40 / 1.7 (75.0)	25 / 4.3 (72.4)	10 / 2.4 (74.3)	3.8 (72.9)	18 / 4.6 (72.1)	30 / 5.2 (71.5)	40 / 8.7 (68.0)
11+00						40 / 3.3 (73.4)	12 / 6.8 (69.9)	6 / 4.2 (72.5)	5.3 (71.4)	15 / 6.5 (70.2)	28 / 8.2 (68.5)	40 / 10.4 (66.3)
+43					Brook Valley	40 / 2.3 (74.4)	28 / 5.1 (71.6)	16 / 7.2 (69.5)	9.2 (67.5)		22 / 10.3 (66.4)	40 / 11.5 (65.2)
12+00						40 / 0.4 (76.3)	25 / 2.7 (74.0)		4.5 (72.2)	8 / 5.1 (71.6)	20 / 6.2 (70.5)	40 / 8.3 (68.4)
T.P. 1	10.84	85.22	2.32	74.38								
13+00						40 / 3.6 (81.6)	27 / 4.2 (81.0)	10 / 8.6 (76.6)	8.3 (76.9)		25 / 10.2 (75.0)	40 / 11.9 (73.3)
+67					Summit	40 / 1.2 (84.0)	25 / 2.6 (82.6)		4.2 (81.0)	10 / 5.1 (80.1)	16 / 8.6 (76.6)	40 / 10.2 (75.0)
14+00						40 / 2.4 (82.8)	20 / 3.1 (82.1)	9 / 4.8 (80.4)	7.6 (77.6)	17 / 8.2 (77.0)	35 / 9.1 (76.1)	40 / 11.3 (73.9)
B.M.4			6.32	78.90	78.92 Established Elev.							

FIG. 9.13.1. CROSS-SECTION NOTES FOR A HIGHWAY.

elevation (H.I.—rod-reading) is recorded in brackets below the rod-reading. These elevations may be computed in the office after the survey has been made, but it is better practice to compute them as the work progresses, since an unreasonable or inconsistent elevation may reveal a mistake in a rod-reading which can be readily corrected while the survey party is still in the field.

Cross-section notes are sometimes so recorded that the stations read up the page, since this is the way they appear to the note-keeper as he stands facing in the direction of increasing stations; i.e., in the direction in which the survey is proceeding. There is less chance of recording readings on the wrong side of the center line, when using this method. When level notes are carried along with the cross-sections, as in Fig. 9.6, the level notes conventionally read down the page; therefore, the cross-section notes are also arranged with the stations reading down the page. Whichever system is used, as the observer faces in the direction of increasing stations, **left and right in the note-book must correspond with left and right in the field.**

A summary of the usual forms for recording cross-section notes is given below. Method *c* corresponds to that used in Fig. 9.13.1.

9.14. Earthwork Cross-Sections.—The simplest earthwork cross-section is one bounded by a level line at the grade of the road (called the "base"), two sloping lines representing the side slopes of the cut (or fill), and the ground surface. (See Fig. 9.14.1.) Railroad and dam cross-sections are usually of this simple type; highway cross-sections do not, as a rule, have level bases but are shaped to conform to the bottom of the pavement, ditches, and shoulders, as shown in Fig. 14.22.1. For preliminary earthwork calculations, the highway cross-section may be conveniently considered to have a level base placed at such a distance below the grade elevation of the pavement that the quantity of earthwork computed to the level line will be equivalent to that indicated by the irregular shape.

SAMPLE FORMS OF CROSS-SECTION NOTES

Left			℄			Right

(a) Numerator is distance "out." Denominator is actual ground elevation.

$\frac{33}{114.8}$	$\frac{20}{113.7}$	$\frac{10}{119.4}$	117.2	$\frac{17}{121.3}$	$\frac{25}{120.0}$	$\frac{33}{124.6}$

(b) Numerator is distance "out." Denominator is rod-reading on ground. Height of Instrument (H.I.) = 125.2

$\frac{33}{10.4}$	$\frac{20}{11.5}$	$\frac{10}{5.8}$	8.0	$\frac{17}{3.9}$	$\frac{25}{5.2}$	$\frac{33}{0.6}$

(c) Combination of (a) and (b). "Rod-readings" being recorded in field, and "elevations" computed either in field or in office.

$\frac{33}{10.4}$	$\frac{20}{11.5}$	$\frac{10}{5.8}$	8.0	$\frac{17}{3.9}$	$\frac{25}{5.2}$	$\frac{33}{0.6}$
(114.8)	(113.7)	(119.4)	(117.2)	(121.3)	(120.0)	(124.6)

Simplified earthwork cross-sections are classed as level, 3-level, 5-level, and irregular sections, depending upon the number and location of the points determined on the ground,

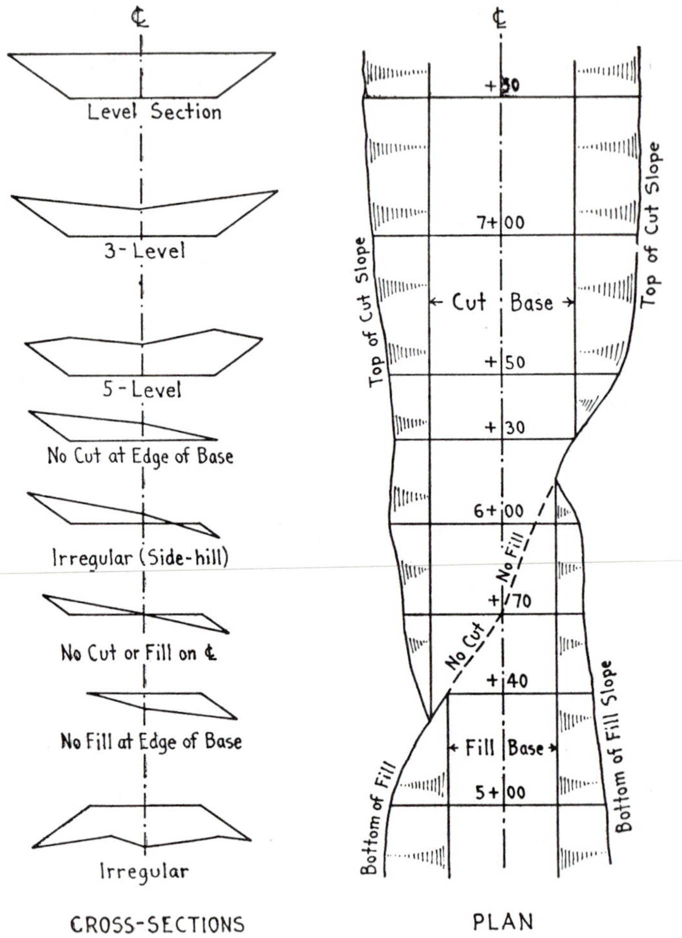

FIG. 9.14.1. PASSING FROM FILL TO CUT.

as shown in Fig. 9.14.1. At the place where the base passes from cut to fill (or vice versa), additional sections are required, as shown in Fig. 9.14.1. These are triangular in shape.

The grade elevation of the subgrade, or base of the section, is determined from the center-line profile of the structure. The difference between the elevation of the surface of the ground and the base elevation at any station gives the cut or fill at that point. Positive differences indicate cut and negative differences indicate fill; commonly cuts are designated as plus and fills as minus.

The side slopes of the section are usually expressed as a ratio, such as 1½ to 1; this means 1½ horizontal to 1 vertical. The 1½ to 1 slope is commonly used in railroad construction; the tendency in highway practice is toward flatter slopes, such as 2 to 1. Dams have still flatter slopes, 3 to 1 or even 6 to 1.

9.15. Setting Slope Stakes.—It is customary to set stakes on the right and on the left of the center line at points where the finished side slopes will intersect the surface of the ground; these stakes are called "slope stakes." Their positions are usually found by trial in the field. For this purpose it is convenient to compute the rod-reading which would be obtained if the rod could be held on the finished grade. The height of the instrument (H.I.) at the site of the work is determined by leveling from a B.M.; the difference between the H.I. elevation and the grade elevation at any station on the center line is the "*rod-reading for grade*," or the "*grade rod*" as it is sometimes called. The rod-reading for grade may be either positive or negative depending upon whether the proposed grade is below or above the H.I. elevation, as illustrated in Fig. 9.15.1(a) and 9.15.1(b) respectively. When the H.I. is above the finished grade, as in Fig. 9.15.1(a), the cut or fill at any point is the difference between the rod-reading for grade and the surface rod-reading. It is the sum of the readings when the H.I. is below the finished grade, as in Fig. 9.15.1(b). Notice that the rod-reading for grade is computed to hundredths of a foot but used to the nearest tenth of a foot when combined with surface readings which are read only to tenths.

In setting slope stakes, such as B in Fig. 9.15.1(a), it is required to find the measured distance AB that equals the computed distance AB. To set the slope stake B in Fig. 9.15.1(a), the surveyor keeps in mind the width of base (40 ft.), side slope (1½ to 1), and the center cut (7.5). At *B* there is actually no cut but it is customary to allude to the cut at *B* as being the vertical distance from the base to *B*. Thus if *B* were level with *A* the distance from the center stake to *B* would be

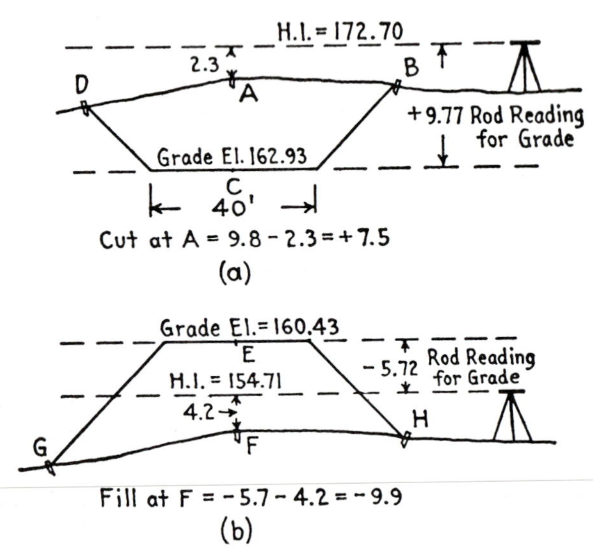

FIG. 9.15.1. DETERMINING CUT OR FILL AT CENTER STAKE.

$20 + 1½ \times 7.5 = 31.2$. This computation is done mentally. The tape is stretched from the center stake, at right angles to the center line. At 31.2 ft. on the tape it is observed that the ground appears to be, say, 2 ft. lower than at *A*, which means that the cut at *B* is about 2 ft. less than at *A*, and the distance out to *B* is therefore $1½ \times 2 = 3$ ft. less than 31.2, or 28.2. The rod is therefore held at the 28.2-ft. point and a surface rod-reading is taken. Suppose the reading is 3.6; the cut at that point then equals $+9.8 - 3.6 = 6.2$. The distance from the center line to the point where the side slope should intersect the ground is $20 + 1.5 \times 6.2 = 29.3$. This is 1.1 ft. farther from the center line than the point first selected. If the ground is level between the 28.2- and 29.3-ft. marks on the tape and the rod is moved to 29.3, the surface rod-reading will still be 3.6, in which case the stake *B* would be driven at 29.3 ft. But if it appears that the ground falls off, say, 0.2 ft., from 28.2 to 29.3, the cut will be 0.2 ft. less and therefore 0.3 ft. less distance out to *B*. Therefore the rod should be held for second trial at $29.3 - 0.3 = 29.0$. After two or three trials the correct position of the stake can be determined; but with experience only one or two trials will be necessary. The stake *B* is then driven into the ground and the cut is marked on its face and station number on its back.

The position of the stake is recorded in the notes as a fraction, with the cut (or fill) in the denominator and the distance from the center line in the numerator. This is shown in Fig. 9.15.2, which is a complete set of notes for the cross-sections shown in Fig. 9.14.1. The fractions at the extreme left and right of the cross-section column show the slope stake locations. The figures in the center are cut or fill on the center line. Intermediate recordings on the cross-sections have been made at critical points on the ground, as at Stas. 5 + 00, 6 + 00 and 6 + 50 in the notes. The intermediate cross-sections at Stas. 5 + 40, 5 + 70 and 6 + 30 were taken where the fill side line, the center line, and the cut side line respectively pass through points of no cut or fill. These sections are utilized in earthwork calculations (Chap. XIV). The positions of these sections must be found by a method of trial in the field. Obviously when there is no cut or fill at a point, the surface rod-reading and the rod-reading for grade must be the same. For example, to determine the station of no cut or fill on the center line, the plus station is first estimated by eye and the grade elevation at that plus station is computed. The difference between the H.I. elevation and the grade elevation at the plus station is the rod-reading for grade. The rod is then held on the plus station and a reading taken. If the reading is the same as the rod-reading for grade, the correct station was assumed; if not, another trial must be made using the result of the first guess as a guide.

Cross-Sections of Millers Road
Westbrook County, Conn.
July 23, 1957

Sta.	Surface Elev.	Grade Elev.	Cross-Sections				
			Bases: Cut 40', Fill 36', Slopes $1\frac{1}{2}$ to 1.				
+50	73.1	67.50	28.4 / +5.6		+5.6		28.4 / +5.6
7	70.5	67.00	27.5 / +5.0		+3.5		28.7 / +5.8
+50	70.1	66.50	25.7 / +3.8	20.0 / +4.8	+3.6	20.0 / +5.8	26.6 / +4.4
+30	68.5	66.30	25.1 / +3.4		+2.2		20.0 / 0
6	67.3	66.00	25.7 / +3.8		+1.3	9.5 / 0	20.7 / -1.8
+70	65.7	65.70	23.9 / +2.6		0		21.6 / -2.4
+40	63.5	65.40	18.0 / 0		-1.9		22.8 / -3.2
5	59.8	65.00	25.5 / -5.0	10.0 / -3.7	-5.2	23.0 / -3.9	24.9 / -4.6

FIG. 9.15.2. CROSS-SECTION FIELD-NOTES.

In Figs. 9.15.1 and 9.15.2, Sta. 5 is an irregular section and Sta. 6 is a special type of irregular section which commonly occurs when part of the road is in cut and part in fill. The position of the point of no cut or fill must be determined on this cross-section so that cut and fill earthwork quantities may be computed separately, as described later in Art. 410. Sta. 6 + 50 is a 5-level section, Sta. 7 is a 3-level section, and Sta. 7 + 50 is a level section.

9.16. Cross-Sections for Borrow-Pit.—The shape of the original ground may be obtained by dividing the area into squares and taking elevations at the corners of each square and at intermediate points where it is necessary to represent the true shape of the surface. These surface elevations are measured to tenths of a foot. The squares may be of any desired dimen-

Cross-Section for Gravel Pit
Brown's Farm, Sanville, N.H.

Ross
Pitman
Campbell

Aug. 7, 1957

Sta.	B. S.	H. I.	F. S.	Elev.
B.M.	8.96	96.38	-	87.42
A 0			8.8	87.6
A 1			6.4	90.0
A 2			4.1	92.3
A 3			3.0	93.4
A 4			4.5	91.9
A 5			7.3	89.1
B 5			6.1	90.3
B 4			5.3	91.1
B 3			2.1	94.3
B 2			3.0	93.4
B 1			5.2	91.2
B 0			8.6	87.8
T.P.			4.88	91.50
etc.				

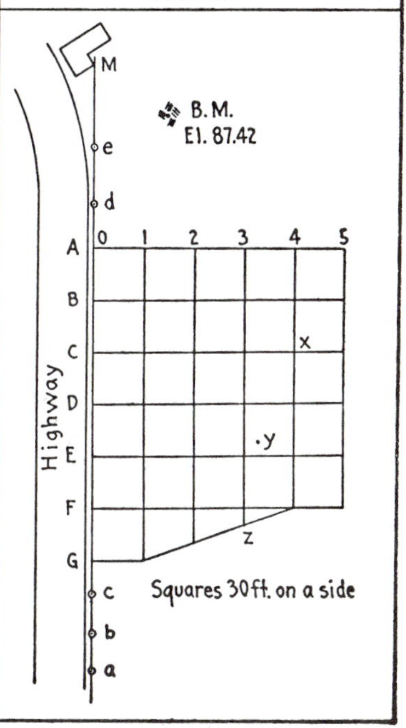

Sketch

FIG. 9.16.1. FIELD-NOTES FOR BORROW-PIT.

sions, usually 10 ft. to 100 ft. on a side, the size chosen depending on the ground slopes. The grid lines should be spaced close enough so that the general slope from one corner to the next will be approximately uniform. The grid is laid out with the transit and tape, and a temporary stake is driven at each grid corner. The corners of the grid are designated by rectangular coördinates, one set of parallel lines being marked by letters and the other by numbers, as shown in Fig. 9.16.1. The point x would be designated (C, 4); and if the squares are 30 ft. on one side the point y would be (D + 20, 3 + 10), and point z would be (F + 10, 3). The notes are kept as in profile leveling except for the designations in the "Sta." column.

When it is desired to determine the amount that has been excavated, the same system of cross-sections is again run out and the new elevations at the corners and at the necessary intermediate points are determined. In borrow-pits, there is usually a steep face of bank left after the material has been removed. Some of the lines of the grid that was laid out before excavation was started will, when reproduced after the pit excavation has been completed, cross the face of the steep slope. It is necessary, therefore, to take a rod-reading at every point where the foot of slope or top of slope crosses a grid line and to record these, as well as readings at all grid corners, in the notes. The base-line from which the system is laid out should be outside the area to be excavated so that the marks will not be obliterated. The bench mark should also be outside this area. For methods of computing the earthwork in borrow-pits see Art. 14.24.

9.17. Cross-Sections for Trench Excavation.—The surface elevations are determined by making a profile of the center line. The grade of the bottom of the trench is obtained either from the plan or by leveling. The width of the trench is measured wherever it changes and the stations of these places noted.

9.18. Use of the Tape Rod in Cross-Section Work.—Where there are many elevations to be calculated, time can be saved by using a tape rod (Art. 4.14). The numbers on this rod increase from the top toward the bottom. The rod is held on a bench mark, and the tape moved up or down as directed by the levelman until he reads the feet, tenths and hundredths which are the same as those of the elevation of the bench mark, e.g., if the elevation of the B.M. is 195.62, the tape will be moved until he reads 5.62. The tape is then clamped firmly to the rod. If the rod is then held on a point 1.61 ft. lower than the bench, the rod-reading will be 4.01, since the readings decrease as the rod is lowered. The elevation of that point is therefore 194.01 ft. or, sufficiently precise for topographic work, 194.0 ft. In this way the elevations are read on the rod to feet and decimals of feet, the tens and hundreds of feet being supplied mentally.

9.19. Leveling to Establish a Grade Line.—The level may be used for setting points at desired elevations as, for example, in establishing the grade line of street or sewer not yet constructed. To set any point at a desired elevation, set up the level and take a backsight on a bench mark, thus determining the height of instrument. Subtract the desired elevation from the height of instrument and the result is the rod-reading for grade. Raise or lower the rod until the horizontal cross-hair indicates this reading. The foot of the rod is then at grade. This is usually set for construction work to hundredths of a foot; for some purposes tenths of a foot will be sufficiently exact. If a target rod is used the target is set at the proper reading, and the bottom of the rod is at grade when the cross-hair bisects the target. A mark is then made at the bottom of the rod on a stake or a batter board. If the grade line such as for a sewer comes beneath the surface of the ground and cannot be reached, a point may be set on a stake or on a batter board where the point is at a convenient whole number of feet above grade and the depth marked. The grade of the pipe may then be set by measuring this distance down from the mark on the grade stake, as described in Art. 10.36. If the construction for which the grade is being set is far above the surface of the existing ground, a grade may be marked on the stake a whole number of feet below the true grade and the height to the true grade noted on the stake. The process of setting grades for curb lines or for resurfacing a street is described in Art. 10.44.

9.20. "Shooting in" a Grade Line.—To save time and to diminish the liability of mistakes, grades may be set by a method known as "shooting in" the grade. First set a mark at the grade elevation at each end of the straight grade line. The instrument is set up 6 or 8 inches to one side (or just behind) of the first point, and the distance from that mark to the axis of the telescope is measured with the tape or rod. Then the rod, which is set at this reading, is carried to the last point on the straight grade line, and, while it is held vertical on the grade mark at that place, the instrument man inclines the telescope until the horizontal cross-hair is on the target, clamping the instrument in this position. The transit instrument is well adapted for this method of setting grades since its telescope can easily be inclined and the cross-hair set on the target held on the distant point by means of the vertical motion clamp and its tangent screw. If a level instrument is used, the horizontal cross-hair may be set on the target by tilting the level telescope by means of the leveling screws. The tilting level, Art. 4.6, may be tilted more advantageously to the desired inclination by means of its micrometer screw which tilts the telescope about a horizontal axis. In each case the purpose is to establish a line of sight that is parallel to the grade line at a fixed height above it. All intermediate points on this grade line may then be set by raising or lowering the rod until the target coincides with the horizontal cross-hair in the instrument. The bottom of the rod is then on the desired grade line.

Obviously the steepness of the grade line that can be laid out by this method is limited by the amount of inclination that can be obtained with the leveling screws on a wye or dumpy level or with the micrometer screw on a tilting level. The method has its most practical application when the transit is being used for both angle measurements and leveling, such as is often done on construction jobs where lines and grades are being laid out.

9.21. To Establish a Datum Plane by Means of Tidal Observations.—Whenever it is necessary to establish a datum from tidal observations it may be determined as follows. Set up a vertical staff, graduated to feet and tenths, in such a manner that both the high and low water can be read for all ranges likely to occur. Read the positions of high and low water for each day for as long a period as practicable. The mean value obtained from an equal number of high and low water observations will give the approximate value of mean sea-level.

The datum obtained in this manner is also called the *high-tide level* and must be carefully distinguished from *mean sea-level.* The former is the plane exactly midway between the planes of *mean high water* and *mean low water.* Mean sea-level is a datum plane so placed that the area between this plane and the curve of high waters is exactly equal to the area between this same datum plane and the curve of low waters. This plane is the datum used by the U.S. Coast and Geodetic Survey, Geological Survey, U.S. Engineers and other governmental organizations when establishing control levels. For a single lunar month, the two data may differ by as much as one foot. If the observations extend over just one lunar month the result will be fairly good, whereas in less than one month a close result cannot be obtained; to determine this accurately will require observations extending over a year or more.

Other common datum planes based upon the tides are *mean high water, mean low water, lower low water,* and *higher high water* (terms used by U.S. Coast and Geodetic Survey). These data are of great engineering importance in interpreting charts and in the design of waterfront developments. For example, if the depths on nautical charts are referred to extreme low water, there is no danger that mariners will run aground through misinterpretation; whereas the extreme high water elevations are essential in the selection of height for piers and wharves.

The place chosen for the gage should be near the open sea where local conditions will not influence the tide. It should be sheltered against heavy seas.

At the beginning of the series of observations the zero of the staff and some permanent bench mark should be connected by a line of levels. The elevation of some point close to the gage, such as a spike driven in a pile to which the gage is fastened, should be determined at the same time, so that the gage can be readily replaced if it should become dislodged. Such a point is called a *Reference Point,* or R.P. These should be tested occasionally to see if the staff has moved. After the reading of the staff for mean sea-level is found the elevation of the bench mark can be computed.

9.22. The Staff Gage.—This is a form of gage (Fig. 9.22.1) which can be easily constructed, and which is sufficient where only a short series of observations is to be made. If made in sections not over 3 ft. long, as described below, it can be packed in a small box for transportation. Each section consists of two strips of wood about 1½ inches square, and 3 ft. long, fastened together at the ends by strips of brass, leaving a space between them of about 1 inch. In this space is placed a glass tube of about ¾ inch diameter and held in place by brass hooks. On one side of the tube a red stripe is blown into the glass. When the gage is set up for observations the sections are screwed to a long vertical piece of joist. The ends of the tube are nearly closed by corks, into which small glass tubes of approximately 1 mm. (inside) diameter have been inserted. When the water rises in the main tube, the red stripe appears to be much wider than it really is on account of the refraction of light by the water. Above the water surface the stripe appears its true width. By observing the position of the wide stripe the height of the water surface can be read within a hundredth of a foot. The heights are red on a scale of feet and tenths painted on the wooden strips. If the size of the small glass tube is properly chosen, the fluctuations of the water surface outside will not disturb the water in the tube, so that the reading is a fair average of the water surface. A gage of this sort may be read to hundredths from a distant point by means of a transit telescope or a field glass.

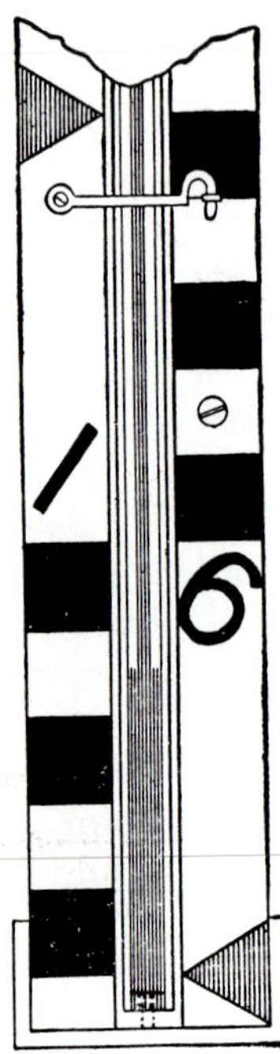

FIG. 9.22.1. STAFF GAGE.

When a long series of observations is to be made, particularly if it is important to obtain a record of every tide, an automatic tide gage should be used. A description of such a gage may be found in "Manual of Tide Observations," Special Publication No. 196 of the U.S. Coast and Geodetic Survey. (See also Vol. II, Arts. 7-23 to 7-26.)

9.23. LEVELING ACROSS A RIVER.—While the effect of curvature and refraction (Art. 4.25) is usually negligible in leveling operations, it may in certain special cases become of great importance to eliminate this error. For example, it is sometimes necessary to carry a line of levels across a river of considerable width, say, half a mile. In this distance the correction for curvature and refraction amounts to about 0.143 ft. under normal conditions, which in a line of bench levels is too large a quantity to neglect. If the correction derived from formulas could be depended upon under all circumstances it would be sufficient to apply it to the rod-reading; but the amount of the refraction correction is variable because of atmospheric changes.

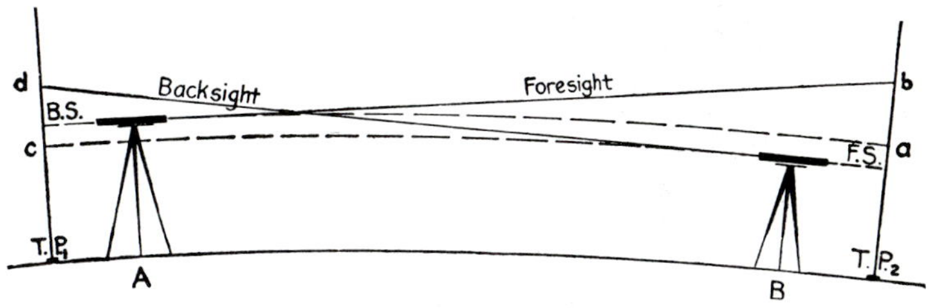

FIG. 9.23.1. LEVELING ACROSS A RIVER.

If it is desired to obtain the difference in elevation between two distant points with great accuracy it will be necessary to use a method that will **eliminate** the effects of curvature and refraction no matter what their actual amount may be. The procedure here described is a form of reciprocal leveling. In Fig. 9.23.1 suppose a backsight were taken on T.P.$_1$ with the instrument at *A* and then a foresight taken on T.P.$_2$. The elevation of T.P.$_2$ as computed from T.P.$_1$ will be too low by the amount *ab*, since the foresight on T.P.$_2$ is too great by this amount. If the difference in elevation is determined by the instrument at *B* the backsight on T.P.$_1$ is too large by the amount *cd*. Hence the H.I. of the instrument at *B* is too great, and consequently the elevation of T.P.$_2$ too great by the amount *cd*. The mean of the two determinations would give the true elevation of T.P.$_2$ if $ab = cd$, but this occurs only when the two sights are taken under the same atmospheric conditions. Therefore it will be seen that two sights must be taken

simultaneously. In order to eliminate the errors of adjustment* in the instrument it is necessary to use the same instrument at both ends of the line. To accomplish both of these results at once it is necessary to take simultaneous readings with two instruments and then to repeat the operation with the instruments interchanged. The magnifying powers of the two telescopes and the sensitiveness of the two spirit levels should be about equal in order to give the best results. It will be noticed that this process is similar to that of the peg adjustment (Art. 4.40).

9.24. Water Levels.—Lines of levels often can be checked satisfactorily by means of a water surface if the line is close to a lake or a pond so that this forms a connecting link between any two benches. The difference between the elevation of a bench mark and the elevation of the water should be determined carefully for the two bench marks selected. Then the water surface elevation is assumed to be the same in both places, and thus furnishes a connecting link between the two benches. The water levels should be taken when there is little or no wind blowing. Or, if this cannot be done, the effect of waves can be partly eliminated by constructing stilling boxes, which admit water slowly through small holes, but check the wave action almost entirely.

9.25. Distributing the Error in a Level Circuit.—Whenever a line of levels is closed upon itself or upon any bench of known elevation some error of closure is sure to be found. In the first instance the error noted is the algebraic sum of the various errors entering into the circuit. When closing on the work of another party the error is composed of the errors of both parties.

A complete discussion of the methods of adjusting level circuits would take us beyond the scope of this book, but a few simple cases will be illustrated. It is always satisfactory to distribute such errors and to know that the elevations are consistent among themselves; this at least avoids having several

*Errors remaining in the instrument, even after adjustment are of unusual importance because the sight is much longer than that used in adjusting the instrument.

values for the same bench mark and the uncertainty of not knowing which one to use.

If the leveling is continuously performed as described in Art. 9.3, then the remaining errors of leveling are chiefly of the class that tend to balance each other, and hence do not increase directly with the distance run, but only as the square root of this distance. The distance itself is not really the measure of the number of set-ups, but it is sufficiently close to it for the purpose of distributing a small error over long lines.

If a line is run in duplicate, forward and back, the difference in elevation is obviously the mean of the two results.

If two bench marks are joined by three routes, all of different lengths, say, 2 miles, 3 miles, and 4 miles, respectively, the best determination should be that from the two-mile line, other things being equal; and the least accurate should be that from the four-mile line. The result obtained from the two-mile line should therefore have the greatest weight in computing the mean, that of the three-mile line should have the next, and the four-mile line should have the least weight. The assignment of weights is made most conveniently by dividing some number (say 12 in this example) by each of the distances in succession. The results are the weights to be used. Dividing 12 by 2, 3, and 4 we obtain the weights 6, 4, and 3. The "weighted mean" is obtained by multiplying each difference in elevation by its weight, adding the results, and dividing by the sum of the weights.

Suppose that three results for a difference in elevation are 41.16 ft., 41.20 ft., and 41.12 ft., then the weighted mean and standard deviation are computed as follows.

Line	M	(Weight) w	$w \times M$	v	$w \times v$	$w \times v^2$
1	41.16	6	246.96	0.00	0.00	0.0000
2	41.20	4	164.80	+0.04	+0.16	0.0064
3	41.12	3	123.36	−0.04	−0.12	0.0048
$n = 3$		$\sum w = 13$	$\sum w \times M = 535.12$		$\sum w \times v = +0.04$	$\sum w \times v^2 = 0.0112$

$$\text{Weighted Mean } M_0 = \frac{\sum w \times M}{\sum w} = \frac{535.12}{13} = 41.16 \text{ ft.}$$

$$\text{Standard deviation of weighted mean} = \sqrt{\frac{\sum w\, v^2}{\sum w(n-1)}} = \sqrt{\frac{.0112}{13\,(2)}} = \pm 0.02 \text{ ft.}$$

This computation can be shortened by omitting the whole number of feet since these are all the same, or by omitting 40 ft. If we omit 41 ft. the calculation becomes

$$\frac{6 \times 0.16 + 4 \times 0.20 + 3 \times 0.12}{6 + 4 + 3} = 0.163$$

$$\text{Elev.} = 41 + 0.163 = 41.16 \text{ ft.}$$

Another convenient way to do this is to deal with the correction instead of with the elevation itself. If we take the first elevation as being a close value, to be used as a first approximation, then its preliminary correction is 0.00; that of the second value is 41.20 − 41.16 = +0.04. That of the third value is 41.12 − 41.16 = −0.04. If we multiply each correction by corresponding weight we have the results shown in the following table:

Line	wt.	corr.	product
1	6	0.00	0.00
2	4	+0.04	+0.16
3	3	−0.04	−0.12
	13		13)+0.04
			+0.003

The final value for the elevation is therefore

$$41.16 + 0.003 = 41.16, \text{ as before.}$$

When there are intermediate bench marks along these routes their adjusted elevations are usually computed by proportioning the correction directly as the distance. That is, if a point is halfway from the start to the second bench, it receives half the correction.

When the circuits are entangled the method becomes more complicated. The exact solution of this problem is one that must be made by the method of least squares. An approximate method used by the U.S. Geological Survey is given in Appendix C, Vol. II.

For general treatment of errors in measurements see Art. 1.10-1.13.

CHAPTER X

CITY AND URBAN SURVEYING*

LINES AND GRADES—CIRCULAR CURVES—SURVEYS OF CITY LOTS, BUILDINGS AND PARTY WALLS

10.1. Instruments Used.—Because of the high value of city and urban land a relatively high precision is demanded in city surveys therein. Furthermore, much of the work of a city engineering department or of the city surveyor in private practice is the establishing of lines and grades for construction work. These classes of surveying work require the use of theodolites, levels, EDM, tapes and rods capable of yielding a high degree of precision. For less exact work, such as the location of contours and physical objects, the plane table or stadia method may be employed. Steel tapes graduated to hundredths of a foot are required for city surveys. Where lines must be measured with a high degree of precision, care must be taken to apply proper pull on the tape and to make the tape corrections described in Arts. 3.25-3.33. Tapes should be standardized by U. S. Bureau of Standards (Art. 3.25-3.26) or frequently compared with standardized tapes. Some cities maintain a *city standard of length*, usually 100 ft. long, located in a convenient place where it will not be exposed to the direct rays of the sun; such as in a corridor of a building. It should be on a flat surface so that the tape may be tested while supported throughout its length.

For city property surveys transits reading to 30″ or to 20″ might be used. With these instruments angles to the nearest 5″ can be obtained by repeating the angles. If much precise work

*See "Technical Procedure for City Surveys," Manuals of Engineering Practice—No. 10, Am. Soc. Civil Eng., revised edition 1957.

is required it will be well to use a 10″ instrument and European theodolites reading to 0′.1 or less should be utilized.

For most bench mark leveling and building layouts, a dumpy or wye level is used with target or self-reading rods. In establishing vertical control for a coördinate system precise levels are required (Art. 4.8 and Chap. 3, Vol. II).

Much city work, such as the staking out of new streets, paving, sewers, or curbs, requires the establishment of both lines and grades. This class of work does not as a rule call for a precision in distances and elevations closer than .01 ft. It is not convenient, for the ordinary surveying party of three men, to carry both a theodolite and a level in addition to the ordinary equipment of lining poles, level-rod, stakes, tape, etc., so the theodolite is extensively used in setting grades as well as in establishing lines. The degree of precision possible with an engineer's transit is satisfactory for ordinary leveling.

ESTABLISHING AND STAKING OUT CITY LINES AND GRADES

10.2. Establishing Street Lines.—It has not been uncommon for the streets in new districts to be accepted by city officials as part of the city's street system without intelligent study of their value or adaptability to the existing street system. Plans of streets presented for adoption should be modified to fit the general scheme of the City Planning Board, if there is one, or to meet the judgment of the City Engineer, before being passed, as he is, or should be, the best qualified official to judge of the merits of such matters.

The plan of proposed streets should show clearly and accurately the mathematical relation of these new street lines to existing monuments on the old streets with which the new ones connect.

When a private way is widened to make it a suitable public way and it becomes advisable to establish new lines, the correct procedure is to make an accurate plan of the present way and abutting properties including all existing improvements. This plan should show clearly the amount of property it is necessary to take from each abutter to accomplish the widening, marking on the plan the dimensions of every taking to

0.01 ft. and its area to the nearest square foot as well as the existing dimensions and areas of all lots affected. These data not only give the local authorities the necessary information for their deliberations but also show each abutter the exact effect the change will have upon his property. Usually such changes are not adopted until a hearing has been given at which all interested parties may express their views.

When the ordinance is finally passed authorizing the widening and the street lines are thus "established" it is the duty of the City Engineer to mark these new lines promptly by monuments, of the legal department to "settle the damages" with the owners who have been deprived thereby of part of their property, and of the public works department to proceed with the construction work necessary to widen the street.

10.3. Establishing Street Grades.—Usually the street grades will be established at the same time that the lines are established (see Art. 10.2), and the legal department of the city will at that time have to settle grade damages as well as land taking damages. In some cases the courts have held that in widening a very narrow street the damage to the abutters is nil since the improvement increased the value of their remaining property to so great an extent. But when a change in street grades is made it rarely happens that improved properties in which their entrances, doorways, driveways and steps already conform to the existing grade are not damaged somewhat by a change in grade. Because of this fact it is the surveyor's duty, when the street grade is to be changed or "established," to show on his plan all approaches to improved property, elevations on driveways and a few elevations in front yards. On the street profile, he should show, not only the profile of the center line, but also that of the curb lines and property lines, the elevations of basements and first floors and steps of all buildings that are near to the street lines so that the effect of change in street grade may be at once apparent.

A description of the proposed street grade is usually put in written form for acceptance by the proper municipal authorities. When this description has been formally accepted by an order of the City Government the grade is said to have been "established." Such an order may refer to the profile by title or recorded number, or it may be a written description of the grade. The profile drawing of each street should also show one or more cross-sections of the street, designating the part of the cross-section to which the profile refers; center of the street, the curb or the sidewalk at the property line.

Following is a description of an established street grade:

"Beginning at station 146 (Maple St.) at the junction of the center lines of Maple St. and Ocean Ave., at grade* 52.00, the grade line falls 0.50 per 100 for 726 ft. to grade 48.37—thence rises 0.82 per 100 for 322 ft. to grade 51.01—thence falls 0.50 per 100 for 122 ft. to grade 50.40—thence falls by a vertical parabolic curve for 100 ft. as follows:

Sta.	Elev.
157 + 70	50.40
157 + 95	50.18
158 + 20	49.76
158 + 45	49.15
158 + 70	48.35

thence falls 3.60 per 100 for 239 ft. to station 161 + 09 at the junction of the center lines of Maple St. and High St., grade 39.75."

10.4. THE DATUM PLANE.—In every city there usually exists some established datum plane to which all elevations may be referred. In the older municipalities, it was customary to choose a datum bearing an intimate relation to the topography of the locality; a town on the seaboard chose mean sea level or mean low water (Art. 9.22); inland cities often selected the mean level of some lake or even some arbitrary datum. Because of this multiplicity of data (there are over 40 in Massachusetts), considerable confusion ensued in providing elevations for such state projects as highways which run from city to city. At present, the network of levels established by the

*The word *grade* is frequently used to mean the *elevation* of a point. Care should be taken not to confuse the meaning of *grade* with *rate of grade*. The latter is sometimes called *gradient*, a word which has some advantages but is not entirely satisfactory.

U.S. Coast and Geodetic Survey and of the U.S. Geological Survey, the datum for both of which is *mean sea level*, is so comprehensive that there is little excuse for a town or city not using this *uniform* and *scientifically* established datum. It is the present practice of both these Federal agencies to establish a permanent bench mark, whenever possible, near the center of the community so that it will serve as a starting point for all the elevations in that town. Information regarding standard bench marks established by these agencies may be obtained by writing the Director of these surveys at Washington, D.C.

10.5. Establishing Bench Marks.—When the datum has been determined, bench marks are established by the method explained in Art. 9.2. The establishment, at the start, of a reliable system of bench marks is of utmost importance, in order that the elevations of all parts of the city shall refer to the same datum. In laying out construction work it is absolutely necessary that reliable bench marks shall be available and sufficiently distributed to be of use in any section of the city without requiring many set-ups of the level to connect a bench mark with the work that is to be done. An advantage gained by having them close together is that they may serve as ready checks on each other as well as on the work at hand. It is not uncommon for a bench mark to be disturbed, and, if the level work is not checked frequently on some other bench mark, an error will surely enter into all of the level work which was started from that bench.

Many of the larger cities run their principal lines of levels with a geodetic level, and wherever possible, utilize the Government bench marks determined by first order leveling. It is advisable to establish the principal bench marks in pairs to diminish the chance of a total loss of a bench mark if there is any disturbance due to construction or other cause. The two B.Ms. preferably should be on opposite sides of the street, but near enough to be seen from one set-up of the instrument.

10.6. Staking Out New Streets.—In staking out a new district or a street widening, the information at hand is usually a plan of the layout which has been approved by the municipal authorities, the street lines as they appear on the plan being the "established lines."

It is the surveyor's duty to stake out these lines on the ground, connecting them properly with the established street lines of the older portion of the city. As soon as the stakes are in place and checked the street lines should be marked by monuments (Art. 10.8), so that there will be no difficulty in retracing the lines as they were originally laid out. If considerable grading work is to be done in building the new streets it may not be practicable to set many of the corner bounds at first on account of the likelihood of their being disturbed. Under these circumstances it is the duty of the surveyor to reference the points by cross transit lines or by some other accurate method before construction work begins; for it is important that the layout, as recorded in the city order, shall be fixed accurately and definitely so that when the streets are brought to the proper grade and the monuments are finally set they will mark the exact position of the established street line.

Just as it is not very wise to use different datum planes instead of referencing all elevations to sea level, it is unwise not to use coördinates in city surveys. If street lines are fixed by coördinates they can be easily reestablished as they are automatically referenced in a system which extends far beyond any construction work.

10.7. Referencing Important Points.—Because important corner stakes can be disturbed before permanent monuments can take their place, it will be highly desirable to tie them in by three or more accurate ties to permanent and well-defined physical objects. (See Art. 5.11.) This should be done whether or not coordinates are used, as it can save a considerable amount of field work.

Since in new districts fixed structures seldom exist, and since objects like trees or fences are liable to be removed at any moment on account of building operations, it will be well to set into the ground additional reference stakes as here explained. This same method applies as well to the referencing of the center line of a highway, a railroad, a dam, a bridge, or the lines of any project where construction activities are likely to disturb the important stakes.

Fig. 10.7.1 illustrates two methods of referencing the stake *A*. In the upper figure a row of stakes *B*, *C*, *D* and *E* are

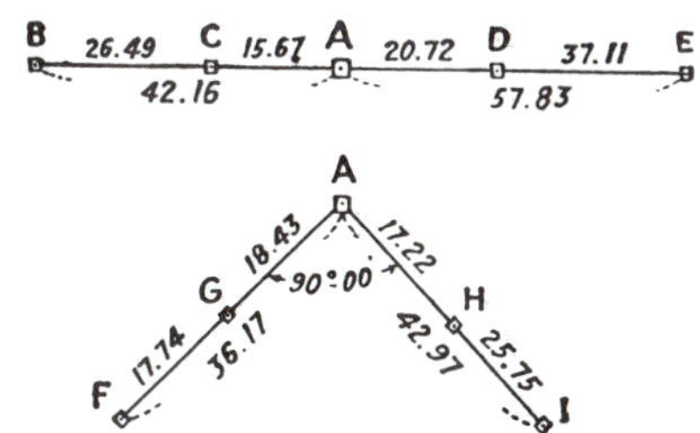

FIG. 10.7.1. REFERENCING IMPORTANT POINTS.

accurately placed on any convenient straight line through *A* that will bring them all well outside of any anticipated construction work. Accurate measurements are taken from *A* to all four stakes and recorded. Point *A* can then be checked or replaced at any time with a minimum of fieldwork.

In the second figure two lines are chosen at right angles (measured with the transit) and points *F* and *G* accurately placed on one line and *H* and *I* on the other. The distances from *A* are measured and recorded as before. This method also permits the location of *A*, if lost, without use of a transit as there are four different pairs of ties that intersect at right angles and which can be used to set *A* and to check its position.

Point *A* could be tied in by taking one accurate foresight on a well-defined distant object such as a church steeple and then placing points on three stakes on that line all within, say, 100 ft. of *A* and measuring from *A* to each of the three points to 0.01 ft. and recording them. The foresight should be described in the notes together with the ties; if all but one of the stakes are disturbed, point *A* can still be accurately reproduced.

In Fig. 10.7.2, for example, line *J A L* was selected intentionally so that it crossed the ledge and was in exact line with the church steeple and so the line *A K* would cross the ledge also. The drill-hole ties are likely to be more permanent than stakes. If stakes M and L are lost point A is still well determined. Even though but one of these reference points remain point A can still be replaced.

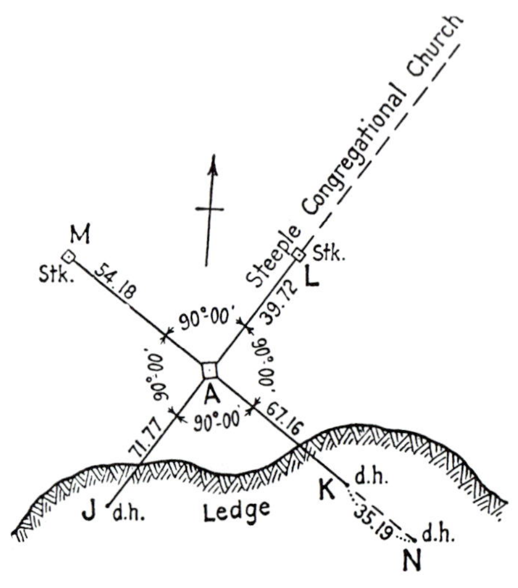

FIG. 10.7.2. REFERENCING IMPORTANT POINTS.

Line ties, involving no measured distances, may also be used to advantage as illustrated in Fig. 10.7.3. Points O, Q, S and U are distant, well-defined and easily described natural foresights. The transit is sighted, for example, on O and then focused on some nearby readily described foresight on the same line; both O and the nearby foresight should be described in the notes. Such ties give precise results only when the near foresights are, say, within 200 ft. of A, and the distant foresights are, say, ten times as far away. If the two points (as P and O) are comparatively near together the method will be useless. The angles between these lines are accurately measured and checked.

10.8. MONUMENTS.—It is important and at the same time customary to define street lines by setting stone bounds, often called *monuments*, at the street corners and at angles in the street lines. The bounds are set sometimes on the side lines, sometimes on the center lines, and sometimes in the sidewalks. The "side lines" of the street are, in fact, called the "street lines." These are the lines marking the property boundaries; from one property line to the opposite property line is

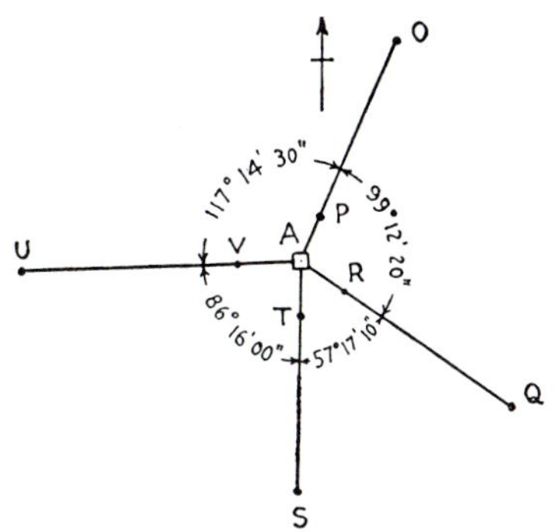

Pt. A - 3" square oak hub, copper tack.
O - c. steeple white Methodist Church with clock.
P - West outer corner east chimney J. A. Knowles & Co. shoe factory.
Q - c. steel chimney L. K. Brown soap factory at ht. of guys.
R - S. side of N. vertical edge of frame of window above big barn door red barn white trimmings about 200 ft. from A.
S - W. side E. corner board, main house of grey 2-story dwelling with French roof and hexagonal cupola.
T - c. conical top of water tank Ideal Laundry Co.
U - Halfway from Nly. vert. outer face of cap on top middle of 3 ventilators on roof of L. E. Howe's dry goods store, Main St. to the Nly. face of main stem of same ventilator
V - Vane rod at gable B. & W. R. R. car shop.

Fig. 10.7.3. Line, or Range, Ties.

the width of the street, no matter what be the roadway width and the sidewalk widths. Almost universally, then, sidewalks and curbs fall within the "bed of the street," albeit the care and upkeep of the sidewalk falls to the abutting property owner.

At street intersections, one monument at the intersection of the center lines will suffice to mark both street lines, but since this point will come in the center of the road pavement where it is likely to be disturbed by traffic or by street repairing it is seldom placed there. The more practicable method is to define the street lines by marking the side lines at the angles or, where these are rounded corners, at the beginning and end of the curves or at the intersection of the two street lines. It is not necessary that all four corners of a street intersection be marked, as a bound on one corner will define the side lines of the two streets and, the width of the streets being known, the other sides can be easily determined. Nor is it necessary to place a bound at one of the corners of every street intersection, provided a street is straight for several blocks, although it is good practice to do so. If stone bounds are placed exactly on the side lines of the streets they are liable to become disturbed by building operations. It is *always* preferable, therefore, to place the bound in the sidewalk flush with the surface of the walk and at such an offset that it will be the least liable to be disturbed by building operations or by the resetting of curb stones. The best place for the bound is at a point about 2 ft. inside the curb stone but at some full number of feet from the street line. In some cities it is the practice to place them at a 2 ft. offset from the property line, but this is usually inadequate; 5 to 10 ft. away is much better practice provided these distances still bring the offset line within the limits of the sidewalk. All monuments should be placed with extreme care as regards both their accuracy of position and their stability.

Monuments are usually squared stone or concrete posts 4 to 8 inches square and 3 to 4 ft. long, the length depending upon the severity of the climate. In New England a monument less than 4 ft. long is likely to be disturbed by frost even though its bottom rests on unfrozen ground. They are carefully squared on top and a drill-hole in this end marks the exact point. This drill-hole may be made either before the stone is set in place, or else after it has been so placed that its center is about in position, when the exact point may be defined by drilling a hole in its top. Sometimes the hole is filled with lead and a copper nail set in the lead is used to mark the exact point. For nice definition of the point, a copper bolt is inserted and two lines scratched across it at right angles; the intersection marks the exact point. When the stone bound is placed at the intersection of the side lines of the streets it is sometimes located entirely in the sidewalk in such a way that its inside corner is exactly on the intersection of the street lines. The three other corners of the bound are usually chipped off so that there may be no mistake as to which corner defines the line, but the line corner frequently becomes worn off and this practice is therefore not recommended. Some surveyors use, instead of stone bounds, a piece of iron pipe or iron plug with a punch-hole in the top of it, driven into the ground or embedded in cement concrete, or a chisel-cut cross in a paved sidewalk. Long heavy stakes are employed temporarily to define intermediate points or points of secondary importance.

10.9. Setting Stone Bounds.—The exact location of a stone bound is usually marked by a tack in the top of an ordinary wooden stake. The permanent monument which is to replace the stake should be set before the frost has entered the ground or before any other disturbance of the stake can have taken place. When the bound is ready to be set the first thing to do is

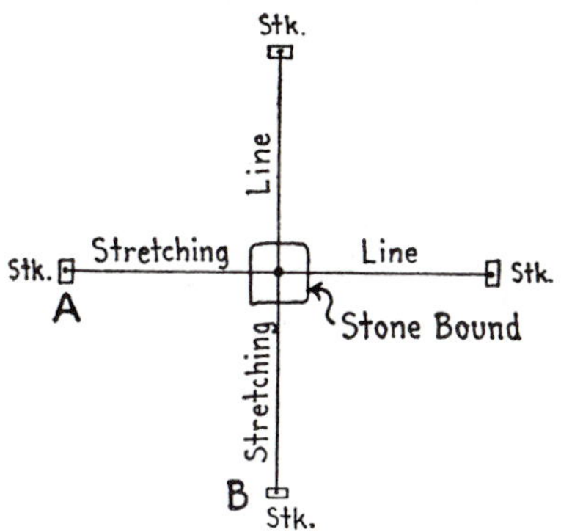

FIG. 10.9.1. SETTING A STONE BOUND.

to drive four temporary stakes in pairs on opposite sides of the bound stake about two feet from it and in such a way that a line stretched from two opposite stakes will pass over the tack in the head of the corner stake (Fig. 10.9.1). Then tacks are set carefully in the tops of these temporary stakes in such positions that a stretching line running between tacks on pairs of opposite stakes will pass exactly over the tack in the bound stake.

Then the bound stake is removed and the hole dug for the stone bound. Care should be taken not to dig the hole any deeper than is necessary so that the bound may be set on firm earth. As to the position of the top of the bound with reference to the surrounding ground, surveyors disagree. Some prefer that the monument should stand out of the ground so that it can be found readily; others believe that if it projects above the surface the bound is likely to become misplaced or worn by traffic, and that it is better therefore to set it just flush with the ground or slightly below the natural surface. If any grading is to be done in the vicinity the bound should be set so that it will conform to the proposed grade. If a cement sidewalk is to be laid the top of the bound must fit exactly and form a part of its surface. When the hole for the bound has been dug to the proper depth it is well to stretch the strings across between the temporary stakes and plumb down roughly into the hole to determine where the center of the bound will come, so that when the monument is dropped carefully into the hole it can be placed so that it will set plumb.

The bound having been set in the hole, the next operation is to fill around it. This should be done with care, the material being thoroughly compacted as the filling proceeds and the bound kept in such a position that the drill-hole in the top of it, if there is one, shall be **exactly** under the intersection of the strings. It is sometimes desirable to put in a foundation of concrete and to fill with concrete around the monument to within a foot of the surface, as shown in Fig. 10.9.2, where a more substantial bound is required, or where the ground is so soft as to furnish an insecure foundation. A bound so placed is not liable to disturbance by frost action. If the top of the bound is plain and the hole is to be drilled after the bound is in place, care should be taken to place the monument so that this hole will come close to the center of the top in order that it may present a workmanlike appearance. After the bound is set exactly in place the temporary stakes are removed.

Some surveyors prefer to use only two opposite stakes and one stretching line, the position of the monument being determined by a measurement along the stretching line from one or both of the temporary stakes. Still another method of temporarily tying in the stone bound, and one which many surveyors use, is to set two stakes such as *A* and *B* in Fig. 10.9.1, and either measure the distance from them to the bound or else set them at some even distance from the bound. This process of using temporary stakes and the stretching line is employed also in setting other types of bounds such as gas pipes or iron rods.

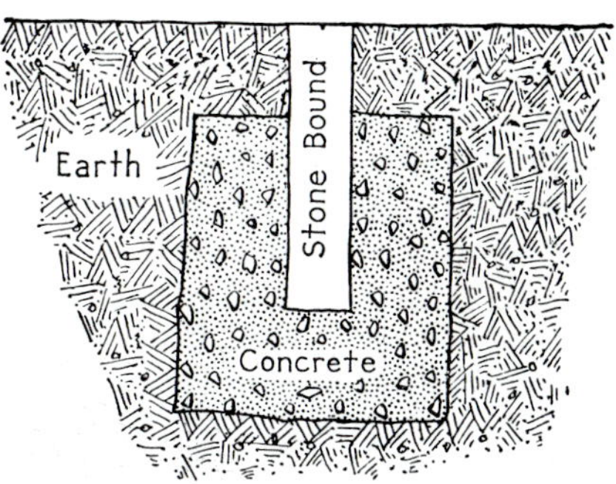

FIG. 10.9.2. SETTING BOUND WITH CONCRETE FOUNDATION.

In the construction of buildings or fences, monuments are frequently disturbed and too often they are reset by the owner of the property without the services of a surveyor. In rerunning a street line, therefore, a surveyor should be on the lookout for such conditions, and he should be cautious in the use of any monument which he has any reason to suspect may have been misplaced.

10.10. STAKING OUT CITY LOTS.—In staking out the lots of a rectangular block, the corners of which have been established, the most direct method is as follows. The theodolite is set up on the S.B. at *A* (Fig. 10.10.1), a sight is taken on *B*, and the front corner stakes of lots 1, 2, 3, 4, etc., are set, with a tack, exactly on line, in the top of each stake. All such work should be done to the nearest 0.01 ft. It will be well first to measure the line *AB*, to see that it is just 600 ft. long. Since it is assumed that the S.Bs. were set exactly, if it is found to be a few hundredths over or under 600 ft., this discrepancy was probably due to the difference between the length of the tape employed

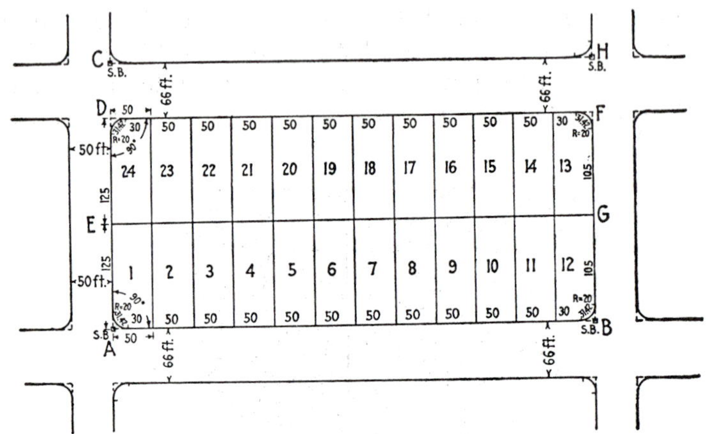

FIG. 10.10.1. RECTANGULAR CITY-BLOCK.

on the present work and that used in the original layout. The twelve lots therefore must be laid out with equal frontages. For example, it may be a hot day when the lots are to be staked out and the tape may read on that day a distance from A to *B* of 599.88 ft. To allow for this each lot should be laid out 49.99 ft. wide. (See also Art. 5.23, for laying out lots.)

With the instrument still at *A* and sighted on *C*, point *D* is set by measuring 66 ft. from *C*, and then point *E* is placed midway between *A* and *D*. Whatever slight discrepancy there may be in the distance between the S.B. at *A* and that at *C* is thrown into the depth of the lots rather than into the width of the street. The distance *DA* should nevertheless be checked.

By setting up the instrument at *B* and sighting on *H*, points *F* and *G* are set. Then by setting up at *F* and sighting on *D* the front corners of lots 13, 14, 15, etc., are determined by giving to each lot its proper share of the measured distance *FD*.

Another set-up of the theodolite at G with the line of sight on E will allow the "back bone" to be run out and the back corners of all the lots established. The depths of the lots can be checked easily by taking direct measurements from the front to their rear corner stakes.

By the method suggested above the street lines are made straight and the slight inaccuracies which may occur in the fieldwork are put into the back and side lines of the lots.

Where the streets are curved, as is frequently the case in new residential districts, before such lines can be established or before the lots can be defined for deed descriptions it is necessary for the surveyor to run out the street lines on the ground and there determine the elements of the curves that will fix them mathematically.

CURVES*

10.11. Curved Alignment.—The horizontal alignment of highways, railroads, canals, many city streets, and property lines consists of straight lines connected by curves. The straight lines between the curves are called "*tangents*." These curves are commonly arcs of a circle. Occasionally they are a combination of circular arcs of different radii, and sometimes they are spirals.

The vertical alignment of highways and railways consists of straight grade lines and parabolic curves tangent to them. The straight grade lines are where the rate of grade is uniform. Where the rate of grade changes, a vertical curve is introduced in order to ease the transition from one grade to the following.

Both horizontal and vertical alignments are usually determined in the office, after a study of plans of the site. The surveyor reproduced on the ground the alignment adopted in the office.

10.12. Elements of a Circular Curve.—Fig. 10.12.1 shows a circular curve connecting the tangents A and B.

*For details on curve layout see standard texts on this subject, such as Route Surveying by C. F. Meyer, International Textbook Co., Route and Construction Surveys by Harry Rubey, The Macmillan Co., Highway Curves by Ives and Kissam, John Wiley & Sons, Inc., and Route Location and Surveys by T. F. Hickerson, McGraw-Hill Book Co., Inc.

Radius of the curve is designated by R.

Vertex, V, or point of intersection, $P.I.$, is the point where the two tangents to the curve intersect.

Intersection angle, I *(or* Δ*)*, is the deflection angle between the tangents, and is equal to the angle at O between the radii.

Point of Curvature, $P.C.$, is the point where tangent ends and the curve begins.

Point of Tangency, $P.T.$, is the point where the curve ends and the tangent begins.

Length of Curve, L, is the length from $P.C.$ to the $P.T.$

Tangent Distance, T, is the distance along the tangent from the $P.I.$ to the $P.C.$ or the $P.T.$

Long Chord, C, is the straight line distance from the $P.C.$ to the $P.T.$

External Distance, E, is the distance from the vertex to the mid-point of the curve.

Middle Ordinate, M, is the distance from the mid-point of the curve to the mid-point of the long chord.

The relations between these various elements are

$$T = R \tan \frac{I}{2}$$

$$E = R \operatorname{exsec} \frac{I}{2}$$

$$M = R \operatorname{vers} \frac{I}{2}$$

$$C = 2R \sin \frac{I}{2}$$

$L = R \times I$, where I is in radians (Arc definition, Art. 278).

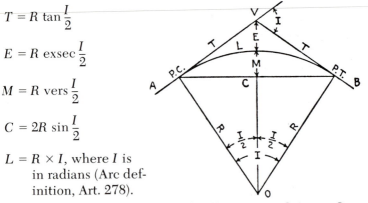

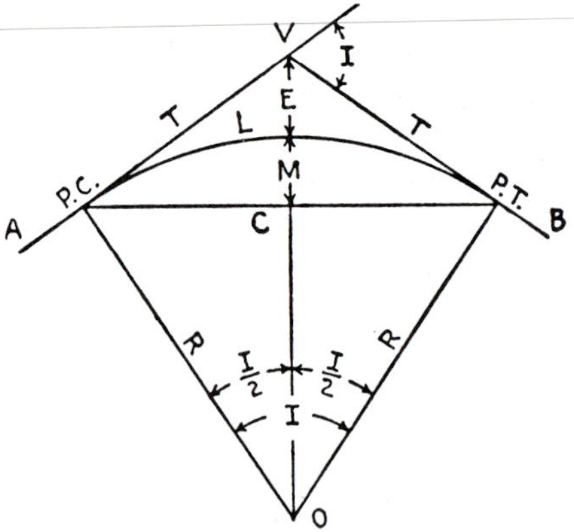

FIG. 10.12.1. ELEMENTS OF CIRCULAR CURVE.

10.13. METHODS OF DESIGNATING CURVES.—In city street and property surveys and in some highway practice a circular curve is designated by its *radius* and the length of curve is the actual arc length. In route surveying the curve is usually defined as the *degree of curve.* This latter definition has two common meanings: (1) by arc definition the curve is defined as the central angle subtended by an arc of 100 ft., and the curve is measured along the actual arc, (2) by chord definition it is defined as the central angle subtended by a 100-ft. chord (Fig. 10.13.1), and the length is recorded as a series of 100-ft. chords and fractional chords. The latter definition is commonly used in railroad practice only.

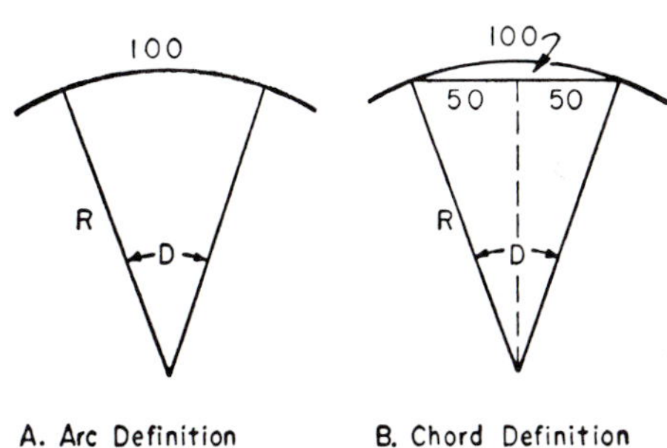

FIG. 10.13.1. ARC AND CHORD DEFINITION OF CURVES.

The relation between radius and degree of curve by arc definition is $R = \dfrac{100}{D_{\text{rad}}}$, or $D_{\text{rad}} = \dfrac{100}{R}$, in which R is in feet and D is in circular measure (radians). The relation can also be obtained by proportion from a circle of radius R, where R is in feet and D is in degrees, as follows:

$$\frac{D}{360} = \frac{100}{2\pi R} \quad \text{or} \quad R = \frac{5729.57795}{D} \quad \text{or} \quad D = \frac{5729.57795}{R}.$$

Then for a curve of $D = 1°$, the radius is 5729.57795. Note also R is exactly inversely proportional to D; thus small values of D are indicative of large radii, and large values of D of short radii.

By chord definition (Fig. 10.13.1), $100 = 2R \sin D/2$, or $R = \dfrac{50}{\sin D/2}$. In this case R is not inversely proportional to D, but is nearly so for small values of D. By this method $R = 5729.65$ when $D = 1°$.

For making rough conversions from degree of curve to radius, or vice versa, it is convenient to memorize the radius of a 1° curve (5729.6), and to divide this number by any other radius to obtain its degree, or to divide the degree of any curve into 5729.6 to obtain its radius.

10.14. RELATION BETWEEN ELEMENTS OF CURVES.—It is apparent from Art. 10.12 that for a given angle I the several elements of a circular curve are proportional to the radius; they are also exactly inversely proportional to the degree of curve *when the arc definition* is used.

$$\frac{T_1}{T_2} = \frac{E_1}{E_2} = \frac{M_1}{M_2} = \frac{C_1}{C_2} = \frac{L_1}{L_2} = \frac{R_1}{R_2} = \frac{D_2}{D_1}$$

These relationships are useful for simplifying computations, particularly when tables are available giving elements such as R, T, M, and E for a 1° curve in terms of increments of central angle.* Values for any other degree of curve can be obtained from these tabulated values by dividing them by the given degree of curve. With the chord definition the proportionality does not apply exactly and corrections have to be made after dividing by degree of curve.

10.15. LENGTH OF CURVE.—When the curve is designated by its radius, the arc for any central angle is equal to the radius times the subtended angle expressed in radians. Since arcs are proportional to radii, the arc lengths for a radius of 1 ft. can be tabulated (as in Table 10.15.1) and the lengths for any other radius obtained by multiplying these values by the radius to be used. Radii of even tens or hundreds of feet are commonly chosen to simplify the calculations. Lacking tables, any arc can be found from the relation that angles are proportional to arcs, as follows:

$$\frac{L}{2\pi R} = \frac{I}{360} \text{ or } L = .0174532925\, RI, \text{ where } I \text{ is in degrees.}$$

*See Textbooks on Route Surveying. For application of chord definition see "Railroad Curves and Earthwork" and "Field and Office Tables," C. Frank Allen, McGraw-Hill Book Co., Inc.

TABLE 10.15.1

LENGTHS OF CIRCULAR ARCS: RADIUS = 1.

Sec.	Length.	Min.	Length.	Deg.	Length.	Deg.	Length.
1	.0000048	1	.0002909	1	.0174533	61	1.0646508
2	.0000097	2	.0005818	2	.0349066	62	1.0821041
3	.0000145	3	.0008727	3	.0523599	63	1.0995574
4	.0000194	4	.0011636	4	.0698132	64	1.1170107
5	.0000242	5	.0014544	5	.0872665	65	1.1344640
6	.0000291	6	.0017453	6	.1047198	66	1.1519173
7	.0000339	7	.0020362	7	.1221730	67	1.1693706
8	.0000388	8	.0023271	8	.1396263	68	1.1868239
9	.0000436	9	.0026180	9	.1570796	69	1.2042772
10	.0000485	10	.0029089	10	.1745329	70	1.2217305
11	.0000533	11	.0031998	11	.1919862	71	1.2391838
12	.0000582	12	.0034907	12	.2094395	72	1.2566371
13	.0000630	13	.0037815	13	.2268928	73	1.2740904
14	.0000679	14	.0040724	14	.2443461	74	1.2915436
15	.0000727	15	.0043633	15	.2617994	75	1.3089969
16	.0000776	16	.0046542	16	.2792527	76	1.3264502
17	.0000824	17	.0049451	17	.2967060	77	1.3439035
18	.0000873	18	.0052360	18	.3141593	78	1.3613568
19	.0000921	19	.0055269	19	.3316126	79	1.3788101
20	.0000970	20	.0058178	20	.3490659	80	1.3962634
21	.0001018	21	.0061087	21	.3665191	81	1.4137167
22	.0001067	22	.0063995	22	.3839724	82	1.4311700
23	.0001115	23	.0066904	23	.4014257	83	1.4486233
24	.0001164	24	.0069813	24	.4188790	84	1.4660766
25	.0001212	25	.0072722	25	.4363323	85	1.4835299
26	.0001261	26	.0075631	26	.4537856	86	1.5009832
27	.0001309	27	.0078540	27	.4712389	87	1.5184364
28	.0001357	28	.0081449	28	.4886922	88	1.5358897
29	.0001406	29	.0084358	29	.5061455	89	1.5533430
30	.0001454	30	.0087266	30	.5235988	90	1.5707963

When the degree of curve is specified:

$$\frac{L}{100} = \frac{I}{D} \quad \text{or} \quad L = \frac{100I}{D}$$

The formula $L = \dfrac{100I}{D}$ applies to *both* arc and chord definitions of D, but in the arc method L is the true arc whereas in the chord method it is a series of 100-ft. chords and fractional

TABLE 10.15.1 (cont.)

LENGTHS OF CIRCULAR ARCS: RADIUS = 1.

Sec.	Length.	Min.	Length.	Deg.	Length.	Deg.	Length.
31	.0001503	31	.0090175	31	.5410521	91	1.5882496
32	.0001551	32	.0093084	32	.5585054	92	1.6057029
33	.0001600	33	.0095993	33	.5759587	93	1.6231562
34	.0001648	34	.0098902	34	.5934119	94	1.6406095
35	.0001697	35	.0101811	35	.6108652	95	1.6580628
36	.0001745	36	.0104720	36	.6283185	96	1.6755161
37	.0001794	37	.0107629	37	.6457718	97	1.6929694
38	.0001842	38	.0110538	38	.6632251	98	1.7104227
39	.0001891	39	.0113446	39	.6806784	99	1.7278760
40	.0001939	40	.0116355	40	.6981317	100	1.7453293
41	.0001988	41	.0119264	41	.7155850	101	1.7627825
42	.0002036	42	.0122173	42	.7330383	102	1.7802358
43	.0002085	43	.0125082	43	.7504916	103	1.7976891
44	.0002133	44	.0127991	44	.7679449	104	1.8151424
45	.0002182	45	.0130900	45	.7853982	105	1.8325957
46	.0002230	46	.0133809	46	.8028515	106	1.8500490
47	.0002279	47	.0136717	47	.8203047	107	1.8675023
48	.0002327	48	.0139626	48	.8377580	108	1.8849556
49	.0002376	49	.0142535	49	.8552113	109	1.9024089
50	.0002424	50	.0145444	50	.8726646	110	1.9198622
51	.0002473	51	.0148353	51	.8901179	111	1.9373155
52	.0002521	52	.0151262	52	.9075712	112	1.9547688
53	.0002570	53	.0154171	53	.9250245	113	1.9722221
54	.0002618	54	.0157080	54	.9424778	114	1.9896753
55	.0002666	55	.0159989	55	.9599311	115	2.0071286
56	.0002715	56	.0162897	56	.9773844	116	2.0245819
57	.0002763	57	.0165806	57	.9948377	117	2.0420352
58	.0002812	58	.0168715	58	1.0122910	118	2.0594885
59	.0002860	59	.0171624	59	1.0297443	119	2.0769418
60	.0002909	60	.0174533	60	1.0471976	120	2.0943951

chords. When the degree of curve is taken in even degrees or half degrees the computation of curve length is quite simple.

10.16. LAYOUT OF CIRCULAR CURVES.—In laying out a curve the tangent distance is computed and measured along the tangents in both directions from the *P.I.* to locate the *P.C.* and *P.T.* If the curve is part of a route survey with stationing, the full stations are set on the curve. If the curve is sharp, intermediate points are often desirable at plus 50- or plus 25-ft. stations. An isolated curve, independent of any stationing, may be laid out by dividing it into an equal number of arcs

(Art. 10.20); or, it may be laid out in arcs which are a full number of feet long with a sub-arc at one end, or points may be set at equal chord lengths apart with a shorter chord at one end.

10.17. Deflection Angle Method of Laying Out Curves.— The most common method for locating points on a curve is the deflection angle method. A deflection angle is one between a tangent and a chord. In Fig. 10.17.1 the angles *P.I. Da, P.I. Db* and *P.I. DE* are deflection angles to points *a*, *b* and *E*, respectively.

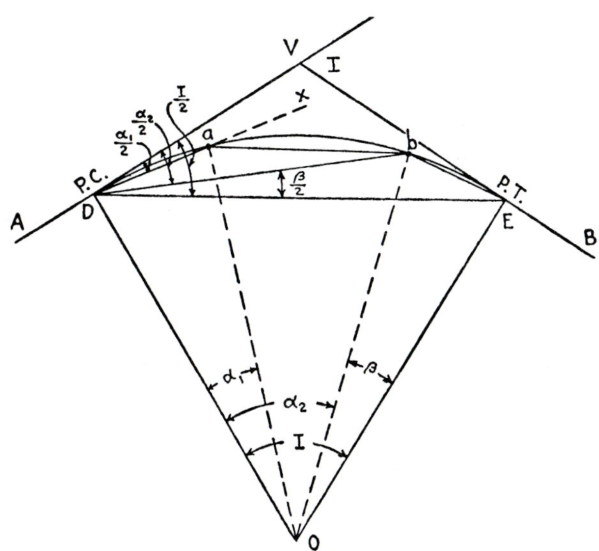

FIG. 10.17.1. DEFLECTION ANGLES FOR LAYING OUT A CURVE.

To lay out the curve, the theodolite is set at the *P.C.*, and sighted on *P.I.*, the circle reading 0°. The first deflection angle *P.I. Da* is set off on the circle of the theodolite; the line of sight will be along *Dx*. Point *a* on the curve is somewhere in the line *Dx*. It is located by setting a stake at the chord distance *Da* from *D*, along the line *Dx*. When point *a* has been established, the deflection angle *P.I. Db* is set off on the circle of the instrument. Point *b* is located by finding the point of intersection of the line of sight from the theodolite and the chord length ab, measured from a. When all points have been set along the curve, a field check is obtained on the linear measurements by measuring the last chord distance bE, and comparing it with the computed length of that chord. A check is obtained on the angles by laying off the deflection angle to the *P.T.* and observing whether the line of sight passes through that point, which was previously set by laying off the tangent distance from *P.I.*

In computing the deflection angles use is made of the following theorems of plane geometry:

1. The angle at the center of a circle is measured by the arc it subtends.
2. The angle between a tangent and a chord the vertex of which lies in the circumference is measured by **half** the subtended arc.
3. The angle between two chords the vertex of which lies in the circumference (an inscribed angle) is measured by **half** the subtended arc.

In Fig. 10.9 the deflection angle $PIDa = \frac{DOa}{2} = \frac{\alpha_1}{2}$; $PIDb = \frac{DOb}{2} = \frac{\alpha_2}{2}$; $PIDE = \frac{DOE}{2} = \frac{I}{2}$; and $bDE = \frac{bOE}{2} = \frac{\beta}{2}$.

To compute the angles DOa, aOb and bOE, the distances along the arc, Da, ab and bE, and the radius or the degree of curve must be known. Using the chord definition, the chord stationing between curve points would be used instead of the actual arcs.

10.18. Laying Out Curve by Arc-Stationing Method.—Referring to Fig. 10.18.1, a 5-degree curve is to be laid out connecting two tangents intersecting at angle I or 24° 15′. The station (prolonged beyond *P.C.*) of the vertex V is 16 + 86.10.

The first step is to compute the tangent distance by formula $T = R \tan \frac{I}{2}$. The radius of a 5° curve is $\frac{5729.57795}{5} = 1145.9156$ and $\frac{I}{2} = 12° 07'.5$. $T = 1145.92 \times \tan 12° 07'.5 = 1145.916 \times$

.2148378 = 246.186. This calculation is often cumbersome, but can easily be done on a hand- or desk calculator. When tables of tangent distances for a 1° curve are used, it will be found that the value of T_1 for $I = 24° 15'$ is 1230.9. This

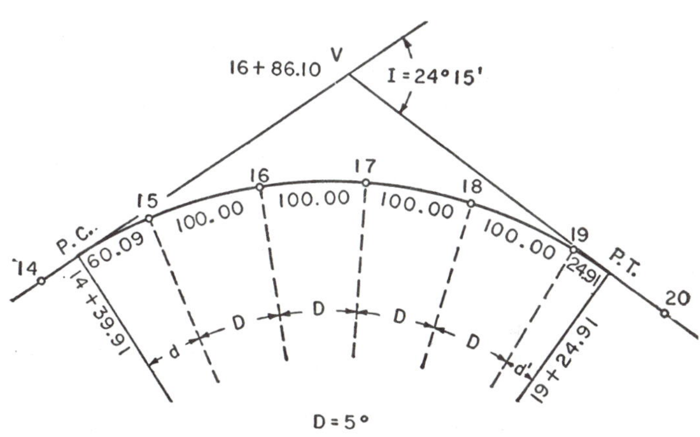

FIG. 10.18.1. ARC-STATIONING METHOD.

divided by 5 gives 246.18. The difference of 0.01 in results is due to uncertainty in fifth place of tabulated values of T_1, and in the tangent function.

The length of curve $L = \dfrac{100I}{D} = \dfrac{100 \times 24.25}{5} = 485.00.$

The station of $P.C.$ = Sta. of $P.I. - T$ = 16+86.10 − 2+46.19 = 14+39.91, and station of $P.T.$ = Sta. of $P.C. + L$ = 14+39.91 + 4+85.00 = 19+24.91.

The curve is then made up of a short arc from $P.C.$ to Sta. 15 of 60.086 ft., four 100-ft. arcs, and a short arc from Sta. 19 to $P.T.$ of 24.91 ft.

The increment of deflection angle between each station point is equal to one-half the corresponding central angle. For a 5° curve the deflection angle increment for each full station is 2° 30′. For the shorter arcs this angle is reduced proportionately. For example, in Fig. 10.10, the deflection angle $P.C.$ − 15 =

$\frac{60.086}{100} \times 2.5° = 1.502°$ or $1° 30' 07''.7$. Similarly, the deflection angle $19 - P.T. = \frac{24.91}{100} \times 2.5° = 0.623°$ or $0° 37'.4$. A convenient formula for computing deflection angles for less than one station length is $\frac{d}{2} = .3lD$, in which $\frac{d}{2}$ is deflection angle in *minutes*, l is length of sub-arc in feet and D is degree of curve in degrees.

The total deflection angles to each station point on the curve are obtained by adding increments, as follows:

Sta.		Deflection Angle
15		$1° 30' 07''.7$
16	$1° 30'.1 + 2° 30'$	$4° 00'.1$
17	$4° 00'.1 + 2° 30'$	$6° 30'.1$
18	$6° 30'.1 + 2° 30'$	$9° 00'.1$
19	$9° 00'.1 + 2° 30'$	$11° 30'.1$
P.T.	$11° 30'.1 + 0° 37'.4$	$12° 07' 30''.0$ Checks $\frac{I}{2}$

The length of chord corresponding to an arc may be found from the formula $c = 2R \sin \frac{d}{2}$ or by subtracting the difference between the arc and chord from the arc. This difference may be obtained by the approximate formula $l - c = \frac{l^3}{24R^2}$* (or from $\frac{c^3}{24R^2}$ if arc is desired from the chord), in which l = length of arc in feet (c = length of chord in feet). Except for very sharp curves this formula will give the difference with sufficient precision by slide rule computation. In example of Fig. 10.18.1, the first chord by formula is

$c = 2 \times 1145.92 \sin 1° 30'.135 = 2291.84 \times 0262143 = 60.079$

This is a cumbersome calculation and does not give accurate results unless the small angle is carried out to at least 5 significant figures.

The difference between arc and chord by approximate formula is $l - c = \frac{\overline{60.09}^3}{24 \times \overline{1146}^2} = .007$, and the chord $c = 60.09 -$

*For derivation see Appendix.

.01 = 60.08. In this case the "approximate" formula gives a precise answer with much less effort than by using the trigonometric formula. The difference between arc and chord for the arc of 24.91 is much less than 0.01 ft., so this chord = 24.91 to nearest 0.01 ft.

For each full station $l - c = \dfrac{\overline{100}^3}{24 \times \overline{1146}^2} = .032$, so each chord = 100.00 − 0.03 = 99.97.

In Table 10.18.1, values of difference between arc and chord are tabulated for 100- and 50-ft. arcs for different degrees of curve. For 100-ft. arc and 5-degree curve the difference is seen to be 0.03 which checks the difference found above.

Station	Points	Description of Curve	Deflection Angles	Chords
21				
20				
	⊙ + 24.91 *P.T.*		12° 07′.5	24.91
19		R = 1145.92	11° 30′.1	99.97
18		L = 485.00	9° 00′.1	99.97
17		T = 246.19	6° 30′.1	99.97
	⊙ + 86.10 *P.I.*	I = 24° 15′		
16		D = 5°	4° 00′.1	99.97
15			1° 30′.1	60.08
	⊙ + 39.91 P.C.		0° 00′.0	
14				
13				
12				

FIG. 10.18.2. FIELD NOTES FOR LAYING OUT CURVE (ARC STATIONING).

The complete data for laying out the curve are found in Fig. 10.18.2. Notice that the deflection angle to the *P.T.* $= \dfrac{I}{2}$. This check should always be applied.

If chord stationing had been used in the above curve layout, the length of curve would have been 485.00 ft., R = 1146.28 ft., T = 246.27 ft., *P.C.* = 14+39.83 and *P.T.* = 19+24.83. The deflection angles would have been 1° 30′.3, 4° 00.3, 6° 30′.3, 9° 00′.3, 11° 30′.3 and 12° 07′.5; and the chords 60.17, 100.00

(each full station) and 24.83. An approximation is introduced in this method by assuming that sub chords are proportional to their central angles. The error is only significant for very sharp curves.

TABLE 10.18.1

RADII FOR DEGREES OF CIRCULAR CURVE

CHORD DEFINITION $R = \dfrac{50}{\sin \frac{D}{2}}$

ARC DEFINITION $R = \dfrac{5729.58}{D}$

Degree of Curve	Radius (Feet)	Log. of Radius	Degree of Curve	Radius (feet)	Log. of Radius	Diff. Between Arc and Chord for Arc of 100 ft.	50 ft.
0-30	11459.19	4.05915	0-30	11459.16	4.05915	—	—
1-00	5729.64	3.75813	1-00	5729.58	3.75812	—	—
1-30	3819.83	3.58204	1-30	3819.72	3.58203	—	—
2-00	2864.93	3.45711	2-00	2864.79	3.45709	0.01	—
2-30	2292.01	3.36022	2-30	2291.83	3.36018	0.01	—
3-00	1910.08	3.28105	3-00	1909.86	3.28100	0.01	—
3-30	1637.28	3.21412	3-30	1637.02	3.21405	0.02	—
4-00	1432.69	3.15615	4-00	1432.40	3.15606	0.02	—
4-30	1273.57	3.10502	4-30	1273.24	3.10491	0.03	—
5-00	1146.28	3.05929	5-00	1145.92	3.05915	0.03	—
5-30	1042.14	3.01793	5-30	1041.74	3.01776	0.04	—
6-00	955.37	2.98017	6-00	954.93	2.97997	0.05	0.01
6-30	881.95	2.94544	6-30	881.47	2.94521	0.05	0.01
7-00	819.02	2.91329	7-00	818.51	2.91302	0.06	0.01
7-30	764.49	2.88337	7-30	763.94	2.88306	0.07	0.01
8-00	716.78	2.85539	8-00	716.20	2.85503	0.08	0.01
8-30	674.69	2.82910	8-30	674.08	2.82870	0.09	0.01
9-00	637.27	2.80433	9-00	636.62	2.80388	0.10	0.01
9-30	603.80	2.78090	9-30	603.11	2.78040	0.12	0.01
10-00	573.69	2.75867	10-00	572.96	2.75812	0.13	0.02
11-00	521.67	2.71740	11-00	520.87	2.71673	0.15	0.02
12-00	478.34	2.67974	12-00	477.46	2.67894	0.18	0.02
13-00	441.68	2.64511	13-00	440.74	2.64418	0.21	0.03
14-00	410.28	2.61308	14-00	409.26	2.61200	0.25	0.03
15-00	383.06	2.58327	15-00	381.97	2.58203	0.29	0.04
16-00	359.26	2.55541	16-00	358.10	2.55400	0.33	0.04
17-00	338.27	2.52927	17-00	337.03	2.52767	0.37	0.05
18-00	319.62	2.50464	18-00	318.31	2.50285	0.41	0.05
19-00	302.94	2.48136	19-00	301.56	2.47937	0.46	0.06
20-00	287.94	2.45930	20-00	286.48	2.45709	0.51	0.06

10.19. When the Entire Curve Cannot Be Run Out from the P.C.—It frequently happens that the entire curve cannot be laid out from the *P.C.* because of some obstruction, as shown in Fig. 10.19.1. When this condition exists, set as many stations as possible from the *P.C.* Then move the transit to the last station set and run the curve from that point out to the end.

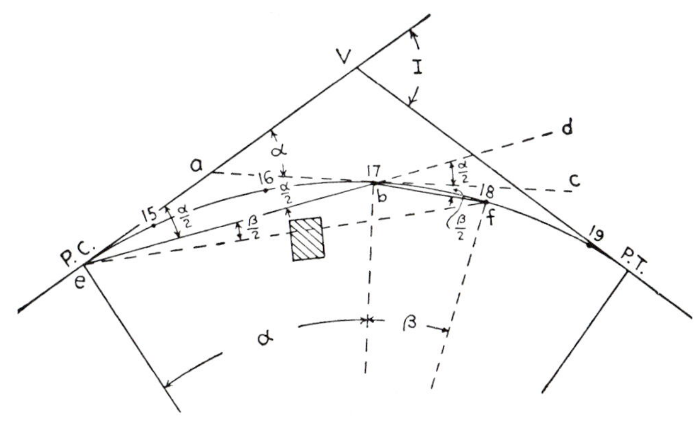

FIG. 10.19.1. OBSTRUCTION ON CURVE.

In Fig. 10.19.1 Stas. *15, 16* and *17* have been set from the *P.C.* In order to set Stas. *18* and *19* and check on the *P.T.*, the instrument is set over Sta. *17*. With the circle reading 0°, invert the telescope and sight the *P.C.* Plunge the telescope so that it is pointing along the line *bd*. Lay off the angle *dbf* = angle *aef*, which is the deflection angle that would have been used if it had been possible to sight Sta. *18* from the *P.C.*

The proof that *dbf* = *aef* is as follows: line *ac* is drawn tangent to the curve at Sta. *17*. Since *ea* and *ab* are tangent to the curve, angle *aeb* = *abe*. The angle *cbf* equals the angle *bef*, since both are measured by one-half the arc *bf*. Therefore *dbf* = *aef*. To set Sta. *19* and check on the *P.T.*, the deflection angles computed for those points from the *P.C.* are laid off from the line *bd*.

Another method for running in the curve from Sta. *17* to the *P.T.* with the transit at Sta. *17* is to backsight on the *P.C.* with the circle reading $\frac{\alpha}{2}$. When the circle is brought to 0°, the line of sight will be along the tangent *a–17–c*. The telescope is then reversed (sighted in the direction of *c*) and a new set of deflection angles are computed and laid out from the *17–c* tangent. The angle to Sta. *18*, for example, is $\frac{\beta}{2}$; and to the *P.T.*, it is $\frac{I - \alpha}{2}$.

In practice some engineers prefer to lay out the curve from the *P.T.* The instrument is set up on the *P.T.* and sighted on the *P.C.* with the circle reading 0°. The same deflection angles are then laid off as if the transit were at the *P.C.* When the deflection angle reads $\frac{I}{2}$, the telescope should then be pointing at the *P.I.*

10.20. Laying Out Curve—Radius Given.—When the radius is given, the arc lengths are found by use of formula $L = .0174533\ RI$ (degrees) or by use of lengths of arc of radius = 1 in Table 10.15.1. It would be possible to convert radius into degree of curve and proceed as in Art. 10.18. The computations would be cumbersome, however, because the degree would have to be carried out to 5 or 6 significant places to obtain the desired precision. For example, the degree of curve

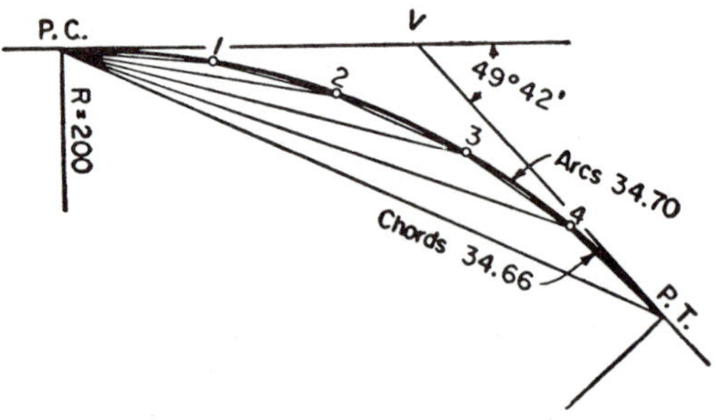

FIG. 10.20.1. LAYOUT OF CURVE IN EQUAL ARC LENGTHS.

corresponding to a radius of 1000.00 ft. is 5° 43′.773. In Fig 10.20.1 an isolated curve of 200-ft. radius is to be laid out to connect tangents intersecting at an angle of 49° 42′. Fou points are to be set on this curve dividing it into 5 equal parts. The length of the curve by formula is $L = 0.017453 \times 200 \times 49.700 = 173.49$ ft. This length is more conveniently obtained from Table 10.15.1, as follows:

$$\begin{array}{rl} \text{Arc for radius} = 1 \text{ for } 49^\circ & 0.8552113 \\ \text{for } 42' & \underline{0.0122173} \\ & 0.8674286 \times 200 = 173.49 \end{array}$$

Each arc length is $173.49 \div 5 = 34.70$ ft. and the deflection angle to point 1 on curve is $49.70^\circ \div 5 \div 2 = 4.970^\circ$ or $4^\circ\ 58'.2$. Subsequent deflection angles are $9^\circ\ 56'.4$ to point 2, $14^\circ\ 54'.6$ to point 3, $19^\circ\ 52'.8$ to point 4, and $24^\circ\ 51'.0$ to *P.T.*, checking one-half *I*.

The chord lengths for each arc are $34.70 - \dfrac{\overline{34.70}^3}{24 \times \overline{200}^2} = 34.70 - .04 = 34.66$.

When the radius is designated and the curve is part of a continuous stationing system, deflection angles and chords must be figured for each station and for sub-arcs at each end of the curve. The central angles subtending these arcs are obtained by proportion or by dividing the arcs by the radius and looking up the corresponding angles in Table 10.15.1.

10.21. Setting Points on a Curve by Offsets from the Tangent.—A curve may be laid out by perpendicular offsets from the tangent and distances to these offsets measured along the tangent from the *P.C.* In Fig. 10.21.1, the offset $a = R$ vers

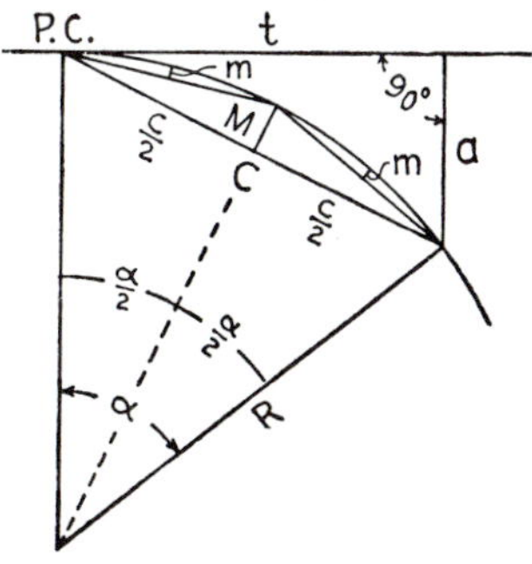

FIG. 10.21.1.

α and the distance from the $P.C.$, $t = R \sin \alpha$. In terms of the radius and chord, from similar triangles

$$\frac{a}{C} = \frac{\frac{C}{2}}{R} \quad \text{or} \quad a = \frac{C^2}{2R}$$

From demonstration in Art. 22

$$t = C - \frac{a^2}{2C} \text{ (Approx.)}$$

10.22. Setting Points on a Curve by Middle Ordinates.—Another method of laying out a curve is by offsets measured at right angles to a chord at its center point, i.e., "*middle ordinates.*" In Fig. 120, $M = R \text{ vers } \frac{\alpha}{2}$, $C = 2R \sin \frac{\alpha}{2}$. In terms of the chord and the radius, $M = \frac{C^2}{8R}$ (approximately). See Art. 14.10 A curve may be laid out by erecting ordinates to successively shorter chords, as shown in Fig. 10.21.1, $m = \frac{\left(\frac{C}{2}\right)^2}{8R} = \frac{C^2}{32R} = \frac{M}{4}$ (approximately).

10.23. Compound and Reversed Curves.—A compound curve is a series of simple circular curves running in the same general direction, one following the other and having a common tangent where they join. A reversed curve is formed by two circular curves having a common tangent where they join but lying on opposite sides of this common tangent.

A compound curve is laid out like a simple circular curve from the $P.C.$ to the point of compound curvature, $P.C.C.$ At the $P.C.C.$ the transit is set up and sighted on the $P.C.$ reading half the central angle of the first section of the compound curve. If the circle is turned to 0°, it should be sighting along the common tangent at $P.C.C.$; the next section of simple curve (of different radius from the first) may be laid off by deflection angles.

In laying out a reversed curve the same general method is used, viz., the first section is laid out like any simple curve. At the point of reversal, *P.R.C.*, set up the transit, back-sight on *P.C.* with proper angle so that the instrument will read 0° along the common tangent, and then lay off the second simple curve on the opposite side of this tangent.

10.24. Spirals.—On railroads and high-speed highways, spirals (or transition curves) are usually introduced between the straight and curved parts of the alignment.

These transition curves are either true spirals, parabolas, hyperbolas or they are a series of compound curves. Where they depart from the tangent they are very flat and gradually grow sharper until they have the same radius as the circular curve to which they are to connect in a common tangent.

In laying out these spirals the tangent distance *Ts* from the vertex to the point where the spiral starts (the T.S.) is computed and laid off on each tangent measuring from the vertex (Fig. 10.24.1). The first spiral may then be laid out by chords

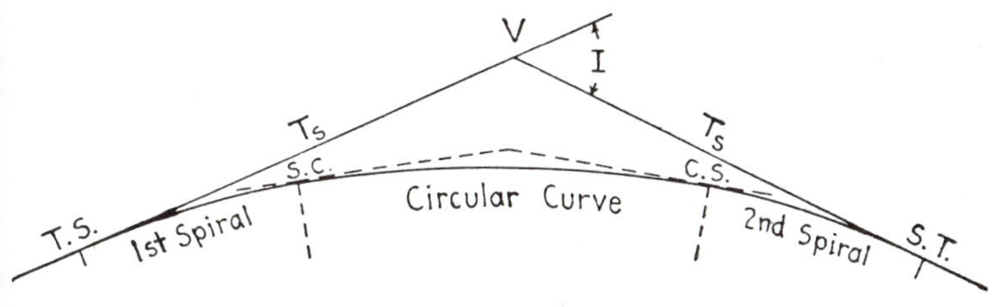

FIG. 10.24.1.

and deflection angles from the main tangent. When the end of the spiral (the *S.C.*) is reached the instrument is moved to the *S.C.* and the circular curve is laid out to its end at the *C.S.* by chords and deflection angles from an auxiliary tangent at the *S.C.* The second spiral may be laid off from the *S.T.* exactly as the first spiral was laid off from the *T.S.* It should check on the *C.S.*

The process of computing *Ts* and the chords and deflection angles of the spiral is too lengthy to include in this text.*

*See Transition Curves for Highways by Joseph Barnett, U.S. Govt. Printing Office.

10.25. CURVED STREET LINES.—In city practice it is common to round off intersecting street lines with circular curves, usually of short radius. Where the intersecting streets are straight the curve may be laid out by one of the methods described above. If the radius is quite short (10 to 50 ft.), it may be an advantage to swing a radius from its center, or if the center is inaccessible, to stretch chords and erect middle ordinates. Where one or more of the intersecting street lines is curved, the problem becomes more complex.

10.26. One Street Line Straight, the Other Curved.—In Fig. 10.26.1 the curved street line DEF intersects the straight street

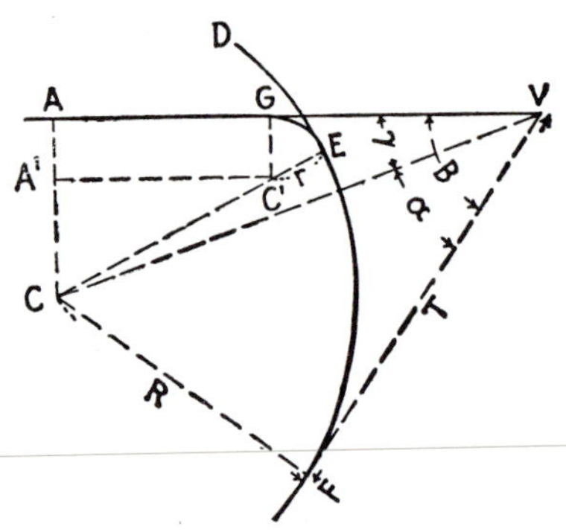

FIG. 10.26.1.

line AV and at this point the circular curve whose center is C' and with a given radius r is to be introduced to round off the corner. It is required to stake out the curve GE on the ground. In the field any tangent line, such as FV, is run off from some known point on the curve and intersected with AV, and the angle β and the distance FV are measured. In the right triangle CFV in which R and T are known, compute angle α and distance CV. In the right triangle CAV, CV and $\gamma = \beta - \alpha$ being known, compute CA and AV. $CA' = CA - r$; $CC' = R - r$. In the right triangle $CA'C'$, CA' and CC' being known, compute $A'C'$ and $A'CC' = GC'E$. Angle $ACF = 180° - \beta$. Angle $ECF = ACF - A'CC'$, from which the length of the arc FE can be readily computed, which locates the point E. $VG = AV - A'C'$, which locates point G of the curve GE, and any intermediate points can be located as explained in the previous articles.

As the radius $C'E$ is often quite short the center of the curve can be located from either its $P.C.$ or $P.T.$ or both, and any intermediate points on the curve can be easily swung in from its center.

If curve EG forms a reversed curve with FE the computations follow the same general method described above.

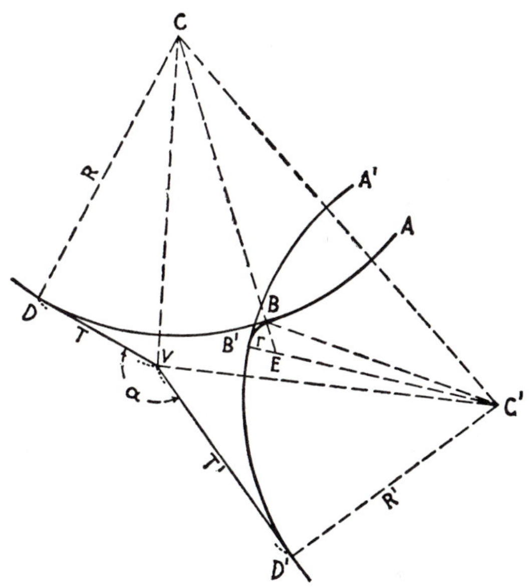

FIG. 10.27.1.

10.27. Both Street Lines Curved.—In Fig. 10.27.1 the two curved street lines ABD and $A'B'D'$ intersect each other, and the curve whose center is E, with a given radius r, is introduced at the intersection of the two street lines. It is required to locate the curve $B'B$ on the ground. In the field the tangent DV is run off from some known point D on the curve ABD and intersected with a tangent $D'V$ from the curve $A'B'D'$ and angle α and distances T and T' are measured. In the right triangle CDV, R and T being known, compute angle CVD and distance CV. Similarly in the triangle $C'VD'$ compute angle $C'VD'$ and distance $C'V$. In the oblique triangle CVC', CV, $C'V$ and angle $CVC' = 360° - (\alpha + CVD + C'VD')$ being known, compute CC' and the angles $CC'V$ and $C'CV$. In the oblique triangle $CC'E$, $CE = R + r$, $C'E = R' - r$, and CC' being known, compute the angle $C'CE$, $CC'E$ and $C'EC$, which is the supplement of the central angle of the curve $B'B$. Angle $DCB = DCV + VCC' - C'CE$, from which arc DB can be computed. Similarly angle $D'C'B' = D'C'V + VC'C - CC'E$, from which arc $D'B'$ is computed. These locate the P.C. and P.T. of the small curve whose center is E.

10.28. Intersection of Circular Arc and Straight Line.—This problem occurs wherever a straight property line cuts a curved street line. In the survey sufficient data must be obtained so that the distance along the curve and also the distance along the property line can be computed to the point where it cuts the curve. The usual method is to lay out a line tangent to the curve at some known station A on the curve, Fig. 10.28.1, intersect that tangent with the property line at D, and measure the distance AD and the angle α.

The problem is to find arc AB and distance BD. The radius R is known. Draw radius OI parallel to the property line and AH and JB perpendicular to OI; then

$$\beta = 90° - \alpha$$
$$AG = DA \cos \beta$$
$$DG = DA \sin \beta$$
$$AH = R \sin \beta$$
$$OH = R \cos \beta$$
$$JB = GH = AH - AG$$
$$\frac{JB}{R} = \sin BOI$$
$$AOB = \beta - BOI$$

Therefore arc AB = Angle $AOB \times R$

Again, $JO = R \cos BOI$

$$BG = JH = JO - OH$$

Therefore $BD = DG - BG$

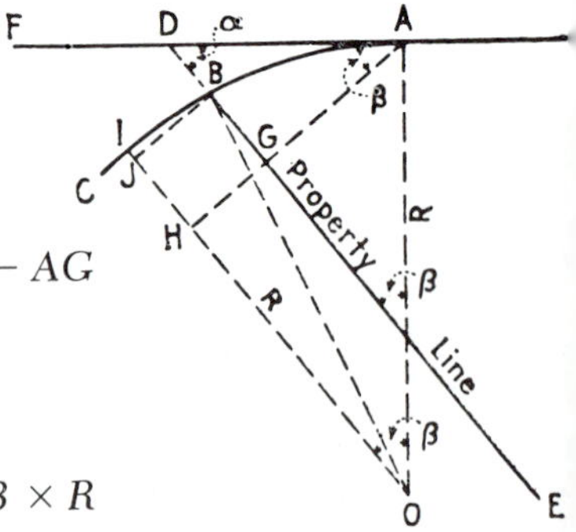

Fig. 10.28.1.

10.29. Staking Out Street Grades.—The fieldwork necessary in setting grade stakes is explained in Arts. 9.20, 9.21. When new streets are constructed the excavation or embankment first is brought to sub-grade, i.e., to the grade of the bottom of the road covering or pavement. The grade stakes set for this work are usually the center and the two side slope stakes, properly marked with the cut or fill, as described in Art. 9.16.

As the work progresses the center stake gets low, being dug out or covered up, and when the construction has progressed nearly to the sub-grade it is customary to set additional stakes, marking the elevation of the sub-grade along the center line and on each side line of the roadway. If sidewalks with curbstones are to be constructed, first the curbs are set to grade, and then the pavement grades are marked on the face of the curb, as well as on stakes in the center of the roadway and at points halfway between the curbs and center. (See Art. 10.33/34)

10.30. Vertical Curves.—In order to avoid an abrupt transition from one grade to another where the rate of grade of a street, highway or railroad changes, it is customary to introduce a vertical curve. The most convenient curve for this purpose is the parabola because the vertical offsets required between the straight grade lines and the curve can be easily computed.

The equation of the parabola is

$$y = 4\,px^2, \text{ or } y = \text{a constant times } x^2$$

From the above equation it is evident that the offsets y vary as the square of the distances x, or $y_1{:}y_2 = x_1^2{:}x_2^2$.

In Fig. 10.30.1, *AV* and *VB* are two grade lines intersecting at V. A vertical parabolic curve *CDE* is introduced, extending equal horizontal distances on both sides of V, with points of tangency at *C* and *E*. The elevations of the stations along the curve are required. *VF* is a diameter of the parabola (a line to which the offsets are parallel). When the parabola extends the same horizontal distance on both sides of V, the diameter *VF* is vertical and therefore all the offsets taken parallel to it are also vertical. These offsets are all proportional to the square of their distances from C along CV. But since the horizontal projection of CV is divided into the same number of equal distances as is CV, these horizontal projections are proportional to the distances along the tangent CV, and the vertical offsets cc', etc., are proportional to the squares of the **horizontal** distances from C.

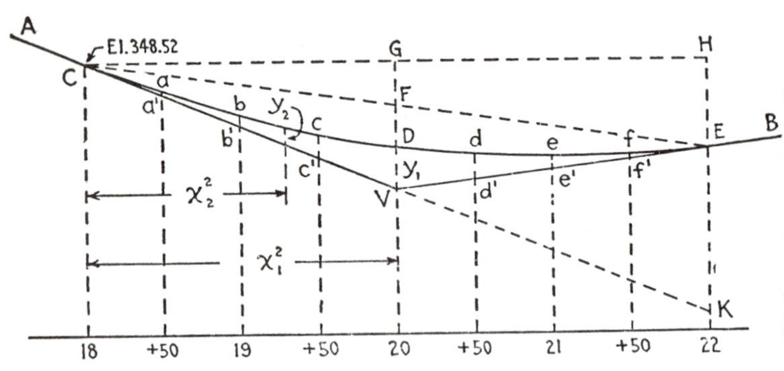

FIG. 10.30.1. VERTICAL PARABOLIC CURVE.

The lines VG and KH are vertical and CH is horizontal. Since the curve extends an equal horizontal distance on both sides of V, $CG = GH$, $CF = FE$ and $EK = 2FV$. From the equation of the parabola $EK = 4DV$. Therefore $FV = 2DV$, so **the point D is midway between F and V.**

If the gradients of AV and VB in per cent are g_1 and g_2,

$$EK = g_1 \frac{CG}{100} - g_2 \frac{GH}{100} = \frac{CH}{200}(g_1 - g_2)$$

in which upgrades are called plus and downgrades minus, and these signs are used algebraically in the formulas.

$$EK = 4DV, \quad \text{or} \quad DV = \frac{CH}{800}(g_1 - g_2)$$

in which CH is the horizontal length of vertical curve in feet.

EXAMPLE: In Fig. 10.19, a -6.00% grade and a $+2.00\%$ grade intersect at V = Sta. 20 +00. The two grades are to be connected by a vertical curve CDE 400 ft. long. The elevation of C is 348.52 ft. Compute the elevation of the curve at each 50-ft. station.

The tangent grade elevations of a′, b', c', V, d', e', f' and E at the 50-ft. stations indicated are first computed. (See Figs. 10.30.1 and 10.30.2.)

$$\text{Elev. } F = \frac{\text{Elev. } C + \text{Elev. } E}{2} = \frac{348.52 + 340.52}{2} = 344.52$$

$$\text{Elev. } D = \frac{\text{Elev. } F + \text{Elev. } V}{2} = \frac{344.52 + 336.52}{2} = 340.52$$

$$VD = \text{Elev. } D - \text{Elev. } V = 340.52 - 336.52 = 4.00$$

Also
$$VD = \frac{CH}{800}(g_1 - g_2) = \frac{400}{800}(-6.00 - (+)2.00) = 4.00$$

$$aa' = ff' = \frac{VD}{16} = \frac{4.00}{16} = 0.25 \qquad bb' = ee' = \frac{VD}{4} = \frac{4.00}{4} = 1.00$$

$$cc' = dd' = \frac{9VD}{16} = \frac{9(4.00)}{16} = 2.25$$

Adding the offsets to the elevations of a', b', c', etc., the grade elevations of the required points along the curve are obtained. (Fig. 10.20.)

Sta.	Straight Grade Elev.	Offsets	Grade Elev. Curve	1st Diff.	2nd Diff.
23					
22 = E	340.52	0.00	340.52		
				−0.75	
+50	339.52	0.25	339.77		+0.50
				−0.25	
21	338.52	1.00	339.52		+0.50
				+0.25	
+50	337.52	2.25	339.77		+0.50
				+0.75	
20 = V	336.52	4.00	340.52		+0.50
				+1.25	
+50	339.52	2.25	341.77		+0.50
				+1.75	
19	342.52	1.00	343.52		+0.50
				+2.25	
+50	345.52	0.25	345.77		+0.50
				+2.75	
18 = C	348.52	0.00	348.52		

FIG. 10.30.2. COMPUTATION OF ELEVATIONS ON A VERTICAL PARABOLIC CURVE.

One of the properties of the parabola is that the second difference should be equal; a check is thus obtained on the calculations. This check is illustrated in the last two columns of Fig. 10.30.2. The first differences must be computed consecutively and given the proper sign + or −. These signs must also be taken into account in computing second differences.

10.31. Highest and Lowest Points on a Vertical Curve.—It is sometimes necessary to find the station of the highest point on a summit vertical curve, or of the lowest point on a sag (valley) curve. The distance from the beginning of the curve to the high (or low) point is $\frac{Lg_1}{g_1 - g_2}$, where L is the total length of the curve in stations; g_1 and g_2 are the rates of grade at the beginning and end of the curve respectively. Upgrades are assumed to be plus and downgrades, minus. Referring to the vertical curve in Fig. 10.30.1, the distance from Sta. 18 to the low point of the curve is $4 \times \frac{-6}{-6-2} = 4 \times \frac{3}{4} = 3$ stations; i.e., the low point occurs at Sta. 21 +00. In many cases the low (or high) point does not fall at a full station, as it happened to do in this example.

10.32. Length of Vertical Curve for Required Sight Distance.—In highway practice the length of the vertical curve over a summit is controlled by the minimum "sight distance" required for safe driving. It is the horizontal distance between the driver's eye in each of two cars approaching the summit from opposite directions.* The vertical curve may be either longer or shorter than the sight distance, depending upon the abruptness of the change of grade and length of sight distance required.

When the vertical curve is longer than the sight distance $L = \frac{S^2(g_1 - g_2)}{8h}$, where L = length of vertical curve in stations, S = sight distance in stations, g_1 and g_2 = per cent of approach and following rates of grade (upgrades are plus and downgrades are minus), and h = height of eye in feet above the pavement at each end of the sight distance, usually taken as 4.5 ft.

*See "A Policy on Geometric Design of Rural Highways," Chap. III American Association of State Highway Officials.

When the vertical curve is shorter than the sight distance the formula becomes

$$L = \frac{2(g_1 - g_2)S - 8h}{g_1 - g_2}$$

The formula to be used may be determined by solving the expression $(g_1 - g_2)S \div 8$ or $(g_1 - g_2)L \div 8$ depending upon whether S or L is known. If either of these terms is greater than h, L is greater than S. Conversely, if either term is less than h, L is less than S.

EXAMPLE: If a + 4% grade is followed by a − 2% grade, and $h = 4.5$ ft., and $S = 500$ ft., find L.

$$g_1 - g_2 = +4 - (-2) = 6$$

$$(g_1 - g_2)\frac{S}{8} = 6 \times \frac{5}{8} = 3.75 \text{ ft., so } L < S$$

Using the formula for L shorter than S,

$$L = \frac{2 \times 6 \times 5 - 8 \times 4.5}{6} = 4.00 \text{ sta., or 400 ft.}$$

10.33. Cross-Section of Street.—For surface drainage the center of a road is raised or "crowned" above the grade of the gutters. Sometimes the surface is formed by two planes, running straight from the gutters to the center of the street; but more often it is shaped in the form of a parabola or a circle. Usually the width of street is great compared with the crown, so the parabola and the circle are virtually coincident.

The curbs are first set to line and grade, then intermediate grade stakes set for finished grade of the roadway.

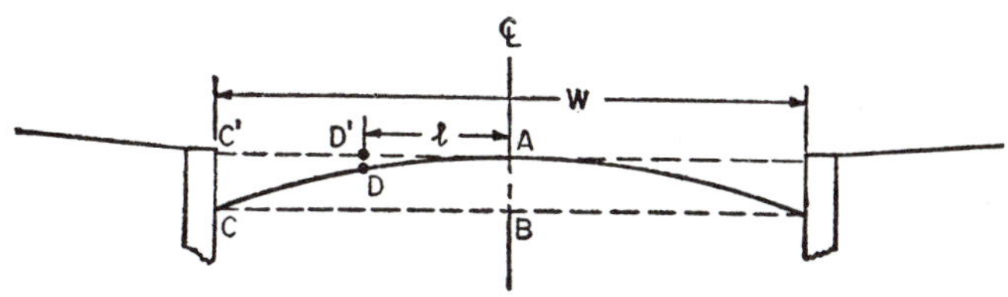

Fig. 10.33.1. Cross-Section of Pavement. Gutters at Same Elevation.

Fig. 10.33.1 represents the cross-section of a pavement. The crown AB is given. The ordinate DD' at any other point on the parabola = $C'C \times l^2/(\frac{1}{2}W)^2$ since in a parabola the offsets from a tangent vary as the square of the distance out along the tangent (Art. 10.30). But $C'C = AB$; hence, if D' is halfway from the center to the curb, $DD' = AB \div 4$.

The sidewalks are sloped toward the curb, as shown in Fig. 10.36.1, so that they will drain into the gutters.

10.34. STAKING OUT CURB* LINES AND GRADES.—If the line stakes that are set for the curbstones are placed directly on the line of the curb they will be disturbed when the trench is excavated. For this reason they are usually set in the sidewalk on an offset line, say, 3 ft. from the outside edge of the curb, and at intervals of about 25 ft. The grade stakes are set at about the same interval, with their tops at grade or at some even distance (6 inches or 1 foot) above or below the grade of the curb. Sometimes the grade stakes are not driven so that their tops bear any relation to the finished grade, but a horizontal chalkmark is made on the side of the stake indicating the proper grade. A stake can be marked much more quickly than the top can be driven to the exact grade.

When new curbstones are being set, or forms for concrete curbs, stakes as a rule cannot be used in old streets because the sidewalks are too hard, and even if stakes could be driven those projecting above the surface of the sidewalk would be a source of danger to pedestrians. It is customary therefore to use heavy spikes about 6 inches long. These are driven into the sidewalk on the offset line and the elevations of their tops determined by leveling. The difference between the elevation of each spike and the elevation of the curb opposite it is calculated. A list of the stations and the distances the spikes are above or below the curb is given to the foreman in charge of the work. These distances always should be transposed into feet and inches (to the nearest ⅛ inch) before being given to the foreman, as it is seldom that the men employed to lay the curbstones know how to use tenths and hundredths of a foot.

Where there are trees or poles in line with the curbs, a nail can sometimes be set in the side of a tree on the line of the curb as well as at its grade. Points like these should be set in

*Called edgestones in some localities.

preference to offset stakes or spikes wherever possible, as there is little liability of the workman misinterpreting such marks. They can fasten their string directly to the nail and set the curb or forms, if the edgestone is to be of concrete, to agree with it.

Before the curbstones are ordered the surveyor usually measures the distances between trees, poles and driveways, and then makes out a schedule of the lengths of straight, of curved and of chamfered stones (opposite driveways) to be used on the job. This schedule is used in ordering the stones, and when they are delivered they should be found to fit the conditions without the necessity of cutting any of them.

10.35. Rerunning Street Lines and Grades.—All kinds of work, such as the construction of fences and buildings, street improvements, and underground construction call for reproducing the street lines and grades. The work of rerunning the line is simple enough if the original bounds are in place. It is not uncommon, however, to find that in excavating for a foundation on a corner lot, the corner bound has been disturbed or that it has been removed entirely. Before the required line can be staked out properly it may be necessary to begin at some reliable bound farther down the street or even on some other nearby street line.

When the line finally has been rerun it is highly desirable to take and to record swing-offsets from the corners of the foundation of several of the buildings located along the street and near to the line. By this record of offsets, this street line can be run out very easily and quickly at any future time, and any disturbance of the bounds at the corners can be detected. **Several offsets to substantial buildings are of more permanent value than stone bounds.** In some offices these offsets to buildings are recorded directly on the street plans. Several offsets from existing buildings give one of the best means of preserving a street line.

Whenever a street line or grade is rerun full notes should be made showing all measurements taken for determining the lines or grades. Sometimes the original street lines have been obliterated so completely that it is necessary to resurvey them and make a new record plan and a new description of them and to have these new lines "established" by a city ordinance.

When a new building is to be constructed the owner usually requests the City Engineer to define the street grade in front of his property. The surveyor who has charge of this work goes to the place and levels from the nearest B.M. to the site of the new building. He has in his possession the established grade of the street and its cross-section. From these he can compute the elevation of the sidewalk at those points along the street line where the grades are desired. On the fence or on stakes set on the side line of the street he marks the grade of the sidewalk at the property line, usually to a hundredth of a foot.

10.36. Staking Out Sewers.—In staking out a sewer, its center line is defined on the surface of the ground and stakes are set every 25 to 50 ft. along the line. In hard surfaces, spikes instead of stakes are used, and in surfaces where spikes cannot be driven, chisel cuts are made in the pavement.

Profile levels of the surface are run along the center line and are plotted on profile paper (Art. 17.20). The proposed grade line of the sewer is then plotted on the same profile, as in Fig. 10.36.1. The grade line of a sewer is the flow line, i.e., the

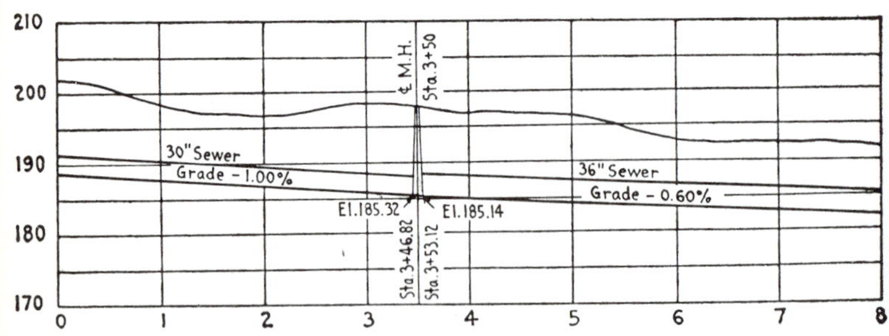

FIG. 10.36.1. SEWER PROFILE.

lowest line on the inside of the sewer. From the profile the approximate cut at each station is readily obtained by computation (subtracting the grade elevation from the ground elevation) or by scaling from the profile; these values should be given to the foreman to guide him in excavating the trench.

As the excavation of the trench progresses, a number of stakes will be displaced. From stakes previously set outside the excavated area, the center line is rerun over the trench as follows: Planks are placed on edge across the trench at each station, the ends of the planks resting with their edges on the surface of the ground (Fig. 10.36.2). To hold them in position,

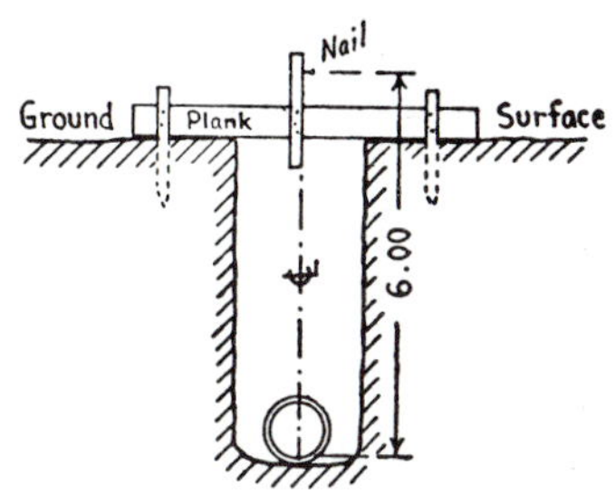

FIG. 10.36.2.
BATTER-BOARDS FOR SEWER.

they are nailed to heavy stakes which are driven in the ground. Sometimes the planks are supported by piling earth around their ends. A strip of wood about 1″ × 2″ × 30″ is firmly nailed to each plank; the strip is in a vertical position with one edge exactly on the center line of the sewer. A small nail is partially driven into the edge of the vertical strip, at a full number of feet (such as 5.00 or 6.00 ft.) above the invert (flow line).

Instead of using a vertical strip nailed to the plank, some engineers prefer to set the top edge of the plank at a full number of feet above the grade of the sewer, and then drive a nail in the top of the plank to mark the center line of the sewer. With this exception, the same procedure is followed as that already explained.

If a stretching line joins the center-line nails, it will be parallel to the grade line of the sewer. In order to set the sewer pipe at the proper distance below the string, a "grade pole" is used. This is a straight wooden pole with saw cuts or notches made in it, every foot throughout its length. Rigidly attached to the bottom of the pole is a right angle metal bracket, which may be extended into and rest on the invert of the pipe. Care should be taken to hold the pole vertically when the pipe is being tested for grade; and to keep the string from sagging between the nails. To obtain a straight grade the string must be kept tight at all times. If the string is left out overnight it is likely to absorb moisture; although it may be fastened taut in the morning, it will begin to sag as soon as it starts to dry out.

10.37. ACCIDENT PLANS.—These plans are for the purpose of showing in great detail the physical features at or near the site of an accident. It may be a railroad wreck, a street railway accident, automobile collision, injury to pedestrian due to faulty sidewalk pavement, or a murder. The surveyor's duty is to make a plan on a large scale so that it may be exhibited on the wall in a court room near the jury and can be readily seen by them. Scales of such plans are usually 1 inch = 4 ft., 1 inch = 8 ft. and 1 inch = 10 ft.; the two former scales are convenient for use by the layman who is familiar only with the use of the ordinary scale divided into inches and eighths. These plans should be shaded and colored so as to show very clearly the physical features which may relate to the case, and the buildings or windows from which witnesses may have claimed to have seen the accident. All trees and poles in the vicinity should be shown and their diameters marked on the plan. Sufficient elevations should be taken so that the surveyor can testify to the rates of grades and amount one part of the pavement is above or below another portion of the street. If the accident occurred after sundown, the location of street lights and store window lights may be important.

In coloring the plan, crayon colors will be suitable. Use the same color for the same physical features, as, for example, dark brown for bituminous-bound pavement, light brown for earth or gravel sidewalks, dark blue for curbstones, as this will define clearly the limits of the roadway, light blue for cement concrete sidewalks, yellow or orange around the fronts of wooden buildings, and red for brick buildings. Sometimes a series of letters A, B, C, lettered beside black crosses plotted at every 25 ft. or 50 ft. along the street will aid the jury in estimating distances on the plan, as well as serving as reference points to assist the witnesses in talking about the plan.

A prominent North point should be drawn on the plan. The measurements should be taken accurately, especial attention being paid to keeping the note-book in neat form as it is sometimes introduced as evidence as well as the plan.

It should be mentioned here that terrestrial photogrammetry (see Chapter XII) provides an excellent tool for the preparation of accident plans, since it minimizes the amount of time spent in the field and maximizes the information content of the survey.

10.38. Setting Batter-Boards for a Building.—One of the most common tasks of the surveyor is to set the batter-boards for the excavation and construction of the cellar of a new building. The dimensions of the building and the elevation at which to set it are usually obtained from the architect, although sometimes the elevation of the ground floor of the building is recorded on the plan itself. In a brick or stone building the lines to be defined are the outside neat lines of the building, and the elevation desired is usually the top of the first floor. In a wooden building the line usually given is the outside line of the brick or stone foundation and the elevation given is the top of this foundation on which the sill of the house is to rest. Sometimes the outside line of the sill instead of the foundation is desired. There should be a definite understanding in regard to these points before the work of staking out is begun.

Generally there is no elevation marked on the plan and the surveyor is simply told to set the top of the foundation a certain distance above the sidewalk or above the surface of some portion of the lot. If there is an elevation referred to City Datum marked on the plan, he should level from the nearest B.M. and set the batter-boards at the grade given.

The location of the building on the lot is given either by plan or by orders from the architect or owner. Not infrequently the surveyor receives the directions to place the building so that its front line is on line with the other buildings on the street and so that it will stand a certain number of feet from one of the side lines of the lot.

His first work is to stake out the location of the building by setting temporary stakes accurately at all corners of the building, e.g., in Fig. 10.38.1, at *A*, *B*, *C*, *D*, *E*, and *F*. A stake also should be set at *G* so that the entire work can be checked by measuring the diagonals *AG* and *FB*, and *GD* and *EC*. These

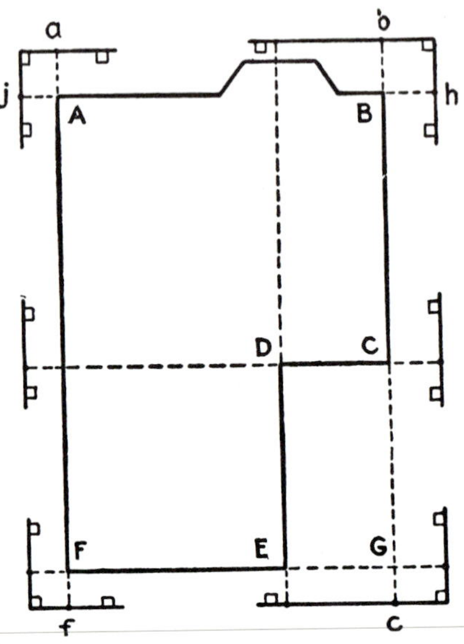

FIG. 10.38.1. SETTING BATTER-BOARDS FOR A BUILDING.

checks **always** should be applied where possible. Then the posts for the batter-boards are driven into the ground 3 or 4 ft. outside the line of the foundation so that they will not be disturbed when the walls are being constructed. On these posts, which are usually of 2 inches × 4 inches scantling, 1-inch boards are nailed. These boards are set by the surveyor so that their top edges are level with the grade of the top of the foundation or for whatever other part of the building he is giving grades. After the batter-boards are all in place they should be checked roughly by sighting across them; they should all appear at the same level. Sometimes, however, on account of the slope of the ground some of them have to be set a definite number of feet above or below grade.

Then the lines are to be marked by nails driven in the top of these batter-boards. The theodolite is set up on one of the corner stakes of the house at *A* (Fig. 10.38.1), for example, and a sight is taken on *F*. This line is then marked on the batter-board beyond (at *f*) and on the one near the transit (at *a*). If the batter-board is so near the transit that the telescope cannot be focused on it, then point *a* can be set within a hundredth of a foot by eye if the surveyor will stand outside of the batter-board and sight point *a* in a line determined by point *f* and the plumb-line on the instrument. Then a sight is taken along *AB* and this line is produced both ways and nails set on the batter-boards at *h* and *j*. In a similar manner all of the lines are marked on the batters. These points should be marked with nails driven into the top edges of the batter-boards and there should be some lettering on the boards to make clear which lines have been given. It is well for the surveyor also to show these marks to the builder or inspector and have it clearly understood just what parts of the structure these lines and grades govern.

It is customary to set batters for the jogs in the building as well as for the main corners; but small bay windows of dwellings are not usually staked out, but are constructed from wooden patterns made and set by the builder.

As soon as the excavation is begun the corner stakes are dug out and the building lines are then obtained by stretching lines between the nails in the opposite batter-boards. These batter-boards are preserved until the sills or first floor are in place, after which they may be removed.

It is good practice to set reference points on the main lines of the building and to place such points entirely outside of the area disturbed by construction of the new building.

10.39. Staking Out a Bridge Site.—The following refers to a small bridge and its supports, such, for example, as a railroad plate girder bridge over a city street, which is supported on two concrete abutments.

A theodolite and tape survey of the general location, including the physical structures in the immediate vicinity and the angle between the side lines of the street and the center line of location of the railroad will be embodied in the "site plan" that usually accompanies the construction plans of such a structure. The base-line of location of the railroad and the street lines should be run with the utmost care so that their relations shall be fixed exactly by the field measurements, because the structure and its supports are designed and located with reference to these data. Points on these lines should be tied in as explained in Art. 10.10.

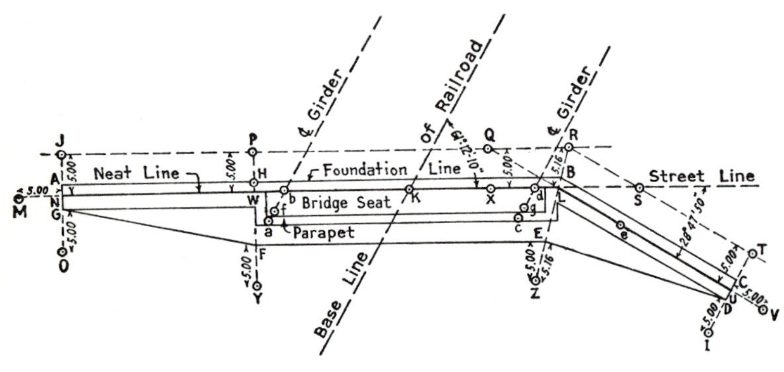

FIG. 10.39.1. STAKING OUT BRIDGE SITE.

Fig. 10.39.1 represents the plan of one abutment of a skew bridge. The outside line of the foundation is *ABCDEFG;* the neat line of the face of the abutment is *NWKLU.* These faces are vertical, but the backs are sloping.

Stakes should be set so that they will define, both in direction and in extent, the faces *NL* and *LU.* They are placed on an offset line at a sufficient distance from these neat lines so that the excavation for the foundation will not disturb them; in this instance 5 ft. from the neat line.

First take from the plan the distances *AN, AG, NW, WK, KL, LU, CU* and *CD.* Then set the temporary point *K* accurately by intersecting the base-line of the railroad with the street line (Art. 6.35).

With the instrument at *K* sight along the street line and set the permanent stake (and tack) *M*, the temporary points *L* and *N* and two temporary points on either side of *S*.

With the instrument at *N* sight on *L*, turn 90° and set permanent stakes *J* and *O*.

With the instrument at *J* sight on a swing-offset of 5.00 ft. from *L* and set the permanent stake *P* opposite *F* and also two temporary stakes on either side of both *Q* and *R*.

With the instrument at *L* sight along the street line and turn the angle 28° 47′ 50″, check it by repetition, set the temporary point *U* and the permanent stakes *V* and *Q*, the latter by the stretching line method, using the two temporary points already placed on line *JR* on either side of *Q*.

With the instrument at *U* sight on *L*, turn 90° and set the permanent stakes *T* and *I*.

With the instrument at *T* sight on a 5-ft. swing-offset from *L* and set permanent stakes *S* and *R* by the stretching line method.

Measure *LS*, *SR*, *RQ* and *QL*. They should all equal 5 ÷ sin 28° 48′ = 10.38. The check distance *LR* = 5.00 ÷ cos 14° 24′ = 5.16.

With the instrument at *R* sight on *J*, lay off 75° 36′ and measure 10.32 + *WF* ÷ cos 14° 24′ = *RZ* and set permanent stake *Z*.

With the instrument at *P*, sight on *R*, turn 90° and set permanent stake *Y*.

By this method stakes *M* and *S* define the face of the head wall and left wing and *J*, *P* and *Q* are intermediate offset stakes used in placing the concrete forms exactly on the street line. Similarly *Q* and *V* define the face of the right wing and *S* and *T* are additional offset points used in setting the forms. Stakes *J*, *O*, *T* and *I* define the face of the end forms.

When construction is about to begin, temporary stakes are set at *A*, *B*, *C*, *D*, *E*, *F* and *G* by measuring from the offset stakes. These stakes will be lost as the excavation progresses. The concrete for the foundation is poured into the hole excavated, or if forms are needed for the foundation their lines are fixed by measuring over from the offset stakes as suggested above. The elevation of the bottom and of the top of the foundation are defined by grade stakes set just outside the excavation near *A*, *H*, *B* and *C*.

When the foundation has been poured to grade and has had a day to "set up," mark on the top of it temporary points at *N*, *W*, *X*, *L*, *e* and *U* by measuring 5 ft. from the offset stakes opposite them. The points give the form carpenter his lines for the face forms of the wall which are erected vertically on these lines. The forms for the back are set at the proper distance from the face forms as given on the plan; these back forms are battered.

Points at elevation of the bridge seat are marked directly on the forms, or else reference points are marked on the face of the forms, and the exact distance from these reference grades to the finished bridge seat is given to the foreman in charge. Since concrete shrinks, care should be taken to make proper allowance for this by having it clearly understood by the foreman that the grades given are for the finished seasoned walls.

After the bridge seat is poured mark upon it the point *K* and lay out the parapet end forms the proper distance from this point. When the parapet has been poured run out the center lines of the girders parallel to the base-line and mark points such as *a* and *c* on the top of the parapet and also points *f* and *b* and *g* and *d* on the bridge seat. Using these lines the bearing plates can be placed accurately on the bridge seat and their position checked by the surveyor before the bridge arrives.

The points on the parapet define the center lines of the girders which must be placed exactly on these lines so that the floor system will fit properly between the girders.

SURVEY OF CITY LOTS

10.40. When the *survey of a City lot* is to be made its purpose and the information required must first be ascertained. The plan is usually made as a basis for an architect's design or for record in the Registry of Deeds or in the Land Court (see Arts. 5.25 and 5.26). Whatever the purpose of the plan the fieldwork methods are substantially alike.

Where the boundaries reasonably admit of more than one interpretation it is customary to register the maximum claim in Land Court cases, whereas if the plan is for other purposes it is advisable to adopt for boundaries only those which are so well fortified by evidence that it is reasonably sure that any Court would validate them.

10.41. Plan for Architect's Design.—The plan is usually drawn on a scale of one-eighth or one-quarter inch per ft. It should show the following:

1. Angles and linear dimensions of lot lines, area, and also measurements in the recorded deed of the property for comparison with measured dimensions.
2. Location of lot lines with relation to existing party-walls and thicknesses of the latter.
3. References to party-wall agreements, by Registry of Deed book and page number. (Art. 5.25.)
4. Any restrictions which prevent free use of either the whole or any portion of the lot, including rights-of-way (all to be found in recorded deeds or other recorded documents). Any evidence that illegal public use has been made of any portion of the lot such as indications that persons have been continuously crossing portions of the property. See Art. 5.23.
5. Street numbers as shown on the buildings.
6. Elevations of present sidewalks and established (official) elevations and widths of sidewalks.
7. Location in streets, size, kind and elevation of inverts (inside of bottom) of sewers.
8. Names of adjacent owners.
9. Plotted location of buildings on lot, including character and number of stories of buildings.
10. Plotted location of buildings on adjacent lots near the party line, and their character. Also any encroachments of cornices, belt course, etc., of adjacent building.
11. Approximate North point. While it is the general rule for engineers to place a plan on the paper so that the North is at the top of the plan, architects usually prefer to have their plans drawn so that the main street is parallel to the bottom of the sheet.
12. Section of city and zoning district in which the lot is located.

In addition to the above information, some architects desire:

13. Elevation of ground and elevation of basements of buildings.
14. Location of existing water pipe, gas pipe, electric conduits and connections, telephone connection and other underground conduits.

10.42. Plan for Record.—A plan for record must be drawn to such a scale as will give the size required by the particular Registry at which it is to be recorded. In some registries plans are redrawn to fit the deed books, this being done by the engineer of the Registry. Other registries have no rules regarding size.

This plan should show:

1. Linear dimensions, angles (or bearings), and area of lot.
2. Location of buildings on the lot, character and number of stories.
3. Existing party walls and their relation to the lot lines.
4. Buildings adjacent and near to party lines, marked wood, stone or brick.
5. Names of adjacent owners.
6. Street numbers as shown on the buildings.
7. References to party-wall agreements, by Registry of Deeds book and page number. (Appendix C.)
8. Any restrictions which prevent free use of either the whole or any portion of the lot, including rights-of-way (all to be found in recorded deeds or other recorded documents).
9. Approximate North point.
10. Section of city in which lot is located.

10.43. Data Required for Fieldwork.—Let us suppose the plan to be made is for the use of an architect.

Data fixing the street lines and the elevation of two or more bench marks should be obtained from the office of the City Engineer.

Next, examine the Land Court records to see if any of this or adjacent property has been put through that Court, because a boundary once fixed by that Court is final. (This of course applies only to States in which Land Courts exist.)

Next, the deeds of the property to be surveyed and usually also deeds of adjacent property should be examined and copied from Registry of Deeds records. The substance of all recorded party-wall agreements and other agreements that may restrict the use of any part of the property should be copied also. Sometimes this information is given by a conveyancer who has been examining the title for the client or for other interests.

With the above information in our possession actual surveying work can be commenced at the site.

10.44. Fieldwork.—The methods used in making the survey should be such as will insure accuracy, reasonable speed, and above all give the least chance of mistakes. Experience has proved that the process followed in the example to be explained gives excellent results.

The parcel here shown (Fig. 10.44.1) is a typical irregular lot containing no right angles, and with party walls on each side;

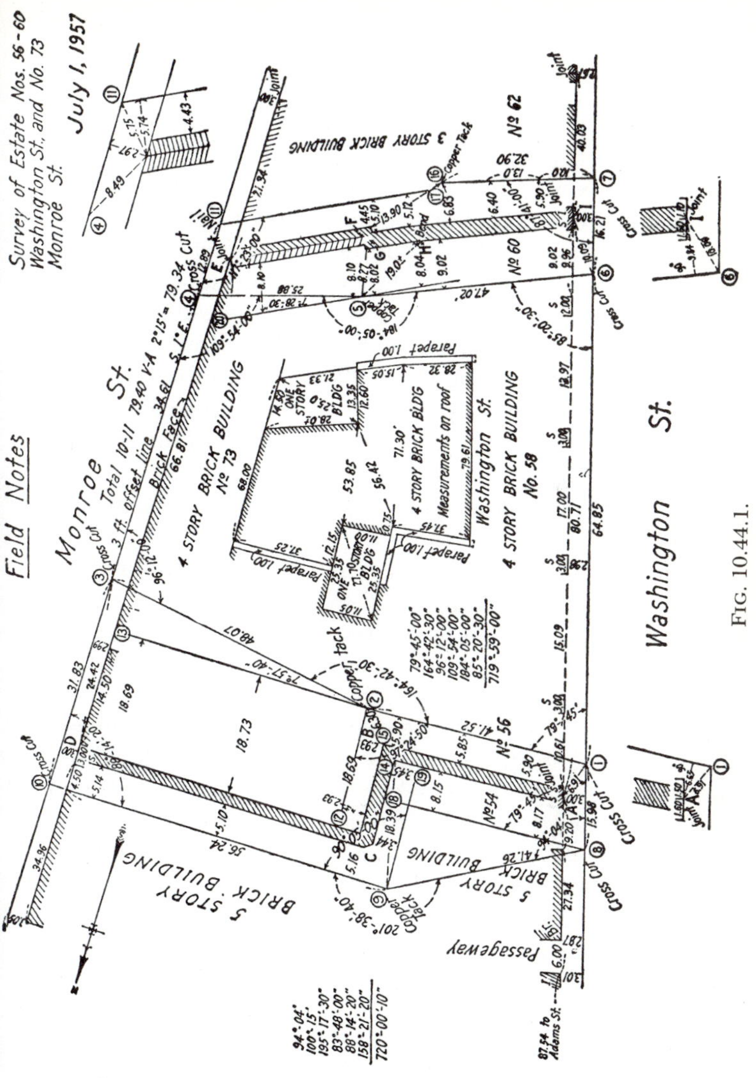
Field Notes
Survey of Estate Nos. 56-60
Washington St., and No. 73
Monroe St.
July 1, 1957
Monroe St.
Total 10-11 79.40 V-A 2°15′= 79.34 Cut
3 ft. offset line
Brick Face
4 STORY BRICK BUILDING
No 73
3 STORY BRICK BUILDING
No 62
No 60
ONE STORY BLD'G
4 STORY BRICK BLD'G
Measurements on roof
Washington St.
4 STORY BRICK BUILDING
No. 58
No 56
No54
5 STORY BRICK BUILDING
Passageway
87.54 to Adams St.
Washington St.
79°-45′-00″
164°-42′-30″
96°-12′-00″
109°-54′-00″
184°-05′-00″
85°-20′-30″
719°-59′-00″
94°-04′
100°-15′
195°-17′-30″
83°-48′-00″
88°-14′-20″
158°-21′-20″
720°-00′-10″

FIG. 10.44.1.

these vary in thickness, thereby presenting in one example most of the usual problems of a city survey of land fully occupied by a building.

Offset lines run parallel to the street lines are first marked on the sidewalks by carefully established points. Joining these sidewalk lines, forming a closed traverse, transit lines are run **within** the property, where sights through doorways or windows can be taken, and extending from street to street; then from these same transit points within the building, lines are run parallel to the party walls and offsets measured from these lines to the interior faces of the walls. Similarly, another traverse is formed by the sidewalk lines and lines run on the **outside** of the building parallel to the exterior faces of these walls, where possible, and offsets measured to the walls from these outside lines.

Fig. 10.44.1 illustrates the specific steps taken in making such a survey.

Let us suppose that the survey starts on Washington St., also that the City Engineer has said that the original monuments defining the street lines have disappeared, and since all of the buildings on that street are very old, occupation will fix the street lines (Art. 5.9).

First set up the theodolite over a point at some convenient distance (say 3 ft.) from the face (at the sidewalk level) of the iron column at the joint between buildings Nos. 60 and 62 (marked *I* in Fig. 10.44.1). The transit is sighted on a swing-offset 3 ft. from the face of the stone post at the wall joint between Nos. 54 and 56. Then cross-cuts (thus +) or nails are put into the sidewalk on this offset line at places convenient for running traverse lines within and without the building, such as at 8, 1, 6 and 7. As it is well to avoid sighting through windows, on account of refraction of the glass, these marks should be placed opposite doors or windows which can be opened. While the instrument is sighted along the line 8-7 the transitman should select two suitable natural sights, one in each direction, preferably at a considerable distance; these he will always use thereafter. The intermediate posts facing on Washington St. and the joints in party walls are then located by offsets and ties as shown, and the frontages of the two adjacent buildings are measured to determine whether these owners have the frontages called for in their deeds. Intermediate distances from door to door are measured.

Similarly, run the offset line 10-11 on Monroe St. The City Engineer has given us a definite line for that street.

While on this offset line the fronts of the building and those immediately adjacent are located and measured as on Washington St. and cuts as 10, 3, 4 and 11 placed at convenient points for running lines into the building or for party-wall locations. It is well to measure between adjacent points and then measure the total distance as a check. There being considerable difference in elevation between points 10 and 11, and the slope of the sidewalk being uniform, the most accurate result is obtained by slope measurement, the distance being measured by the tape lying directly on the sidewalk, and the vertical angle taken by measuring the height of the axis of the telescope above point 10 and sighting on a point at the same height above point 11.

Obviously, permission to enter and take measurements in a store or occupied building always should be obtained. Managers and tenants are usually willing to allow privileges of this sort within reason, but naturally resent intrusion without permission. It is unwise, however, to give tenants information regarding the purpose of the survey. Giving the name of the engineering firm represented, denying knowledge of the purpose of the owner and referring inquiries to him is a safe policy unless one has been otherwise instructed.

Next run line 1-2 parallel to the inside face of the party wall. If the wall bends slightly make the two end offsets equal and measure one in the middle to record the bend. The brick wall is the object to be surveyed and where it is covered with sheathing it is imperative to know the thickness of that sheathing. This can be obtained by measuring through a small gimlet-hole or drill-hole made with a Yankee drill. If this cannot be measured, note its character; soft sheathing is usually 2 inches thick and plaster on the brick is usually from ¼ to 1 inch thick.

Now place point 2, a copper tack on the floor, at a convenient point with due consideration also of the position of line 2-3. The angle at 1 is read and doubled, and distance 1-2 is measured.

Set up on point 2 and measure the swing-offset to the wall joint on Washington St. (6.65). This will show directly its position with relation to the inside face of the party wall. Run line 2-12 parallel to the short wall, measure the angle at 2, the offsets, and the distance 18.60 to the wall face. Similarly, line 2-13 is laid out parallel to the party wall and the angle and swing-offsets measured. Being unable to run this line 2-13 through to Monroe St. on account of the wall, point 3 opposite a window is selected on the offset line 10-11 which was run on the Monroe St. sidewalk and the distance 2-3 and angle at 3 are measured.

Run line 6-5 parallel to the inside face of the wall and measure swing-offset 9.84 to the joint on Washington St. Set up on point 5 and run line 5-20 parallel to the wall; the small jog and the flat bend in the wall are located approximately as indicated in Fig. 10.44.1, but their exact location will be found later by calculation. Line 5-4 is run and angles measured at 5 and 4.

To the closed traverse 1-2-3-4-5-6 we have now tied the interior faces of the party walls, their street joints and the street face of the building. We must now obtain a survey of the outer faces of these party walls.

Set up on point 8 and run a line parallel to 1-2. If the offsets taken to the outer face of the wall are approximately equal, the presumption is that both angles (at 1 and 8) are correct, as walls are usually built with parallel faces, but if there is a difference in the outside offsets an angle error is probable and should be investigated.

The perpendicular distance between lines 1-2 and 8-18 is easily calculated and the sum of the offsets on both sides subtracted from the computed perpendicular distance gives the thickness of the wall. Brick walls are built in multiples of 4″ that is, they are 8″-12″-20″, etc. in thickness. If the result gives a 10″ or an 18″ wall it is well to investigate as a mistake is proba ble. It is also advisable to measure if possible the size of the bricks in the wall being surveyed, as those laid in the 70's are usually about 7½" instead of 8" long—a peculiarity which may account for an apparent error.

Similarly, set up on points 9-10-11 and lay off lines parallel to the interior faces of the party walls, measure the offsets to their outer faces; these give the wall thicknesses and also check the angles previously turned. When possible, measure the swing-offset to party-wall joint, and when it is not possible measure the distance along the street line from the transit line to the joint. It is well also to have tie measurements, when possible, to all angle points in walls, as from point 9 to 14, and from point 2 to 15. As it is impossible to run a line parallel to the wall through point 7, a line 7-16 is run approximately parallel to it and several points on that transit line are marked from which the shortest distances to the wall have been measured.

As the plan should show the limits of buildings of different heights and area ways, it is well also to take roof measurements as shown in the subsketch in Fig. 10.44.1. Roof measurements along the party walls, fronts and a few ties across the roof not only will give the sizes of the various buildings but also serve as rough checks on the survey made at the street level.

10.45. Calculations and Plotting of the Survey Traverses.— 1*st.* Calculate the traverse by coördinates—Arts. 15.9 and 15.10. It should close with 1:10 000 or better. The error is distributed according to (Art. 15.7). The balanced traverse is made a basis for the second step.

2*nd.* Plot the traverse on a large scale, preferably ¼ inch to a foot. Plot all walls and other details of the survey on this plan. See Fig. 10.45.1.

3*rd.* Add to this plan all party-wall and deed data, placing the deed dimensions of the lot within the lot and the deed dimensions of adjacent lots within their respective lots.

4*th.* From a study of all recorded agreements and field data determine the division lines from the evidence supporting them. In deciding which lines to select one must bear in mind that since a costly building may be erected on this lot it is well to submit all questionable boundaries to the owner, or to his attorney, and obtain from him a decision with respect to them.

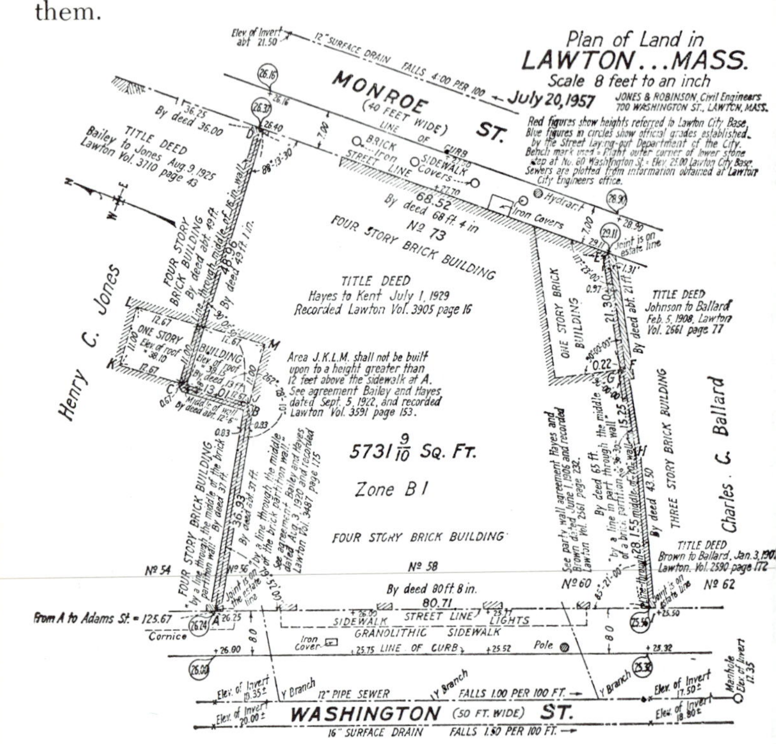

FIG. 10.45.1.

5th. Calculate the coördinates of all angle points and then the azimuths and lengths of the sides adopted as the boundaries. Check the calculated azimuths by the azimuth of the field survey.

6th. Using the lines and angles as actually recorded on the finished plan, and adopting one street line as an axis, compute the entire property boundary again by the coördinate method as a final check on the finished plan.

10.46. Calculation Details.—Traverse 1-2-3-4-5-6 is calculated by coördinates. Since Washington St. is a convenient base-line, we assume that the azimuth of line 6-1 is 0° and that azimuths increase clockwise. Assume the coördinates of point (1) are $x = 0$, $y = 200.00$, with the x's increasing toward the east and y's increasing toward the north. By doing so we avoid negative coördinates in future surveys of property in the same block.

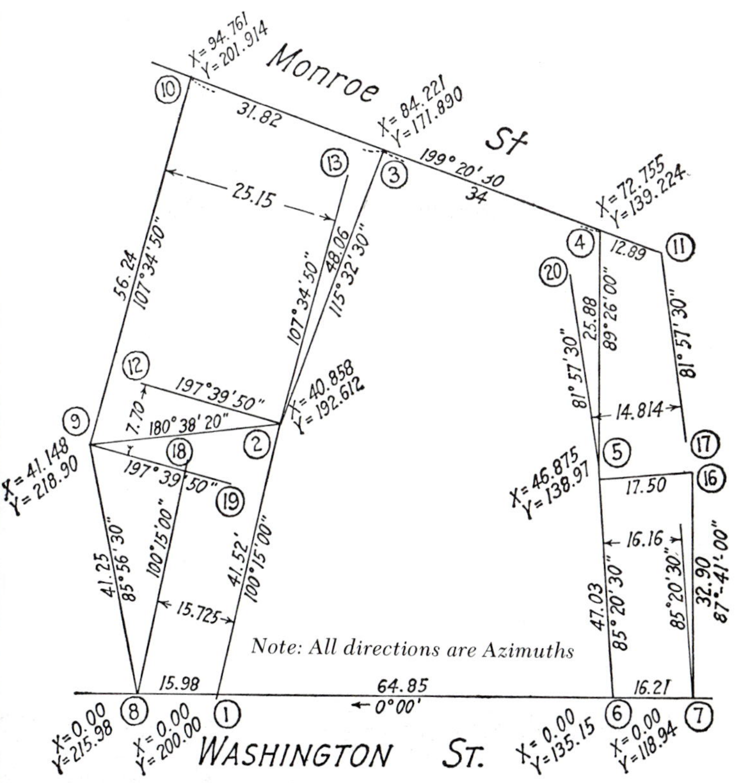

FIG. 10.46.1. ADJUSTED TRANSIT TRAVERSES.

The sum of the angles was 719° 59′. Since the angles at 4 and 5 were the most liable to error, being adjacent to short sides, 30″ is added to each angle.

It will be found that this traverse closes within 0.02 east and west and 0.005 north and south. If we increase the distance 5-6 by 0.01, calling it 47.03, and decrease the distance 2-3 by 0-01 calling it 48.06, the traverse will be well balanced.

Now calculate the coördinates of points 8-9-10-11-7-16; also the perpendicular distances between lines 1-2 and 8-18, between lines 2-12 and 9-19, between 9-10 and 2-13, between 5-20 and 11-17, and from point 16 to the line 6-5. These distances form an excellent check on the two traverses and the plot, and also permit of computing the thickness of walls. The results of the above are shown in Fig. 10.46.1.

The survey then should be plotted; and if this is done accurately on a scale of 4 ft. to the inch, errors of 0.15 ft. or greater can be detected readily by scale.

10.47. Determination of Boundaries from Analysis of Evidence.—The information obtained from the Registry of Deeds is now put on the plan. (Fig. 10.45.1).

Our title deed is the conveyance of Hayes to Kent. The names of the grantor and grantee are lettered in the middle of the plan; also the date of the transfer and the volume and page where it is recorded at the Lawton Registry.

In this instrument the line *A–B* is described as being about 37 ft. in length and passing through the middle of the brick partition wall. This record is lettered on the plan as shown. Line *C–B* is described as being 13 ft. in length, but no mention is made of its passing through a party wall, hence the record "By Deed 13 ft." is all that is necessary. Similarly the deed information relating to lines *C–D*, *D–E*, *E–I* and *I–A* is added.

Our title deed mentions two party-wall agreements, one referring to line *G–I*, the other to *A–B;* also to an agreement restricting the height of any building to be placed on area *J–K–L–M*. These agreements are very important and the substance of them should be lettered on the plan as shown in Fig. 10.45.1.

Similarly the information contained in the deeds of the adjacent owners is added to the plan. With these data before us we can now study the evidence of the boundaries.

It should be remembered that when any property line is defined in the deed by distance and also by a physical object, the latter rules in case of disagreement of the two. (See Art. 5.23.)

Line *G–I* is a line through the middle of the party wall; that was fixed by agreement, but this only applies for a length of about 43.50 ft. Ballard, the present owner of No. 62, acquired his property in two separate parcels at different times, but only the one acquired from Brown carries a party-wall agreement.

As there are two separate walls which back up on the line *E–F* (as indicated by the cross-hatching in Fig. 10.45.1) that line is undoubtedly the property boundary. Each party is occupying to that line and these conditions have apparently existed for a sufficient time to make it the boundary by "adverse possession" even if there were no other evidence of its validity. (See Art. 5.23.)

The line of Monroe St. as stated is that given by the City Engineer and is authentic.

In line *C–D* we have more of a problem. The wall is a party-wall. It is used by both parties and there is no wall joint in the front. The sum of the two frontages from fieldnotes is 104.77, or 0.44 ft. greater than by the deeds. Furthermore, if we lay off Jones' deed distance from his northerly corner it will fall about 3 inches north of the center of the party wall, and if from the joint at *E* we lay off our deed distance along Monroe St. it will fall 2¼ inches south of the center of the wall.

As the center of the wall which is being used by both parties is the fairest boundary, it giving each party a greater frontage on Monroe St. than that called for in their respective deeds, the wisest thing to do is to assume it to be the boundary and call the attention of the owner to the discrepancy. As party-wall rights without a recorded agreement may prove troublesome, it is wise at this time for the two adjacent owners to sign a party-wall agreement in accord with the above decision before the proposed new building is commenced.

Line *C–B* should be fixed in the same manner as *C–D* as conditions are similar.

Although line *A–B* is described as passing through the middle of the brick partition wall (which is 20 inches thick), reference to the fieldnotes shows that the street joint *A* is actually 0.08 ft. south of the center line of the wall.

It may be better to accept the wall joint as the corner at *A* and run to the center of the wall at point *B*, thus sacrificing a small triangular strip of land, for the purpose only of avoiding a controversy which might arise if 0.08 ft. of the front of building No. 54 were apparently cut away. But, as in the other cases, the decision of this question should be made by the owner.

On consulting the agreement between Bailey and Hayes dated Sept. 5, 1950 we find the statement "and it is further agreed that the area hereinafter described which is now covered by a one-story building and indicated as *ABCD* on a plan by Jones & Robinson, surveyors, dated Aug. 20, 1950 and herewith recorded shall not be built over to a height greater than 12 ft. above the sidewalk on Washington St. where the land of the party of the first part adjoins the land of the party of the second part."

On consulting that plan we find that the area restricted is the one indicated by the letters *JKLM* on our plan. A note giving the substance of this restriction is therefore added to our plan.

10.48. Calculation of the Adopted Boundaries of the Property.—As the joints at *A* and *I* are at the lot corners we first calculate their coördinates using the ties from points 1 and 6 and the offsets from line 1-7. With 6.91 as a hypotenuse and 3.00 as a perpendicular we find the distance from 1 along line 1-7 to the foot of the perpendicular from A. This added to the "*y*" of 1 gives us the "*y*" of point *A* whose "*x*" is obviously 3.00. In the same way using tie 10.09 and offset 3.00 we find the coördinates of point *I*. The distance *AI* calculated from these coördinates (80.71) agrees with the measurement taken in the field.

The thickness of wall 4–*B* = 15.725 − (5.90 + 8.17) = 1.655 ft. = 20 inches. Similarly the computed thickness of wall *BC* is 16 inches.

As point *B* is at the intersection of the center line of walls *A*–*B* and *C*–*B* it is 5.90 + 0.83, or 6.73, from line 1-2 and 2.93 + 0.66, or 3.59, from line 2-12.

The coördinates of point *B* may be calculated as indicated in Fig. 10.48.1.

Draw *B–c* parallel to 2-12. Having the perpendicular 3.59 and the opposite angle at *c*, calculate the distance 2-*c* and then having the coördinates of 2 and the azimuth of line 2-*c* we can obtain the coördinates of *c*. Similarly having the perpendicular 6.73 and the angle at *c*, calculate *B–c;* and then the coördinates of *B* from the azimuth and distance of line *B–c*.

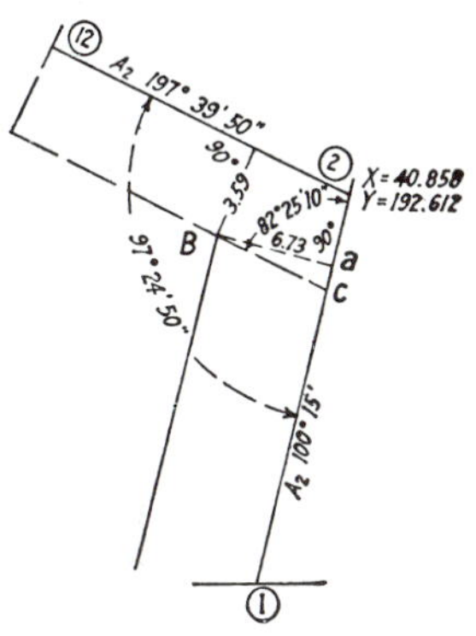

FIG. 10.48.1.

An alternative method is to calculate *a-c* and then 2-*a*. We can obtain now the coördinate of *a*, and as *a–B* is at right angles to 2-*a* we can obtain its azimuth readily and likewise the coördinates of *B*. By this second method it is only necessary to look up the sine and cosine of one azimuth, namely 100° 15′, whereas by the first method it is necessary to look up the sine and cosine of two angles, 100° 15′ and 197° 39′ 50″.

In a similar way obtain the coördinates of *C* and *D*, the latter being 5.80 from line 9-10 and 3.00 from line 10-3.

To obtain the coördinates of *E* we use the tie 8.49 from 4 and the swing-offset of 2.97. This gives us the coördinates of a point 0.03 ft. into Monroe St. which we will call *e*. We know that point *E* is 3.00 from Monroe St. traverse line and also lies on the line *E–F*, which is parallel to 11-17. As the angle between Monroe St. and the line 11-17 is 117° 23′, to obtain the distance from *e* to *E* we divide 0.03 by the sine of 117° 23′. With the coördinates of *e* and the azimuth and distance from *e* to *E*, we can calculate the coördinates of *E*.

A convenient graphical method of finding the difference in coördinates between *e* and *E* is illustrated in Fig. 10.48.2. Draw a line (using any convenient large scale) 30 units from and parallel to *D–E* and intersecting line *E–F* at *g*. Then through g draw a line g–h parallel to the axis of y and another through point E parallel to the axis of x intersecting line g–h at f. Using the same scale we will find that E–f is approximately 34 units in length and f–g 5 units in length. If now we assume that 30 units = 30 thousandths of a foot the difference between the x's of e and E is 34 thousandths and the difference between the y's of e and E is 5 thousandths of a foot.

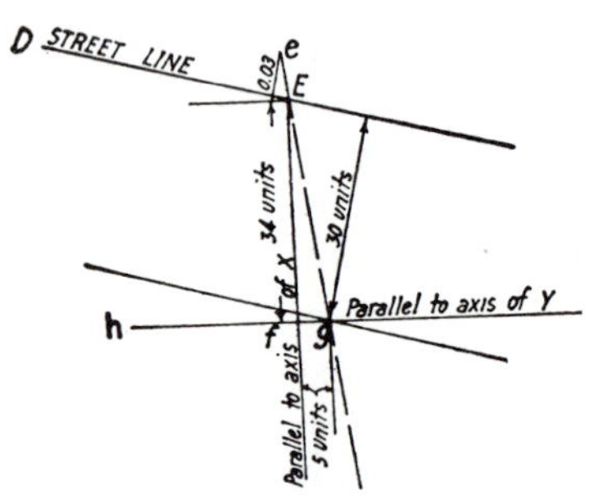

FIG. 10.48.2.

Distance EF is obtained from the roof measurement of 21.33 − 0.03 = 21.30 that is, the distance from Monroe St. to the end of the party-wall. With this distance and the azimuth of E–F, which is the same as 11-16, we now calculate the coördinates of F. An exact value for the length EF is not important because if the jog F is 0.1 to 0.2 ft. nearer to or farther from Monroe St. it matters little.

The computed perpendicular distance between lines 11-17 and 5-20 is 14.81. If distance 5.74 at joint E is subtracted from 14.81 it gives 9.07 as the distance from E to line 5-20. Since G is the center of the 20 inch wall G–H, it is 8.85 ft. from line 5-20. The difference between 9.07 and 8.85, or 0.22, gives the length of the line F–G, which is so short that we are safe in assuming it to be at right angles to E–F, and then compute the coördinate of G.

As indicated by the offsets, line G–H is so nearly parallel to E–F that it is better to assume it to be so and to ignore the small variation of 0.02 ft., in calculating the coördinates of H

and the length of *G–H*. Having the coördinates of *G* and *I* the azimuth and length of *GI* are computed readily. With azimuths of *G–H*, *H–I* and *G–I*, calculate the angles of the triangle *GHI* and with the side *G–I* known we can find the other two, and then the coördinates of *H* and *I*.

From the coördinates of points *A*, *B*, *C*, *D*, *E*, *F*, *G*, *H* and *I* we can find the azimuths and lengths of the property line. This gives a balanced traverse (Fig. 10.48.3).

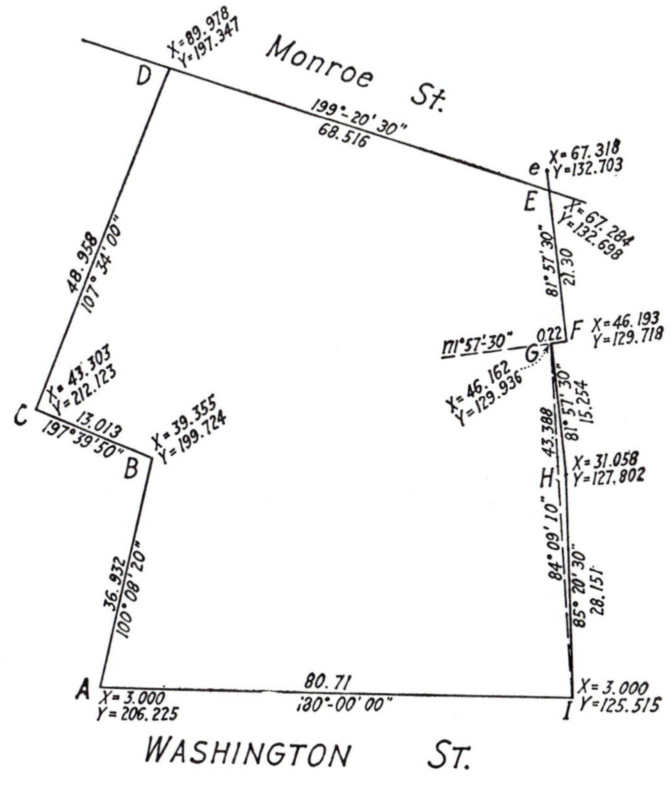

Fig. 10.48.3.

10.49. Final Adjustments of Property-Line Dimensions.—Using the traverse of boundaries with lengths calculated to the nearest thousandth and the azimuth to the nearest 10 seconds, a plan of the boundary lines can then be produced that should balance.

Structural companies and architects usually calculate structural members to the nearest 1/16 inch. They often prefer to have the boundary dimensions recorded on the plan, therefore, only to the nearest 0.005 ft. To accommodate this situation the lengths in the calculated boundary traverse can now be modified by adding a thousandth or two to certain lines and subtracting from others so that most of the lines will read to a full hundredth of a foot, but in a few cases to half hundredths. Using these new lengths and the old angles, compute the boundary traverse anew by coördinates, using this time a different axis. Then adjust any slight error of closure that may exist by making slight changes preferably in the angles; or by making a small change in a short side if that seems to be the more practical.

This was done in producing the final plan (Fig. 10.45.1).

10.50. Other Data Shown on Final Plan.—When sewers, water pipes and various other utilities have to be shown on a plan confusion will result unless each object is designated clearly. For that reason it is well to use not only different colors but also different characteristics in the lines representing the various pipes, etc. Satisfactory results will be obtained by using the following representations:

Water pipes—Full blue lines thus
Sewers—Brown lines thus
Surface drains—Brown lines thus
Gas pipes—Green lines thus
Elec. conduits—Red lines thus
Tel. and Tel. pipes—Red lines thus

Present elevations are shown in red.
Official elevations are shown in blue within circles.
Deed measurements and quotations from records are shown in green.

The Bench Marks used should also be shown on the plan.

Modern copying machines are employed a great deal to copy sketches and plans; though they will copy colored lines they generally reproduce them in black and white. It is very helpful, therefore, to code the various utility lines, elevations, etc. with the symbols here suggested, in addition to color-coding them.

10.51. Profile and Section of Building Walls—Fieldwork.— In obtaining the profile and sections of wall *E–F–G–H–I* the following method is used. On Washington St. (see Fig. 10.32) a mark is made in the vicinity of point 7 on line 1-7, so selected that when a line parallel to *I–H* is laid off and projected upwards it will pass through the windows of all the floors of building No. 62 and also, when run over the roof, this line can be produced to Monroe St. without meeting obstructions, such as penthouses or chimneys. Let us suppose that point 7 fulfills these conditions.

Since it is impossible to place a point on the roof of No. 62 sighting directly from point 7 another point 7*a* is marked on this line on the opposite sidewalk of Washington St. and a foresight 7*b* is placed on the face of the building beyond point 7*a*.

The transit is then set up on 7*a* and sighted on 7 and a point 7*c* is marked on the roof of No. 62, far enough back from the parapet of the building to form a suitable place to set up a transit. The transit telescope should now be plunged (instrument reversed) and point 7*c* again marked, and, if there is any variation between the two sights, points 7*c* should be shifted to conform to the average between the two positions. (See Art. 6.9.)

The instrument is then set up over point 1 and sighted on 7. If a leveling rod is now held horizontally with its end at point 7*c* the right-angle offset from 7*c* to line 1-7 can be read on the rod, and the distance 7-7*c* computed.

The transit is now set up on the roof-point 7*c*, sighted on the point 7*b*, and double reversed to set 7*d* near the angle at *H*. Offsets are measured to the wall *H–I* and the distance 7*c*-7*d* measured.

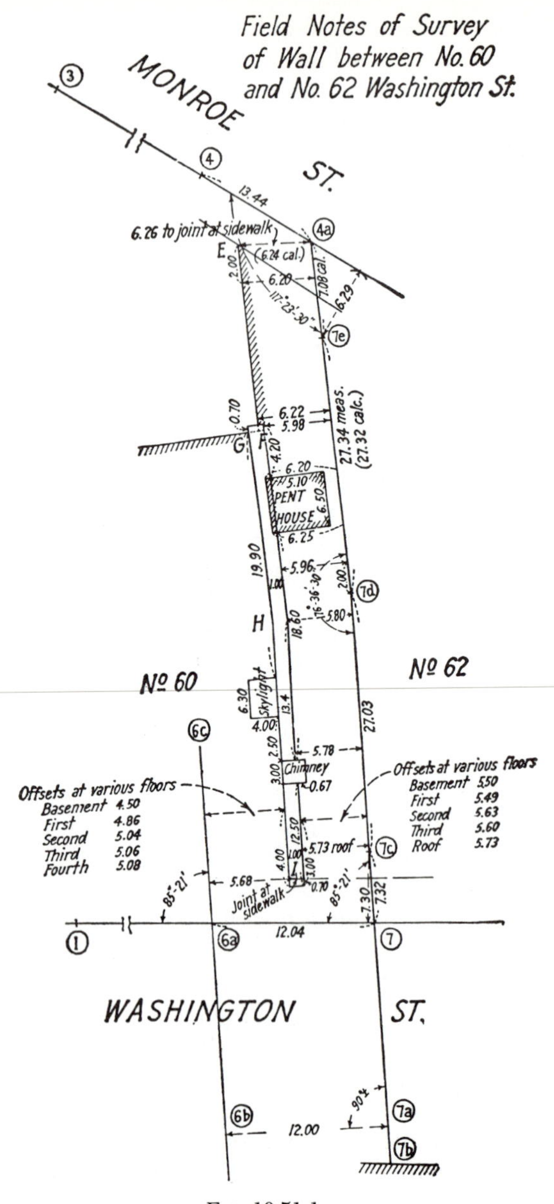

Field Notes of Survey of Wall between No. 60 and No. 62 Washington St.
MONROE ST.
6.26 to joint at sidewalk
13.44
(6.24 cal.)
6.20
7.08 cal.
6.29
117°-23'-30"
2.00
6.22
5.98
0.70
27.34 meas. (27.32 calc.)
4.20
6.20
5.10
PENT HOUSE
6.50
6.25
19.90
5.96
176°-36'-30"
2.00
5.80
18.60
H
Nº 60
Nº 62
Skylight
6.30
4.00
13.4
27.03
2.50
5.78
Chimney
3.00
0.67
Offsets at various floors
Basement 4.50
First 4.86
Second 5.04
Third 5.06
Fourth 5.08
Offsets at various floors
Basement 5.50
First 5.49
Second 5.63
Third 5.60
Roof 5.73
12.50
5.73 roof
4.00
3.00
5.68
0.70
Joint at sidewalk
85°-21'
85°-21'
7.30
7.32
12.04
WASHINGTON ST.
90°
12.00

FIG. 10.51.1.

The transit is now set up on point *7d*, sighted on *7c*, and the angle 176° 36′ 30″ (from Fig. 10.51.1) turned, and point *7e* is marked on the roof near the Monroe St. wall. It is now pointed on a line parallel to *G–H* and *F–E* and offsets are measured to the face of these walls.

Line *7d*-*7e* is produced to Monroe St. and point *7f* is marked on the easterly sidewalk of that street, by double reversal of the transit- the distance *7d*-*7e* is measured.

The transit is now set up over *7f*, the line *7e*-*7f* intersected with line 3-11 at point *4a* and the right-angle offset from point *E* to line *7e*-*7f* measured.

The angle between lines *7e*-*7f* and 3-11 is measured, the right-angle offset from *7e* to line 3-11 measured with the rod; and the objects on the roof (wall thicknesses, penthouse, skylight and chimney) measured and tied onto the roof traverse and the survey on the roof is completed.

The profile of the wall should now be taken. First establish a temporary bench mark or reference mark on the face of building No. 60 from which vertical tape measurements can be made to the roof of No. 60, the higher building.

The transit (used as a level) is now set up on the roof of No. 60 and a vertical reading taken with the tape on the bench mark established on the face of the building.

Readings should now be taken either with the tape or rod at every change in slope on the roof of No. 60 where it adjoins No. 62, at every change in the slope of the roof of No. 62 where it adjoins No. 60, on the roof of the one-story building at *F* and *E* at every change of slope in the parapet wall *G–H–I*, and on top of the penthouse, the chimney and the skylight.

Employing the same methods used in ascending from Washington St. another bench mark can be established on Monroe St. and the levels checked.

The elevations of the basement floors of buildings No. 60 and No. 62 should now be obtained. The use of flashlight, acetylene light or candle will be needed in dark parts of the building. With the above information the profile is plotted, and also the roof plan of the division wall as shown in Fig. 10.51.2. The next step is to survey for the wall sections.

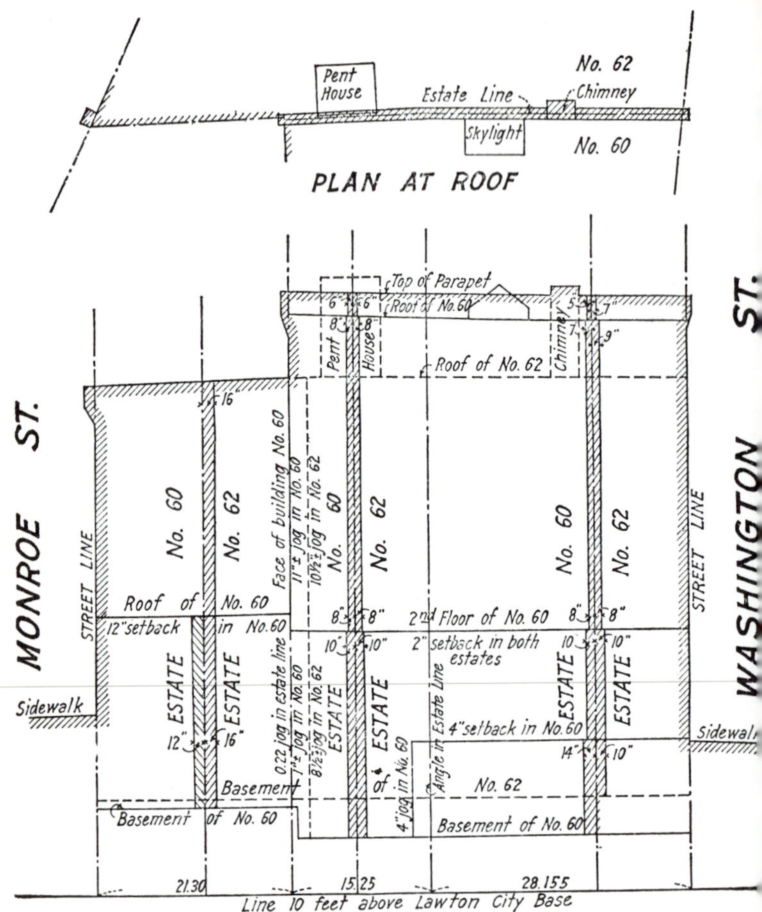

Profile and Sections of Wall between Estates
No. 60-62 Washington St., Lawton, Mass.

Scale 8 feet to an inch

August 15, 1957

Jones and Robinson, Civil Engineers
700 Washington St., Lawton, Mass.

Fig. 10.51.2.

A section of this wall at Washington St. is obtained by measuring the offsets from line 7*a*-7 to the southerly face of the party-wall through the window at each floor of No. 62.

Another point 6*a* near point 6 should be marked where a line parallel with *H–I* will pass through the windows of all the floors of No. 60 Washington St. Set up on point 6*a*, establish a point 6*b* on the westerly line of Washington St. making 6*a*-6*b* parallel to 7-7*a*. Offsets can be measured from line 6*b*-6*a* to the northerly face of the wall through the window at each floor of No. 60.

Similarly offsets to the same wall can be measured from transit lines run parallel to the wall from the Monroe St. side of the building.

The section at Washington St. being different from the section at Monroe St. it is evident that intermediate sections should be obtained to locate where the changes, or jogs, occur. These can be obtained by locating and measuring the jogs (ignoring chimneys) which occur in each floor in both buildings.

For instance, in the basement of No. 60 we find that at Washington St. the face of the wall is 14 inches north of the party line. Inspection shows that there is a 4 inch jog 30 ft. from Washington St. The wall east of that point is presumably 10 inches north of the party line. Now the face of the wall running in from Monroe St. is 12 inches north of the party line but there is a jog of 0.22 ft. in the party line at *FG*. Hence the face of the wall east of *G* is 0.078 ft. north of line *H–G* produced, since previous measurements show that the northerly face of the wall west of *G* is 10 inches north of *G–H*. If both solutions are correct there should be a jog of approximately ⅝ of an inch at point *G*. The jog is actually 1 inch, for wall *EF*, which is intended for a 12 inch wall, actually measured 0.97 ft.

With the location of all of the jogs at all the floors in buildings No. 60 and 62 the entire sections can be obtained and verified.

As the measurements through the window indicate that the face of the wall in No. 60 is 2 inches south at the second floor from where it is at the first floor, and in No. 62 the face of the wall at the second floor is 2 inches north of the face of the wall at the first floor, a setback occurs on both sides of the wall. The position of these setbacks must be obtained by leveling.

With all this information shown the profile and sections can be completed as shown in Fig. 10.51.2.

In plotting the profile it is assumed that one is standing *within* the property. This places Washington St. on the right-hand side of the profile and Monroe St. on the left.

In plotting sections it is customary to assume that one is standing on the principal street, which is Washington St. in this case; hence the sections are drawn to show No. 60 on the left and No. 62 on the right.

If the sections are drawn in red on the profile, which is in black, their location being indicated by the dot and dash line (marked party line) they will be clearly distinguished.

10.52. Giving Lines and Grades for Large City Building.— In giving lines and grades for new buildings which abut directly upon street lines and which occupy nearly all of the land, batter-boards are impracticable. For a large city office building which is to have a steel frame or reinforced concrete columns and floors, the services of a surveying party may be required throughout much of the construction period.

Before any excavation work begins the surveyor should take levels carefully, and also offsets from accurately defined lines at well-defined points on all existing buildings and structures abutting on or near the property, and as the work progresses he should rerun the levels frequently so as to discover any settlement of these structures while it is incipient. It is important that the B.M. shall be outside of any area that can possibly be disturbed by the construction.

As the work on the new building progresses he will set grades and lines for all rough excavation work, for masonry foundations of basement walls and column footings, mark all columns center on the completed footing, set lines and grades for forms for concrete walls and piers, check their position before the concrete is poured, mark grades for each floor level at all columns, check the grade of forms for floors before they are poured, check up the elevation at many points throughout each floor before and after the forms supporting the floors have been struck and for a month or more thereafter so as to determine whether the green floor slab is sagging.

The grid lines connecting all column centers should be referenced on the architect's plan to the property lines shown on the surveyor's plan (Fig. 10.52.1). The surveyor accurately reproduces these column lines on the ground and reference marks on each column line are marked on the opposite sidewalks on a base-line parallel to the street line and on the party walls, so that each line of columns is referenced at both ends (Fig. 10.52.1). As far as possible these marks should be well outside the limits of actual construction so that they will not be disturbed. It is advisable also to set reference marks for levels at the ends of each line of columns, preferably by setting the level up, by trial, so that the middle cross-hair will be an exact number of feet above the City Datum. Then the reference marks are cut in the walls surrounding this lot and on the walls of the buildings that face the site of the new building. This insures a consistent set of reference marks readily seen from all parts of the new building site and which are dependable because all were cut at the height of the line of sight and in plain view of the levelman.

A sketch always should be given to the superintendent of construction showing all reference marks made. In construction work more mistakes are made by misunderstanding than by errors in measurements. It is advisable, therefore, to give all dimensions to the superintendent both in feet and hundredths and also in feet and inches and fractions (usually to $1/16$ inch).

In laying out large plants involving many buildings, connecting pipe lines, and tracks the entire layout may be referred to a rectangular coördinate system. By this means the property lines, buildings, and all points needed for purposes of construction are tied together, and points of known coördinates may be set by measuring from any convenient stations in the system.

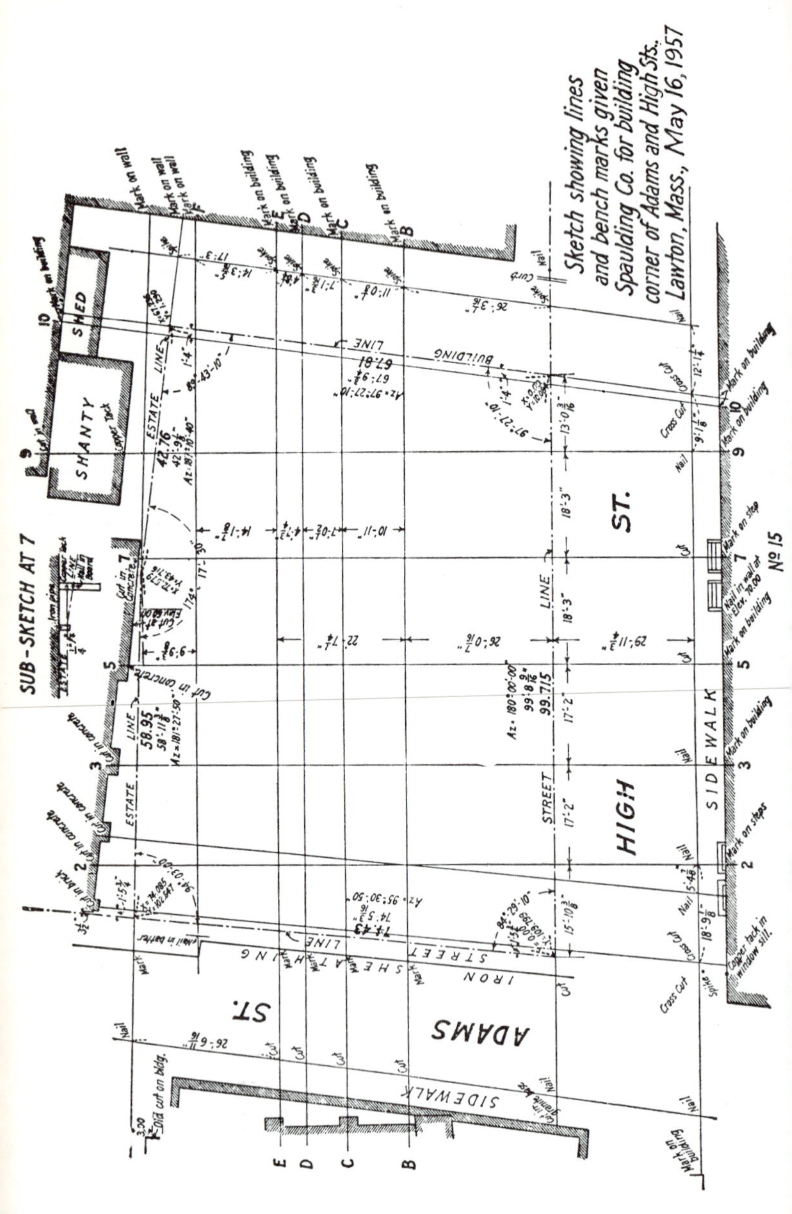
Sketch showing lines
and bench marks given
Spaulding Co. for building
corner of Adams and High Sts.,
Lawton, Mass., May 16, 1957
SUB-SKETCH AT 7
SHANTY
SHED
HIGH ST.
ADAMS ST.
SIDEWALK
ESTATE LINE
STREET LINE
BUILDING LINE
IRON SHEATHING
No 15

RECTANGULAR COÖRDINATE SYSTEM FOR SURVEYING CITIES

10.53. General Description.—In the survey of a city or a metropolitan district it is customary to refer all points to a system of plane rectangular coördinates. Some point within the area, preferably a station already established by triangulation, is selected as the fundamental point of the coördinate system. All distances and all coördinates are computed as though they were in a plane tangent to the earth at this point. The effect of disregarding the curvature of the earth is found to be extremely small for such areas. The true meridian through this tangent point is taken as the axis of Y for the system, and the line at right angles to it as the axis of X. All points within the area may be located with reference to this starting point. In order to avoid negative values of the coördinates, the initial point may be assigned the coördinates 10 000 N, 10 000 E, or any other convenient values. In such a system all azimuths, or bearings, refer to the initial meridian or Y axis. Consequently at all points not on the initial meridian the local meridian is not used. The direction of the Y axis is sometimes called "grid North" to distinguish it from the geographical meridian.

In the survey of the city of Baltimore (Fig. 10.53.1) the origin was taken at the Washington Monument in the center of the city. The city map is divided into squares 1000 ft. on a side. It is also divided into blocks one mile square, each square being shown on a page of the atlas. These squares are designated by numbers north or south, east or west of the origin, as, 1S2W, 3N4E, etc. The position of a point is shown by its distance in feet from the coördinate axes, as 2020.5 N, 1105.0 E.

Among other cities that have been surveyed by this coördinate method the following may be mentioned: New York, Atlanta, Rochester, Cincinnati, Pittsburgh, and London, Ontario.

One of the advantages of a coördinate system is that if any point of known coördinates is lost it can be replaced by means of the known coördinates of other points in that district. If the stations from which it was originally located have been lost, it may be relocated from any other points whose coördinates are known. If a large area is destroyed by fire the entire district may be laid out again from points outside the devastated area.

It is of utmost importance, therefore, that all stations be marked permanently.

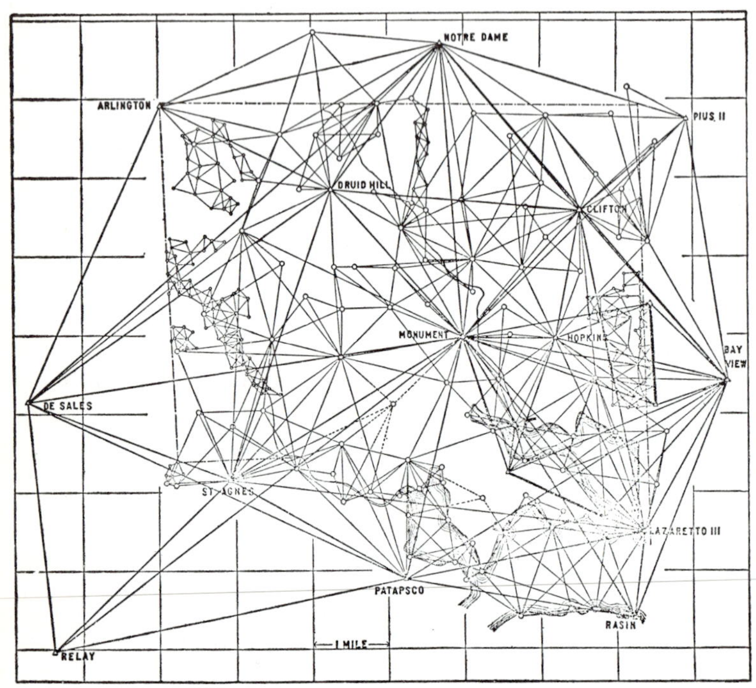

FIG. 10.53.1. TRIANGULATION SCHEME FOR SURVEY OF THE CITY OF BALTIMORE.

(Printed by permission of Major Joseph W. Whirley, Chief Engineer of the Topographical Survey Commission, Baltimore, Md.)

10.54. Triangulation Scheme.—The principle points of the survey (control points) are usually located by triangulation, trilateration or accurate traverses. In a triangulation system the length of one side of some triangle must be measured, then if sufficient angles are taken (with great precision) all other sides may be calculated by plane trigonometry. The shape of the triangles is important. It is difficult to lay down rules for the shape of acceptable triangles without going into the discussion from a mathematical standpoint. In general, it is necessary to avoid small angles which occur opposite the base, or known side, or opposite the side being calculated. It is not correct to say that small angles should never be admitted in the system; they should not be admitted when they enter the system in such a way as to weaken it. (See Coast and Geodetic Survey Special Publication No. 247.) In beginning such a survey it is usually possible to utilize such triangulation as already exists in the locality in question. For a method of converting the latitudes and longitudes of such triangulation stations into plane coördinates see Hosmer's *Geodesy*, and Coast and Geodetic Survey Special Publication No. 235. The triangulation may be precise enough to furnish base-lines upon which the entire system of triangles may be built up. But if the triangulation in this locality is a long distance from the base from which it was computed the errors in length may have accumulated to such an extent that the lines will not be sufficiently accurate to serve as base-lines for the city survey. It may be possible however to utilize the triangulation to furnish the general position of the survey as a whole, but it will be necessary to measure one or two special base-lines with great accuracy and to make special determinations of azimuth. Land values in cities are high, and are increasing. The precision required in a city survey is fully as high as that of any kind of survey that is made, and it is a mistake to begin such a survey with work of low precision, for if the precision of the triangulation is barely sufficient for the present purposes it is only a question of time when the whole system must be resurveyed. Trilateration is triangulation with many distances measured, usually by EDM (see Art. 3.14 and Art. 10.55).

There will in general be two classes of triangulation stations: (1) those occupied with the instrument and located with first-order precision; and (2) such points as flagpoles, church spires, or water tanks, not occupied with the instrument, but located by intersection. These latter points will be useful, in locating the theodolite by means of the three-point problem. (See Vol. II, Art. 1-57.) It is of special importance in city surveys to center instruments and signals precisely over station marks.

10.55. Measurement of Base-line.—The lines chosen as bases should be placed for convenience on fairly level ground, but for accurate results they should be such as will give well-shaped triangles, and the former consideration should never outweigh the latter. The base is measured with invar tapes or wires, a spring balance, and thermometers (as described in the publications already cited). The tapes should be tested at the U.S. Bureau of Standards both before and after the measurements. The measuring stakes, or the measuring tripods, should be lined in carefully with a transit, and the grade of the tape determined carefully by leveling. If two independent measurements of the base agree within about one part in 500,000 the base will be satisfactory for this work. It is not too difficult, however, to secure a precision of about one in one million.

Base-line for second or third order control surveys may be measured using electronic measuring systems described in 3.16-3.18. Where line-of-sight intervisibility is possible, a base-line may be measured using only one pair of setups, thus eliminating repetitive tape measurements and overcoming difficulties of rough terrain. Measurements of many lines of a triangulation network are usually feasible, rather than only one base-line for many triangles. When all distances are measured, the network may be extended by trilateration, the angles being computed from the sides. Checks may then be made with measured angles.

10.56. Measuring the Angles.—If a direction theodolite, reading to seconds by means of a micrometer, can be secured for this work it will give the best results. If a repeating transit (10″ vernier) is used the angles must be measured with a sufficient number of repetitions to give angles with a standard deviation, say, 0.1″. Each set of angles should consist of six angles with telescope erect and six with telescope inverted. The particular order in which the direct and reverse angle are taken may be varied somewhat. About five sets will be required on each angle. In order to distribute the readings over the circle and thereby eliminate errors of graduation each time a new set is begun the initial setting should be advanced by an amount equal to 180°/n, where n is the number of sets to be measured. Careful attention must be given to the centering and the leveling of the instrument. The signals should be such as to give no error of "phase" (side illumination) and should be exactly centered over the station mark.

10.57. Adjustment of the Principal Triangulation.—In a system of triangulation of such importance it is advisable to have the entire main scheme of triangles adjusted by "Least Squares." Once done it will not have to be repeated. This adjustment removes all inconsistencies and prevents accumulation of error such as would occur if errors were allowed to remain in the system. For individual triangles it will be sufficient to divide the error of closure by 3 and to apply this correction to each angle. This may be done for the preliminary computation of all triangles. The test of accuracy most easily applied is to see if the sum of three angles of each triangle is equal to 180° + any spherical excess which may be required. The spherical excess is only 1″ for each 76 square miles of area of the triangle, so, for the triangles discussed here the spherical excess will be quite small, but in some instances it will not be negligible. In first-order triangulation the error of closure of a triangle should be about 1″ or less. In city work it should not fall below this accuracy.

10.58. Azimuth.—Before beginning any other work it is important to obtain the azimuths of two or more lines, preferably connected with, or near to, the initial station. These azimuths must be determined with an accuracy comparable with that of the triangulation, that is, with a standard deviation of about 0″.1 (For the special methods required see Vol. II, Chap. 2, and Coast and Geodetic Survey Spec. Pub. No. 237.) These azimuths must refer to the initial meridian. If they are observed at any station other than the initial point the amount of the convergence of the meridians due to the curvature of the earth must be determined and allowed for.

10.59. Subsidiary Triangulation.—After the principal triangulation has been completed other stations may be located, the triangles being smaller and the accuracy somewhat less, than that of the main scheme.* The object of this is to place stations in more convenient locations from which to run traverses for locating street lines, or lot lines. In using these stations, it is desirable to complete a closed traverse with the network to obtain a check on intermediate work.

10.60. Traverses.—After the triangulation has been completed, traverses may be run from one triangulation station to any other, connecting all street corners, lot corners, and other points whose location is desired. Since the coördinates of the points from which these traverses start are of a high order of precision and since the traverses close on other accurate points, it is not necessary to make the traverse measurements themselves with so great accuracy as would be necessary if there were no controlling system of triangulation. It is for this reason that the triangulation is really economical. When a traverse is closed on a fixed station the error of closure is computed and the error thrown into the traverse itself. The positions of the triangulation stations are not altered, as these are considered to have been finally fixed in the general adjustment.

10.61. Method of Locating Street Lines, Property Lines and Buildings.—The coördinates of all traverse stations are calculated after the traverse has been adjusted. To locate street corners, points on offset lines, or lot corners, it is only necessary to turn angles and measure distances from the traverse stations to points on the lines in question. The azimuths or bearings of all these lines may be calculated readily (always referred to the Y axis of the system). When the distance and the bearing are known the difference in coördinates (Δy) and Δx) can be found by the same procedure as used to find latitudes and departures in Art. 15.5.

Buildings, property lines and underground conduits may be located by the various methods discussed and demonstrated in

---

*For specifications for 2nd and 3rd order triangulation, see Classification and Standards of Accuracy for Geodetic Surveys, U.S. Bureau of the Budget, published by American Congress on Surveying and Mapping. Also see Vol. II, Art. 1-7.

Chap. VI. All important points should be located from at least two different instrument stations as a check against mistakes.

EXAMPLE

Coördinates of point *a* from station 3 (Fig. 10.61.1) would be calculated as follows:

Azimuth 3 − *a* = 41° 10′ 30″
Dist. 3 − *a* = 41.29 ft.

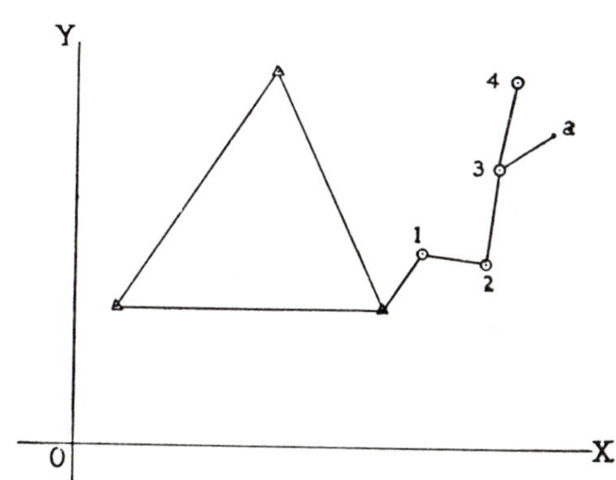

FIG. 10.61.1.

By logarithms

lat. (Δy)	31.08
log diff. lat	1.49246
log cos az.	9.87662
log dist.	1.61584
log sin az.	9.81846
log. dep.	1.43430
dep. (Δx)	27.18

	North	East
Coörd of 3	2160.41	4209.20
Diffs.	+ 31.08	+ 27.18
Coörd of *a* =	2191.49	4236.38

(See Art. 15.5)

By calculator:

lat (Δy) = 41.29 (cos 40° 10′ 30″)
= 41.29 (0.7527·22) = 31.08

dep. (Δx) = 41.29 (sin 40° 10′ 30″)
= 41.29 (0.6583611) = 27.18

10.62. Level Control.—Vertical control involves providing precise elevations referred to one well-defined datum so that their accuracy is assured. Vertical control is produced by running first-order levels (Vol. II, Chap. 3) and establishing permanent bench marks placed in pairs at about one-half to one mile apart. The bench marks are placed in pairs so that both are visible from one set-up of the level instrument, thus giving an immediate check on settlement of the bench marks. The levels should be run in loops of about one mile in length on a side; the error of closure in each loop is a valuable check on the observations. The entire net should be adjusted by the method of Least Squares to give the most probable elevations and to remove all inconsistencies. From these first-order elevations, levels of low order may be conveniently run to provide grades for construction. (See Art. 9.25.)

CHAPTER XI

MINE SURVEYING

11.1. General Remarks.—In this chapter the usual limitations and difficulties met with in surveying a mine are pointed out and some of the instruments and common methods described. A brief description is given also of the methods of establishing the boundaries of mining claims in the United States.

Mine surveys are made for the purpose of determining the relation of mine workings to the boundaries of the property, obtaining data from which to establish direction for proceeding with the workings, determining the amount of material taken from the mine and the probable amount of available ore that can be worked, obtaining data from which graphical representation of mine workings may be made.

Most of the principles of mine surveying apply also to tunnel surveys.

MINING INSTRUMENTS

11.2. Mining Theodolites.—In modern mining all of the accurate angle measurements are taken with a theodolite, several forms of which are designed for this purpose. The essential features are lightness and adaptability for measuring accurate azimuths of nearly vertical or of very short sights. If the telescope is of low power the illumination of the field is better and its focal length is usually shorter, both of which are conveniences in mine surveying. The theodolite should be provided with a full vertical circle which should be enclosed in a metal case to protect the graduations from dripping water.

With an ordinary transit one cannot take a downward sight more steeply inclined than 55° or 60° to the horizon. Various attachments have been devised for sighting more steeply inclined lines, the object being to permit a sight to be taken over the edge of the horizontal circle of the instrument. This is usually accomplished by attaching an auxiliary telescope to the side or to the top of the main telescope of the engineers' transit, so that the instrument will afford all the advantages of the ordinary transit and also will make it possible to sight even down a vertical shaft.

FIG. 11.3.1. MINING TRANSIT WITH INTERCHANGEABLE SIDE AND TOP TELESCOPE.

(Courtesy, C. L. Berger & Sons, Inc.)

11.3. Eccentric Telescopes.—Some mining transits are constructed so that an auxiliary telescope may be attached to an end extension of the horizontal axis, such an instrument being known as a *side telescope transit*. In another type of transit, called the *top telescope transit*, the auxiliary telescope is mounted above the main telescope. The distance between the centers of the main and auxiliary telescopes is called the *eccentricity*. Some instruments (Fig. 11.3.1) are made with an interchangeable telescope which can be attached at either the top or the side of the main telescope.

As these telescopes are detachable, a transit need not be encumbered with one when the main telescope alone can be used, which is the case in most surveying required in mines.

In comparing the merits of the various forms of attachment it must be remembered that the object to be accomplished is to transfer the meridian accurately from one station to another, these stations sometimes being close together in plan but distant in elevation.

With the side telescope the effect of eccentricity in the measured azimuths may be eliminated by the reversal method described below. When a top telescope is attached, the main telescope cannot be inverted. With such a transit the correction for eccentricity must be applied to all altitude readings.

When it is not important to double the angle azimuths may be carried more rapidly by means of the top telescope than with the side telescope, but it is good practice to double all of the angles of a traverse.

With the side telescope vertical angles are read directly, and the average of the direct and reversed readings gives the true angle, thus vertical control is carried more rapidly.

The interchangeable side and top telescope is in common use. The side telescope is preferred by most surveyors for measuring both horizontal and vertical angles, because by taking an equal number of direct and reversed readings, no correction for eccentricity or index error need be applied. For example, if four angles are read, the result of the fourth horizontal angle reading divided by 4 will give the angle whose vertex is at the point over which the instrument is set; while the average of the direct and reversed vertical angle readings gives the true angle.

When using an auxiliary telescope for a steeply inclined sight, the main telescope may be employed for the backsight or foresight if the vertical angle is small enough to permit its use.

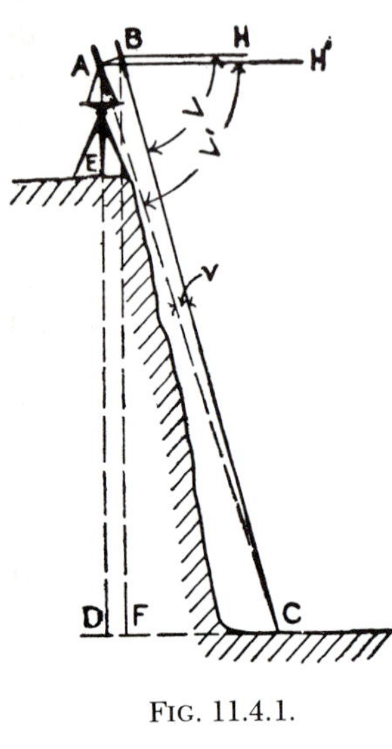

FIG. 11.4.1.

11.4. Vertical Angle Correction for Eccentricity of Top Telescope.—If the top telescope is used for measuring vertical angles, a correction for eccentricity must be made. In Fig. 11.4.1 the vertical angle V has been taken by means of the top telescope to point C in the bottom of a shaft. The distance AC was measured, A being the horizontal axis of the main telescope. The distances desired are DC and AD. HB and $H'A$ are both horizontal,

$$\text{then } V' = V - v$$

where V is the angle measured by the top telescope, v is the angular correction for eccentricity, and V' is the corrected angle.

$$\text{Since } \sin v = \frac{AB}{AC} = \frac{\text{Distance between telescopes}}{\text{Distance measured}}$$

we may construct a table for any instrument giving the values of v for different measured distances.

The vertical and horizontal components of AC are then

$$AD = AC \sin V' = AC \sin (V - v)$$

and

$$DC = AC \cos V' = AC \cos (V - v)$$

11.5. Horizontal Angle Correction for Eccentricity of Side Telescope.—In Fig. 11.5.1, the center of the small circle represents the center of the instrument, and the line of sight of the eccentric telescope is tangent to this circle. Angle H has been measured with the side telescope and H' is the angle which would have been obtained had it been measured with the main telescope. The difference in direction between the lines from the F.S. to the side telescope and to the main telescope is the angle α, and the difference in direction between the lines from the B.S. to the side and to the main telescope is the angle β. By the construction indicated in the figure the angle formed between the two dash lines is equal to the angle measured

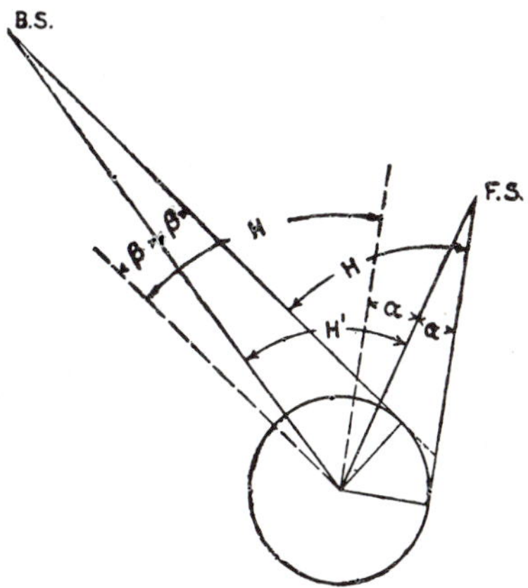

FIG. 11.5.1.

with the side telescope. The required true angle H' therefore is equal to $H - \beta + \alpha$. Since the usual practice is to measure all horizontal angles as azimuths in a clockwise direction, the true horizontal angle will equal the angle measured with the side telescope minus the correction angle for the B.S. plus the correction angle for the F.S. These correction angles α and β obviously balance when the horizontal distances from the center of the instrument to the F.S. and the B.S. are equal.

The cotangent of the correction angle for either case equals the horizontal distance from the center of the instrument to the point sighted divided by the distances between the main and side telescopes.

When the main telescope is used for either the backsight or the foresight, only one correction for eccentricity is applied.

11.6. ADJUSTMENTS OF MINING TRANSITS.—It is assumed that all ordinary adjustments of the transit have been made. When an external focusing transit is used, the adjustment of the objective slide is of importance, because often in mining work the azimuth has to be transferred through very short horizontal sights, or inclined sights with short horizontal component. Exceptional care must be taken to make the horizontal axis of the telescope truly horizontal and the line of sight exactly perpendicular to it.

Most auxiliary telescopes are provided with a screw adjustment which allows alignment of the vertical hair of the top telescope or the horizontal hair of the side telescope with the corresponding hair of the main telescope. The cross-hairs of the auxiliary telescope are made parallel to those of the main telescope by the manufacturer. If parallelism is lost, it may be regained by shifting the individual cross-hairs.

If angles are measured by repetition, taking equal numbers of direct and reversed sights, it is not essential to have perfect alignment of the main and side telescopes as the errors compensate for one another. If single angles are taken with the side telescope, or if the top telescope is used, the instrument should be in perfect adjustment.

11.7. Compasses used in Mines.—Compasses are useful in most mining operations for performance of rapid or preliminary surveys, both on the surface and underground. They may also be used to align development openings and drill rods, or to determine strike and dip of geological structures. The Brunton compass, or pocket transit, is widely used. This instrument may be used to determine bearings, or vertical angles may be determined by the use of the clinometer.

In using a compass, particularly underground, it must be kept in mind that there are many attractions for the compass needle that may affect its accuracy.

11.8. OTHER INSTRUMENTS.—The best kind of level for use in mines is a dumpy level having a low power telescope. It should be equipped with a reflector for illuminating the cross-hairs. Leveling rods for mine work are made similar to ordinary leveling rods, except that they are shorter, the 3-ft. and the 5-ft. lengths being the commonest. These rods are usually of the Philadelphia pattern, the target being employed when there is difficulty in reading the graduations directly. A good form of target for rods used in mine surveys has a slit silhouetted against an illuminated white background. This can be made by cutting a horizontal slit across half the face of the level target so that the light can be held behind it.

The objects sighted at in underground surveying must be illuminated. An ordinary plumb-bob hung in front of a white card or tracing cloth illuminated by a light held in front of and at one side of the plumb-bob makes a good signal. Even when the point is close to the instrument and the atmosphere is clear a white card held behind the plumb-bob string is of great aid to the transitman.

The equipment should contain at least two tapes, a 50- or 100-ft steel tape graduated to hundredths, and another tape of such length that the longest slope measurements required, such as those from level to level, can be made in one measurement. Tapes 200 or 300 ft. in length are suitable for this purpose. It is frequently impracticable to hold a tape level and obtain the measurement by plumbing because of darkness in the mine. Inclined measurements accompanied by vertical angles are taken to speed up the field work. A short pocket tape is convenient for measuring the height of instrument, height of point and offsets. The open form of reel shown in Fig. 3.3.1 should be used because tapes used in mines often become wet and dirty.

Modern instrumentation increasingly used in mining surveying includes gyrotheodolites, optical plummets, and lasers. Based on the same principle as the gyro-compass for navigation at sea, accurate gyro attachments provide azimuths directly to an accuracy of a few arc seconds. This solves the rather difficult task of transferring a line from the surface into a mine. Nevertheless, for transferring a line down a vertical shaft, the optical plummet can greatly simplify the task and increase the plumbing precision. The laser provides a very powerful and narrow light beam which can be used directly as a reference line for alignment of shafts, horizontal tunnels, or inclined tunnels and for line and grade measurements in a variety of mining and tunneling applications.

UNDERGROUND SURVEYING

11.9. Transferring a Direction into a Mine.—There are three general methods of transferring a direction into a mine: first, by means of the main telescope or an auxiliary telescope on the theodolite, second, by means of plumb-lines hung in a shaft, and third, by means of a gyroscopic instrument.

If the mine is entered by an adit* or a tunnel the passageway is usually so nearly level that the direction can be transferred into the mine by setting up the transit at the mouth of the adit and, after taking a backsight on a fixed station on the surface, taking a foresight into the adit and establishing a point within the mine. If the entrance is by a shaft which is highly inclined, but not vertical, the same general method is employed except that the auxiliary telescope will be required.

When the shaft is vertical it is still possible to use the transit and its auxiliary telescope, but a more accurate method is to use long plumb-lines hung in the shaft. If the mine has but one shaft it is necessary to suspend two plumb-lines (sometimes three or four) in the shaft. If the mine has two or more shafts a point may be located at the bottom of each shaft by plumbing down from the top. Then a traverse can be run in the mine connecting these underground points and another traverse on the surface connecting the corresponding points at the tops of the shafts; in this way a closed traverse is formed connecting all of the points.

11.10. Plumbing Down a Vertical Shaft.—To the mine surveyor the plumb-line is an instrument of precision, excelling even the transit under some conditions, and the work of transferring the azimuth down a vertical shaft usually can be accomplished more accurately by the plumb-line than by any other method.

The method usually followed is to suspend two bobs (sometimes three or four) from points located at the collar of the shaft and whose positions have been accurately determined from

---

*An approximately horizontal underground passageway running from the surface into mine workings and used for drainage and ventilation.

the surface surveys. The azimuth of the line joining the two points at the collar of the shaft is the azimuth of the plane defined by the plumb-lines. If the instrument is taken underground and set up at *C* (Fig. 11.10.1) and "jiggled"; that is, moved until the line of sight *CA* and the two wires *A* and *B* are in the same vertical plane then the azimuth of the line *CAB* becomes known. The distance *BA* should be measured and should be equal to the distance between the two points established at the collar. The distance from the wires of the plumb-lines to the transit should be chosen so that it will be unnecessary to move the objective far when focusing. Since this condition is seldom possible, the object glass slide should be in excellent adjustment. When the instrument is in the same vertical plane as the wires a point is established over or under the instrument and an angle turned to another station as *D* further along the level, thus furnishing a permanent line in the level whose azimuth is readily computed. By measuring *CD* the position of *D* may be calculated. It is good practice to check the position of *D* by "jiggling in" at a second point *C*′.

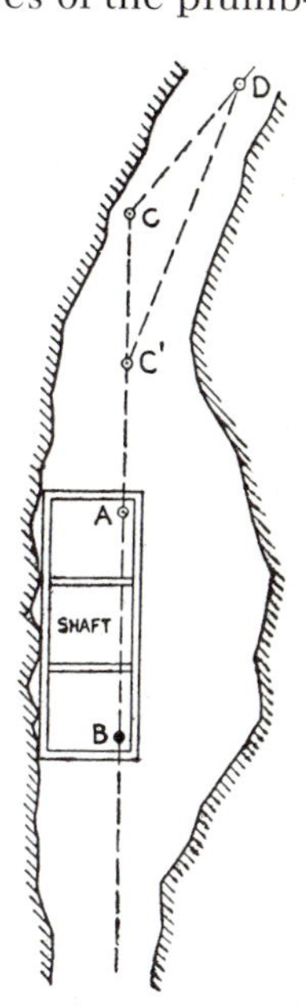

FIG. 11.10.1.

The plumb-lines should be as small in diameter as the weight will permit. They are usually of copper, annealed iron or piano wire. With regard to the size of plumb-bob and wire there is a great difference of opinion. The plumb-bobs should be heavy enough to straighten out all bends in the wire. They should be immersed in oil or water or an oil emulsion, so as to stop their swinging in the shortest possible time, and the receptacle in which the plumb-bobs are immersed should be covered in order to protect the surface of the liquid from water dripping down the shaft. All air currents should be checked so far as possible, since experience has shown that in deep shafts they have considerable effect upon the plumb-lines. Great care should be taken to see that no part of the wires comes in contact with the shaft, or is splashed by water dripping down the shaft.

When once the plumb-lines are hung the meridian may be transferred to all the levels of the mine once and for all time, so that a little extra precaution and time given to this operation are worth while. The surveyor should always keep in mind the fact that in plumbing the azimuth down the mine the direction of the meridian is of much more importance than the actual position of the points themselves, because an error due to an incorrect direction of the meridian may be multiplied many hundreds of times in carrying the traverse through the mine.

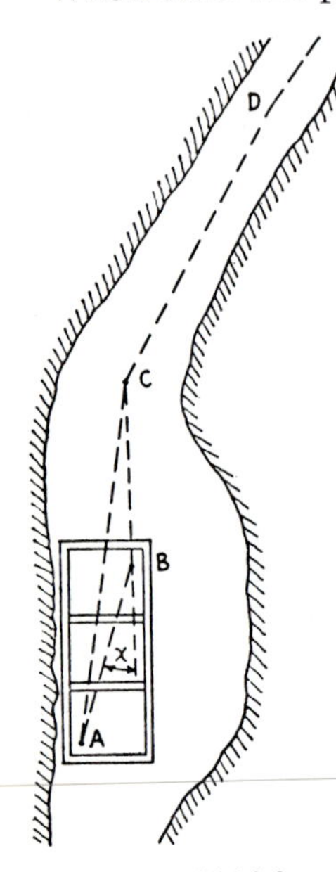

FIG. 11.10.2.

A second method of shaft plumbing which is widely used is the Weisbach triangle method (Fig. 11.10.2). The transit is set up at C, slightly to one side of the vertical plane created by the two wires. The angle ACB should be kept very small, less than one degree, and the sides AB and BC should be roughly equal in length. The angle ACB is measured by repetition to obtain the maximum accuracy possible with the transit employed. The distances AB, BC, and AC are measured to the nearest 0.001 foot. Due to the small angles, $AB + BC = AC$ very closely. The value of the angles may be calculated using the law of sines.

$$\sin ACB : AB = \sin CAB : BC = \sin x : AC$$

As a check on the accuracy of the work, the sum of angles ACB and CAB should equal angle x.

A point should be established over or under the transit and an angle turned to a second point at least 50 feet away, establishing a permanent line.

The methods described may be employed in establishing the location of the wires at the collar of the shaft as well as in transferring the meridian from the wires at the shaft bottom.

11.11. The Use of Gyro Instruments.—It is quite obvious that the physical dimensions of a shaft limit the accuracy of the transferred azimuth. A slight error in the plumbed position of one of the points will cause a large angular error due to the short distance between the points. Furthermore the "jiggling in" of the transit it not a precision approach either. It is therefore advisable to utilize a gyro instrument for precision work. With the gyroscopic instrument the direction is directly transferred to within a few seconds of arc into any part of the mine without the direct physical connection to the surface, although it will be necessary to drop a point vertically down the shaft to link up the surface traverse with the underground traverse. One must also employ the gyro instrument on one or more stations of the surface traverse—directly overhead, to the extent possible—to be able also to correlate the azimuths of both traverses.

11.12. Measurements of Distances Down Vertical Shafts.—If the depth of the shaft is no greater than the length of the tape, then the zero end of the tape may be lowered by having a small weight attached to it, and a vertical measurement taken from a point at the top of the shaft to a level line of sight established by the transit set up in the bottom of the shaft. When the distance is greater than the tape-length it may be measured along the guides for the cage or skip, or it may be measured by means of a long wire. The difficulty in measuring deep vertical shafts by means of long tapes or wires is that they stretch on account of their weight. The amount the tape stretches may be determined as follows. If a tape is suspended from a spring balance the pointer will register the total weight of the tape; this is twice the average tension in the tape. Then if the pull is increased by attaching a weight W to the end of the tape the average tension is $W + \frac{1}{2}$ (weight of tape). Knowing this average amount of pull the actual length of tape can be found by stretching the tape out on a floor and testing it as described in Art. 3.29, giving it the same amount of pull.

11.13. Transferring a Direction into a Mine When There Are Two Shafts.—The above methods presuppose that the mine has so far been opened only by one shaft. If there is a second shaft or an adit, it is, of course, only necessary to use one line to plumb or otherwise transfer the position down each shaft; the computed distance between these points then becomes a base-line of substantial length. In Fig. 11.13.1 the traverse *ABCD* is run out on the surface to connect the two shafts at *A* and *D*. The points *A* and *D* are plumbed down the

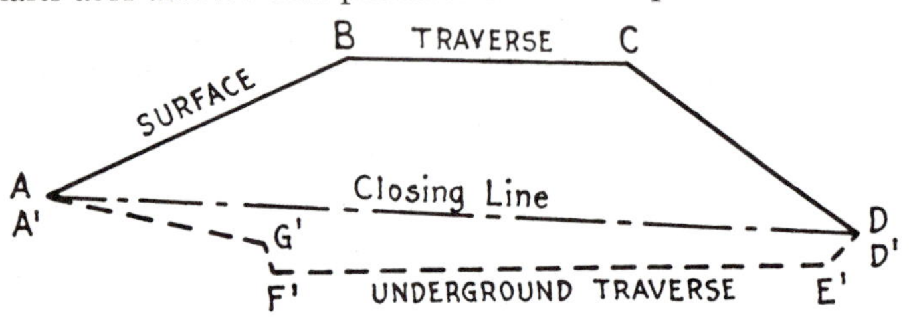

FIG. 11.13.1.

shafts and the corresponding points *A′* and *D′* established at the bottom. An underground traverse *A′G′F′E′D′* is then run out. In the surface traverse the length and azimuth of *AD* and in the underground traverse the length and azimuth of *A′D′* are missing. The horizontal length and azimuth of each of these lines can be determined from their respective traverses as explained in Art. 15.12. The azimuth of *AD* in the surface traverse is referred to the true, and, since nothing is yet known in regard to the direction of the meridian in the mine, the underground traverse is referred to an assumed meridian. The correct azimuth of *A′D′* is the same as the azimuth of *AD*, provided the plumbing down the shaft has been done accurately. The difference between the true and assumed azimuths of *A′D′* is a correction then to be applied to the azimuths of all of the lines of this underground traverse.

11.14. UNDERGROUND TRAVERSES.—Surveying in a mine is necessarily a process of traversing, for only the working passages are available for lines of survey. The line of traverse is often run so that the longest possible sight may be taken. In the tortuous passages of a mine it is frequently necessary to take very short sights on the main traverse and since the azimuth is transferred to distant connections through these short lines great care should be exercised; and instrumental errors should be eliminated by reversing the telescope and using the mean of the two results. After the main traverses have been run, the surface boundaries may be established accurately underground and the stopes* and working places surveyed by more convenient and less accurate methods, from the traverse stations already established.

Accurate traverses should be carried into all important workings of the mine; these include the shafts, levels, cross-cuts, raises and winzes. Permanent transit points should be established in the shaft at every level; in an inclined shaft it is good practice to put in one or two points on line between the levels to hold the direction of the shaft. Offset distances called "rights" and "lefts" are measured on each side of the traverse line within the mine workings for the purpose of locating the limits of the workings. The floor of passages is defined by determining the elevation of the necessary number of points, and the roof by measurements from points located on the floor. Elevations are defined within the mine either by direct leveling, as in leveling out on the surface, or by means of inclined distances and vertical angles. Secondary openings, such as stopes, are surveyed by running the lines from the main traverse through a chute or mill-hole, from which the necessary dimensions and angles are taken to define the shape and extent of the stope holes. Sometimes these stopes are surveyed by use of the compass and clinometer. Where long stopes are to be surveyed a transit line is run from a station in a level up through a chute at one end of the stope, carried along through the stope, and then down through another chute at the other end closing on another station in the level.

The field-notes of these details are usually kept in the form of sketches which show the traverse line that forms the skeleton of the survey and the measured offsets and angles which locate the points needed to define the boundaries of the hole. It is often necessary to draw sketches showing the opening

---

*_Stopes_ are rooms formed by the excavation of ore above or below a level.

both in elevation and in plan and to add descriptive remarks sufficient to make the sketch clear to any surveyor even though he is unfamiliar with that particular place in the mine.

11.15. Traverse Stations and Station Marks.—Stations should be located where they are not likely to be disturbed, and where they can be of the maximum value to the surveyor. Usually the most satisfactory method of installing a station mark consists of drilling a short hole in the roof rock, driving a tight-fitting wooden plug into the hole, and placing a spad in the plug (Fig. 11.15.1). Station marks can be attached to the caps of mine timber, but since timbers are likely to be moved, multiple auxiliary marks are advisable.

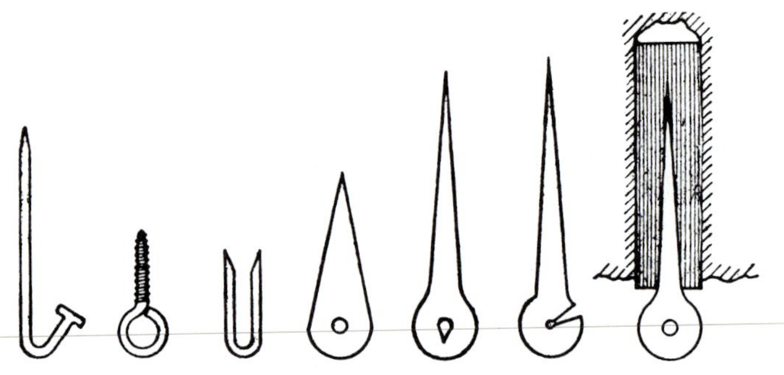

FIG. 11.15.1. STATION MARKS IN ROOF.

Owing to the workings in the mine these overhead marks sometimes cannot be employed; then the stations are established on the floor. Floor station marks are commonly a short stout spike or a sharpened boiler rivet driven into a tie of the track. In the heads of these spikes a small drill-hole is made to distinguish them from other spikes. For a temporary mark a punch mark on the rail, a mark on a block of lead or iron or a chalk mark is often used.

In sighting overhead stations, a plumb-line is suspended, and the angle and distances taken to a mark on the plumb-line, usually to the top of the plumb-bob. When the station is on the floor a pencil may be sighted, and if the station mark has a small drill-hole in it a small finish nail can be inserted, upon which to sight.

All such points must be illuminated. This may be done by holding a piece of tracing cloth or thin paper back of the station mark, behind which a light is held; thus causing the nail or pencil held on the mark to be silhouetted against a white surface. Often a light held to one side of the object sighted will illuminate it sufficiently, especially if a sheet of white paper is held behind it.

In most underground surveying it is necessary to provide additional illumination for the cross-hairs. This is accomplished by holding a light forward and to one side of the objective end of the telescope.

Besides the devices which mark the exact station point a white ring is often painted around them to aid in finding the point. Another method is to paint an arrow on the wall, pointing toward the station. It is good practice to paint the number of the station near by, either on the hanging wall or on the foot wall. In some mines a small copper, brass or aluminum tag upon which the station number is stamped is attached near the station mark.

The method of numbering stations sometimes depends upon their positions and sometimes it does not. Where the station number relates to its position in the mine one of the methods is to number all points on the first level from 100 to 199, on the second level from 200 to 299, and so on. Another method is to designate the station as 1206 N or 1318 S, if the workings are about North and South; the first number would mean station 6 on the 12th level north of the shaft. Where the numbering of the stations has nothing to do with its location they are usually numbered consecutively. A book containing the number, location and description of all stations should be kept up to date, and if the mine has more than one shaft this fact should be indicated in the notes or in the station numbers.

11.16. Setting up the Transit.—If the point to be occupied is on the floor the transit is set up in the usual manner; if it is in the roof a temporary point can be marked on the floor by plumbing down from the roof point, and then the transit set up over the temporary floor point. If the point is in the roof, however, the usual method is to suspend the plumb-line from the point and set the instrument up under it. Most instruments have a mark on the telescope barrel over the intersection of the center lines of the horizontal axis and the telescope. This point is set approximately under the plumb-bob and the instrument leveled; the telescope is brought to a horizontal position by the telescope bubble, and then by means of the adjustable head the point is brought directly under the plumb-bob.

11.17. TRAVERSING.—In running the traverses in levels, drifts and horizontal, or nearly horizontal passages, the work is done the same as on the surface and under these conditions the main telescope of the transit can be used. The angles are measured as direct angles, and doubled as a check. Vertical angles are read in the ordinary manner, but care should be taken to determine the index correction. Whenever the vertical circle is read with the telescope both direct and inverted, the index correction should be determined and applied to each reading to eliminate errors produced by plate bubbles. The mean of these two corrected angles is the most probable value. Where it is necessary to use an eccentric telescope the proper correction for eccentricity must be applied as described in Arts. 11.4 and 11.5.

11.18. Elevation Differences.—Since nearly all traverse measurements are made as inclined measurements, and since vertical angles often are used together with the distances for obtaining elevations, it is frequently necessary to measure the H.I. *(height of instrument)* and the H.Pt. *(height of point)* above or below the station mark. If the transit is set **above** the station the H.I. is recorded + (plus), and if the transit is set **under** the station the H.I. is recorded − (minus); but if the point sighted is **below** the station the H.Pt. is recorded + (plus), and if it is **above** the station the H.Pt. is recorded − (minus).

The vertical distance is obtained by computing the "sine distance," which is the inclined distance × the sine of the vertical angle, and then applying to that distance the proper correction, depending upon the amount and the sign of the H.I. and H.Pt. distances. For example, Fig. 11.18.1 indicates a set-up of the instrument under a point in the roof; the vertical angle was taken to the top of the plumb-bob hung from another point in the roof. Here the H.I. is minus and H.Pt. is plus.

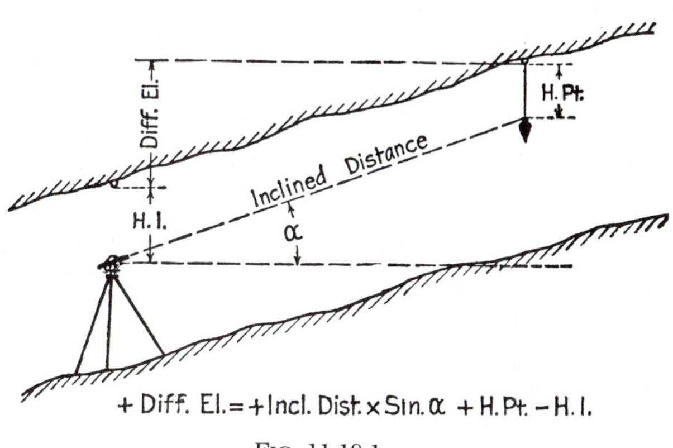

Fig. 11.18.1.

Fig. 11.18.2 illustrates a set-up where the instrument station is in the roof, the station ahead is in the floor, and the vertical angle was taken directly to the point in the floor; here the H.I. is a minus quantity. There are several different combinations

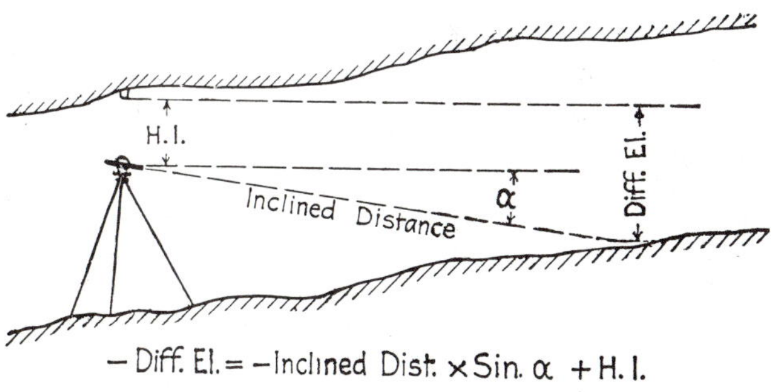

Fig. 11.18.2.

which arise, but no difficulty will be experienced if the relative position of the points is visualized before the computations are made.

The horizontal distances are usually obtained by computing the "cosine distance," that is, the inclined distance × the cosine of the vertical angle.

11.19. Traverse Notes.—As in all classes of surveying the form of notes differs with local practice and the personal preference of the surveyor. Probably the most common note form has the following column headings: height of instrument, instrument station, backsight, foresight, horizontal angle, slope distance, vertical angle, height of point, and remarks. Columns for distances right, left, up, and down may be included.

Great care must be exercised to take notes in such a manner that they cannot be misinterpreted later, particularly when measuring angles by repetition or when using auxiliary telescopes.

11.20. Station, Computation, and Coördinate Books.—Besides the field note-book, most mining engineers keep three other survey record books, a book containing a description and location of all traverse stations, one containing the computations of all traverses and other surveys, and a coördinate book containing the computed values of the coördinates of station points.

11.21. MINE MAPS.—The maps of a mine are frequently the only means of clearly conveying information regarding the shape and extent of the workings. They should be made so as to show in the clearest possible manner the passages and stope-holes, their extent as well as their location. Since in different mines the veins of deposit have different amounts of dip and different directions of strike, considerable judgment must be exercised in the choice of planes upon which to project the plans of the workings so that their location and extent will be shown in the clearest possible manner.

The usual plans prepared for a mine are

1. Surface plan.
2. Underground workings in plan.
3. Underground workings in longitudinal section.
4. Underground workings in traverse section.
5. Map of stopes.

6. Assay map.
7. Geological data.

1. The Surface plan should show the boundaries, roads, joining railroad, buildings, mine dump, drains and water supply.

2. The underground plan is a projection on a horizontal plane. It should show all underground workings such as the shafts, levels, cross-cuts, winzes, raises, stopes, faults, ore shoots, mill-holes, and sometimes the geology; the progress should be shown by different colors, by difference in character of lines or by dates lettered on the plan.

In mines where the dip is 70° or greater the plan is often complicated because lines on one level overlap lines representing other levels. To avoid this confusion each level should be plotted in plan separately. In Fig. 11.26.2 all levels are shown on one plan because the stopes are not plotted on that plan and hence there is no overlapping of lines on different levels.

When the dip is less than 60° to 70° the plan of the entire mine can usually be plotted on one sheet, and if the levels are colored differently the plan is more easily understood.

3. The longitudinal section may be either a vertical section, as in Fig. 11.26.2, or a section lying in the plane of the dip, in which case it is called a "plan on the vein." In the former the true vertical distance between levels is shown, whereas in the latter the levels are shown at distances apart equal to the slope distances along the shaft. Even though the levels are usually constructed on a grade of about 1 per cent for drainage and haulage they are often plotted on the longitudinal section as level passages.

This longitudinal section should show all workings shown on the plan [see 2. above] and should show the progress of the work in a clear manner, especially in the stopes.

If the vein is fairly regular and the coördinate planes are taken with the assumed north and south along the strike, then the longitudinal section represents the levels in their true lengths.

4. The transverse section is usually plotted in a plane at right angles to the plan and longitudinal section. It is well in some cases to plot several transverse sections, each one a separate sheet, in order that their lines may not overlap and produce a confusing drawing. If the longitudinal plan is in a vertical plane through the strike, then the transverse plane is at right angles to this plane and on it the shaft would show in its true length.

5. In some mines stope maps are kept showing in detail the size and shape of each stope as a separate map and indicating upon the plan the progress made. These working plots may be either vertical, horizontal, or parallel to the vein or seam. In any case, the thickness of the deposit is recorded at frequent intervals together with other particulars, such as thickness of waste or value of ore. These thicknesses are all measured at right angles to the plane of the working plan, so that when multiplied by the area on the plot, the cubic contents of any section is obtained. Where the ore occurs in irregular masses, not conforming particularly to any one plane, the above system does not apply and some other method must be devised by the surveyor. The best way of estimating amounts not mined is to sketch their probable extent on such a chart from the data available and to make use of the area and thickness method as suggested above.

6. Assay plans are made from the plan and longitudinal section of the mine on which is plotted the location from which assay samples have been removed.

7. A plan showing geological data should show the boundaries of the various formations, the planes of bedding and the foliation, the fault planes with their displacement, and all veins and dikes encountered. These data could of course be shown to some extent by three coördinate planes as in solid geometry, but it is usually better to plot the geological data taken in each level. Any inclined rock surface is then represented by a series of contour lines corresponding to the levels from which the information is actually obtained. Strikes, dips, and intersections may then be determined by use of a protractor, a scale and a table of cotangents.

In a metal mine a plan of each level, when filled in with all the geological data, will have as much detail as can be shown conveniently. It is usual to make a geological plan of each level separately on thin tracing paper so that any two or possibly three consecutive ones may be superimposed. The particular position, strike, and dip of any ore shoot or surface may then be found as easily as though they were all plotted on the same piece of paper.

The scale of the surface map is usually 100 or 200 ft. to an inch. The general plans of underground workings are often 50 ft. to an inch and the detail stope plans are 20 ft. or 10 ft. to an inch. The general underground plans can best be plotted by the coördinates of the station points. The coördinates can be laid out on the plan and then if it changes scale by shrinking or swelling it does not affect the accuracy of the plan, for the new station points located in workings as they progress are plotted by coördinates from the nearest of the coördinate points originally plotted on the map. Since these maps are used so frequently and for so long a period, it is advisable to plot them on the best quality of mounted paper available.

11.22. Mine Models.—These may be constructed of wood and glass. A wooden box without a top is built first. Horizontal strips of wood are fastened to the inside of the box at distances apart equal to the vertical distances (to scale) between the levels. On each of these horizontal strips a sheet of glass is placed on which has been accurately painted the horizontal projection of that particular level. The glass should be treated first with a coat of copal varnish or gelatine so that it will take the paint or ink. Strips of glass are inserted between the horizontal sheets to represent a shaft or winze; more conveniently, sheet plastics can be used to build up the entire model.

11.23. Laying out Mining Work.—Drifts or cross-cuts* are laid out by putting in two spads in the roof, not too near together, from which the miner can hang two plumb-lines and sight the center of the heading he is to run.

---

*A *drift* is a horizontal passageway following a vein. A *cross-cut* is a horizontal passageway across or approximately at right angles to a strike. The latter is the direction of a horizontal line in the plane of a deposit.

Vertical shafts are carefully plumbed on the inside of the frames, and frame by frame, as these are put in. It is best to hang the plumb-line from several frames above the bottom one, as these upper ones are more likely to have ceased to move. Hang the line an even fraction of an inch each way from the true position of the corners and note any accidental variation in the last frame set, so that in future work, if it is desired to hang the plumb-line from this frame, its error of position can be allowed for. The dimensions of a shaft or drift are given either "in the clear," meaning net measurements inside all timbers, or "over all," meaning gross measurement outside all timber and lagging.

11.24. UNDERGROUND SURVEYING PROBLEMS.—In the practice of mine surveying, problems are constantly arising which tax the ability and ingenuity of the surveyor. Most of these problems involve the connection of an existing mine opening with another point located at a distance and perhaps at a different elevation; e.g., drifts may be driven toward one another and must meet head-on, two mine levels must be connected, a drift must be driven to intersect a drill hole from the surface, etc. The solutions of problems of this type are not difficult when the coördinate system of mapping is employed. The bearing of the connecting lines is found by the method described in Art. 15.12.

SURFACE SURVEYING

11.25. SURFACE SURVEYING IN RUGGED MOUNTAIN REGIONS.—In accurate work, such as the surveying of mining claims for patent,* the ordinary mining transit may be used. Measurements are made with a steel wire tape, 300 to 500 ft. long and marked every 10 ft. (or 20 ft.); it is used with a short auxiliary steel ribbon tape which is divided to hundredths of a foot. Over rough ground it is not practicable to measure horizontal distances by use of plumb-lines. Slope measurements are taken from the horizontal axis of the instrument to the

*By patent proceedings is meant the proceedings necessary to obtain from the government a fee simple deed to the mining claim.

point at which the telescope is sighted, care being taken not to over-stretch the tape or to kink it. (Art. 3.23.) The most accurate work is done by stretching the tape with a tension handle (a spring balance) which can be attached by a clamp to any part of the tape. Where it is feasible, just enough tension is given so that the stretch of the tape compensates for the shortage due to sag. Sometimes assistants will have to hold the middle point, or the points at one-third and two-thirds the length of the tape, up to the line of sight, giving at the same time enough pull to make the sag equal in different sections of the tape.

In making general maps of a mining district, only monuments and important locations need be shown accurately. This accurate work, which is the first to be done, forms a skeleton on which to make a general map. The topography can be filled in by using a transit fitted with stadia hairs and a compass.

Considerable success has been achieved by use of the subtense bar (see Art. 3.34) in rugged terrain, as also by use of self-reducing tacheometers (see Art. 3.28).

The best topographical data in mountainous country are obtained by running traverses along the ridges and valleys; these are also usually the best places to travel. Much sketching is necessary and the work should be plotted by the surveyor himself each day as the work proceeds. In this work a rough determination of the topography is sufficient, since the plans are usually plotted to the scale of 1/10000 or smaller, and therefore such instruments as the hand compass, clinometer, and aneroid barometer can be used. With such instruments one man can do the entire work. The plane-table cannot be used to advantage in wooded mountain or in mine surveying, but photographic surveying may often prove useful in filling in details of topography.

11.26. Mine Boundaries—Location Survey—Lode Location.*—In most countries mineral rights are defined by vertical planes through lines marked out on the surface. Title

---

*For further information with regard to this subject see the Manual of Instructions for the Survey of the Mineral Land of the United States, issued by the Bureau of Land Management, Dept. of Interior, Washington, D.C.

to metalliferous lands, however, as granted by the United States, conveys the right to all minerals included in the downward prolongation of the portions of veins cut off by the vertical end bounding planes, i.e., a vein can be worked in the dip indefinitely, but in the direction of the strike it is limited by the end bounding planes of the claim.

The federal law allows a claim to cover 1500 ft. measured along the direction of a vein and 300 ft. on each side of it. These dimensions, which constitute the maximum, can be reduced by local laws. Fig. 11.26.1 shows the measurements

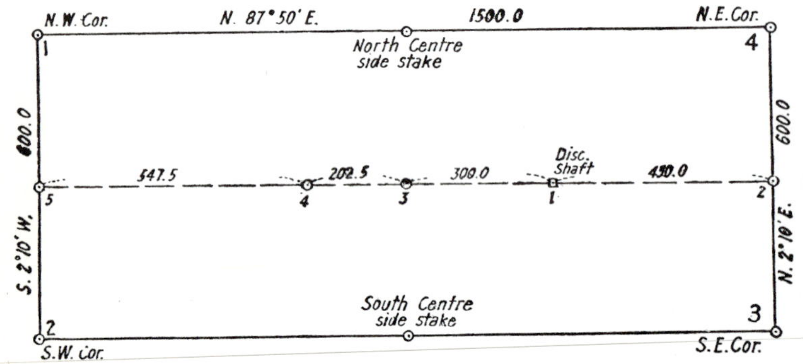

FIG. 11.26.1. PLAN OF LODE CLAIM.

made in staking out a claim 1500 ft. long and 600 ft. wide, the ends being perpendicular to the center line. The discovery was made at the point marked "discovery shaft." The direction of the center line follows the supposed direction of the strike of the vein. The claimant decided to limit his claim on the east end 450 ft. from the discovery shaft. The center line was therefore extended 1050 ft. westward, for he is entitled to a claim 1500 ft. long. At stations 2 and 5 right angles were turned off and the four corners set. When measuring from the discovery shaft westward the center point (station 3) was set as a means of establishing the center side stakes. The corner stakes and the center side lines of a claim are marked with the number of the corner and the name of the claim. Bearings and ties are taken to patent corners of the other (adjoining) claims, or, bearings from one corner of the claim are taken to mountain peaks or other permanent points. A location certificate is then made out stating the name of claimant, date, place, extent and description of claim (metes and bounds); this is sent to the county recorder.

The end lines of claims are not necessarily perpendicular to the center line; they may make any angle, for example, to fit an adjoining claim, but the two end lines must be parallel. Claims not infrequently have angles as shown in Fig. 11.26.2, or the

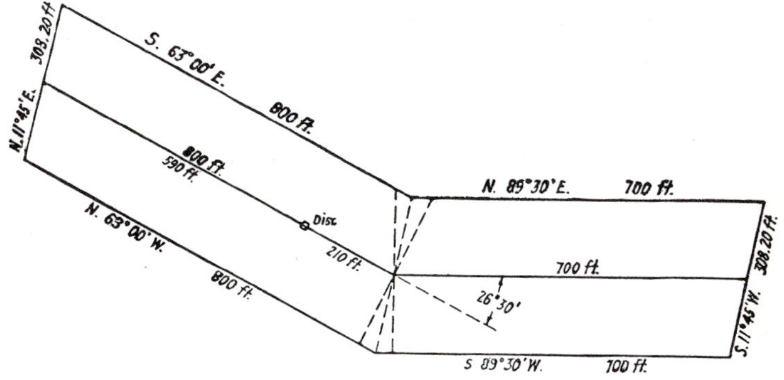

FIG. 11.26.2. PLAN OF LODE CLAIM.

side line may be broken lines. Where a claim conflicts with another it may stop short of the full distance allowed by law. In order to guard against troublesome litigation, an effort is sometimes made to surround a valuable claim with others, thus forming a "group." The more valuable claim is then protected as regards all "extralateral rights."

11.27. SURVEYING FOR PATENT.*—The surveying of claims for patent from the United States Government can only be obtained by those who have received appointment of United States Deputy Mineral Surveyor and they must have an order from the Surveyor General of the state or territory in which the claims are located before making any such survey.

*For more detailed information see "Mineral Land Surveying," by James Underhill (published by John Wiley & Sons).

In surveying for patent much more accurate work has to be done than when merely locating a claim. Before applying for an order for a patent survey the claim survey should be examined to see that it is correct and closes accurately. Any inconsistencies should be removed, and the description made to agree with the lines intended in the claim. If the dimensions on the ground are subsequently found to differ slightly from those in the certificate an error of a few feet or a few minutes of an angle will usually be permitted. The claimant then deposits the fees for the Surveyor General's office, and makes application for the survey, accompanied by certified copies of the certificates of location. If the location certificates are found correct the order is issued. Otherwise the certificates are returned for correction.

The mineral surveyor then executes the survey just as for a claim location survey except as to accuracy of measurements and manner of marking corners. An error of one part in two thousand is permitted by the Surveyor General. The corners are monumented with stones, rock in place, or by marked trees. Sometimes a witness corner is set on line when it is not possible to set a regular corner monument. The survey is tied in (from Cor. No. 1) to a Section corner or to some U.S. locating monument. The workings and improvements (such as shafts, tunnels, buildings, etc.) are tied in to the corners and are shown on the plan. An estimate of the value of these improvements must be made and must be shown to be at least $500 per claim.

Placer claims may be taken in 20-acre tracts, the bounding lines of which must conform with the general system of survey lines established by the Government, but if such survey has not been extended to the district, they must be bounded by true meridian and east and west lines. The survey of coal land is subject to somewhat similar rules.

REFERENCES

Introduction to Mine Surveying.
by W. W. Staley, Stanford University Press, 1939.
Practical Tunnel Driving.
by Richardson and Mayo, McGraw-Hill Book Company, Inc., New York, 1940
Mine Surveying.
by E. B. Durham, McGraw-Hill Book Company, Inc., New York, 1913.

CHAPTER XII

PHOTOGRAMMETRY*

12.1. PHOTOGRAMMETRY.—Photogrammetry is the science or art of obtaining reliable measurements from photographs taken from the air or from the ground.

Aerial photographs and the maps prepared from them by the application of photogrammetry are used for many types of projects, such as airport, highway, railway, waterway, pipe line, and transmission line locations; for irrigation, water supply, and flood control; for harbor and river improvements, city zoning and planning, traffic studies, parks, grade separations, and assessors' tax plans; for soil conservation studies, soil identification and mapping, and timber estimates; for military surveys and explorations.

Aerial photographs may be classified as either *vertical* or *oblique.* Occasionally combinations of verticals and obliques are employed.

Vertical photographs are taken with the camera approximately vertical; as a result the plate or film is horizontal. While vertical views look like a map, the scales of the individual photographs are usually far from uniform. Although all photographs taken with the optical axis vertical are called "verticals," they are at best really perspective views of the terrain produced on horizontal planes.

In taking *oblique* photographs, the camera axis is purposely inclined from the vertical. The apparent horizon may or may not appear in the picture depending upon the angle of inclination. The image plane, where the film is located during exposure is always, however, fixed at right angles to the axis of the

*See Higher Surveying by Breed and Hosmer, Vol II, 8th Ed., Chaps. 4, 5 and 6, for complete text on this subject.

camera. From a given altitude an oblique photograph covers a larger area than a vertical photograph.

12.2. Cameras.—Cameras used for taking vertical photographs are usually mounted in gimbals and sighted through a hole in the floor of the plane. The axes of the gimbals are supported on vibration absorbing mounts fastened securely to the structure of the airplane. By means of an attached level, the camera axis may be held vertical or nearly so at the instant of exposure.

Oblique photographs for reconnaissance may sometimes be taken with a small, hand-held camera. Oblique photographs for mapping, however, are normally taken with cameras mounted in the aircraft with a fixed angle of inclination.

Cameras used for aerial surveying are of the fixed focus type. At the instant of exposure the film is held flat in the focal plane of the camera by means of a vacuum back or a glass pressure plate. The coördinate axes on the photograph are defined by fiducial marks in the focal plane. These are photographed on the negative at the instant of exposure.

The principal point (where the optical axis intersects the film plane) is defined by the fiducial system.

A common focal length is 6 inches, while others might be used occasionally.

12.3. View Finder.—In taking vertical photographs a view finder is used in conjunction with the camera. It serves to determine when an object to be photographed is vertically below the camera. It is also used to determine the angle of crab of the airplane; i.e., the angle between the longitudinal axis of the airplane and the direction of flight when the airplane is flying against a cross wind. To correct for crab, the camera is rotated about its vertical axis so that successive photographs will line up with the true direction of flight. The view finder can also be used to determine intervals between exposures necessary to obtain the desired overlap of successive photographs.

12.4. Taking the Photographs.—When taking the photographs of the area to be mapped the airplane is flown back and forth over parallel straight lines, called "flight lines," usually at a pre-determined altitude. The proper distance between the flight lines is pre-computed and the lines are plotted on a map before the flight, as an aid to the pilot in keeping the plane on its course. This is called flight planning. The altitude is given by an "altimeter" installed in the plane. If vertical photographs are to be used for compiling maps, exposures are so timed that the overlap between successive pictures in the direction of flight will be 60 per cent; this overlap is sometimes called *forward lap* or *end lap.* To avoid an hiatus between adjacent strips of photographs, the flight lines are so spaced that the photographs will also overlap; this is called *side lap* and it is usually from 15 to 45 per cent, averaging 30 per cent.

The best time to take photographs is when the weather is bright and cloudless. For photogrammetric mapping the best time of year is when there is a minimum of foliage and no snow cover. Photographs to be used for forestry studies, however, are taken when the trees are in full foliage. The three-hour periods after sunrise and before sunset should be avoided because of excessive length of shadows.

12.5. Relation Between Photoscale and Altitude of Camera.—Strictly speaking, there is no such thing as the scale of a photograph because the ground usually has varying elevations; it is only when the ground is perfectly level that a vertical aerial photograph could have a definite and uniform scale. However, any elevation above sea-level may be selected as a datum and if the height is known at which a photograph was taken the photo scale may be computed for the datum chosen.

There are two methods of expressing the scale of air photographs: (1) By the *Representative Fraction* (R. F.) and (2) By the *Scale Factor.* The R. F. is the usual fractional scale in which the numerator is unity; it is the ratio of a distance on the map to the corresponding distance on the ground and is expressed here as s. The scale factor is the usual expression of "so many feet to an inch," in which an inch on the map represents the specified number of feet on the ground and is expressed here as S. This notation is seldom used for the Photoscale.

If a distance D on the ground (datum) is represented by a distance d on the photograph, then the scale expressed as a fraction is $\frac{d}{D} = s$. From similar triangles (Fig. 12.5.1) it is seen that

$$\frac{d}{D} = \frac{f}{H} = s$$

where f = the focal length of the lens and H is the elevation of the nodal point of the lens above the datum plane both expressed in the same units.

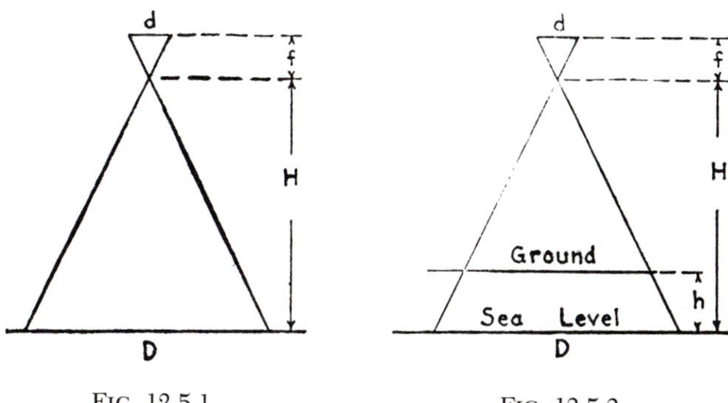

FIG. 12.5.1. FIG. 12.5.2.

Another form of this equation is is as follows (Fig. 12.5.2): If H is the elevation of the camera above sea-level and h is the elevation of the ground, then

$$s = \frac{d}{D'} = \frac{f}{H - h}$$

From this relation we may find the altitude required to give a photograph the desired scale s, when the focal length of the lens and the average elevation of the ground are known.

12.6. Altitude of Airplane.—When preparing for a photographic mission the pilot must be informed of the altitude at which the pictures are to be taken. Since the photo scale depends upon the average elevation of the plane above the ground, the flying altitude must be increased by the average height of the terrain, in order that the pilot may fly according to his altimeter which refers to mean sea-level.

EXAMPLE: Scale of photograph 1/20000; focal length of lens 8.25 inches; average altitude of ground 1500 feet.

$$s = \frac{f}{H - h} \qquad H = h + \frac{f}{s}$$

$$H = 1500 + \frac{8.25}{12}(20{,}000) = 1500 + 13{,}750 = 15{,}250 \text{ ft.}$$

Any change in the altitude of the airplane will alter the scale of the photography. For a focal length of 8.25 inches the change in scale of photograph would be about 12 feet per inch for each 100-foot change in altitude.

12.7. Number of Negatives Required to Cover a Given Area.—When the dimensions of the area to be photographed, the photoscale, the image format, and the end and side overlaps are known, the approximate number of photographs required to cover a given area may be computed.

EXAMPLE: Size of area 310 square miles; format 9 inches × 9 inches; scale 1/12000; end lap 60%; side lap 30%.

Total area covered by one photograph =

$$\frac{9 \times 12000}{12 \times 5280} \times \frac{9 \times 12000}{12 \times 5280} = 2.90 \text{ sq. miles.}$$

Net area covered by one photograph =

$$2.90\,(1.00 - .60)\,(1.00 - .30) = 0.812 \text{ sq. miles.}$$

Approximate number of photographs required to cover area =

$$\frac{310}{0.812} = 382 \text{ photographs.}$$

Two additional exposures are usually taken at the end of each flight line. Also the boundaries of the area photographed parallel to the flight lines are overlapped by at least 25% of a photograph width.

A more accurate estimate of the number of photographs required may be obtained by computing the number of flight lines required and the number of photographs per flight line.* Additional pictures may be required to insure complete

*See Higher Surveying by Breed and Hosmer, Vol. II, 8th Ed., Arts. 4-17, 18.

coverage when the path of the airplane deviates from the flight plan.

12.8. Displacements and Errors in Aerial Photographs.—Image points on an aerial photograph are displaced from their correct perspective position as a result of lens distortion, changes in dimensions of photographic materials, such as by film and print paper shrinkage and others.

The true perspective positions of points on a photograph differ from their true orthographic or map positions due to the inability of the pilot to maintain flight at constant altitude, changes in scale caused by relief, and displacements caused by tilt of the camera axis.

12.9. Distortions in Cameras, Film and Printing Paper.—The cameras used in aerial surveying are ruggedly built to withstand rough handling and temperature changes. They are factory adjusted. High precision photographic lenses are essentially distortion free. Corrections may be made in the photogrammetric plotting process for any known distortion of the lens if needed. Changes and distortions in film and paper can be controlled by the proper selection of materials and care in handling and processing. The dimensional change in topographic aerial film base is very small. Printing paper will shrink, but if the shrinkage is uniform, the effect of the change can be corrected; in double weight printing paper the shrinkage is practically uniform. Changes in film and printing paper may be determined by comparing the distances between reference marks set in the camera and the corresponding marks photographed on the film and appearing on the print.

12.10. Scale Changes.—When taking aerial photographs for mapping, the plane is usually flown at a precomputed altitude. Because the altimeter, which is used for indicating the altitude, is not a precise instrument, and because meteorological conditions vary, the altitude will vary, thus causing a change in the photoscale during flight.

Another cause affecting the scale is the varying elevation of the ground. The scale near the summit of a hill would be larger than the scale near its base.

If a day is selected when the barometer is changing slowly and the wind is fairly uniform in direction and intensity, the flying altitude ought not to vary more than 5% from that specified.

In making scale corrections use is made of the following terms: The *Nadir* or *Ground Plumb Point* is the point on the ground vertically beneath the lens **at the instant of exposure.** Its image on the print is called the *photograph nadir. Principal Point* is the point at which the optical axis of the camera intersects the film. If this point is not shown on the print, it is taken at the intersection of the lines between the fiducial marks (which are at the corners or along the sides of the print) on the margin of the print. When the optical axis is truly vertical, the nadir and principal point coincide. When the plate has some tilt, which is usually the case, these two points do not coincide. A point approximately midway between the nadir and principal point in a tilted photograph is called the *isocenter.*

12.11. Displacement Caused by Relief.—Because all photographs are perspectives, all objects in the photograph except the plumb point and those in the datum plane are displaced. In Fig. 12.11.1 the top of the chimney A will be shown on the datum at A'. The displacement on the datum is $A'' A'$, and on the photo, $a''a$. The displacement of B is $B''B'$. The

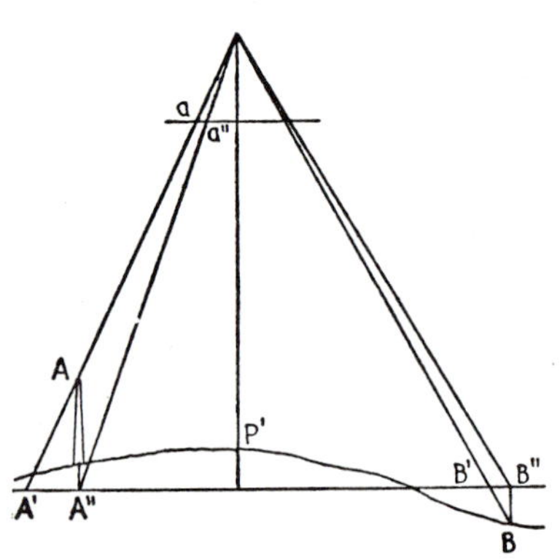

FIG. 12.11.1. DISPLACEMENT DUE TO RELIEF.

nadir point P' has no displacement. It will be observed that the displacement of points below the datum is inward. Since the displacement is in a vertical plane, the **error on the print is radial from the photograph plumb point,** which in a truly vertical photograph coincides with the principal point.

The displacement on the photograph is calculated as follows: The hill A (Fig. 12.11.2), the height of which above the

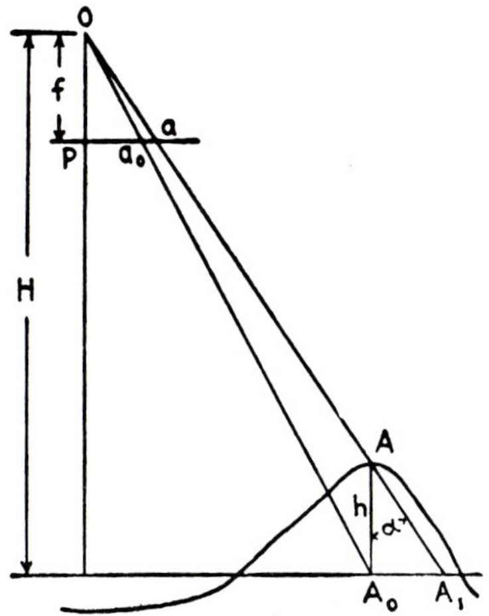

FIG. 12.11.2. RELIEF DISPLACEMENT.

datum is H, appears on the photo at a; the ground displacement is A_1A_0, and the displacement on the print is aa_0. By measuring the distance Pa on the print and dividing it by f, the angle α = angle A_0AA_1 may be found.

$$a_0a : A_0A_1 = f : H$$

$$a_0a = \frac{f}{H} A_0A_1 = s \cdot h \tan \alpha.$$

Another formula for computing the displacement is derived as follows: In the last equation $s = \dfrac{f}{H}$ and $\tan \alpha = \dfrac{Pa}{f}$. Substituting we obtain

$$a_0a = \frac{f}{H} \cdot h \cdot \frac{Pa}{f} = Pa \cdot \frac{h}{H}.$$

EXAMPLE: $H = 10000$ ft., $f = 6$ in., $h = 500$ ft., Pa measured on the photograph = 3″.075.

Then by the first formula the displacement is

$$a_0a = \frac{6}{100000 \times 12} \cdot 500 \cdot 12 \cdot \frac{3.075}{6} = 0.154 \text{ in.};$$

and by the second formula

$$a_0a = 3.075 \times \frac{500}{10000} = 0.154 \text{ in.}$$

Thus by computing the displacement of a point on a photograph, we can, if desired, plot the correct position of the point on the photograph.

12.12. Tilt.—In taking vertical photographs the ideal position for the optical axis, at the instant of exposure, is truly vertical; i.e., the principal point and plumb point coincide. This condition, however, seldom exists. The angle of inclination of the optical axis and the vertical is called the *tilt*. Experience shows that the tilt may be kept within 1 degree, and seldom exceeds 3 degrees. When tilt exists, the film, which is at right angles to the optical axis, is not truly horizontal. This causes the image of the object on the film to be displaced. If the angle of tilt is known, the amount of displacement may be computed; but there is no simple way of determining the angle of tilt. To determine the tilt of a photograph the images of at least three ground points, the positions of which have been determined by ground surveying methods, must appear on the photograph. Then the tilt may be determined analytically, graphically, or by rectifying instruments. Displacement caused by tilt is radial about the isocenter.

12.13. Effect of Displacement on Plotting Caused by Relief and Tilt.*—Since the displacement caused by relief is radial from the plumb point, and that caused by tilt is radial from the isocenter, the positions of these points are indeterminate unless the tilt is known. In graphical plotting, however, relief and tilt are both assumed to be **radial** from **the principal point.**

---

*See Higher Surveying by Breed and Hosmer, Vol. II, Eighth Ed., Arts. 4-24—4-28 for more complete text on displacements caused by relief and tilt.

With this assumption, the resulting error in graphical plotting is negligible when both relief and tilt are small.

12.14. Ground Control.—Ground control is the term applied to those points that are identifiable in the photographs and whose horizontal positions and/or elevations referred to a selected datum are known. Their locations and/or elevations are determined by ordinary ground methods of surveying. The amount of ground control may consist of many or few points according to the accuracy required and the mapping method to be employed.

In deciding how the control points should be located it should be observed that a minimum of three points known in position are necessary to fix independently the position of a vertical photograph from which a planimetric map is to be compiled. Where stereoscopic processes are to be used, at least two points known in position and three points known in elevation are required in the area of common overlap. A fourth point of known elevation is also desirable for stereoscopic mapping to provide a check on the elevation of the other points.

The location of control points is further influenced by the method to be used in compiling the map. It should be remembered that triangulation stations are usually elevated points and that these are subject to large relief displacements in the pictures. Traverses, on the other hand, may be run along highways, in valleys, near to the level of the selected datum, and therefore be nearly free from this objection. Differences in relief are of minor importance in using stereoscopic instruments, since they are compensated for in the stereoscopic plotting process.

A control traverse with levels has the advantage that it may be run by an ordinary surveying party with the usual equipment, but a triangulation system of the order of accuracy required usually takes less time and is more economical provided the terrain is suitable for triangulation.

In taking levels, points should be selected which are easily identified in the pictures and which are placed so as to give the best determination of tilt. For plotting with stereoscopic machines, the points of known elevation should be chosen so that they will fall near the four corners of the stereoscopic model (Art. 12.20).

The photographs are often taken before the ground control has been established, though sometimes traverses are run before the flight is made and the picture points tied in afterwards. The requirements for control point locations are sometimes such that the best locations cannot be selected until after the pictures have been inspected.

12.15. INTERPRETATION OF AERIAL PHOTOGRAPHS.*—In a single aerial photograph there is a wealth of detail which can be interpreted by direct inspection or by association with other familiar objects. The interpretation is greatly assisted, however, when a pair of photographs is viewed stereoscopically (Art. 12.20). Whenever possible the use of a stereoscope is recommended.

Man-made features, such as roads, railroads, canals and buildings, usually stand out quite clearly in aerial photographs. Roads of concrete or gravel appear as white bands. Bituminous roads show darker in color. The importance of a road is not always indicated by its prominence in the photograph. An unsurfaced local road may be more striking than a dark-surfaced paved road, particularly if the latter is in a wooded area. On large-scale photographs, oil stains on concrete pavements, and to a lesser extent on bituminous types, define the paths of heavy traffic flow. In areas laid out by Public Lands system roads will appear in a rectangular pattern usually following section lines; most of these roads are secondary highways. The more important highways are usually laid out along direct routes cutting diagonally across section lines.

Railroads appear as regular scars across the landscape, consisting of long tangents connected by flat curves. The track will appear as a dark band, and at some scales the rails can be distinguished. Since railroad gradients are quite flat there will

*See Manual of Photographic Interpretation, American Society of Photogrammetry, 1960.

be frequent cuts and fills. Railroad bridges will usually be narrower than highway bridges and will be found over or under highways as well as over streams. Sidings and station buildings will appear along the track and trains or standing freight cars may appear in some photographs.

Streams and other bodies of water usually appear black but they also appear white if the photograph was taken in such a position that the rays of the sun were reflected from the water into the lens. A body of water may appear black in one part of the picture and white in another. If the water is muddy it has a tendency to photograph lighter than if it is clear. Water also appears light in color when a bright sky is reflected. Deep water usually appears darker than shallow water. In clear, still water features on the bottom, such as boulders, shoals, logs, etc., may show in the photograph. Streams are usually well defined by black twisting lines and by the presence of dense vegetation. Drainage ditches are detected by their straightness and regularity. Canals are also regular in shape. In hilly country a canal or irrigation ditch will follow along the contour of the terrain. A stereoscope is of great assistance in tracing drainage patterns and drainage areas.

Shadows play an important part in photo interpretation. In order to obtain the most natural effect from a photograph it should be viewed so that the shadows are falling toward the observer; otherwise the relief may appear in reverse, the valleys looking like ridges and the ridges like valleys. Vegetation looks dark on photographs partly because its green color reflects less light than other colors and partly because the upper branches of the tree cast a shadow on those below. Evergreens appear somewhat darker than deciduous trees. In winter the evergreens stand out and the other trees almost disappear. For preparing contour maps photographs taken in the winter or early spring have the advantage that the ground surface is most exposed.

Tall grasses or crops cast longer shadows than short ones and therefore appear darker on the photographs. Different types of crops and those cut and uncut can be distinguished by differences in color. These color differences delineate the boundaries of fields and also serve to identify property boundaries. Fence lines may be detected at field boundaries or at the edges of cleared areas. The height of trees is indicated by the length of shadow, and the presence of low bushes is evidenced by the absence of shadows.

Some idea of the shape and types of bridges, water tanks, towers and buildings is revealed by their shadows. A transmission line location, for example, will appear on the photographs as a cleared swath with an occasional pair of small squares indicating tower foundations. The line itself will be practically invisible except for the shadows cast by the towers.

Shore lines with beaches appear prominently in photographs, but shore lines in marshy areas where both the water and marsh vegetation print in dark color may be difficult to trace. Gravel pits and strip mines appear as irregular white patches of extreme prominence.

An observer skilled in geology and land forms can tell much of the geology of the region which he sees in the photograph. Also he can ascertain the kind of soils that exist in different areas and the presence of sand and gravel materials useful in construction. Soil maps have been prepared of large areas of the country. Color photography greatly enhances photographic interpretation. See Manual of Color Aerial Photography.

12.16. PHOTOGRAMMETRIC MOSAICS AND MAPS.—A *mosaic* is a continuous representation of the ground obtained by piecing together individual photographs into a composite picture. A mosaic can be assembled to almost any desired degree of accuracy. The cost, however, increases as the standard of accuracy is raised.

Some of the advantages of mosaics are low cost, rapidity of reproduction, completeness of detail and ease of understanding. They are more readily understood by the non-technical man than engineers' plans. In estimating timber, in traffic studies, and in many other applications, the mosaic is often preferable to a line map.

The following terms may be used to define the different types of photographic mosaics and maps.

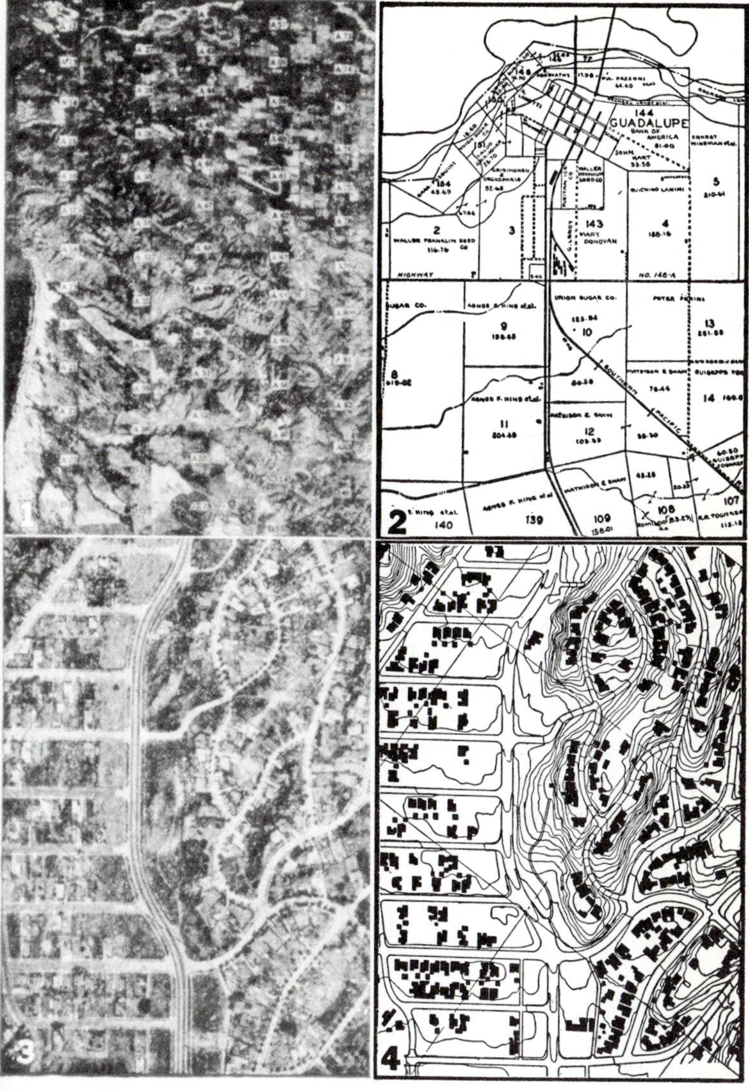

FIG. 12.16.1. MAPS MADE FROM AERIAL PHOTOGRAPHS. (1) INDEX MAP, (2) PLANIMETRIC MAP, (3) MOSAIC, (4) TOPOGRAPHIC MAP.

(Courtesy, Fairchild Aerial Surveys, Inc.)

Uncontrolled Mosaic.—A representation of the ground made by matching aerial photographs without reference to ground control points.

Controlled Mosaic.—A representation of the ground made from aerial photographs by bringing them to a uniform scale and fitting them to ground and picture control stations (Art. 12.18). Inaccurately called a *photo map* (Fig. 12.16.1-3).

Planimetric Map.—A map showing the natural or cultural features (or both) in plan only, sometimes called "line map" (Fig. 12.16.1-2).

Topographic Map.—Map made from data derived from aerial photographs showing contour lines and planimetric data (Fig. 12.16.1-4).

Orthophoto Map—An orthographic projection obtained by differential rectification directly from the stereoscopic projection of overlapping photographs. It has the appearance of a photograph with the complete information content thereof without being a perspective image. It is made by one of several types of orthoprojectors recently developed.

The *index map* (Fig. 12.16.1-1) is a photograph of the individual prints arranged in the same relative positions as taken during the flight. Numbers on the negatives identify the prints in the order taken. The index map is useful for finding the particular pictures covering a given area, and for planning the construction of a map.

12.17. Orienting Photos and Compiling Maps by Graphical Methods.—All photographs are true **perspectives,** whereas maps are **orthographic projections.** The perspective, or photograph, is a view in which the rays of light from the object points converge to a point, the lens of the camera, and diverge from the lens to the negative. With the exception of the ray of light that lies in the optical axis of the lens, the rays are not perpendicular to the negative. An orthographic projection is a map on which all objectives are represented as they would appear if the eye were on a line everywhere perpendicular to the plane of the map. The map compiler makes the orthographic projection map from the perspectives (photographs).

12.18. Transferring Details to the Tracing.—After the skeleton of the map has been prepared, the detail can be filled in from the original photographs. In each case the photograph is placed under the tracing cloth and oriented with respect to those points the positions of which are fixed. Detail is taken off along the line joining any two of these points. The print is so oriented that one point on the photograph lies under its map position and the second point lies on the line joining the two points on the map. Differences in scale between print and plot are then corrected by estimation during the compilation process, due regard being taken for any known relief differences along this line.

In the case of controlled mosaics, there will be no apparent difference in scale between the positions of the control points on the print and on the plot. Detail can be taken from the mosaic or from rectified prints, and a planimetric map drawn. The operator must use his best judgment, based on knowledge of the terrain, in selecting the positions of those details where mis-matches occur as a result of relief.

The tracing of detail is greatly simplified by the use of photographic apparatus such as the *Vertical Reflecting Projector, Vertical Sketchmaster* and *Rectoplanigraph.* These instruments project an image of the photographs on to the drawing. By adjusting the machine or the table top, the image may be brought into the scale of the map by fitting images of control points to their plotted positions on the map.

12.19. Contours.—The contours on a map are obtained by means of the stereoscopic principle applied to two overlapping photographs (Art. 12.28). In order to do this there must be sufficient ground control to enable the stereoscopic pairs to be set up in the stereoscopic plotting machine at proper scale and orientation.

Aerial contact prints, enlargements and precise mosaics are often used as plane table sheets, especially in reasonably flat country where the picture or mosaic if properly rectified becomes a true map. A photographic plane table sheet has the great advantage that all of the planimetry is already on the map sheet in correct scale and relationship. The sole task remaining for the topographer is to add the contours and complete the planimetry in spots where it is not properly recorded by the photograph (such as under trees) and adding geographic names, political boundaries and similar features. The photograph is frequently prior-mounted, either on composition board or on an aluminum sheet.

12.20. STEREOPHOTOGRAMMETRY*—is the science of producing maps from photographs by the application of the principles of stereoscopy. The method involves the study of stereo-pairs of photographs by means of the stereoscope, an instrument which shows the terrain in three dimensions.

In aerial stereo-surveying it is customary to take the photographs from an air base of such length that about 60% of the terrain shown in one picture will show again in the adjacent one. Photographs taken in the above manner are referred to as *overlapping* or *stereo-pairs.*

A close examination of a stereo-pair of photographs would show that while in general the overlapping portions appear alike they are actually slightly different because they were taken from different viewpoints (from each end of a short base).

In natural or unaided binocular vision the interocular distance is the base; therefore when an object is observed a slightly different view of the object is seen by each eye. These two views of the same object seen by two different eyes are resolved into a single view in which there is an appearance of solidarity. Thus a sense of relief or depth results which is not apparent when the object is seen with one eye only.

The same impression of binocular vision may also be obtained artificially by viewing with a stereoscope two photographs each similar to the view seen by each separate eye in natural vision. The two photographs seem to merge or fuse into a single model of the object which stands out clearly in three dimensions.

---

*See Higher Surveying by Breed and Hosmer, Vol. II, 8th Ed., Chap. 5 for principles of stereophotogrammetry.

12.21. Stereoscopes.—When overlapping aerial photographs are to be viewed stereoscopically, it is often impractical to do so with unaided eyes. An instrument designed to assist an observer in combining the images from a stereo-pair of pictures into a single view having the effect of solidity or depth is called a *stereoscope.* Stereoscopes are commonly of two types: lens or mirror. Sometimes the two types are combined.

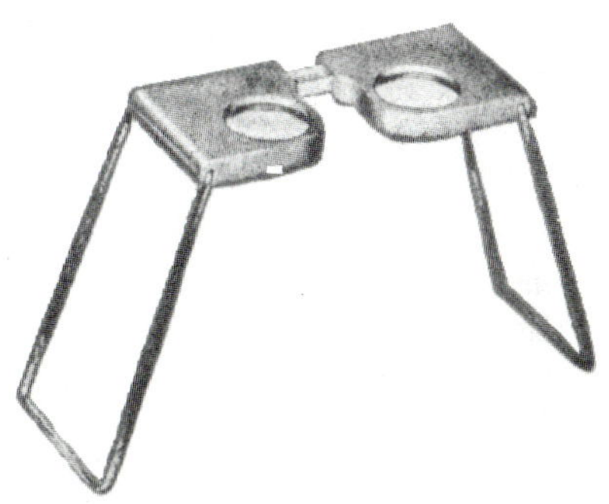

FIG. 12.21.1. LENS STEREOSCOPE.
(Courtesy, Abrams Instrument Corporation.)

Fig. 12.21.1 illustrates a folding, pocket-size lens or "direct vision" stereoscope and Fig. 12.21.2 illustrates a magnifying mirror type stereoscope. In aerial photographs 9 by 9 inches in size with a 60% overlap (60% of the area of one photograph included in the other photograph of a stereo-pair), the entire stereoscopic model may be viewed in the mirror stereoscope, but only a portion of the overlap may be viewed with one setting of the lens type.

For most satisfactory use, the alignment of stereo-pairs of pictures should be done carefully. Corresponding image points should be separated by the amount for which the instrument is designed. The lens stereoscope shown in Fig. 12.21.1 calls for a separation of 2½ inches and the mirror stereoscope shown in Fig. 12.21.2 requires a separation of 9¾ inches. The alignment of the photographs is best done through use of the perspective centers and their conjugate images. In an aerial photograph taken with the optical axis truly vertical, the perspective center is at the center of the photograph. This point (principal point) is found through use of fiducial marks

FIG. 12.21.2. MAGNIFYING MIRROR STEREOSCOPE.
(Courtesy, Fairchild Camera and Instrument Corporation)

placed either near the corners or along the sides. In Fig. 12.21.3 p_1 is the perspective center of the vertical aerial photograph I and p_2 is the perspective center of photograph II. p_2' is the image on photograph I of p_2, and p_1' is the image on photograph II of p_1. p_2' and p_1' are often called conjugate images of the principal points.

A satisfactory method of aligning these photographs for stereoscopic examinations is to take a sheet of drawing paper, say of size 11″ by 22″, and draw a bisecting line parallel to the long dimension. The principle to be followed is to have the perspective centers and their conjugate images fall along this line and to have corresponding images such as p_1 and p_1' separated by the amount required by the stereoscope. The axis of the stereoscope should be placed parallel to the edge of the sheet of drawing paper. The middle view in Fig. 12.21.3

shows photographs aligned for examination by means of a lens stereoscope and the lower view Fig. 12.21.3 shows the same pair adjusted for a mirror stereoscope.

Both lens and mirror stereoscopes provide a means of viewing aerial photographs stereoscopically; each has different

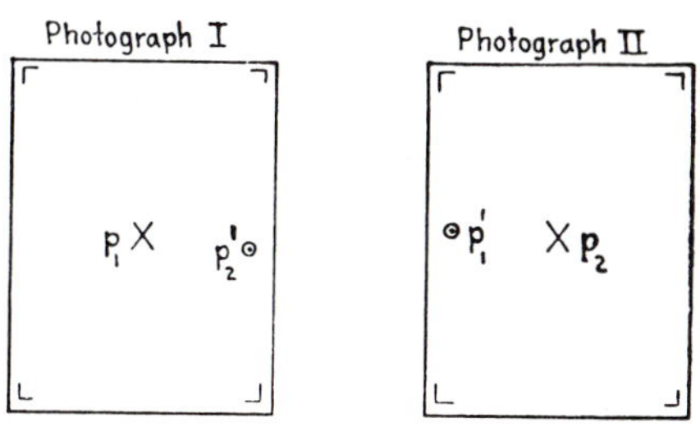

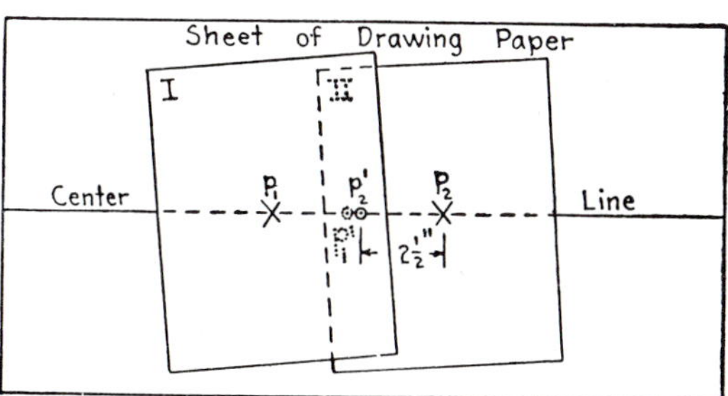

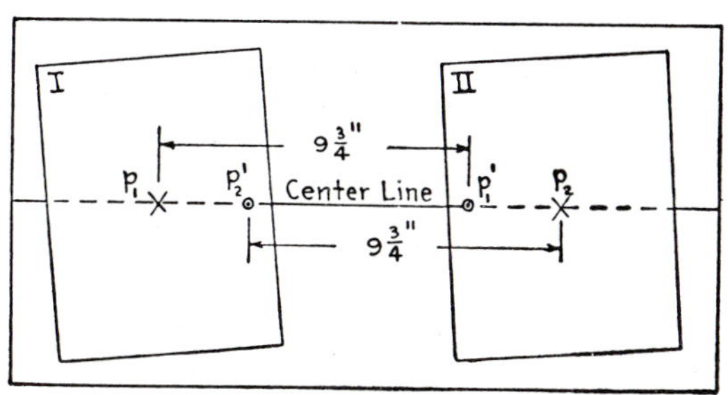

FIG. 12.21.3. ARRANGEMENT OF PHOTOGRAPHS FOR VIEWING IN STEREOSCOPES.

practical advantages. Lens stereoscopes cover a somewhat limited field, but the details are more closely defined. This type is to be preferred when a minute examination of the features is desired, such as when selecting ground control stations. Mirror stereoscopes yield a wider field of view and are preferable when such features as geological forms are being studied on the photographs. Mirror stereoscopes, however, have attachable binoculars to provide any desired magnification.

12.22. Stereoscopic Plotting Instruments.*—The stereoplotter is an optical and/or mechanical instrument of high precision for preparing maps from aerial photographs. Its function is to (1) create a precise three-dimensional scale optical model of the area common to two overlapping aerial photographs (2) provide a means for the operator to view this spatial model (3) provide a means for measuring the model and plotting its orthographic or map projection. In forming the model, the plotter re-creates the exact relative and absolute orientation of the two photographs. Such factors as tilt, scale variations, and conversion of perspective projection to orthographic projection are compensated for automatically in the stereoplotting process.

There are a number of different types and makes of plotters available to the engineer, each performing the same basic function.

Instead of plotters, some of the major plotter manufacturing companies are named, since most of them produce several types and models of plotters to meet a variety of purposes and accuracies. This list can only be considered as a representative sample and is rather incomplete. For more details, the reader is advised to contact the firms or their U.S. representatives directly. Most of the commonly used instruments are produced by Bausch and Lomb (U.S.A.), Kelsh (U.S.A.), Wild (Switzerland), Kern, (Switzerland), Zeiss (W. Germany), Jenoptic (E. Germany), OMI-Nistri (Italy), Galileo-Santoni (Italy), SOM (France), Hilger & Watts (U.K.)

---

*See Manual of Photogrammetry, 3rd Ed., Chaps. XII to XIV, American Society of Photogrammetry, 1965, for description of stereoscopic instruments.

Practically all topographic mapping done by photogrammetric methods is accomplished by means of stereoplotters. These instruments have been refined to give a high degree of precision. For many purposes topographic maps can be prepared by photogrammetric methods in less time and at less cost than by ground survey methods. The better plotters are capable of producing large scale maps with one, two or five-foot contours suitable for use as a basis for design and construction drawings.

12.23. The Kelsh Plotter.—This plotter is typical of those applicable to topographic mapping for civil engineering projects. Its operation is illustrated in Fig. 12.23.1. The two pro-

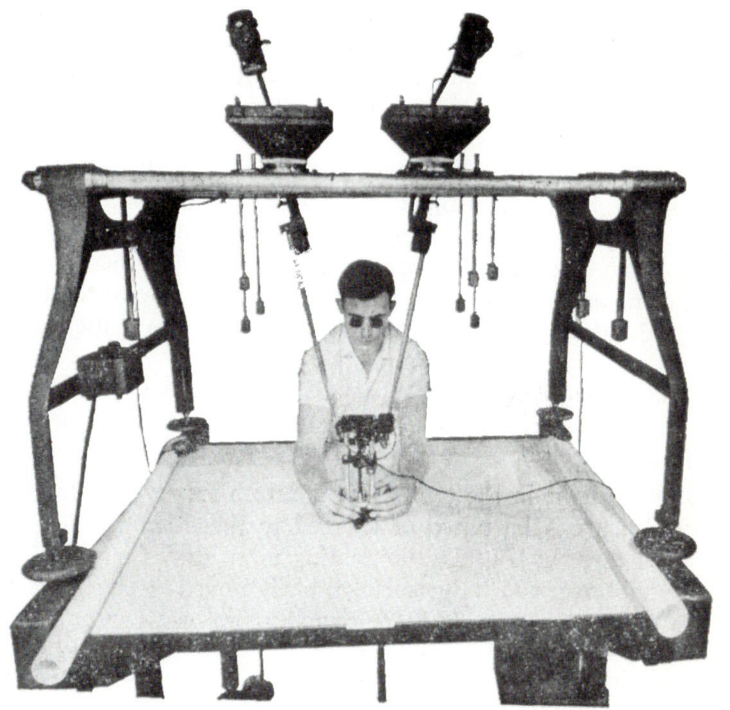

FIG. 12.23.1. KELSH PLOTTER.
(Courtesy, Michael Baker, Jr., Inc.)

jectors, which are functionally identical to the aerial camera, are adjusted so as to have the same spatial orientation as existed at the time of two successive exposures of the taking camera in the air, except that all ordinates on the model are reduced in the ratio of the plotting scale to the natural scale. The complete orientation of the spatial model is achieved in two stages known as relative and absolute orientation: first, the two projectors are brought into the same relative angular relationship as existed in the aerial camera when the two successive exposures were made; secondly, absolute orientation is obtained by adjusting the pair of projectors as a whole so that the scale, position and orientation of the spatial model agree with the plotted positions of ground control points.

The projectors receive positive transparencies on glass plates called diapositives, which are prepared by contact printing from the original full-sized aerial film negatives. The distance from the diapositive to the interior perspective center of the projector lens is made the same length as that of the taking camera. An adjusting cam, which is activated by the arms which run from the lens to the tracing table, automatically raises or lowers the lens minutely to compensate for lens distortion. The light sources are also actuated by the connecting arms so that the light is directed to the tracing table. In Fig. 12.8 the operator is shown holding the base of the tracing table.

The Kelsh Plotter projects its conjugate images on to the small screen on top of the tracing table in complementary colors, such as red and blue. The operator receives the stereoscopic impression by wearing a pair of glasses with red and blue lenses, which insures that each eye will receive the projected images from only one of the projectors. Under these conditions the mind will fuse the separate sets of images received by each eye into one spatial model.

A dot of light in the center of the screen serves as a floating mark which can be raised or lowered to the desired elevation. A pencil directly beneath the floating mark draws on the map manuscript over which the tracing table is moved by the operator as he causes the floating mark to follow the contour or planimetric feature.

A dial indicates the exact elevation (in feet) of the floating mark, so that when the operator raises or lowers the tracing table to touch the floating mark to the ground, he can read the elevation on the dial. So, too, when he wishes to draw a contour line at a given elevation he sets the tracing table to the correct height and moves it about, carefully keeping the floating point in contact with the ground while the pencil below draws the contour line on the map base.

A pantograph attachment may be added to the Kelsh Plotter which permits a choice of drawing scales.

12.24. Testing Topographical Maps Made from Aerial Surveys.—A commonly accepted specification for vertical accuracy is as follows: "90% of all contours shall be within ½ contour interval of their correct position. The remaining 10% shall not be in error by more than one contour interval."

A commonly used specification for horizontal accuracy is as follows: "90% of all identifiable horizontal features shall be within .025 inches, their correct position as determined from the nearest grid lines and the remaining 10% shall not be more than .05 inches in error."

Contour maps made from aerial surveys are usually tested by running a traverse and profile between two of the control points upon which the map was constructed. A check plot is then compiled showing the ground-surveyed profile as a horizontal line and the map profile as a departure therefrom. The limits of timbered areas are shown on this profile. The quality of the map is judged by measuring the increments of the profile which are in excess of the one-half contour intervals.

On small scale maps it is rather unusual to encounter difficulty in obtaining the required horizontal accuracy. On very large scale maps, the horizontal accuracy of the map is generally checked during the "field completion" survey which is made to check the map for such things as building locations obscured by vegetation, fence lines, mine symbols and the details of road intersections. In the course of this survey the horizontal accuracy and vertical accuracy are concurrently checked by sample traverse and level lines usually run between control stations.

12.25. APPLICATION OF AERIAL PHOTOGRAMMETRY TO ENGINEERING.—During the early stages of its development, the application of photogrammetry to engineering was usually limited to the preparation of reconnaissance and preliminary location maps. With the greatly increased accuracy of modern equipment and methods, photogrammetric surveys yield sufficient accuracy for many of the necessary design and construction measurements. Very large scale design maps such as 1″ = 20′ to 1″ = 100′ with one or two foot contour intervals are practical and economical. Precision methods will produce a map from which design measurements may be accurately scaled. Such mapping is more difficult than smaller scale mapping for location studies due to the difficulties of eliminating the effect of ground cover, and the necessity of flying the airplane at low altitude.

Photogrammetric methods are also extensively used for making cross-section surveys for the determination of earthwork quantities. A series of elevations and offsets from a baseline may be determined directly with a stereoplotter. Advantage is taken of the fact that greater accuracy is possible in reading the elevation of a discrete point than in tracing a contour. The cross-section data may be recorded manually or automatically by suitable instrumentation. By rephotographing a construction project area after the work has been completed, a new set of cross sections may be obtained and construction quantities computed for payment purposes.

By linking a photogrammetric stereoplotter with an electronic computer, the data from the stereomodel can be converted to numerical or digital form for acceptance by the computer. One such concept, known as the Digital Terrain Model (DTM), has been developed in the Photogrammetric Laboratory at the Massachusetts Institute of Technology, and is available commercially. Such a system is illustrated in Fig. 12.25.1. The principal components are a stereoplotter for forming a spacial image of the terrain, a scanning device (co-

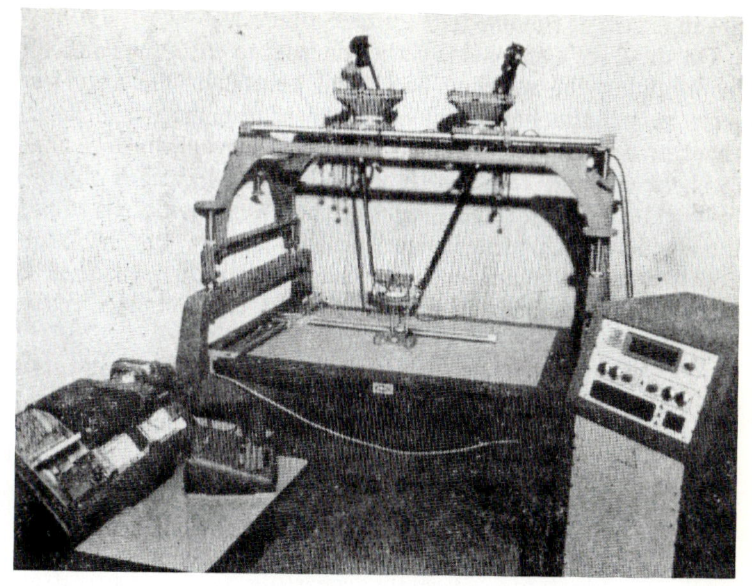

FIG. 12.25.1. STEREOPLOTTER AND COMPUTER OUTPUT SYSTEM.
(Courtesy, Keuffel & Esser Co.)

ördinatograph) for locating x and y coördinates, a digit counter attached to the plotter tracing table for determining z coördinates (elevations), a digital control unit for digitizing input, and a card punch output unit. Other systems also have typewriter and punched or magnetic tape output. Where accurate contour maps are available, these may be used as the source of digital data.

The band of terrain for which topography is desired is represented in digital form by obtaining the xyz coördinates for a large number of points in the area. Essentially the same data normally represented by contours is expressed in digitized form. With the stored digital terrain model and appropriate computer programs, the engineer can make numerical evaluations of problems involving terrain without drawing a map.

This concept has been further developed in both software and hardware, leading to such instruments as the analytical plotter, a computer plotter combination.

12.26. Other Engineering Applications of Photogrammetry.—For many engineering projects, the two main advantages of photogrammetry, namely (1) instant collection of a complete set of information and (2) no physical contact with the object, are used in a variety of fields. It is far beyond the scope of this book to go into details of these mostly terrestrial applications of photogrammetry.

The research in this area has progressed to such an extent, that even an ordinary camera can be utilized for precision work, provided that sufficient object control and a computer for compilation are available. For many projects, however, special stereo cameras exist, rigidly inter-connected and with a fixed relative orientation. With such a stereo camera, the advantages of recording a traffic accident photogrammetrically are obvious, because there is virtually no time required and the scene of the accident can be cleared immediately while any evidence can be obtained at a later time from the photography.

Objects which are not easily accessible, such as excavations in open pit and underground mining, architectural facades, structures, sculptures etc. are often surveyed by photogrammetry, as also fast moving and changing events in laboratories, such as vibrations, deformations due to loading, liquid surfaces, and many more.

12.27. Analytical Photogrammetry.—In the analytical approach to photogrammetry, the true positions of points are determined mathematically from measurements made on the photographs. For example, the coördinates of a selected image point are measured with great accuracy by means of a precise stereo-comparator (Vol. II, Art. 5-24), and the true coördinates then found by geometric relations determined from control points on the photographs. The equations, however, are often quite complex and require considerable time to compute.

With the increasing availability of electronic digital computers, the calculations can be made rapidly and economically. The data are punched on tape or cards for input to computer. The spatial coördinates of the desired object points are then determined by the computer according to stored instructions. The high accuracy capabilities of the analytical approach make it especially applicable to such photogrammetric problems as aerotriangulation for carrying control through a series of overlapping photographs, and for tracking missiles and aircraft.

12.28. Aerotriangulation.—This is a procedure of obtaining coördinates of points by photogrammetric means covering large areas. Although historically developed for providing the necessary ground control for photogrammetric map compilation, analytical photogrammetric procedures and modern instrumentation made it possible to achieve accuracies high enough for use in 2nd to 3rd order geodetic control densification. Aerotriangulation procedures include methods such as radial triangulation, slotted templet layouts, cantilever extension, bridging, analogue, semianalytical and analytical strip and block adjustments. These vary greatly in approach, instrumental, computational and control requirements as well as resulting accuracies. The reader is advised to consult with special literature and experts in this field before getting involved in aerotriangulation.

PART III

COMPUTATIONS

CHAPTER XIII

COMPUTATIONAL TOOLS AND THEIR USE IN SURVEYING

13.1. General Remarks.—The ultimate purpose of most surveys is to obtain certain numerical results. In the section on Surveying Methods it has been pointed out that in all surveys there should be a proper relation between the precision of measurement of the angles and distances. To secure final results to any given degree of precision, the measurements in the field must be taken with sufficient precision to yield such results. In computing from a given set of field-notes the surveyor should first determine how many digits or figures he should use in the computations, the aim being to obtain all the accuracy which the field measurements will yield without wasting time by using more significant figures than are necessary, thereby arriving at misleading results.

Final results should be carried to as many significant figures as the data will warrant and no more. In order to insure the desired precision in the last figure of the result it will usually be necessary to carry the intermediate work one place further than is required for the final result.

13.2. Significant Figures.—The number of significant figures in the result of an observation is the number of digits that are known. For instance, if a distance is recorded as 24,000 ft. when its value was obtained to the nearest thousand feet only, it contains but two significant figures. The zeros are simply put in to show the place of the decimal point. If, however, the distance has been measured to the nearest foot and found to be 24,000 ft. there are five significant figures, for the zeros are here as significant as the 2 or 4. Similarly a measurement such as 0.00047 contains but two significant figures, the zeros simply designating the position of the decimal point, for, had this same value been recorded in a unit 1/100000 as large the result would have been 47.

Again, if a series of measurements is taken between two points to hundredths of a foot and three of the results are 4.88, 5.05 and 5.00 it is evident that each of these distances contains three significant figures; if each one is multiplied by 1.246 the results are 6.08, 6.29 and 6.23, respectively. But had the measurements been taken to the nearest tenth of a foot and found to be 4.9, 5.1, and 5.0, these values when multiplied by 1.246 should appear as 6.1, 6.4 and 6.2. This illustration indicates the proper use of significant figures. Since the measurements 4.9, 5.1 and 5.0 are good to only two significant figures (maximum error = 0.05) the factor 1.246 should be taken as 1.25. Similarly in the use of such a constant as 3.141592654 (π) it is a waste of time to use more significant figures in the constant than are known in the number with which it is to be combined.

Conversely, if in using a computational device (a large computer or a hand calculator) which works with many more digits than the input data warrants, one must be aware of the need to lop off and round off at the end of the computation to give a correct answer. For example to find the area of a rectangular lot which measures 80.083 ft. × 99.963 ft., a result of 8005.336929 ft.^{2} or 0.1837772481 acres is misleading. Correctly it is 8005.3 ft.^{2} or 0.18378 acres. The area of a circle whose measured radius is 37.75 ft. should not be given as 4476.965881 ft.^{2}, but rather as 4477 ft.^{2} Of course speaking of a theoretically exact circle of exactly 37.75 ft. in radius, one could give a theoretically exact area of 4476.965881 ft.^{2} or as many more digits as one may care to use.

13.3. Classification of Computational Tools.—In increasing order of capability for surveying calculations, the following types of tools can be listed:

1) Graphs and tables
2) Logarithmic tables
3) Slide rule
4) Mechanical calculators
5) Electro-mechanical calculators

6) Electronic calculators
7) Programmable electronic desk calculators
8) Electronic computers.

13.4. Graphs and Tables.—The simplest computational tools are graphs and tables where a certain type of arithmetic computation has been presented in graphical or numerical form for further use.

As an example for a graphical problem solution, a hyperbola sheet can be used directly for the determination of the area of a triangle. If the coördinate axes of a hyperbola coincide with its asymptotes then the product x · y is constant. Figure 13.4.1 shows a series of hyperbolae with different constants.

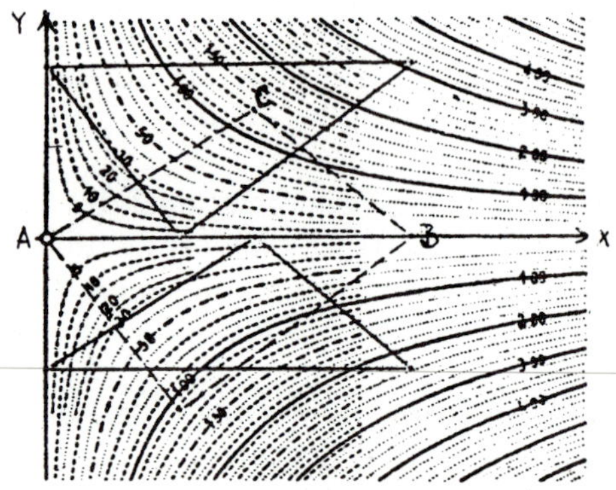

FIG. 13.4.1. HYPERBOLA GRAPH.

The area of the triangle ABC is directly read at point C, provided that AC is parallel to the X-axis, B is on the X-axis and A on the Y-axis.

Tables are widely used, particularly in connection with calculators. In these tables certain arithmetic, trigonometric or other functions are numerically represented for a complete set of arguments. Such tables include the trigonometric functions sine, cosine, tangent, cotangent, arithmetic functions, such as the square of the argument, or its square root, arc lengths and many others.

13.5. Logarithmic Tables.—The concept of logarithms is based on a few mathematical properties such as

$$\log (a \cdot b) = \log a + \log b$$
$$\log (a/b) = \log a - \log b$$

Virtually any number can be used as a base for logarithms, but in practice either logarithms of the base ten or logarithms with the exponential function are used, the latter being considered as "natural" logarithms.

The use of logarithms reduces multiplication and division to simple addition and subtraction, but cannot handle addition and subtraction as such. The logarithms were in common use prior to the age of electronic calculators and computers and were the main computational tool for the surveyor and many other professions. Even up until the advent of pocket size calculators, logarithms were frequently used for field computations. However, within the past few years they have become obsolete because their use is rather time consuming. For this reason, no further details about logarithms are given.

13.6. The Slide Rule.—For fast computations and lower accuracy (three significant figures) the slide rule has been exclusively used, but is now also being replaced by pocket-size calculators.

Based on logarithms, the slide rule presents a mechanical means of adding and subtracting distances which represent logarithmic values. It is therefore extremely convenient and fast for multiplications and similar operations. Due to its limited size, which is designed to fit into pockets and briefcases, the reading accuracy of the ordinary 10-inch slide rule is limited to 3 and 4 significant figures. The 5-inch slide rule is accurate to 2 or 3 figures, and the 20-inch slide rule to 4 and 5 figures.

This limited accuracy and the fact that the decimal point has to be placed correctly for the result by rough estimation requires a certain practical experience for the user. Again, the pocket calculator is replacing the slide rule because it is easier to use.

13.7. Mechanical Calculators.—Up to the mid 1960's mechanical (rotary) calculators were the prime tool for surveyors. They are still used to a fair extent, especially in the field when there is no electrical power supply. Used with tables of natural trigonometric functions they are capable of solving most surveying problems. Without distinguishing the different types and designs, it can be stated that these machines usually have three registers, namely setting register, rotation register and accumulation register.

A number can be entered directly into the setting register by sequentially depressing buttons for the desired numerals or by shifting levers to the numerals. One positive (usually clockwise) rotation with the crank handle will put this number into the accumulation register while the rotation register will indicate the number "one" for one rotation. If the accumulation and rotation registers had been shifted by one place to the left or right, the same turn of the crank handle would have resulted in a multiplication by 0.1 or by 10 respectively, with these factors appearing in the rotation register. The decimal place has to be set manually. Any digit that is wrongly entered into the setting register can be altered as long as the handle has not been turned.

All four basic arithmetic operations can be considered as some form of addition, which is exactly how they are performed. No other operations are possible, although one can obtain the square root, for instance, by selective division.

Addition is achieved by setting one number into the setting register and subsequently placing it into the accumulation register by one positive turn of the handle. A second number is then set into the setting register and added to the first one by another positive rotation of the handle. The resulting sum is shown in the accumulation register, while the rotation register will show the number two, indicating two positive rotations. This procedure can be continued at will, to add more numbers.

Subtraction is nothing else but a negative addition. Therefore once the first number has been placed in the accumulation register the second number is set into the setting register and subtracted from the first one by executing one negative turn of the handle. Again the resulting difference will appear in the accumulation register, while the rotation register will show a value "zero", as the positive and negative turn of the handle cancel the rotations. If a third number is to be subtracted from the result, it is again set into the setting register and operated by one negative turn. The accumulation register will show the new result, while the rotation register should indicate a "minus one." However, since the calculator does not display signs, the decadic supplement will appear, which in this case would be a "nine" in every digit.

It can also happen that the result in the accumulation register becomes negative. In this case again the decadic supplement is displayed (e.g. 999999732 would mean −268). Usually a bell rings, when the accumulation register changes the sign.

Multiplication is only a sequential addition of the same number. Once the number is set in the setting register, each positive turn of the handle will result in a multiplication by one. For example, to multiply 3421 by 824, one would have to set 3421 into the setting register and turn the handle 824 times. A better way would be to turn the handle four times, then shift the accumulation and rotation registers by one place to the right and rotate twice, shift again and rotate eight times. With a little experience one would realize that 8 = +10 −2 and therefore multiply by ten and subtract twice for the same result, thereby reducing eight positive rotations to one shift and three rotations (one positive and two negative).

Division is merely a successive subtraction of the divisor from a given number. Once the result is zero the division is complete. Rather than working from right to left as in multiplication, one works from left to right.

The first number is set and transferred to the accumulation register by one positive turn of the handle. Then both setting and rotation registers are cleared and the second number is placed in the setting register. Then the first numbers of the accumulation register and of the setting register are aligned by shifting and the handle is cranked anticlockwise (negative) until the bell rings. Then one positive rotation brings the accumulated result to a positive value before the register is shifted by one digit to the left. This procedure is repeated until the absolute value of the accumulation register is smaller than the value in the setting register. The result is displayed in the rotation register, with a possible uneven fracture indicated in the accumulation register. For example, 101 divided by 3 would have 101 in the accumulation register, and 3 in the setting register. When aligned, a negative rotation would result in a negative value after one rotation, as 300 > 101. Therefore a shift to the left would give a subtraction of 30 from 101, which would cause a sign change at the 4th rotation (120 > 101). One positive turn would establish 30 in the rotation register and leave 11 in the accumulation register. After a shift, the bell will ring again at the 4th negative rotation, and after one positive rotation result in the value 33 in the rotation register and the value 2 in the accumulation register. Therefore, the final result would be: $101 \div 3 = 33 + 2/3$. If more decimal places are desired, both numbers can initially be set further to the left, thereby giving a result like $101 \div 3 = 33.666 + 2/3000$.

13.8. Twin Hand Calculators.—There are some mechanical calculators which have two setting and accumulation registers operated with one handle and rotation register. A small lever is used to select the operation mode, which can be

a) same direction
b) opposite direction
c) one machine only.

This arrangement has the advantage that the same operation can be administered to two different numbers simultaneously (e.g. in traversing the cosine and sine of an azimuth can be multiplied by the same distance to immediately give latitude and departure).

13.9. Electro-Mechanical Calculators.—These machines are a further development of the mechanical calculators whereby the hand operation motion is replaced by electrical operation. The machines remain noisy and slow and have little added features. Some are equipped with a printer to eliminate readout and manual recording.

13.10. Electronic Calculators.—In the early 1960's nonprogrammable electronic calculators appeared on the market. This type of calculators is virtually replacing all the tools mentioned in the previous articles. They work noiselessly and with great speed, display a visual result, and are very reliable and accurate. Some have features like $\sqrt{x}$, trig functions, logarithms, π, e^x, x^y, working storage, etc. (see Figs. 13.10.1 and 13.10.2).

FIG. 13.10.1. HEWLETT-PACKARD'S LOW-COST HP-25 PROGRAMMABLE SCIENTIFIC POCKET CALCULATOR.

FIG. 13.10.2. ELECTRONIC DESK CALCULATOR.

(Courtesy, Casio, Inc.)

The input of numbers is achieved with a 10 key setting, with the decimal point being set manually for each number.

By depressing the appropriate key, the arithmetic operation is performed within a fraction of a second and the result replaces the last input number in the visual display. The arithmetic sign and the decimal point are automatically placed either in fixed point notation at the correct position after an arithmetic operation or in floating point rotation, with the decimal point being placed after one digit, while numbers to indicate powers of ten are also displayed (e.g., +3.6257 +02 means 365.27). By using the latter notation the machine can handle very large numbers or very small decimals without losing significant digits.

It should be noted that some machines have limitations concerning the number of digits to be operated. On some machines there will be a shift into floating point notation if the capacity of the machine would be over-run, as $(987654.321)^2$ would show as $9.754610578 \times 10''$. On others the machine would refuse to perform the operation and will give zero as an answer. Most models have a warning system (flashing light, zero output, etc.) if the range is exceeded or if an illegal procedure such as division by zero, square root of a negative number etc. is attempted. Some calculators always round correctly, others do not.

Most non-programmable electronic calculators have one or more storages or memories which can be utilized for storing certain numbers for recall at a later time. This is especially convenient for multiplication with a constant and similar operations. The content of a storage register is usually recalled to the display to be then used in arithmetic operations.

All common surveying problems can be computed using non-programmable electronic calculators, provided they are sufficiently large to handle the requisite number of digits. Certain smaller machines do exist, however, that give trig functions to only 5 decimal digits and will not properly calculate for precise work. For example a multiplication like 1062.817 (sin 31° 15′ 37.2″) would give 551.517, whereas the correct value is 551.525 as found by a larger-capacity machine.

13.11. Programmable Electronic Desk Calculators.—This type of computer is the most suitable one for the surveyor, as it can handle most of the surveying problems very efficiently. By programming a certain series of arithmetic operations, this type of calculator requires no intermediate recording and therefore reduces the possibility of human error.

The difference between these machines and non-programmable ones is storage (or memory) capacity and the ability to automatically perform a series of subsequent instructions. The memory can be used as storage for both the instruction sequence (program) and numerical data such as intermediate results. To properly write a program, experience and knowledge of the particular computer is required. Although a detailed description exceeds the scope of this book, it can be stated that there is nothing magic about programming. A program is simply a listing of logical steps in their proper sequence. An attempt to classify these machines on the basis of their capacity is shown in Table 13.11.1.

TABLE 13.11.1

GROUPING OF PROGRAMMABLE DESK CALCULATORS

Group	Instructions	Storages	Remarks
I	0-100	0-5	With or without program reading facilities
II	100-200	5-20	Few hard wired functions (usually square root)
III	200-500	15-50	Hard wired trigonometric and logarithmic functions
IV	500-2000+	50-250+	Interchangeable blocks of hard wired functions.

By reason of the rapid development of these machines, no particular ones are described in detail, but rather a brief overview is given for each group.

13.12. Simple Programmable Desk Calculators (Group I).—With the restricted amount of program instructions available, only simpler problems can be handled effectively, such as area determination, volumes (cut and fill), curve data, solution of a triangle, stadia reductions, small traverse, resection (see Fig. 13.12.1).

The use of trigonometric functions is extremely limited, as they are not hard wired and their program takes up a major part of the memory space. Any program is entered manually into the machine either with a simple example or as a sequence of instructions. If there is a program recording unit it can be recorded onto a punched card or magnetic card or some similar device after it has been checked. Then it can be read in from this card for further use. The remarks given in article

FIG. 13.21.1. SIMPLE PROGRAMMABLE DESK CALCULATOR.
(Courtesy, Litton Monroe)

13.10 about rounding-off, accumulation, decimal point, etc. apply for these machines as well and are not repeated here.

13.13. Small Programmable Desk Calculators (Group II).—Although these machines have a larger memory capacity, they have only a few hard wired functions, often the square root only. Calculations for traverses and resections, which are rather difficult to achieve with group I calculators can easily be carried out, however. Other survey computations are possible, especially those involving trigonometric functions, since after their inclusion in the programming there is still sufficient memory space available for arithmetic operations.

All these machines have program recording units, and several are equipped with print-out devices, which give permanent records of all entries for checking purposes (see Fig. 13.13.1).

13.14. Medium Programmable Desk Calculators (Group III). —These machines have many hard wired functions, and are

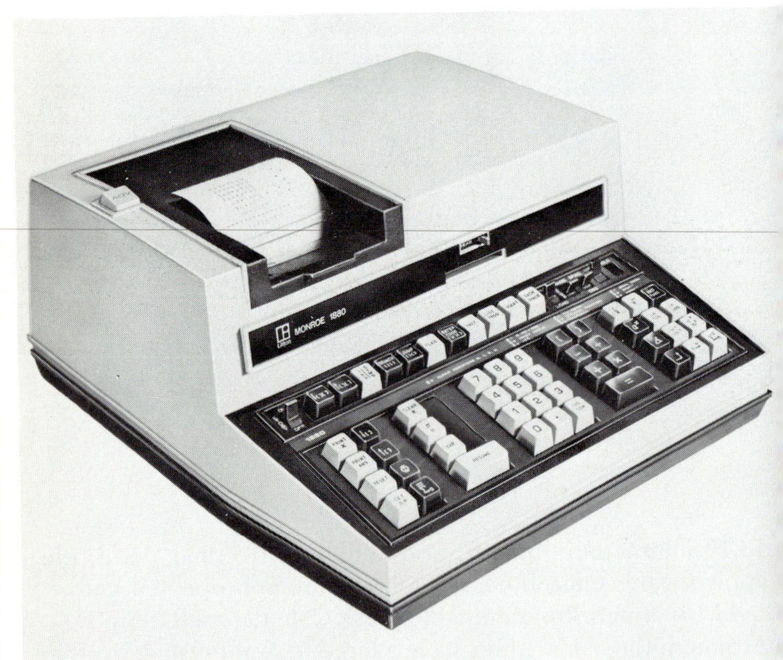

FIG. 13.13.1. SMALL PROGRAMMABLE DESK CALCULATOR.
(Courtesy, Litton Monroe)

therefore capable of handling all common surveying problems, even small adjustment problems. The trigonometric functions are hard wired, usually based on radians. However, a few simple programming steps convert degrees, minutes and seconds (DMS) into radians without difficulty. Other functions include logarithms, although these are not all that necessary for surveyors. Some having hard wired transformations from polar coördinates to rectangular coördinates and vice versa are very useful. Coördinate transformations, all types of coördinate computations and the solution of small systems of linear equations can be achieved with these calculators (see Fig. 13.14.1).

FIG. 13.14.1. MEDIUM PROGRAMMABLE DESK CALCULATOR.

(Courtesy, Toshiba)

13.15. Large Programmable Desk Calculators (Group IV). —The main difference between these calculators and electronic computers is their memory and storage capacity. Interchangeable blocks of hard wired functions permit the same

FIG. 13.15.1. LARGE PROGRAMMABLE DESK CALCULATOR.
(Courtesy, Hewlett-Packard)

basic machine to be used for a wide variety of applications. Printer and plotter attachments make these machines very versatile, and some can be used in direct interaction with electronic computers.

Advanced problems, such as statistical evaluations, small adjustments, photogrammetric strip formation and many others can be solved with these machines quickly and economically (see Fig. 13.15.1).

13.16. Electronic Computers.—Every aspect of our life is influenced to some extent by electronic computers, the field of surveying being no exception. Any type of computation can be handled, whether it involves one or 100,000 unknowns, and numerous program packages are available for all kinds of surveying problems.

13.17. **Evolution of Electronic Computers.**—The stages of technological development of computers are often referred to as "generations". The first computers were of enormous size and low reliability with relatively small, slow memory; they utilized vacuum tubes and were very expensive. In the second generation computers, vacuum tubes were replaced by transistors and diodes, the speed and size of memory were increased and much of the wiring was replaced by printed circuits, thus reducing the physical size of the computer as well as the heat generated. Reliability was increased while costs were reduced. The technological advances incorporated in the current (third) generation of computers utilize micro-miniature circuits and provide almost unlimited storage capacities. The speed of these machines has been increased to such an extent that about 200 remote consoles may be operated simultaneously, as if each were a separate computer. This mode of operation is known as *time-sharing*. Remote time-sharing consoles currently consist of either a typewriter or a teletype unit with the necessary circuitry for communicating with the central computer. The programs and data are typed in on the keyboard, and the results are returned on the same keyboard.

13.18. **Computer Languages—Programming.**—Computer operations are controlled by a "Program" spelling out the procedure necessary to solve a problem. Programs may be written in a variety of languages but all of them must be translated into the language understood by the machine.

The earliest programs were written in machine language

describing exactly, in minute detail, every step to be taken. The first improvement permitted the use of meaningful, easily remembered symbols instead of strings of rather confusing numbers. This improvement also assigned to the computer the task of keeping track of location of data in memory.

More recent improvements included programming languages in which computational operations are specified in everyday terms. Some of these languages are FORTRAN, ALGOL and COBOL. Many very useful programs have been written in FORTRAN to solve geometric problems that are well defined and reoccur frequently in the same form. Although all these languages are suitable for developing classical programs, they do not provide the level of communication between engineer and computer needed to achieve the full capabilities of the computer in engineering practice.

Still later improvements are the development of more understandable, problem-oriented computer languages which use words and terms familiar to the user. For example, one such language written in terms familiar to the surveyor or civil engineer is COGO, for *CO*ordinate *GeO*metry. This language uses such words or commands as Locate, Line, Angle, Azimuth, and Bearing. COGO is a general purpose program that may be used to solve many of the problems encountered by the surveyor. Programs written in any of the other languages are less flexible, and may be thought of as one-purpose programs.

Problem-oriented languages have greatly improved communications, and remote time-shared consoles have reduced problem turn-around time. These improved techniques now make it possible and economically feasible for surveyors to take advantage of the speed, accuracy and economy of computers.

CHAPTER XIV

SURVEY COMPUTATIONS

14.1. Arrangement of Computations.—All surveying computations should be kept in a special computation book. At the head of the page should appear the title of the work, the number and page of the field note-book from which the data are copied, the names of the computer and checker, and the date. The work should be arranged neatly and systematically so that every part of the computations can be traced by anyone who is familiar with such work. Where possible the work should be so arranged that numbers will have to be written but once. Each important value, each column, etc., should be labeled so that it can be readily found. In large offices it is found more satisfactory to use separate sheets for computations. These may be bound in covers, each job being a book by itself. It is important that each separate sheet contain enough information to identify it if it becomes separated from the rest.

When programmable machines are used, special care should be taken to obtain the print-out in a format comparable to the computation book, with proper headings etc. Then all outputs can be kept on record.

If this is often neglected, only the person who performed the computation knows which number represents a desired result. The print-out is then just about as useful as a scribble pad, and retracing of certain steps is next to impossible.

14.2. Checks.—It is very important that all calculations should be checked, not merely at the end of the computation but also at as many intermediate steps as possible. In this way a great waste of time may be prevented and serious mistakes avoided. One good method of checking is to perform the operations when possible by two independent methods, for example. Very often two men do the computing, one man's work acting as a check on that of the other. The two may work by the same or by different methods, and the results may be compared at intervals. **Every part of the work should be done independently, from the copying of data out of the note-book to the final results.** Each computer also should check his work roughly by estimating approximately what the result should be.

The use of programmable calculators greatly eliminates the need for computational checks. Once a program has been checked, the operations will be correct for any subsequent work. It is, however, important to check all input quantities and use an accuracy indicator within the program as a safeguard against blunders.

14.3. REDUCING THE FIELD-NOTES FOR COMPUTATIONS.—Before any of the computations are made the measurements must be corrected for systematic and constant errors, such as temperature of the tape, sag, errors in length of tape, and alignment errors. The errors in the angles are balanced by altering the value of those angles which were taken from short sights since the angular errors are most likely to occur in these. Whenever distances have been measured on a slope, these distances are reduced to horizontal by multiplying them by the versed sine of the vertical angle and subtracting the result from the **corrected** slope distance, the correction for error in the tape being made **before** this is done. Sometimes instead of a vertical angle the slope distance and the difference in elevation between the points are the data contained in the field-notes.

The reduction of field-notes is often performed together with the actual computations as a preliminary step, especially when using electronic computers.

14.4. CURVED BOUNDARY BY OFFSETS.—The offsets to the brook (Fig. 5.5.1) were taken at regular intervals in one portion of the survey and in another portion offsets were taken at the points where the direction of the brook changes. The offsets which were taken at regular intervals give a series of trapezoids with equal altitudes the area of which can be obtained by one computation. Although there are several approximate rules for this computation the two most common are those known as the *Trapezoidal Rule* and *Simpson's One-Third Rule.*

14.5. Trapezoidal Rule.—If the figure is considered as made up of a series of trapezoids, all having the same base, the total area can be found by the following rule:

$$\text{Area} = d\left(\frac{h_e}{2} + \sum h + \frac{h_e'}{2}\right)$$

where d = common distance between offsets,
h_e and h_e' = end offsets of the series of trapezoids,
and $\sum h$ = sum of the intermediate offsets.

14.6. Simpson's One-Third Rule.—In the development of this formula the curved line is assumed to be a parabolic curve. It is claimed by some that this affords results more nearly correct than the Trapezoidal Rule. The latter is probably sufficiently exact for most problems of this kind, where the offsets at best can give but an approximate location of the crooked boundary, which is frequently a brook or a crooked wall the middle of which must be estimated.

Simpson's One-Third Rule is as follows:

$$\text{Area} = \frac{d}{3}(\mathrm{h}_e + 2\sum h_{\text{odd}} + 4\sum h_{\text{even}} + h_e')$$

where d = common distance between offsets,
h_e and h_e' = end offsets of the series, 1st and last offsets,
$2\sum h_{\text{odd}}$ = twice the sum of all the odd offsets (the 3d, 5th, 7th, etc., from the end)
$4\sum h_{\text{even}}$ = four times the sum of all the even offsets (the 2d, 4th, 6th, etc., from the end).

This rule may be derived by taking any three consecutive offsets and passing through their extremities a parabola whose axis is parallel to the offsets. The area of the parabolic segment formed by the chord and the curve is known to be two-thirds the area of the (circumscribed) parallelogram bounded by the chord, a parallel tangent, and the end offsets extended. The area of this segment added to the area of the trapezoid formed by the first and third offsets is the area of the two strips. Combining several such pairs of strips the above expression is obtained.

In order to apply this rule it is evidently necessary to have an **even** number of strips; if the number of strips is odd, all but one end strip may be computed by this rule and the extra strip treated as a trapezoid and computed separately. If there is a triangle or trapezoid at the end of the series which has a base greater or less than d it must be computed separately.

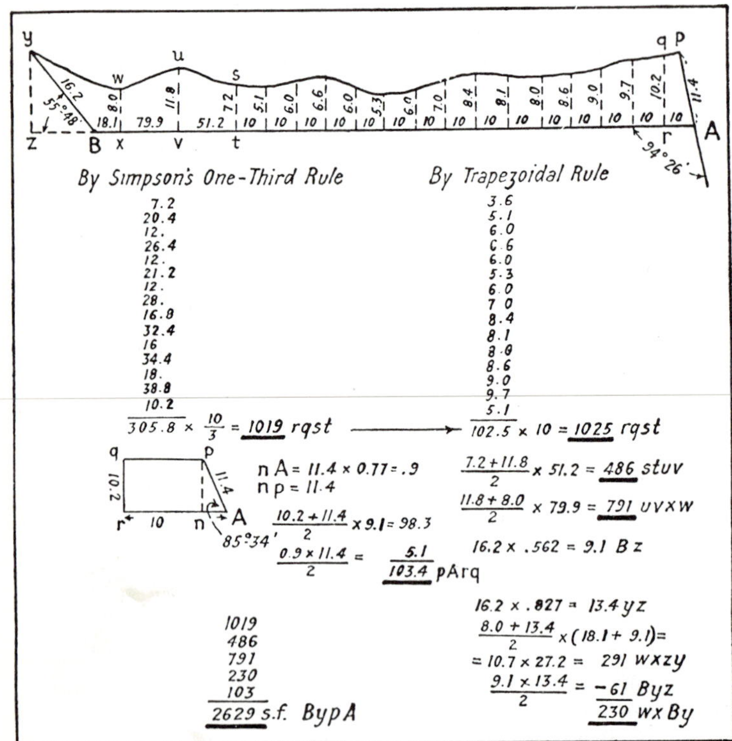

FIG. 14.6.1.

Fig. 14.6.1 shows the computation of a series by both methods and also the computation of several trapezoids and triangles at the ends of the series. The data are taken from the field-notes in Fig. 5.5.1.

14.7. Straightening Crooked Boundary Lines.—In Fig. 14.7.1, *AEFGH* represents a curved boundary between two tracts of land, and it is desired to run a line from *A* so as to make the boundary a straight line and to leave each tract of the same area as before.

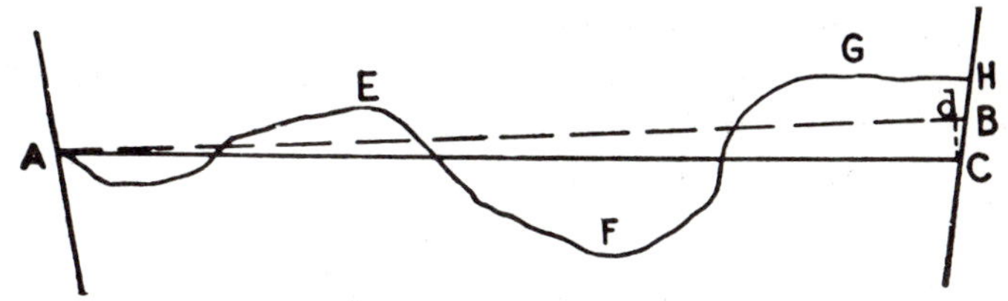

Fig. 14.7.1. Straightening a Crooked Boundary.

The trial line *AB* is first run, and the distance *AB*, the angles at *A* and *B*, and the necessary offsets to the curved boundary are measured in the field. Then the areas of the property between this trial line and the curved line are computed as explained in the previous articles. The sum of the fractional areas on one side of the trial line and the sum of the areas on the other side of it should be equal. If not made so by the trial line, the difference between these sums is the area of a correction triangle *ABC* which must be taken from one tract and added to the other because the trial line has taken this difference from one of the tracts and it should therefore be restored. The area and the base *AB* being known the altitude *dC* can be computed. Then in the triangle *ABC*, the lines *BC* and *AC* and the angle at *A* are calculated; and the line *AC* is staked out, its calculated length being checked by measuring the line *AC* in the field and the angle at *A* being checked by the measured distance *BC*.

14.8. Area by Triangles.—If the field has been surveyed by setting the transit in the middle of the field and taking angles between the corners (Art. 5.7), the areas of the triangles may be found by the trigonometric formula:

$$\text{Area} = \tfrac{1}{2}\, a\, b \sin C,$$

where *C* is the angle included between the sides *a* and *b*.

If all three sides of any of the triangles have been measured or if the field has been surveyed with the tape alone (Art. 5.8), the area of the triangles can be found by the trigonometric formula.

14.9. AREA OF A QUADRILATERAL BY TRIANGLES.—Most city lots have four sides, and while some standard coördinate method is often employed in computing their areas, it is not uncommon to divide these quadrilaterals into simple triangles and to use this for computing the area and checking the geometric exactness of the fieldwork.

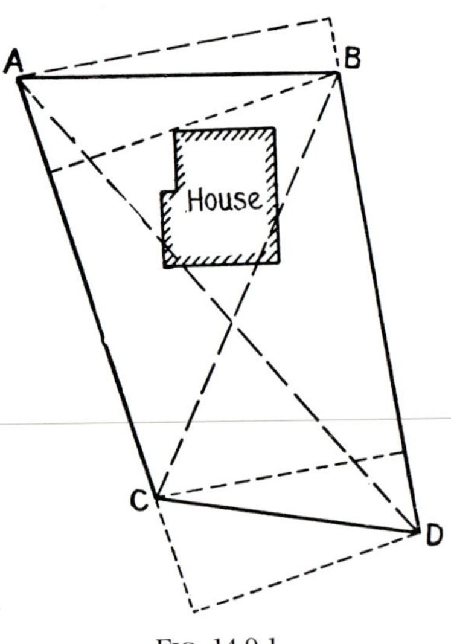

FIG. 14.9.1.

In Fig. 14.9.1, *ABCD* represents an ordinary city lot in which all the sides and angles have been measured. It is evident that the diagonal *BC* can be computed either from *BC*, *CD*, and the angle *D*, or from *AB*, *AC*, and the angle *A*. These two determinations of *BC* should check each other. Similarly two independent determinations of *AD* can be found. These evidently check all the fieldwork and calculations as far as they have gone. In computing these triangles the best way is to resolve all the work into right triangle calculations, as suggested by the dotted lines on the figure which show two pairs of perpendiculars, one pair for triangles ABC and BCD, and another pair for triangles ADC and ABD. Not only is this method more simple than to use the oblique triangle formulas, but it gives at the same time altitude distances which are useful in computing the area of the lot. The area can be obtained by calculating the area of one pair of triangles and checked readily by calculating the other pair.

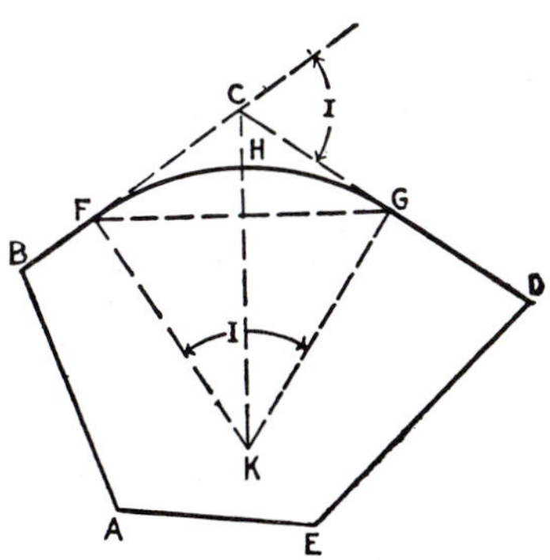

FIG. 14.10.1.

14.10. AREA OF CURVED CORNER LOT.—In Fig. 14.10.1 $ABFHGDE$ is the boundary of a corner lot, all the angles and distances of which have been determined in the field. The area of $ABCDE$ can be computed easily (see Art 14.17). Then the area of $FCGH$ must be subtracted from the traverse area. The angle I is known and the radius KF of the curve is given or can be computed from data such as CH or CF obtained in the field (Art. 10.12).

$$KFHG = \frac{FHG \times HK}{2} = \frac{I^\circ \times 0.0174532925 \times (HK)^2}{2}$$

$$\begin{aligned} KFCG &= FC \times FK \\ FCGH &= KFCG - KFHG \\ &= KG\,(CG - FHG \div 2) \\ &= \text{Radius (Tangent} - \tfrac{1}{2}\text{ Arc)} \end{aligned}$$

The length of the arc of curve whose radius is 1 and whose central angle is 1° is 0.017 453 2925 which will give results up to eight significant figures, provided I and R are correct to eight significant figures.

The total arc can be found readily as

$$L_c = \frac{I^\circ}{360^\circ}(2\pi R)$$

(and this may be needed anyway in the survey, if not for the area).

The area of *FCGH* could have been calculated by computing the area of the triangle *FCG* and then subtracting the area of the segment *FHG* from it. The area of this segment, however, cannot be calculated accurately by any short formula. An approximate formula for the area of a segment is

Area of Circular Segment = ⅔*MC* (approximately), where *M* is the middle ordinate and *C* is the chord length.

$$M = \frac{C^{2*}}{8R} \text{ (approximately).}$$

Another method for calculating this area would be

$$\text{Area of } KFHG = \frac{I^\circ}{360}(\pi R^2)$$

$$\text{Area of } KFCG = KC \times FG \quad or \quad FC \times FK$$

(whichever of these values are readily available).

---

*In FIG. 14.10.2, *OB* = Radius of circular curve.
CH = Middle Ordinate for chord *AB*.
CD is drawn tangent to the curve.

DB = Tangent Offset for chord *CB*.
OE is drawn perpendicular to *CB*.
In the two similar triangles *OEB* and *CBD*,

$$DB : CB = BE : OB$$

$$DB : CB = \frac{CB}{2} : OB$$

$$DB = \frac{CB^2}{2OB}$$

$$\textbf{Offset from Tangent} = \frac{\textbf{(Chord)}^2}{\textbf{2} \times \textbf{Radius}} \qquad (1)$$

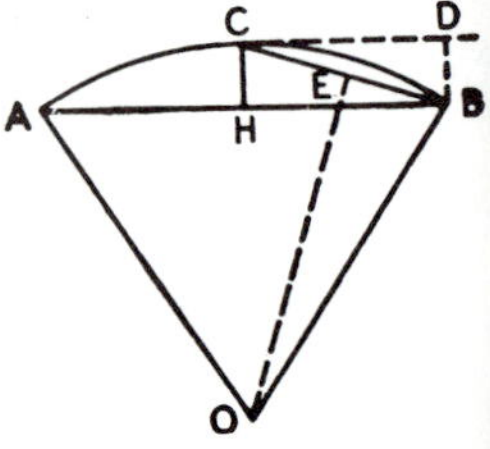

FIG. 14.10.2.

But $DB = CH$, and $AB = 2 \times CB$ (approximately).

Then subtraction of these two areas gives directly the area of *FCGH*.

Expressed in terms of *C* and *R*,

$$\text{Area of Circular Segment} = \frac{C^3}{12R} \quad \text{(approximately)}$$

These formulas are fairly accurate when *M* is very small as compared with *C*. They are most useful, however, as a check on computations made by the preceding method.

14.11. Rough Checks on Areas.—If the traverse has been plotted to scale, it can be divided easily into simple figures such as rectangles or triangles, their dimensions scaled from the plan, and their areas computed, thereby giving an independent rough check on the area.

A piece of tracing paper or cloth ruled into small squares can be placed over the plot of the area and a count made of the number of squares and fractional parts included within the area desired. Then, the area of one square being known, an approximate area of the figure may be obtained. This method is particularly applicable to areas with irregular boundaries.

14.12. Planimeter.—One of the commonest ways of checking the area of a traverse is to obtain its area by means of an instrument called the *planimeter*. It is a small instrument consisting of an arm, carrying a tracing point, which is fas-

---

$$\therefore CH = \frac{\left(\frac{AB}{2}\right)^2}{2\,OB} = \frac{AB^2}{8\,OB} \quad \text{(approximately)}$$

$$\textbf{Middle Ordinate} = \frac{\textbf{(Chord)}^2}{8 \times \textbf{Radius}} \quad \text{(approximately)} \quad (2)$$

The following will give some idea of the accuracy of this formula:

When radius = 20 ft. and chord = 10 ft.,
M = 0.625 (correct value is 0.635).
When radius = 100 ft. and chord = 25 ft.,
M = 0.781 (correct value is 0.784).
When radius = 100 ft. and chord = 100 ft.,
M = 12.500 (correct value is 13.397).
When radius = 1000 ft. and chord = 100 ft.,
M = 1.250 (correct value is 1.251).

It is evident from the above that this formula will not give accurate results when the chord is large in comparison with the radius.

tened to the frame of the instrument; the arm is either fixed or can be adjusted to any desired length. The frame touches the paper at only three points: the anchor point (pole), the tracing point, and the circumference of a small wheel which is free to revolve. On the rim of this wheel is a scale and beside it is a vernier which is used in reading the scale. For some planimeters the length of the arm can be regulated by setting it at the proper reading on a scale which is marked on the arm, so that a unit on the wheel scale will represent any desired unit area on the plan, such as a square inch or a square centimeter.

In using the instrument the anchor point is set at some convenient position on the drawing **outside** of the area to be measured and then the tracing point is run around the perimeter of the area to be determined. The reading on the wheel is recorded when the tracer is at the starting point. The tracer, in passing around the perimeter, should be kept as **closely as possible** on the boundary line and should return **exactly** to the starting point. Then the scale is again read, and the difference between the two readings is the area that has been traced out, expressed in some unit depending on the length of the arm. The result can be transposed easily into the unit of the scale of the map.

Usually the settings for the scale on the arm are furnished by the marker for various units of area. It is safer to test this setting by running the instrument around a known area, such as 4 square inches and determining the interval passed over by the wheel by making several tests and by setting the anchor point at different positions. This interval divided by 4 will be the value of one square inch of plan area and this is equivalent to a certain number of square feet of surface, depending upon the scale of the map. It is important that the sides of the trial square should be laid off so that they agree with the present scale of the map which, owing to swelling or shrinking of the paper, is frequently not quite the same as when it was first drawn (Art. 18.13).

14.13. The Amsler Polar Planimeter.—The most common planimeter is the *Amsler Polar Planimeter* (Fig. 14.13.1). This

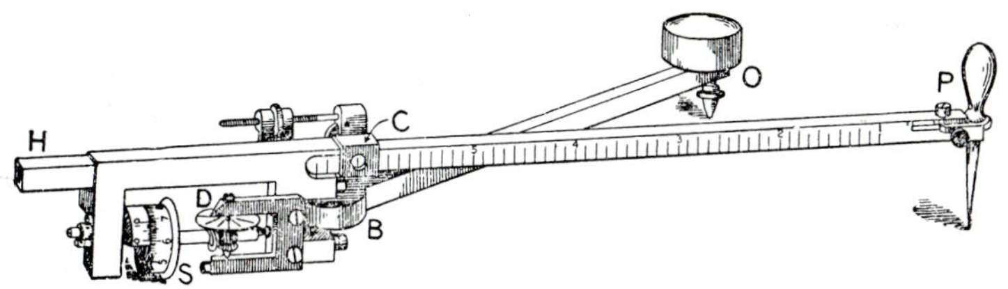

FIG. 14.13.1.

instrument has two arms, *BO* and *HP*. The arm *BO* is of fixed length; it is anchored at *O* by a needle point which sticks into the paper and is held in position by a small weight which is detachable. At *B* it is connected by a pivot to a collar, *C*, through which the tracing arm *HP* can slide. At *P* is a tracing point which is moved along the outline of the area to be measured; the distance *CP* being adjusted to conform to the scale of the map. The graduated wheel *S*, whose axis is parallel to *HP*, records the area in units dependent upon the length of the arm *CP*.

The planimeter rests then on three points, the anchor, the tracing point, and the periphery of the wheel. As the tracing point is moved around the given area, the wheel drags along, sometimes slipping and sometimes rolling, and the difference between the reading of the scale on the wheel at the beginning and end of the circuit represents the area of the figure. Besides the scale on the wheel there is a small disk *D* which records the number of full revolutions of the wheel. The result of reading the disk, the wheel, and its vernier will usually give four digits in the number representing the area.

Since the length of the anchor arm is fixed and the point *O* stationary, the pivot *B* moves on the circular arc whose center is *O* and whose radius *(R)* is the distance *OB*. The wheel, however, does not follow the arc of a circle, but the instrument must be so constructed that the wheel will always lie somewhere on the line *dC* or on *dC* produced (Fig. 14.14.1).

If in moving the tracing point its arm be maintained in such a position with reference to the anchor arm that the plane of the wheel will always pass through the anchor point, it is evident that the wheel will not revolve at all on its axis but will slip on the paper without changing its reading. The tracing point can therefore be started at a given point and moved about in the path of a circumference, returning to the same point again without recording any reading of the wheel. This circumference is called the *zero circumference*, the *correction circle*, or the *zero circle* (Fig. 14.14.2).

14.14. Theory of the Amsler Polar Planimeter.—The following proof has been taken from Cours de Mécanique, by Édouard Collignon.

Let A (Fig. 14.14.1) be the area to be measured. Conceive cd (corresponding to the tracing arm) to be a straight line of con-

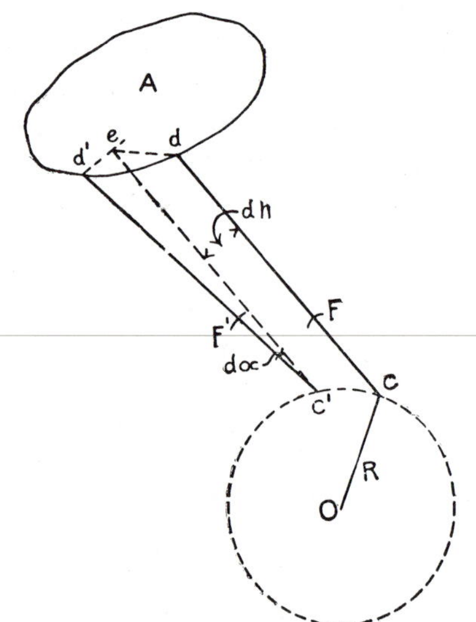

FIG. 14.14.1.

stant length moving so that one end d is always upon the outline of A and the other end c is always upon a given curve cc' (in general a circle described from O).

Let cd and $c'd'$ be consecutive positions of the moving line, and let an expression be obtained for the elementary area $cdd'c'$ generated by the line in moving from the first position to the second. This movement may be considered as composed of two parts; a translation from *cd* to a parallel position *c'e*, and a rotation from *c'e* to *c'd'*, the first generating a parallelogram *cdec'*, and the second a sector *c'ed'*.

Let dA' = the elementary area $cdd'c'$
L = the length of cd
dh = the width of the parallelogram
$d\alpha$ = the elementary angle of rotation

Then $$dA' = L \cdot dh + \tfrac{1}{2}L^2 \cdot d\alpha \qquad [1]$$

Now suppose a wheel F fixed upon cd, its plane perpendicular to that line, so that in the displacement of cd the wheel rolls when the point F moves perpendicularly to cd, and glides without turning when F is displaced in the direction of cd. Let $d\theta$ be the angle through which the wheel turns upon its axis in passing from F to F'. If r is the radius of the wheel, $rd\theta$ is the length of arc applied to the paper. This length is equal to dh (the rotation of the wheel in the translation from cd to $c'e$ corresponding to the normal displacement only) + the arc $L'd\alpha$ (letting $cF = L' = c'F'$).

$$\therefore r \cdot d\theta = dh + L'd\alpha \qquad [2]$$

With the wheel beyond c on dc produced, $r \cdot d\theta = dh - L'd\alpha$
Eliminating dh from equations [1] and [2]

$$dA' = r \cdot L \cdot d\theta + \left(\frac{L^2}{2} - LL'\right) d\alpha \qquad [3]$$

$$\int dA' = \int r \cdot L \cdot d\theta + \int \left(\frac{L^2}{2} - LL'\right) d\alpha \qquad [4]$$

Conceive now the point d to traverse the entire outline of A, the elements dA' being reckoned positively or negatively according to the direction in which they are generated. Two cases are to be noticed:

(a) When the directing curve cc' is exterior to (but not including) the area A (Fig. 14.14.1). The algebraic sum, $\int dA'$, will be the difference between the sum of the positive and the sum of the negative elementary areas, and will equal the area area A.

$\int r \cdot L \cdot d\theta = rL\theta$

$= Lu$ (where $u = r\theta$ = algebraic sum of arcs applied to paper by wheel).

$\int d\alpha = 0$, since cd returns to its original position without having made a circuit about O.

$\therefore$ Integrating expression [4], $A' = A = Lu$.

(b) When the directing curve cc' is within the area A (Fig. 14.14.2). The line cd now makes an entire revolution in order to return to its primitive position, and $\int d\alpha = 2\pi$. Also the area

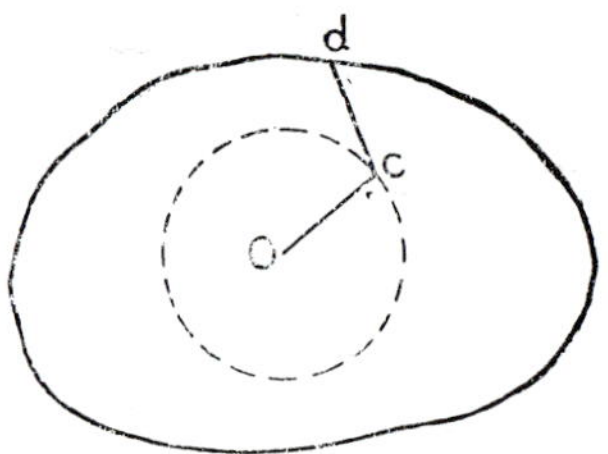

FIG. 14.14.2.

$$A = \int dA' + \text{area of circle described by } Oc$$

By integrating expression [4]

$$A' = Lu + 2\pi\left(\frac{L^2}{2} - LL'\right)$$

$$\begin{aligned} A &= A' + \pi R^2 \\ &= Lu + \pi(L^2 - 2LL' + R^2) \\ &= Lu + \text{the area of a circle of radius } \sqrt{L^2 - 2LL' + R^2}. \end{aligned}$$

The sign of $2LL'$ is $-$ if the wheel be between tracing point and pivot point; otherwise it is $+$.

This circle is the so-called "circle of correction" or "zero circle" and its value may be found by measuring with the planimeter a circle or other figure of known area inclosing the directing curve cc'.

It will be seen that the radius $\sqrt{L^2 \pm 2LL' + R^2}$ is the distance from anchor point to tracing point, when the plane of the wheel passes through the anchor point; in other words, is the radius of the zero circle.

If C = circumference of wheel
and n = number of revolutions made in a given measurement
$Lu = LnC$

If L and C be given in inches A will be found in square inches. By varying L the area A corresponding to one revolution of the wheel ($n = 1$) may be varied at pleasure. Commonly, if the area is sought in square inches the length L is made such, by adjustment of the tracing arm, that one complete revolution of the wheel corresponds to 10 square inches of area.

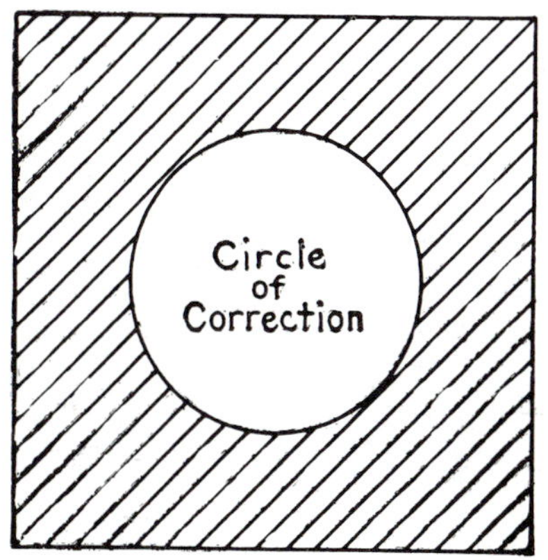

FIG. 14.14.3.

Since, for the anchor point **outside** the area to be measured, $A = LnC$, it appears that for any setting of L, A is directly proportional to n. So that L may be set at random and n' determined for a known area A' (say a circle or rectangle) in whatever unit the area is desired, then $\dfrac{A}{A'} = \dfrac{n}{n'}$. But this process evidently does not apply to the case of anchor point **inside** the area to be measured.

In finding the area of the circle of correction, the instrument gives directly only the **difference** between circle and known area.

If the perimeter lies entirely outside of circle (Fig. 14.14.3), then the record of instrument gives the shaded area only, and this **subtracted** from known area will equal the circle of correction.

If the known area (Fig. 14.14.3) does not lie entirely without the circle then record of instrument must be **added** to the known area if the circle of correction is the larger, otherwise **subtracted.**

14.15. Use of Polar Planimeter.—In measuring a closed area the anchor point is pressed into the paper at a position **outside** of the area of the figure, if it is not too large, and the tracing point is started from a definite point on the periphery of the area, preferably such as will bring the two arms approximately at right angles to each other. The wheel is then read. The tracing point is then moved around the outline of the area, being careful to follow the line accurately, until the starting point is reached again. The wheel is again read and the difference between the two wheel readings gives the area in the unit depending upon the setting of the arm. The disk should also be read when the wheel is read if the area is large enough to require a full revolution of the wheel. Care must be taken to bring the tracing point **exactly** back to the point from which it was started.

While some instruments have a tracing arm of fixed length, so that all areas recorded by the wheel are in the same unit, square inches for example, many planimeters have adjustable tracing arms which can be set by means of a clamp and slow-motion screw at whatever reading of the scale on the tracing arm is desired. Usually it will be necessary to use a reading glass to make this setting accurately. The arm is sometimes marked by vertical lines and beside these lines are letters and figures indicating the unit of area to which the setting corresponds. In some instances, however, the scale is marked as a continuous scale and the proper settings for given area units are supplied by the instrument maker.

A way of avoiding the use of the correction circle in measuring large areas is to divide the area into smaller ones by light pencil lines and to determine each fractional area separately. If, however, the anchor point is placed inside the area the value of the correction circle must be applied to the reading of the wheel as explained in the previous article.

A planimeter can be readily used even though the setting of the arm is not known. A square, say four inches on a side, can be accurately drawn on the map and its area determined with the planimeter in the usual manner, making several independent determinations with the anchor point in different positions so that if there are any irregularities in the paper over which the wheel passes which are affecting one result, this error will not enter into the other determinations. The mean of these results divided by the number of square inches in the given area will be the wheel reading for one square inch. This being determined the area in square inches of any given figure can be obtained by the planimeter and this can be easily converted into the desired units by using the scale of the map.

It is well before beginning to trace out the figure to run the tracing point around the figure, keeping approximately on the line so as to be sure that the anchor point has been placed in a satisfactory position. To insure accuracy the area should always be measured by at least two independent determinations with a different position of the anchor for each measurement for the reason explained above. Furthermore, it is of extreme importance to check the area roughly by observation or by scaling and rough calculations. If the paper has shrunk since the drawing was made the amount of this change should be determined and allowed for in arriving at a correct value for the area. By the use of a polar planimeter a result which is not in error more than one per cent is easily obtained, except in the case of very small areas.

Current models usually are equipped with both a magnification window over the vernier and a measuring wheel dial graduated in black on a white background for easy observation.

14.16. The Rolling Planimeter.—The rolling planimeter,

unlike the polar planimeter, is not anchored to the drawing. It has a tracing point at the end of an adjustable pivoted arm which is fastened to a frame which is supported on two rollers. In using this planimeter the whole instrument moves forward or backward in a straight line while the tracing point traverses the outline of the area to be measured.

With the rolling planimeter it is possible to obtain a remarkable degree of accuracy, results correct to a tenth of a per cent being easily reached.

14.17. CALCULATION OF AN AREA WHEN THE X, Y COÖRDINATES OF ALL OF THE CORNERS ARE KNOWN.—Assuming that the rectangular coördinates of the corners have been found we may obtain the area by means of these X and Y coördinates alone. It is not essential that all the coördinates should have been found from the same closed traverse, except for the check given by such a survey. If the coördinates have been determined accurately by any means the area will be accurate.

In Fig. 14.17.1 the area of the field, 1, 2, 3, 4 is equal to the algebraic sum of the trapezoids

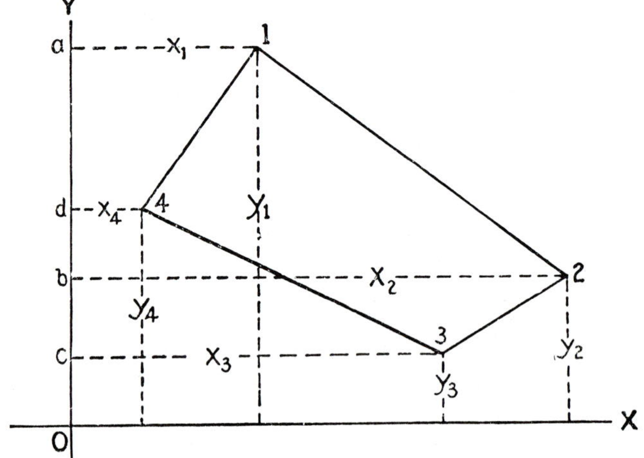

FIG. 14.17.1.

$$(a, 1, 2, b) + (b, 2, 3, c) - (a, 1, 4, d) - (d, 4, 3, c)$$

Expressed as an equation in terms of the coördinates the area is

$$1, 2, 3, 4 = (y_1 - y_2)\frac{x_1 + x_2}{2} + (y_2 - y_3)\frac{x_2 + x_3}{2} - (y_4 - y_3)\frac{x_4 + x_3}{2} - (y_1 - y_4)\frac{x_1 + x_4}{2} \quad [1]$$

$$= \tfrac{1}{2}\{y_1(x_2 - x_4) + y_2(x_3 - x_1) + y_3(x_4 - x_2) + y_4(x_1 - x_3)\} \quad [2]$$

From this last equation is derived the following rule for obtaining the area of a closed field from the coördinates of its corners:

1. Number the corners consecutively around the field.
2. Multiply each $\begin{Bmatrix}\text{abscissa}\\\text{ordinate}\end{Bmatrix}$ by the difference between the following and the preceding $\begin{Bmatrix}\text{ordinates}\\\text{abscissas}\end{Bmatrix}$, always subtracting the following from the preceding (or always subtracting the preceding from the following), and take **half the sum** of the products.

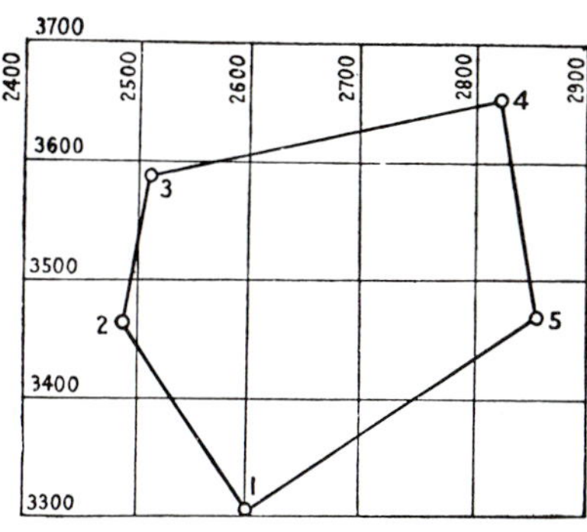

FIG. 14.17.2.

Fig. 14.17.3 illustrates the computation of an area (Fig. 14.17.2) from the coördinates of its corners. For convenience each Y has been diminished by 3000 ft. before multiplying by the corresponding difference in X's. Observe that the correctness of the subtractions may be checked by finding the sum for the fourth column, which should equal zero.

Corner	North (Y) (Feet)	East (X) (Feet)	Diff. Adj. X's	+ Areas	− Areas
1	3304.29	2601.71	+ 372.91	113,473	
2	3462.91	2486.21	+ 90.55	41,917	
3	3591.20	2511.16	− 333.79		197,337
4	3651.81	2820.00	− 347.96		226,804
5	3470.19	2859.12	+ 218.29	102,638	
				258,028	424,141
					258,028
					2 | 166,113
					83,056 Sq. Ft.

FIG. 14.17.3. AREA OF LOT FROM COÖRDINATES OF ITS CORNERS.

Equation (1) may be developed into the following form:

$$1, 2, 3, 4 = \tfrac{1}{2}\,(x_2y_1 - x_1y_2 + x_3y_2 - x_2y_3 + x_4y_3 - x_3y_4 + x_1y_4 - x_4y_1) \quad [3]$$

When this formula is to be used the coördinates may be arranged in the following simple manner:

$$1, 2, 3, 4 = \tfrac{1}{2}\left(\frac{x_1}{y_1} \times \frac{x_2}{y_2} \times \frac{x_3}{y_3} \times \frac{x_4}{y_4} \cdots\cdots \times \frac{x_1}{y_1}\right) \quad [4]$$

From equation (3) it will be seen that the area is equal to the sum of the products of the coördinates joined by **full** lines in (4) **minus** the sum of the products of the coördinates joined by **broken** lines. This formula involves the multiplications of larger numbers than in (2), but does not require any intermediate subtractions.

14.18. Area Calculation by Double Meridian Distances and Double Parallel Distances.—Since the coördinate method, described in Article 14.17 is very easily programmed on electronic calculators, most calculations follow this approach. However, it might occur that the coördinates are not known, and only azimuths or bearings and distances are available. This can be easily arranged into a temporary coördinate system by either projecting the lines onto a meridian, usually the one passing through the most westerly point of the survey, or by projecting the lines onto an East-West Axis, preferably the one passing through the most southerly point of the survey. Then the same trapezoid approach is used to determine the area. For the first option this leads to the "Double Meridian

Distance Method" (DMD) while for the second it is called "Double Parallel Distance Method" (DPD).

As they are very rarely used nowadays, no further details are given and the reader is referred to previous editions.

14.19. Volumes.—Determining the amount (volume) of earthwork and masonry required for construction projects is an important branch of practical surveying. In projects involving the movement of earthwork such as pipe lines, sewers, cellars, borrow-pits, grading, highways, and railroads, a knowledge of the volume is necessary to compute cost. Other volume problems require yardage in piles of sand, gravel or coal, and capacity of reservoirs.

Masonry volumes may be computed by solid geometry from dimensions on the plans, but it is common practice to check the pay quantities by recomputing the volume from dimensions of the actual structure taken in the field. These solids are usually prisms, pyramids, cones, or portions of these solid figures.

Earthwork volumes are computed from the cross-sections taken in the field, as described in the previous articles. These cross-sections define parallel areas at definite stations and at definite distances apart along the center line of the project. The solid between these approximates a prismoid and is considered as such in computing earthwork volumes.

Two formulas are used for computing the volume of a prismoid:

(1) End-Area Formula
(2) Prismoidal Formula.

14.20. End-Area Formula.—In the end-area formula the volume is assumed to be equal to the average of the areas of the two ends multiplied by the distance between them, i.e.,

$$V = \frac{A_1 + A_2}{2} \times l$$

in which A_1 and A_2 are the areas of the two end bases and l is the distance between them. Although this formula, because of its simplicity, is almost universally used in computing earthwork volumes, it is not exact and in the majority of cases it gives a volume somewhat **larger** than the exact value.

14.21. PRISMOIDAL FORMULA.—A prismoid is a solid bounded by planes, the end faces being parallel and having the same number of sides. The correct volume of a prismoid is given by the prismoidal formula, i.e.,

$$V = \frac{l}{6}(A_1 + 4A_m + A_2)$$

in which l is the distance between the two bases A_1 and A_2; and A_m is the area of the section **midway between them.** Notice that A_m is not defined as the mean of A_1 and A_2. To find the area of the middle section, the dimensions of the middle section are first found by **averaging the dimensions** of the end sections; from the values the middle area is computed.

The average end-area method is amost universally used for computing earthwork quantities. Although approximate, its precision is usually consistent with the field methods used in determining the cross-sections upon which the calculations are based. The unit cost of ordinary excavation is relatively low and refinements in earthwork calculations are not justified in most cases. In computing masonry quantities, however, more precision is required, since a cubic yard of concrete costs many times as much as a cubic yard of earth excavation. The prismoidal formula is very useful in determining accurately the volumes in certain parts of masonry structures.

The greatest difference between the results of the two formulas will be found for tapering solids of greatly different end areas, as illustrated in the following example.

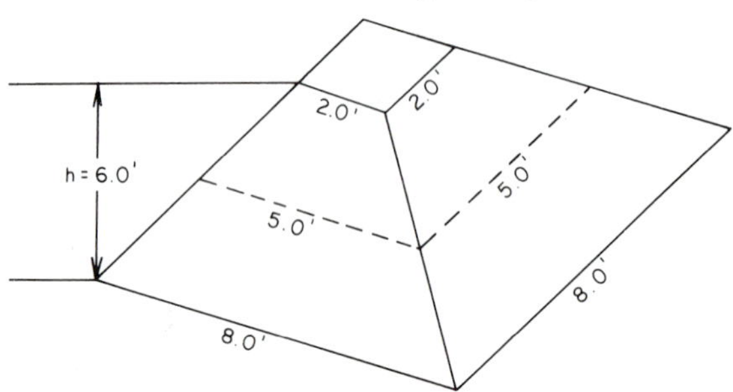

FIG. 14.21.1.

A prismoidal base for a statue is to be built of masonry or poured concrete, with dimensions shown.

The volume by average end-area method:

$$V = \frac{(2.0 \times 2.0) + (8.0 \times 8.0)}{2}(6.0) = 204 \text{ ft}^3$$

and by prismoidal formula:

$$V = \frac{6.0}{6}\left[(2.0 \times 2.0) + 4(5.0 \times 5.0) + (8.0 \times 8.0)\right] = 168 \text{ ft}^3$$

This shows a result 21% high for the end-area method.

$$\text{In Fig. 14.21.2,}\ A_1 = \left(\frac{2.0 + 2.5}{2}\right) 2 = 4.5 \text{ s.f.}$$

$$A_2 = \left(\frac{2.0 \times 5.0}{2}\right) 12 = 42.0 \text{ s.f.}$$

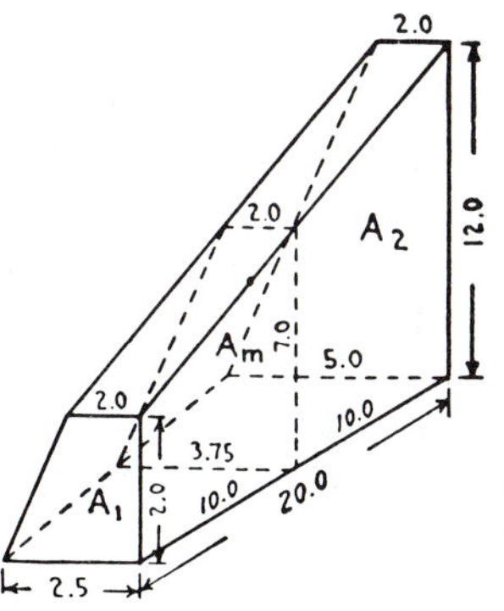

FIG. 14.21.2.

The approximate volume by average end-area method = $\left(\frac{4.5 + 42.0}{2}\right) 20 = 465$ c.f.

The area of the mid-section = $\left(\frac{3.75 + 2.0}{2}\right) 7 = 20.1$ s.f.,

and the exact volume by prismoidal formula = $\frac{20}{6}(4.5 + 4 \times 20.1 + 42.0) = 423$ c.f. In this example the average end-area formula gives a result that is about 10 per cent too large. This example illustrates why the end-area method is not used in practice in computing the quantities of masonry.

14.22. EARTHWORK COMPUTATIONS.—A common method for computing the quantity of earthwork is to plot the cross-sections to scale and obtain the areas from this plot with a planimeter (Art. 14.12) or by other graphical methods.

The cross-sections are plotted to the same horizontal and vertical scale; 1 in. = 5 ft. and 1 in. = 10 ft. are commonly used. Special cross-section paper is available; this is ruled with heavy lines into one-inch squares and with lighter lines into one-tenth-inch squares. A heavy vertical ruling is selected to represent the center line for a series of cross-sections; it should be one that will admit of plotting the cross-sections so that they will not run off the edge of the paper. The elevations are then plotted at their respective distances to the left and right of this line. A heavy horizontal ruling near the bottom of the sheet is arbitrarily marked with an elevation (such as 70.00) and this line is used for plotting the elevations of first cross-section. The next cross-section is plotted above the first and its elevations are referred to another convenient horizontal line (as 70.00). Each cross-section has a separately designated datum line. (See Fig. 14.22.1 which is a plot of a portion of the cross-section notes shown in Fig. 9.13.1.) The plotted points are connected by a heavy freehand line which is drawn to represent the shape of the ground.

For estimating earthwork quantities, the final cross-section is superimposed on the original ground line, as shown by light lines in Fig. 14.22.1. The reference elevation for plotting the final cross-section is the elevation of the finished grade at the center line, shown by the short horizontal dash line, such as 73.49 for Sta. 11 + 43. The area between the original and final cross-sections may be determined by planimeter (Art. 14.12), or by dividing the area into triangles and trapezoids and computing the area in parts, using dimensions scaled from the plot. In using the planimeter it is often convenient to measure the area in square inches and then to convert it to square feet according to the scale of the plot.

Another rapid way of obtaining cross-section areas is to divide the area into vertical strips each one foot wide to the scale of the plot, and to add the lengths of these strips cumulatively. In Fig. 14.22.1 the width between the fine grid lines represents one foot. The area can be obtained by scaling the average length of each grid strip, starting at the left edge of the section and working across to the right edge. The lengths may be recorded cumulatively by ticks marked on the edge of a long piece of paper, such as adding machine paper, or measured cumulatively with a pocket tape graduated in inches. The sum of the strip lengths in inches multiplied by the plot scale in feet per inch will give the area in square feet.

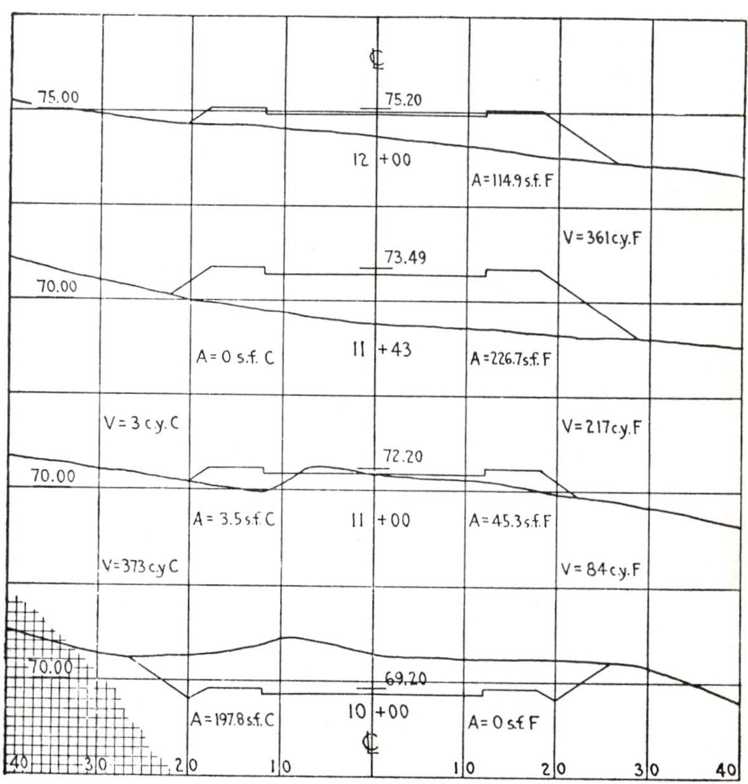

FIG. 14.22.1. CROSS-SECTIONS FOR HIGHWAY.

Cut and fill areas are computed and recorded separately, as at Sta. 11 + 00 in Fig. 14.22.1 the cut area = 3.5 s.f. and the fill area = 45.3 s.f.

The volume of earthwork between adjacent stations is computed by the average end-area method and the cut and fill are recorded separately between sections, as in Fig. 14.22.1: between Stas. 11 +00 and 11 +43, the cut volume = 3 c.y. and the fill volume = 217 c.y.

14.23. Computing Earthwork from Cross-Section Notes.—Earthwork quantities may be computed directly from the cross-section notes when they are kept in the form illustrated in Fig. 9.15.3. The areas of the sections are found mathematically, and the volumes by average end-area method. Referring to Fig. 9.15.3, the areas at Sta. 6 + 50 and Sta. 7 + 00 would be computed by dividing the figures into pairs of triangles, as shown in Fig. 14.23.1.

Area at Sta. 7 +00 equals

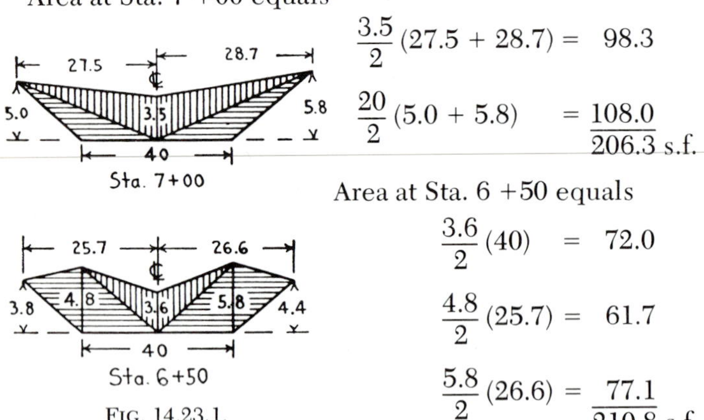

$$\frac{3.5}{2}(27.5 + 28.7) = 98.3$$

$$\frac{20}{2}(5.0 + 5.8) = 108.0$$

$$206.3 \text{ s.f.}$$

Area at Sta. 6 +50 equals

$$\frac{3.6}{2}(40) = 72.0$$

$$\frac{4.8}{2}(25.7) = 61.7$$

$$\frac{5.8}{2}(26.6) = 77.1$$

$$210.8 \text{ s.f.}$$

FIG. 14.23.1.

Volume between 6 +50 and 7 +00 equals

$$V = \frac{50}{27}\left(\frac{206.3 + 210.8}{2}\right) = 386.2 \text{ c.y.}$$

The areas of other types of sections may be obtained by a similar process. Cut areas and volumes are kept separately from fill areas and volumes. In passing from cut to fill (or vice versa), as in the notes shown in Fig. 9.15.3, the zero area of fill at Sta. 6 +30 and the zero area of cut at Sta. 5 +40 must be included in the calculations of volumes. For example, the area of cut at Sta. 5 +70 is $\frac{2.6}{2}(20) = 26$ s.f., and at Sta. 5 +40 the cut is 0; therefore the volume by average end-area method between Stas. 5 +40 and 5 +70 is $\frac{30}{2 \times 27}(0 + 26) = 14.4$ c.y.

14.24. Computing Borrow-Pit Excavation.—Borrow-pit cross-sectioning and the method of determining the amount of cut at each corner are explained in Art. 9.16.

The volume of the excavated material removed from the borrow-pit is computed by assuming that the solid within each of the grid squares, rectangles, or triangles is a truncated prism, the volume of which is equal to the area of a right, or horizontal, section multiplied by the mean of the cuts (heights) at the corners. Expressed as a formula, the volume of a square or rectangular truncated prism is:

$$V = A\left(\frac{h_1 + h_2 + h_3 + h_4}{4}\right)$$

where A is the area of the right section and h_1, h_2, h_3, and h_4 are the cuts at the four corners of the prism; i.e., the cuts at the four corners of the cross-section grid squares (or rectangles).

Similarly, the volume of a triangular truncated prism is:

$$V = A\left(\frac{h_1 + h_2 + h_3}{3}\right).$$

When a number of prisms have the same right area, these may be combined and their volume computed in one operation by the "assembly of prisms" method. If the volumes of a contiguous group of prisms of the same right area are computed separately, it will be seen that the right area and the divisor 4 are common to all the calculations. Likewise, some of the heights will be common to one or more of the calculations. Such computations may be combined and simplified by applying the "assembly of prisms" method, expressed as follows:

$$V = A\left(\frac{\sum h_1 + 2\sum h_2 + 3\sum h_3 + 4\sum h_4}{4}\right)$$

where A = area of one right section in the assembly
$\sum h_1$ = sum of all heights applying to one square
$\sum h_2$ = sum of all heights common to two squares
$\sum h_3$ = sum of all heights common to three squares
$\sum h_4$ = sum of all heights common to four squares.

In Fig. 14.24.1 the volume of the borrow-pit bounded by *ABCDEF* may be computed by the "assembly of prisms" method as follows:

$\sum h_1$	$\sum h_2$	$\sum h_3$	$\sum h_4$
3.2	4.0	5.8	5.1
3.8	4.4		6.6
4.6	4.4		8.8
2.1	7.3		7.1
3.6	3.6		9.4
	5.0		6.6
	4.9		
	6.2		
	5.4		
	4.2		
17.3	49.4	5.8	43.6

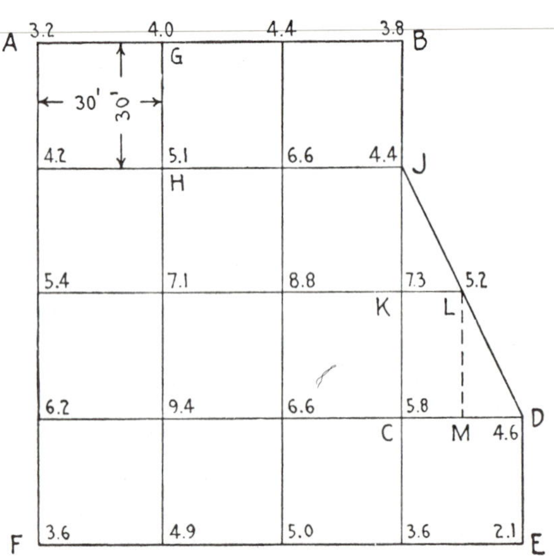

FIG. 14.24.1.

$$V = \frac{900}{4}\left[17.3 + 2(49.4) + 3(5.8) + 4(43.6)\right] = 69{,}278 \text{ cu. ft.} = 2566 \text{ cu. yds.}$$

The volume bounded by *JKL* is a triangular truncated prism and may be computed as follows:

$$V = \frac{225}{3}(4.4 + 5.2 + 7.3) = 1268 \text{ cu. ft.} = 47 \text{ cu. yds.}$$

In computing the volume of the trapezoidal prism whose area is *KLDC*, it is customary to divide it into two triangular prisms by drawing diagonals or into a rectangular prism and triangular prism, depending on the slopes of the surface. The method to use should be indicated in the field-notes; it depends on the shape of the surface of the ground. If the latter method is indicated, the cut at $M = 5.2$ is found by interpolation.

The volume of *KLMC* is

$$V = \frac{450}{4}(7.3 + 5.2 + 5.2 + 5.8) = 2644 \text{ cu. ft.} = 98 \text{ cu. yds.},$$

and the volume of *LDM* is

$$V = \frac{225}{3}(5.2 + 4.6 + 5.2) = 1125 \text{ cu. ft.} = 42 \text{ cu. yds.}$$

Total volume of the borrow-pit

$$V = 2566 + 47 + 98 + 42 = 2753 \text{ cu. yds.}$$

14.25. Computing Earthwork from Contours.—Estimates of earthwork quantities may be made directly from contour maps, such as those prepared from aerial surveys (Art. 12.19). From a study of the terrain as represented by the contours, a horizontal alignment (as for a highway) can be chosen. The profile along this line and the cross-sections at right angles to it can be plotted from elevations obtained from the contours (Art. 7.6). A construction grade line and grading cross-section is then selected and the computation of earthwork made as described in Art. 14.22. Several locations can be studied on the topographic map before the most desirable line is chosen. In some work the contract quantities are obtained from contour plans based on aerial surveys, and lines and grades are not run out in the field until construction is ready to start. Field surveys are needed, however, to establish ground control points for aerial survey.

14.26. Estimates for General Grading.—Estimates for grading may be made conveniently by means of a topographic map. On this map will appear the contours of the original surface. The contours representing the finished surface are also sketched upon the map; the smaller the interval between the contours the more precise will be the result. In Fig. 14.26.1 the full lines represent the contours of the original surface, which is to be altered so that when the necessary cutting and filling have been done the new surface will have the appearance indicated by the dash contours. At contour 20 and at contour 25 no grading is to be done. On the plan, first sketch the lines *ABCDEF* and *AGHIJB* which are lines of "no cut" and "no fill," i.e., lines which enclose areas that are either to be excavated or filled. The amount of excavation and embankment must be computed separately. In sketching such lines the lines *AB*, *ED*, and *HI*, as will be seen, follow the

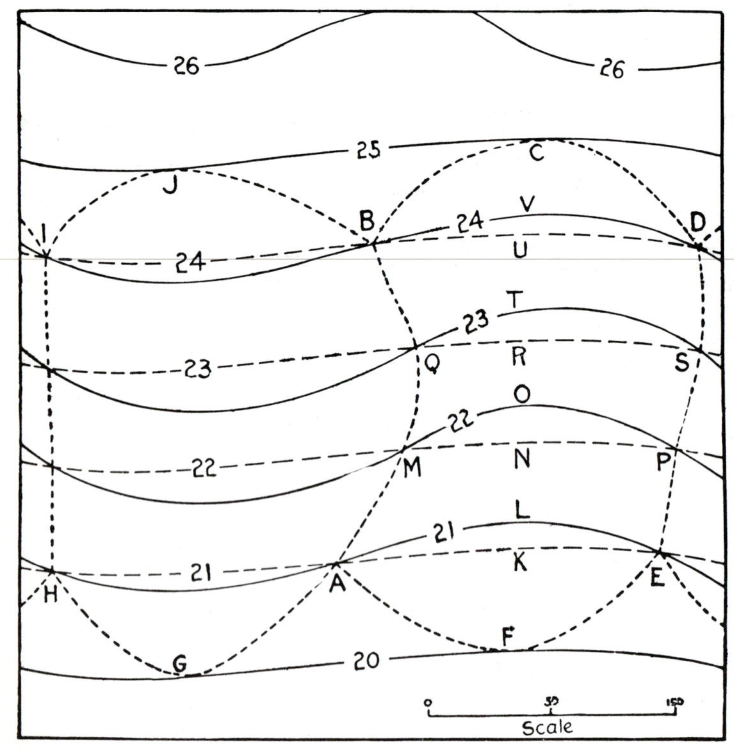

FIG. 14.26.1.

intersection of the original contours with the new ones, since at these points there is no cut or fill. There are no direct data on the plan which define where the earthwork ends at *C* but the assumption is here made that the fill will run out to meet the original surface at about the next contour at *C*. In this example the fill must run out somewhere between the 24-ft. contour and the 25-ft. contour, for if it ran beyond the 25-ft. contour there would be another new 25-ft. contour shown on the plan. Therefore the line *BCD* has been sketched to represent the limits of the fill in that vicinity; similarly *EFA*, *AGH*, and *IJB* have been sketched.

There are three general methods of computing the earthwork from the data given on a plan; (1) by computing directly the amount of cut or fill between successive contours, (2) by assuming a horizontal plane below the lowest part of the earthwork and computing the volume of earth between this plane and the original surface, then computing the volume between the same plane and the finished surface; the difference between these two volumes will be the amount of earthwork, or (3) by drawing on the plan a line of no cut or fill, a line representing, say, 5-ft. cut or fill, a line representing 10-ft. cut or fill and so on. Then compute the volume between these successive 5-ft. layers.

14.27. Cut and Fill between Contours.—(1) Referring to Fig. 14.26.1 and applying the first method, the volume of the solid *AMPE* is that of a solid having two parallel end planes *AKEL* (a plane at elevation 21) being the lower, and *MNPO* (a plane at elevation 22) being the upper plane. The altitude between these two end planes will be the difference in elevation between 21 and 22, or will be 1 ft.

The areas of the horizontal planes *AKEL*, *MNPO*, *QRST*, and *BUDV* may be obtained by planimeter (Art. 14.12) or otherwise, and the volume of the solid *AKEL-MNPO* may be obtained by the End Area Method (Art. 407, p. 486), its altitude being 1 ft. If it is desired to obtain the volume by the use of the Prismoidal Formula the volume of the solid *AKEL-QRST* may be found by using *AKEL* as one base, *QRST* as the other and *MNPO* as the middle area, the altitude, or length, of the solid being the difference between 21 and 23, or 2 ft. While neither of these methods is mathematically exact, they are the ones usually employed for this kind of a problem. The solid *AKEL-F* may be considered to be a pyramid with a base *AKEL* and an altitude of 1 ft.

EXAMPLE

In Fig. 14.26.1 the amount of fill on the area *ABCDEF* is computed below.

Area *AELK* = 900 sq. ft.	$900 \times \frac{1}{3} =$	300 cu. ft. (Pyramid)
" *MNPO* = 1000	$\frac{900 + 1000}{2} \times 1 =$	950.
" *QRST* = 1020	$\frac{1000 + 1020}{2} \times 1 =$	1010.
" *BUVD* = 680	$\frac{1020 + 680}{2} \times 1 =$	850.
	$680 \times \frac{1}{3} =$	227 (Pyramid)
	3)3337	cu. ft.
	9)1112	
	124	cu. yds. Total Fill.

14.28. Cut and Fill Referred to Horizontal Plane.—(2) Referring again to Fig. 14.26.1 and applying the second method, the area of *ABCDEF* is found (by planimeter); this is the area of a plane at, say, elevation 20, since none of the fill extends below contour 20. Then the area of *ABCDEL* is found, which is the area of the plane cutting the original ground at elevation 21. Similarly the areas of *MBCDPO*, *QBCDST*, and *BCDV* are found. The volume of the solids between these planes may be computed by the End Area Method or by use of the Prismoidal Formula, alternate contour planes being used as middle areas, as explained in the preceding paragraph. The volume of solid whose base is *BCDV* is a pyramid whose altitude is the vertical distance between the 24-ft. contour and point *C*, which is 1 ft.

By the same general method the areas of *ABCDEK*, *MBCDPN*, etc., which refer to the new surface of the ground, may be obtained, and the volume of the solids between successive contour planes computed. The difference between this quantity and the quantity between a plane at elevation 20 and the original surface will give the amount of fill.

While in this particular problem the first method is the shorter, still there are cases where the second method will be somewhat simpler. It is particularly useful when the actual amount of cut or fill is not desired but when it is required to know if the proposed alterations will require more or less earth than can be easily obtained on the premises and, if so, about how much the excess will be. Under these circumstances the portions of cut and fill will not have to be computed separately. A line is drawn around the limits of the entire area where the grading is to be done, the volume between an assumed plane and the original surface is found, and

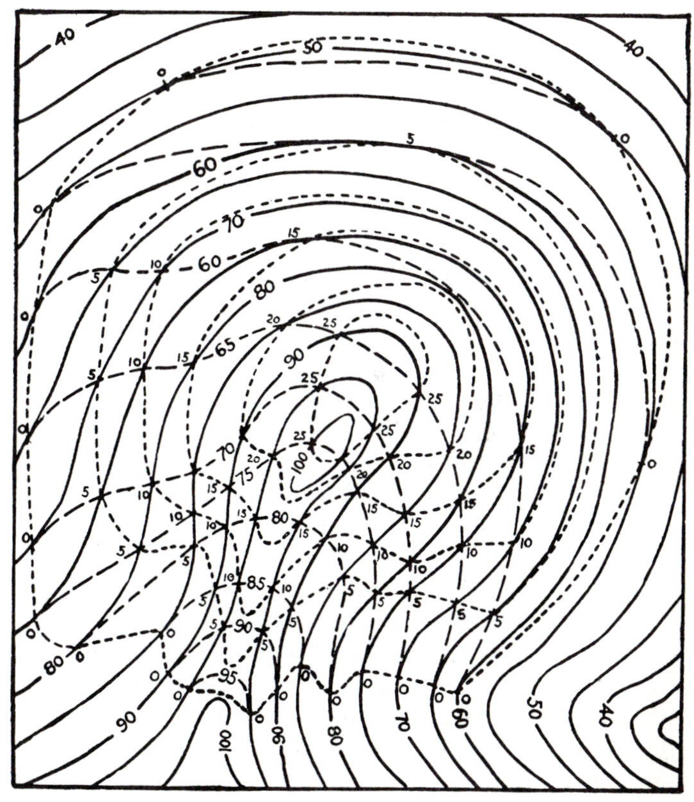

Fig. 14.29.1.

then the volume between the same plane and the proposed surface. The difference between the two values will give the amount of excess of earthwork.

14.29. Lines of Equal Cut or Fill.—(3) Fig. 14.29.1 illustrates a third method of computing earthwork from the data given on a topographic map. The original contours are shown in full lines and the contours of the proposed surface in dash lines. Through the intersection of the new contours with the original ones is drawn the line of "no cut" (zero line), the line where the cut is just 5 ft. (marked 5), the line of 10-ft. cut (marked 10), etc. These dotted curves enclose areas which are the horizontal projections of irregular surfaces which are parallel to the final surface and at 5 ft., 10 ft., 15 ft., etc., above the final surface. The solids included between these 5-ft. irregular surfaces are layers of earth each 5 ft. thick, and their volumes may be computed by either the End Area Method or by the Prismoidal Formula as explained in the preceding paragraphs. The areas of these horizontal projections are obtained from the map and the vertical dimensions of the solids are the contour intervals.

14.30. ROUGH ESTIMATES.—Rough estimates of the quantity of earthwork are often required for preliminary estimates of the cost of construction or for monthly estimates of the amount of work done. For preliminary estimates of road construction, very frequently the notes of alignment and the profile of the center line are the only information at hand. From this profile the center cuts or fills can be obtained, and the cross-sections can be assumed to be level sections and computed by the End-Area Method. The slight errors resulting will be corrected in the final estimate.

In obtaining the required data from which to make an approximate estimate of the quantity of earthwork, the engineer has an opportunity to exercise his judgment to an unusual degree. Rough estimates do not, as a rule, call for a large amount of fieldwork. It is important that as few measurements as possible should be taken and that these should also be at the proper places to give complete data and to allow simple computations.

CHAPTER XV

SIMPLE COÖRDINATE CALCULATIONS, TRAVERSES

15.1. The First Geodetic Problem.—Surveying coördinate computations depend upon the simple notion of transforming polar coördinates to rectangular coördinates and vice versa. The former concept may be called the first geodetic problem:

If the rectangular coördinates of a point A are given (X_A, Y_A) along with the length and direction of AB, find the rectangular coördinates (X_B, Y_B) of point B.

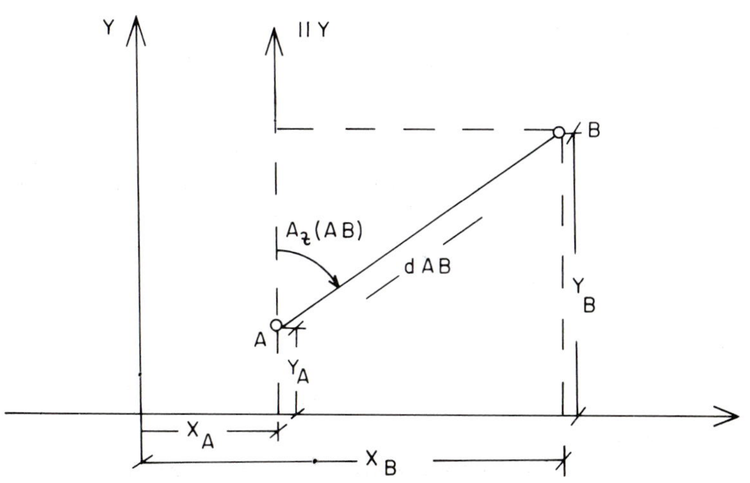

Fig. 15.1.1. Basic Coördinate Relations.

According to Figure 15.1.1 we obtain in triangle ABC:

$$\sin(\text{azimuth AB}) = \frac{X_B - X_A}{\text{distance AB}}$$

$$\cos(\text{azimuth AB}) = \frac{Y_B - Y_A}{\text{distance AB}}$$

Then

$$X_B = X_A + (\text{distance AB}) \times \sin(\text{azimuth AB})$$

$$Y_B = Y_A + (\text{distance AB}) \times \cos(\text{azimuth AB})$$

15.2. THE SECOND GEODETIC PROBLEM.—In the reverse case, the second geodetic problem, the coördinates of both points A and B are given, while the distance and direction between them are unknown. Again from Figure 15.1.1 it follows that:

$$\tan(\text{azimuth AB}) = \frac{X_B - X_A}{Y_B - Y_A}$$

and

$$\text{azimuth AB} = \text{arc tan}\left(\frac{X_B - X_A}{Y_B - Y_A}\right)$$

Once the azimuth has been determined, the distance follows from simple trigonometric relations.

$$\text{distance AB} = \frac{X_B - X_A}{\sin(\text{azimuth AB})} = \frac{Y_B - Y_A}{\cos(\text{azimuth AB})}$$

Due to the propagation of errors, it is advisable to depend more upon the fraction with the larger denominator. One may also use the well known Pythagoras formula

$$\text{distance AB} = \sqrt{(X_B - X_A)^2 + (Y_B - Y_A)^2}$$

although one must be cautious about capacity of the calculator used, lest needed significant digits be chopped off at the right end of the squared values.

15.3. CALCULATION OF TRAVERSE.—Traverses are run for a definite purpose which in turn determines the precision required and hence the methods that must be employed to obtain that precision. No survey is mathematically perfect, because no measurement is ever perfect. In a compass traverse, the bearings may be several minutes in error and the distances are usually measured without great precision; such a closed traverse in all probability would not meet the geometric requirements of a closed figure. Even in closed traverses observed with high precision there will be small errors in both angles and distances, and these precise surveys will not mathematically close until adjustments are made to the dimensions. These adjustments will be explained later.

Before "calculating the traverse" all linear distances should be corrected for all constant and systematic errors, and if slope measurements have been made, they should be reduced to the horizontal. (See Chap. III.)

Before leaving the field the angles should be added to see that the sum equals the proper number of right angles (if interior angles are measured their sum should equal 180° $(n - 2)$, in which n is the number of angles), or that the algebraic sum of deflection angles equals 360°, or is as near the correct value as is consistent with the precision used in measuring the angles. If, for example, the angles are measured to the nearest half-minute, the maximum uncertainty in any angle is $\pm 15''$, which is a compensating error because it is as likely that the true angle is $15''$ too small as $15''$ too large. On this basis, the allowable error in the sum of the angles for a nine-sided traverse is $\sqrt{9} \times 15'' = 45''$. Hence if the sum of the angles comes within $45''$ of the correct value, then the angles have been measured within the limits to be expected. If the difference were 2 or 3 minutes, the angles have probably been carelessly measured; if the difference is still greater, then a mistake is indicated which should be located and corrected before leaving the field.

In order to avoid large errors and mistakes linear distances are checked by measuring the lines forward and backward; the angles should be at the least doubled for a check. By these means a field check is obtained on each individual measurement.

For traverses that do not close, the angles may be checked by astronomical methods (Chapter VIII) or the methods explained in Art. 15.23.

The calculation of traverses is usually carried out by means of rectangular coördinates. Before these coördinates can be found it is necessary to adopt some meridian, or axis of Y, and then to find the bearings, or the azimuths, of all the lines of the traverse with reference to this meridian. The best reference meridian to use is the **true meridian,** or geographical meridian, passing through some point of the survey. This gives bearings which are unchanging, and also permits showing the survey in its correct relation to all other surveys in the locality that also show the true meridian. (Methods of determining the true meridian are given in Chapter VIII.) If, however, the survey was made with a magnetic compass, or if the direction of the true meridian cannot be ascertained, then the magnetic meridian may be used. If the direction of neither the magnetic nor the true meridian is known, any direction may be assumed arbitrarily as a meridian, or Y axis, and bearings or azimuths referred to this direction.

Unless the azimuths have been obtained in the field by the method of Art. 5.16, they must be calculated from the measured angles. If the deflection angles have been measured the azimuths may be found quickly by starting with one known (or assumed) azimuth and **adding the deflections algebraically,** right deflection angles being taken as positive, and left deflections as negative. This method is illustrated in the example in Fig. 15.4.1.

15.4. Balancing the Traverse Angles.—The first step in calculating a closed traverse of a transit survey is to balance the angles, i.e., the angles should be made to fulfill the geometric conditions of the figure. While the angles of a traverse may have all been read with the same care, they may not all be equally accurate, because of difficulty that may exist in centering the instrument over the station or in sighting on points requiring plumbing. On long lines the errors from the above causes are small, but on short lines they are large. Where the conditions under which the angles have been measured are known, the angular error should be distributed on the angles that were difficult to measure. If all the angles are assumed to be measured with the same accuracy the error should be placed on the angles dependent on the short side or sides.

The following shows the balancing of the angles of the survey shown in Fig. 5.5.1. It will be noticed by referring to Fig. 5.5.1 that there is a very short line running between sta. *D* and *E*. It is therefore more likely that small errors in angle occurred at the angle *D* and *E* because a slight inaccuracy in setting up the instrument at these stations and in taking the short sight to the adjacent station is likely to produce a small error in the measured angle. This survey should be balanced as follows.

Sta.	*Defl. Angles*		*Corrected Dist.*
B	87° 30′ L		179.2
C	91 09 L		164.6
D		91° 29′ R	99.7
E	92 00 L		169.3
F	95 17 L		286.2
A	85 34 L		299.2
Sums	451° 30′ L	91° 29′ R;	Diff. 360° 01′

This survey has an error in angle of only 01′ indicating an accurate survey. To be geometrically correct the algebraic sum should be 360°. Inasmuch as the angles are measured to minutes no smaller value than a minute in the adjustment of angles should be made. The angle at Station *D* (adjacent to shortest side) should be changed to read 91° 30′ R. If, in this same survey, the angles had been read to half minutes and the

Line.	*Bearing.*	*Azimuth.*	
A—B	North	180° 00′	
B	87° 30′ L	87 30 L	
B—C	N 87° 30′ W	92° 30′	
C	91 09 L	91 09 L	
	178 39		
C—D	S 1° 21′ W	1° 21′	
D	91 30 R	91 30 R	
	92 51		
D—E	N 87° 09′ W	92° 51′	
E	92 00 L	92 00 L	
	179 09		
E—F	S 0° 51′ W	0° 51′	
F	95 17 L	95 17 L	
	94 26		
F—A	N 85° 34′ E	265° 34′	
A	85 34 L	85 34 L	
A—B	N 00 00	180° 00′	(Check)

Fig. 15.4.1. Bearing and Azimuth Calculation.

total error was still one minute, then a correction of one-half minute would be added to the angle *D* and one-half minute subtracted from angle *E*, both of which lie next to a short line.

Using the balanced angles, bearings or azimuths are then computed as in Fig. 15.4.1, for the survey on Fig. 5.4.

The bearings and azimuths of the first line should be recomputed from the last bearing and azimuth; this is a check which is essential.

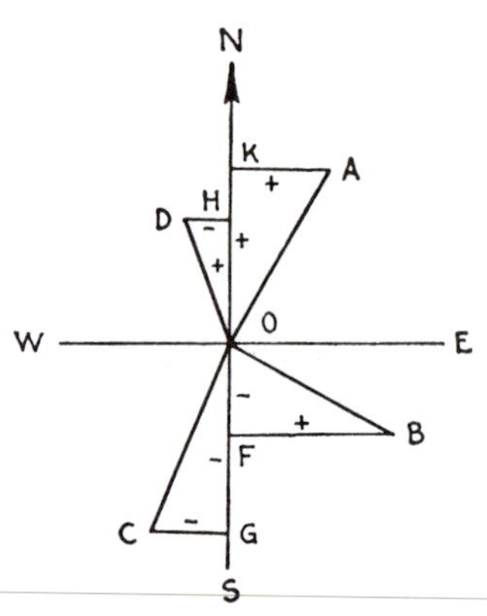

FIG. 15.5.1.

15.5. CALCULATING THE DIFFERENCE OF LATITUDE (LAT.) AND DEPARTURE (DEP.).—After the bearings or azimuths have been computed, the next step in obtaining the coördinates is the calculation of the *latitudes* and *departures*. The latitude of any line is its projection on a north and south line and the departure of the line is its projection on an east and west line. Latitudes are also variously termed *latitude difference* or *difference of Y's* and departures are termed *departure difference*, or *difference of X's*. In Fig. 15.5.1 the latitude of *OA* is *OK*, of *OB* is *OF*, of *OC* is *OG* and of *OD* is *OH*. The departure of *OA* is *KA*, of *OB* is *FB*, of *OC* is *GC* and of *OD* is *HD*. North latitudes are designated N or + and south latitudes S or −. East departures are designated E or + and west departures W or −; thus for a line with a northeast bearing both latitude and departure would be positive; if the bearing were southwest both latitude and departure would be negative.

The computation of latitude and departure is, therefore, a direct application of the first geodetic problem (see Art. 15.1). The result will have the correct arithmetic sign, if the azimuth is used. When bearings are available, it may therefore be advantageous to convert them into azimuths before computing latitudes and departures. If one uses bearings, however, the plus sign must be associated with N and E, and the minus sign with S and W.

15.6. Computation of the Error of Closure of a Traverse.— Since all the measurements in a survey are subject to error, any traverse which supposedly returns to its starting point will not do so in reality, but there will be a short line of unknown length and direction connecting the first and last points of the traverse; this short line is caused by the accumulation of errors.

If the calculated traverse returns mathematically to the starting point then the sum of the + Lats. and the − Lats. should be exactly equal and, similarly, the sum of the + Deps. and the − Deps. should be exactly equal. The amounts by which they fail to balance are the Lat. and Dep. of the line of closure. This distance is sometimes called the "linear error of closure" and reflects the algebraic sum of all the accidental errors of measurement both in angles and distances.

Suppose the difference of latitude and of departure between the first and final point is Lat. + 0.14 ft. and Dep. − 0.06 ft. This means that the calculated position of the final point is 0.14 ft. N and 0.06 ft. W of the first point. The distance between the first point and the computed positions of the last point, i.e., the "error of closure," is found thus

$$\sqrt{0.06^2 + 0.14^2} = 0.15 \text{ ft.}$$

This closure distance, however, does not indicate the precision of the measurements until compared with the total distance around the field. Therefore a more useful and convenient measure of the accuracy is the ratio of the two, that is, the linear error of closure divided by the perimeter of the figure. This is often expressed as a fraction whose numerator is unity, as 1/5000. Such a fraction states that the error of the survey is one part in 5000 parts and tells at once the quality of the work. This ratio is sometimes referred to as the "*error of closure*."

After the sums of the N and S Lats. and of the E and W Deps. have been found they should be examined to see if they differ sufficiently to indicate a large **mistake** in the fieldwork.

If a large difference is discovered the first step would be to check all the calculations to make sure that the mistake is not in the calculations themselves. If after this recomputation the error of closure still does not come within the required limits the distances must be remeasured in the field. (See Art. 15.13.)

15.7. Distribution of the Errors of Latitude and Departure, or "Balancing" a Closed Survey.—Before any quantities such as coördinates, dimensions, or areas, are calculated from the data of a closed survey the errors must be distributed so that a mathematically closed polygon is obtained. There are several rules which may be used for accomplishing this result.

In any closed figure the algebraic sum of the Lats. must equal zero and the algebraic sum of the Deps. must equal zero. If they do not, then corrections should be added or subtracted from the Lats. and Deps. so as to make these sums zero.

1. If the survey has been made by use of a compass and a chain the error in position of a point is assumed to be due as much to the error in bearing as to the error in distance, a reasonable assumption for these instruments. Hence the "compass rule."

The correction to be applied to the $\left\{\begin{matrix}\textit{latitude}\\\textit{departure}\end{matrix}\right\}$ *of any course is to the total error in* $\left\{\begin{matrix}\textit{latitude}\\\textit{departure}\end{matrix}\right\}$ *as the length of the course is to the perimeter of the polygon.*

This rule is mathematically correct for the assumption made, that is, it fulfills the requirements of the method of least squares.

2. If the survey is made with the transit and tape in such a way that it is reasonable to assume that the resulting error is due more to taping than to transit work, the "transit rule" may be used. First the angles are balanced as explained in Art. 15.4 before the Lats. and Deps. are computed. The errors in the sums of Lats. and Depts. are then distributed as follows:

The correction to be applied to the $\left\{\begin{matrix}\textit{latitude}\\ \textit{departure}\end{matrix}\right\}$ *of any course is to the total error in* $\left\{\begin{matrix}\textit{latitude}\\ \textit{departure}\end{matrix}\right\}$ *as the* $\left\{\begin{matrix}\textit{latitude}\\ \textit{departure}\end{matrix}\right\}$ *of that course is to the sum of all the* $\left\{\begin{matrix}\textit{latitudes}\\ \textit{departures}\end{matrix}\right\}$ *(without regard to algebraic sign).*

This rule is not mathematically correct. It is supposed to alter the distances more than the bearings, but it will not do so under all conditions. It will, of course, give a closed figure.

If it is known that certain measurements are less reliable than others, the adjustment may be made by *arbitrary distribution,* in which large increments of error are assigned to those measurements which are suspected of being of doubtful accuracy.

The compass rule, although originally devised for adjusting compass surveys, is quite commonly used to adjust transit and tape traverses, particularly in cases where there is no reason to believe that the distances are less reliable than the angles. (See example in Fig. 15.24.2.)

With reference to the selection of a rule for balancing a traverse it should be observed that it is more important from a practical standpoing that the polygon should be closed than that it should be closed by any particular method. If the survey is made to close, then all the resulting (computed) dimensions will be consistent, and errors will not accumulate from the computations.

In regard to what is and what is not revealed by the error of closure it should be observed that *the constant errors are not reflected* in the error of closure. If, for example, the tape is 0.01 ft. too long and a correction for this constant error has not been applied to the distances the computed closure is not affected thereby. If the temperature of the tape is constant throughout the survey and a temperature correction is not applied to the recorded distances, similarly, it will not affect the closure. If, however, the tape temperature is recorded at a certain amount and the distances corrected accordingly but, as a matter of fact, the temperature changes during the work, then the effect of the variations in temperature enter as a part of the error of closure. This error of closure is made up chiefly of errors of an accidental nature. Notice that small accidental errors in determining the corrections for temperature, sag, pull, etc. remain in the final result, and appear in the error of closure.

In each of the preceding rules it is assumed that all measurements are equally well made and therefore that all lines are to be given the same weight in balancing. Such conditions rarely occur, however, in practice. One line may have been measured over level ground with the tape supported throughout its length; another may have required plumbing over steep slopes using short lengths of tape; and still another line may have been measured through dense bushes where it was impossible to keep the tape exactly straight. Such conditions may justify assigning most, if not all, of the error of closure to the Lats. and Deps. of these difficult lines. At any rate, wherever there is definite knowledge concerning the difficulties of measuring particular lines it is advisable to assign greater corrections to such lines than to others, that is, to assign arbitrarily to such lines the proportional part of the error that it is believed they should have. The rules given above for balancing may always be used as guides even when they are not applied strictly. They may be used in distributing any remaining error not arbitrarily assigned. With any method the corrections are applied so as to diminish the sum in the Lat. (or Dep.) column that is too large and increase the sum in the column that is too small.

During the process of making the field survey it is valuable to make notes in the field book of places where it was difficult to measure the angles accurately and also the lines where difficulty was met in making the measured distances. These facts aid greatly in determining how to distribute errors of closure in the calculated Lats. and Deps.

Another fact to be considered in the balancing of closed traverses is that such errors as that of "lining in" the intermediate points, judging whether the tape is horizontal, measuring around trees or bushes, all tend to make the recorded distances too long. When tapemen are careless about pulling the tape taut the recorded distance will be too long from this cause also. It is considered better, as a general rule, to shorten the Lats. and Deps. in the "long" columns than to lengthen those in the "short" columns. If, however, the lines are clear and an excessive tension is used then the recorded lengths may be too short.

It should be noticed that a strict adherence to any rule requires that the total error, which may be but a few hundredths (or a few tenths), shall be designated in thousandths in order to express the correction for each line. As it is inadvisable to carry such small fractions the whole matter often comes down to a more or less arbitrary distribution of the error of closure.

When calculating the corrections it should be noted that it is unnecessary to work out the complete proportion for each correction; if the total error is divided by the perimeter (compass rule) or by the sum of the latitudes or departures (transit rule) the result is the constant *correction per foot or per tape-length*, according to the unit used. If this is multiplied by the distances (or latitudes and departures) in succession the corrections are obtained. This may be done mentally or by slide rule. The number of feet per 0.01 ft. correction may be obtained by inverting the ratio of the error of closure and dividing by 100.

15.8. Graphical Adjustment of Error of Closure in a Traverse.—The graphical method of adjustment affords a simple means of making traverse adjustments. If a traverse *ABCDEA′* (Fig. 15.8.1) has been plotted by bearings and distances (or developed graphically by the plane table method) there will be a certain error of closure, *AA′*. To adjust the traverse graphically, construct a straight line *AA′* representing the length of the perimeter of the traverse with the points *B*, *C*,

D, and *E* plotted on this straight line in proportion to the distances of these points from the starting point A. At point *A'* construct line *A'a* parallel to the error of the closure and equal to this error at any convenient scale. Draw *Aa;* this gives the

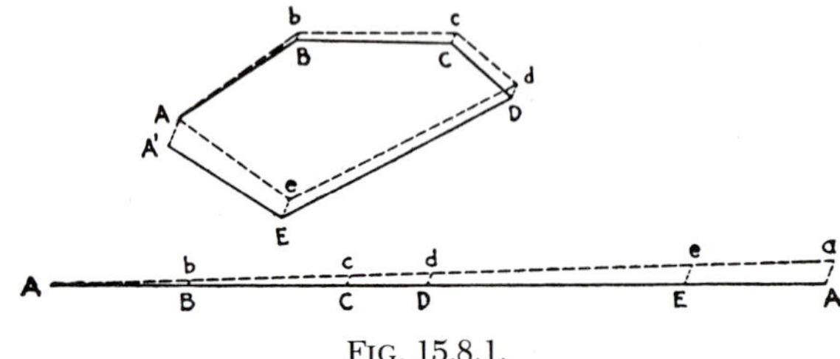

FIG. 15.8.1.

correction triangle AA'a. Lines *Bb*, *Cc*, *Dd*, and *Ee* are drawn parallel to *A'a*, and each one of these lines represents the correction (direction and distance) to be applied at points *B*, *C*, *D*, and *E* on the original traverse to give the adjusted positions *b*, *c*, *d*, and *e*, thus closing the traverse.

The correction triangle may be constructed to adjust any closed traverse. The base of the triangle may be plotted to one scale and the error of closure may be magnified by plotting to another. If the base of the triangle *AA'* is drawn as an E and W line, then the altitude of *a* above the base is the error of closure in latitude and the perpendicular distance of *a* to the left or right of *A'* is the error in departure. Similarly the errors in latitude and departure may be scaled from the diagram at points *B*, *C*, *D*, and *E*.

15.9. CALCULATION OF RECTANGULAR COÖRDINATES OF THE STATIONS OF A SURVEY.—The coördinate system of surveying is widely used for state and municipal surveys and for large scale construction projects. In this system the position of each survey point is fixed by *X* and *Y* coördinates referred to two lines at right angles to each other, usually some meridian and a line perpendicular to it. The advantage of this system lies in the fact that as all surveys refer to the same origin they are tied to each other, and any lot may be re-located by means of the known coördinates of its corners, even if all points in the immediate vicinity have been lost.

Usually the coördinate lines run north and south and east and west. For some construction surveys it may be more desirable to have one coördinate parallel to a direction along which the principal features are oriented, such as parallel to a street line.

Whenever practicable the coördinates should be tied in to an established state or municipal plane coördinate system. Otherwise the origin may be taken through one of the survey stations. The X coördinates of the successive points are obtained by adding algebraically the corresponding departure, or difference of X. The Y coördinates are obtained similarly from the latitudes of the successive lines. These X and Y coördinates are often called the **total departures** and **total latitudes** of the survey points.

Example: In the example shown (Fig. 15.12.1) the coördinates of the points A, B, C, and D referred to A would be computed as follows:

Point	Y	X
A	0 (assumed)	0 (assumed)
B	−242.3	−319.4
	+301.9	−430.9
C	+ 59.6	−750.3
	+318.6	+226.3
D	+378.2	−524.0

Negative coördinates, as at B, can and sometimes are avoided by adopting some other station as an origin, as, for instance, B itself or a point 1000 ft. south and 1000 ft. west of A.

15.10. Calculation of the Coördinates of the Corners of a Closed Survey.—The origin may be one of the survey stations, such, for example, as the most westerly or the most southerly station, or it may be at the intersection of a meridian through the most westerly and a perpendicular through the most southerly points. All of the coördinates will be positive if this last suggestion is adopted. The coördinates are calculated exactly as in the preceding article. There is an opportunity here to check the calculations by recomputing the coördinates of the first point from those of the last and observing whether the original value is again obtained. For an example of this calculation see Fig. 17.9.1.

15.11. Calculation of the Length and Bearing which are Consistent with the Balanced Latitudes and Departures.—In describing property in a deed it is important to use bearings and distances which represent the best that can be obtained from the measurements. Since we have adjusted the traverse so as to increase the accuracy of the Lats. and Deps. it is of almost equal importance to calculate distances and bearings which are consistent with these new Lats. and Deps.

This is easily obtained using the formulae for the second geodetic problem (Art. 15.2) with either the newly adjusted coördinates, or the coördinate differences, which represent Lats. and Deps.

15.12. Calculation of Closing Side from Lats. and Deps. of Several Lines.—Suppose that *A*, *B*, *C*, and *D* are connected by angles and distances and that the Lat. and Dep. of each line have been computed, as shown in Fig. 15.12.1.

LINE	BEARINGS.	DIST., FEET.	LATS.		DEPS.	
			N.	S.	E.	W.
A–B	S 52° 49′ W	400.9		242.3		319.4
B–C	N 54 59 W	526.1	301.9			430.9
C–D	N 35 23 E	390.8	318.6		226.3	
		sums	620.5	242.3	226.3	750.3
			242.3			226.3
		diffs.	378.2			524.0

FIG. 15.12.1. CALCULATION OF CLOSING SIDES.

It is desired to find the distance from *A* to *D* and the bearing of this line. The total difference of latitude of *A* and *D* is the difference in the sums of the two latitude columns, or 378.2. Similarly the total difference of departure is 524.0. These two coördinates determine the direction and distance from *A* to *D* exactly as though *AD* **were one of the lines of the traverse.**

Applying the second geodetic problem we obtain:

$$\tan(\text{az. AD}) = \frac{524.0}{378.2} = 1.3855$$

$$\text{azimuth AD} = 305°\,49.2'$$

$$\text{Dist. AD} = \frac{-524.0}{0.81086} = \frac{378.2}{0.58524} = 646.2 \text{ ft.}$$

15.13. Detecting Mistakes in Distances.—Mistakes in fieldwork may often be detected by means of the calculations. One of the easiest mistakes to make in surveying is to omit a whole tape-length in counting. If such a mistake were made and the latitudes and departures were computed, the linear error of closure of the survey would prove to be about a tape-length. In order to find in which line this mistake probably occurred compute the bearing of this linear error of closure and examine the traverse to find a line having a bearing the same or nearly the same. The error in departure divided by the error in latitude equals the **tangent of the bearing of the line** that represents the error of closure of the traverse. The errors of the survey, of course, will prevent these bearings from agreeing exactly. If two mistakes have been made it may be difficult and sometimes impossible to determine where they occurred. When an error of this sort is indicated by the computation the line should be remeasured. It is bad practice to change an observed measurement merely because it is found by calculation to disagree with other measured distances.

It may, and frequently does, happen that there are several lines in the traverse that have about the same bearing. In such a case it is impossible to tell in which of these lines the mistake occurred. But if a cut-off line is measured, as was suggested in Art. 5.17, and one portion of the survey balances, then the other part will contain the mistake. By proceeding in this way the number of lines in which the mistake could occur is reduced so that its location can be determined and checked by field measurement.

15.14. Detecting Mistakes in Angles.—If the angular misclosure indicates that a blunder has occurred during the angular measurements, then the following procedure may be used to locate the station at which a remeasurement is required. Compute or plot the traverse in both clockwise and anticlockwise direction starting from the initial point. The traverse stations will be at their correct position up to the point where the wrong angle was measured and from there on the positions will be wrong. Since this point is approached from both directions, it will be the only one with matching coördinates (or matching position plot), while each of the stations will have two different positions or two different sets of coördinates. Thus the station at which the angle blunder occurred is detected. If there is more than one blunder in the angular measurement, the method does not work and it is advisable to remeasure all angles.

15.15. SUPPLYING MISSING DATA.—If any two parts of a traverse, either distance or bearings, are missing, it is possible to supply these missing parts by computation and thus obtain a complete set of notes for the traverse. Since the algebraic sum of the latitudes must be zero we may write an equation of the forms $\sum l \cos B = 0$; similarly for the departures, $\sum l \sin B = 0$. In each equation there may be (the same) two missing parts. A solution of these equations simultaneously will give the missing parts. In practice this method is seldom applied but geometric solutions are preferred.

The solutions of the different cases may be classified according to whether the two missing parts are in the same line, in any two adjacent lines, or in two non-adjacent lines.

15.16. When the bearing and length of the same line are missing, then the two parts are computed at once from the latitude and departure of this course, according to the 2nd geodetic problem (15.2). This case is of frequent occurrence in the subdivision of land.

When the missing parts occur in adjacent lines we may draw a line joining the extremities of the known part of the traverse, leaving a triangle containing the unknown parts. The solution of this triangle gives the required parts. For example, suppose that a lot *ABCDEFGH* (Fig. 15.16.1) is crossed by a line *XY* and the distance *Dm* and the angle *DmY* have been measured. *ABCDmnA* is a closed traverse in which the two distances *mn* and *An* are missing. If we draw *Am* we have a continuous traverse *ABCDmA* in which *mA* is unknown both in direction and distance. These two missing parts may be computed by the method of Art. 15.2. Then in the triangle *Amn* we know *Am* and all of the angles and may therefore compute the distances *An* and *mn*. If the angles were missing instead of the lengths the solution of the triangle would give these angles. This case occurs frequently when a new (proposed) line is projected across the boundaries of an existing lot, as when taking land for a highway or a pipe line.

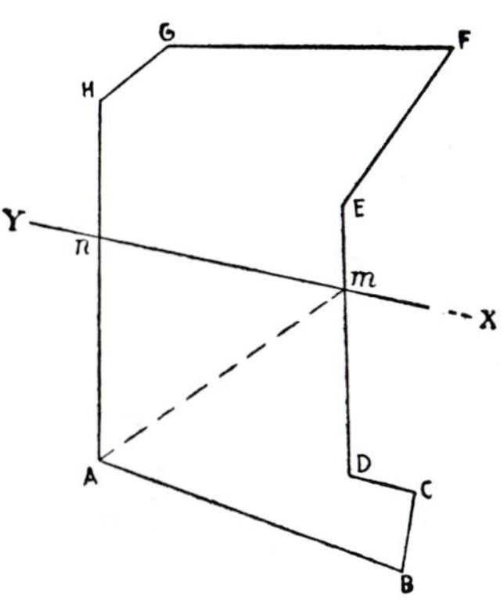

FIG. 15.16.1.

If the missing parts occur in non-adjacent sides, such for examples as *HG* and *EF*, we may suppose *G* to be moved to *H* and *F* to be moved an equal and parallel distance *FF'*. Then we may compute the traverse *HABCDEFF'H* as in the preceding case (triangle *EFF'*) and find the length of bearing of *FF'*, which will be the same as for *GH*.

15.17. Other Cases.—The solutions of the *other cases* of missing data are not so simple, as they involve the use of simultaneous equations; they will not be discussed here.

15.18. One Side, Two Angles Missing.—Besides the three cases mentioned above there are some special cases which can be solved. The following example is typical of this kind of problem. In Fig. 15.18.1 the lines and angles measured are shown by full lines. The bearing of *AB* is given. Here one side and two angles are missing. The solution is as follows. In the triangle *EAB* find *EB*, *EBA*, and *AEB*. In the triangle *EDC* find *EC*, *DCE*, and *DEC*. Then in the triangle *EBC*, in which *EC*,

EB, and *EBC* are known, find *ECB*, *CEB*, and *BC*. All the angles and sides are then known.

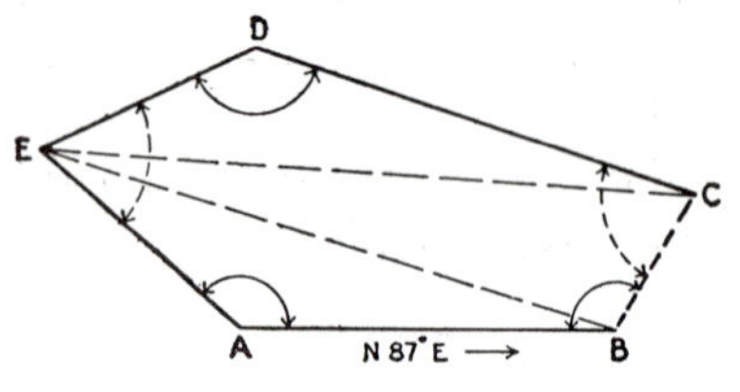

FIG. 15.18.1.

15.19. THE SUBDIVISION OF LAND.—There are many different problems which may arise in the subdivision of land and which may be solved simply by the application of the principles of trigonometry. A few of these problems are so common and so frequently involved in the working out of more complicated cases that their solution will be given.

15.20. To Cut Off from a Traverse a Given Area by a Straight Line starting from a Known Point on the Traverse.—In Fig. 15.20.1, *ABCDE* represents the traverse which has been plot-

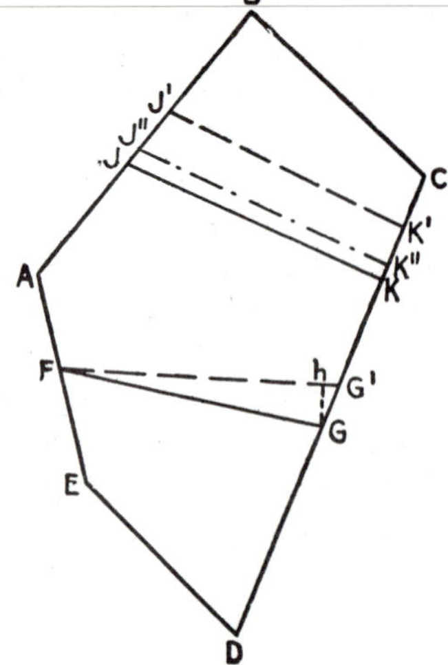

FIG. 15.20.1.

ted and whose area has been computed. It is desired to cut off a certain area by a line running from *F*, which is at a known distance from *A* or *E*. The line *FG′* is drawn on the plan so as to make the area *FG′ DE* approximately equal to the desired area. The line *DG′* is scaled off and the scaled distance used as a trial length. Then the side *FG′* and its bearing can be found by the method explained in Art. 15.16, and the area *FEDG′* computed in the usual manner. The difference between the required area and the area of *FEDG′* is the amount to be added to or subtracted from *FEDG′*. If this correction area is a minus area then *GFG′* will represent the correction triangle. In this triangle the base *FG′* and its area being known the altitude *hG* and the distances *GG′* and *FG* can be readily computed. In the traverse *FGDE*, which is the required area, the length of the missing side *FG* and its bearing can be supplied.

Instead of using the trial line *FG′* the line *FD* might have been first assumed and the correction triangle would then be *FDG*. This method has the advantage of containing one less side in the first trial area, but the correction triangle is large, whereas in the method explained above the correction triangle is small, which may be of advantage in that part of the computation.

15.21. To Cut Off from a Traverse a Given Area by a Line running in a Given Direction.—In Fig. 15.20.1, *ABCDE* represents a closed traverse from which a given area is to be cut off by a line running at a given angle *(BJK)* with *AB*. On the plot of the traverse draw the line *J′*K′ in the given direction cutting off *J′BCK′* which is, as nearly as can be judged, the required area. Scale the distance *BJ′* and use this trial distance in the computations. Then compute the distance *J′K′* and the area of *J′BCK′* by the method suggested in Art. 15.18, i.e., by dividing *J′BCK′* into two oblique triangles. The difference between this area and the required area is then found; this is a correction trapezoid to be added to or subtracted from *J′BCK′*. It will be assumed that it is to be added to *J′BCK′*.

In this correction trapezoid the area and one base *J′K′* are known; also the base angles, *J′* and *K′*. From these data an approximate value for the altitude of the trapezoid can be obtained and the length of the other base $K''J''$ of the trapezoid computed from this altitude and the length of $J'K'$. Then the area of this trapezoid $J'K'K''J''$ can be determined accurately; the difference between this and the required correction will be small and the dimensions of the second correction trapezoid $J''K''K'J'$ can probably be computed readily from its area, and the length of $J''K''$, which are known. By successive trials, probably not more than two, the correct line JK can be found. If lines AB and CD are approximately parallel the trapezoid is nearly a parallelogram and then its correct altitude can be quickly determined.

15.22. To Find the Area Cut off from a Traverse by a Line running in a Given Direction from a Given Point in the Traverse.—This problem may be readily solved by drawing a line from the given point in the traverse to the corner which lies nearest the other extremity of the cut-off line. The area of the traverse thus formed is then computed, and this area corrected by means of a correction triangle.

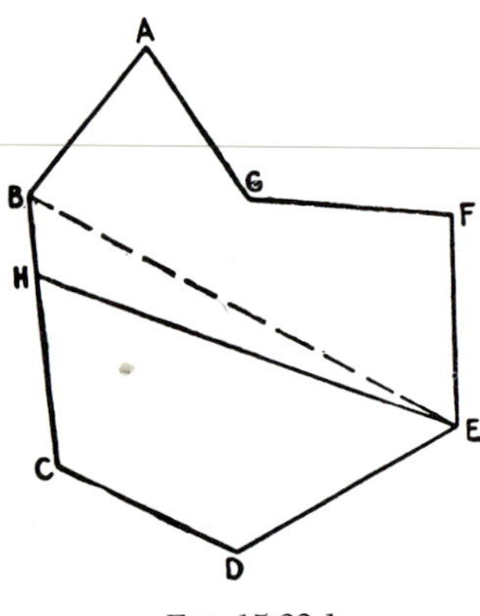

FIG. 15.22.1.

In Fig. 15.22.1, *ABCDEFG* represents a plot of a field. It is desired to run the line from *E* in a given direction *EH* and to compute the area *HEFGAB* cut off by this line. The latitude and departure of points *B* and *E* being known the bearing and length of *BE* and the area of *ABEFG* can be computed. Then the area and the remaining sides of the triangle *BEH* can be obtained from *BE* and the angles at *B* and *E*.

It is obvious that the solution of such problems as these is greatly facilitated by plotting the traverse before attempting the computations. This type of problem would be greatly simplified by the strict use of coördinates.

15.23. COMPUTATION OF AZIMUTHS WHEN CHECKING ANGLES TO A DISTANT OBJECT.—In this kind of problem the coördinates of all the points along the traverse can be computed with reference to some coördinate axes. At A and B (Fig. 15.23.1) angles have been taken to S, and from these angles the coördinates of point S, referred to AB and a line perpendicular to AB as axes, can be computed (Art. 15.9). Coördinates of S referred to the same axes should have the same value when figured from BC as a base when calculated from the base CD, and so on. If, however, when computed by means of angles at D and E, the point falls at S', and angle E and F give its locations also at S' there is evidence of a mistake in the traverse at station D. If the two locations of S and S' are such that a line between them is parallel to either CD or DE, the mistake was probably made in the measurement of the length of the line parallel to SS' and the distance SS' should be approximately equal to the amount of the mistake in measurement. If, however, SS' is not parallel to either CD or DE the mistake probably lies in the angle at D.

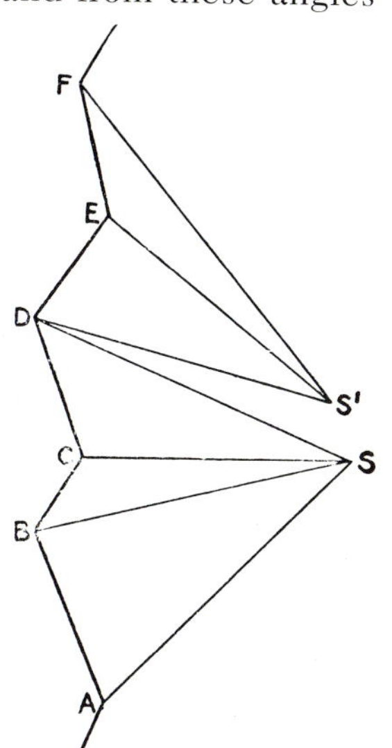

FIG. 15.23.1.

15.24. Other Types of Traverses.—It is not necessary that a traverse returns to its initial point. It could be an open traverse extending from a known point. In this case however, there is no checking possibility and no accuracy measure possible.

Therefore, particularly when using coördinates, traverses are extended until they end at a known coördinated point. If at this terminal point, a reference direction can be measured, the traverse is fully controlled, as the initial azimuth is propagated through the traverse angles to the closing azimuth, while direct coördinate closure is obtained with the coördinated end points. Figures 15.9 and 15.10 show such a traverse and the pertinent calculations.

Sometimes it might not be possible to end at a known point. The astronomical azimuth (sun-shot) on a direct azimuth observation with a gyro-instrument (compass for lower accuracy) would at least provide a means for angular closure.

Traverse From P_1 to P_2

Bearing of P_2 to P_1: S15-20-52W

Coördinates of P_1: N 10,596.76, E 18,601.42

Coördinates of P_2: N 10,248.37, E 19,097.73

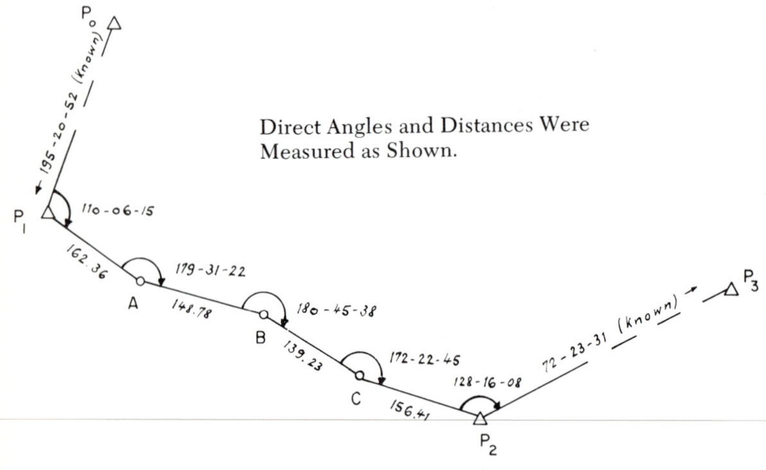

FIG. 15.24.1. TRAVERSE WITH COÖRDINATE CONTROL AT BOTH ENDS.

(a) Verify azimuth closure, and adjust angles to achieve mathematically correct azimuths.

(b) Calculate Lats. and Deps. and balance traverse if within permissible 1/5000 closure.

(c) Calculate coördinates of all stations, using adjusted Lats. & Deps.

(d) Recalculate bearings and distances from adjusted final coördinates. The calculations and results are illustrated in Fig. 15.24.3, while a more detailed azimuth calculation is presented in Fig. 15.24.2. In Fig. 15.24.4 the adjusted bearings are shown. Two complete adjustments performed on a WANG 500 PT are presented in Figures 15.24.5 and 15.24.6.

The calculations and results are illustrated in Fig. 15.24.3, while a more detailed azimuth calculation is presented in Fig. 15.24.2. In Fig. 15.24.4 the adjusted bearings are shown. Two complete adjustments performed on a WANG 500 PT are presented in Figures 15.24.5 and 15.24.6.

195-20-52	P_0-P_1
−180	
15-20-52	P_1-P_0
+110-06-21	
125-27-13	P_1-A
+180	
305-27-13	A-P_1
+179-31-28	
484-58-41	
−360	
124-58-41	A-B
+180	
304-58-41	B-A
+180-45-44	
485-44-25	
−360	
125-44-25	B-C
+180	
305-44-25	C-B
+178-22-51	
484-07-16	
−360	
124-07-16	C-P_2
+180	
304-07-16	P_2-C
+128-06-15	
432-23-31	
−360	
72-23-31	P_2-P_3

FIG. 15.24.2. CALCULATING AZIMUTHS OF TRAVERSE.

Point	Traverse Angle	Azimuth ϕ	Distance d	Sin ϕ Cos ϕ	ΔX $= d \sin\phi$	ΔY $= d \cos\phi$	X	Y	Pt.
P_0		195-20-52							
P_1	110-06-15 +6	125-27-13	162.36	+0.81458 −0.58004	+132.26	−94.18	18601.62	10596.76	P_1
A	179-31-22 +6	124-58-41	148.78	+0.81937 −0.57326	−8 +121.91	+4 −85.29	+132.18 18733.60	−94.14 10502.62	A
B	180-4[illegible]-38 +6	125-44-25	139.23	+0.81167 −0.58411	−7 +113.01	+4 −81.33	+121.84 18855.44	−85.25 10417.37	B
C	178-22-45 +6	124-07-31	156.41	+0.82781 −0.56100	−7 +129.49	+3 −87.74	+112.94 18963.38	−81.30 10336.07	C
P_2	128-16-08 +7	72-23-31			−8	+4	+129.41 19097.79	−87.70 10248.37	P_2
P_3	972-23-00 −900-00-00 72-23-00		606.78		+496.67 −496.37 + 0.30	−348.54 +348.39 − 0.15	$X_{P_1}-X_{P_2}=496.37$	$Y_{P_1}-Y_{P_2}=-348.39$	

Corr. $= \frac{31}{5} = 6''.2$ From coordinates: $\phi P_0P_1 = 195°\,20'\,52''$ and $\phi P_2P_3 = 72°\,23'\,31''$.

FIG. 15.24.3. TRAVERSE COMPUTATION WITH COMPASS RULE.

	AZIMUTH	BEARINGS	CORRECTIONS USED (SEC.)
P_0P_1	195-20-52	S15-20-52W	00
P_1A	125-27-13	S54-32-47E	06
AB	124-58-41	S55-01-19E	12
BC	125-44-25	S54-15-35E	18
CP_2	124-07-16	S55-52-44E	24
P_2P_3	72-23-31	N72-23-31E	31

These are azimuths and bearings already corrected for closure in azimuth—the 31″ have been distributed, about 06″.2 per angle.

FIG. 15.24.4. CALCULATION OF ADJUSTED BEARINGS.

On WANG 500 PT
TRAVERSE COMPUTED and
BALANCED BY COMPASS RULE
June 12, 1975

10596.760	Y
18601.420	X
2543247.000	B
2. 54 32 47	B
162.360	D
10502.584	Y
18733.676	X
2550119.000	B
2. 55 01 19	B
148.780	D
10417.294	Y
18855.582	X
2541535.000	B
2. 54 15 35	B
139.230	D
10335.968	Y
18968.591	X
2555244.000	B
2. 55 52 44	B
156.410	D
10248.230	Y
19098.076	X
.1380	Y
−.2860	X
1910.35	E
4. 64 14 05	B
.317	D
10596.760	Y
18601.419	X

Despite a closure precision of only 1/1910.35, the adjustment is made in this instance for illustrative purposes.

10596.760	Y	P_1
18601.419	X	
2. 54 32 29	B	
162.276	D	
10502.621	Y	A
18733.599	X	
2. 55 01 02	B	
148.703	D	
10417.364	Y	B
18855.435	X	
2. 54 15 16	B	
139.158	D	
10336.070	Y	C
18968.379	X	
2. 55 52 28	B	
156.329	D	
10248.368	Y	P_2
19097.790	X	
5869559.48	F	AREA*
134.746544	A	

*In an open traverse this result is meaningless, but is printed by the computer in this case anyway. It is meaningful in the case of a closed traverse.

FIG. 15.24.5. COMPASS RULE ADJUSTMENT ON WANG 500 PT CALCULATOR.

On WANG 500 PT
TRAVERSE COMPUTED and
BALANCED BY TRANSIT RULE
June 12, 1975

```
10596.760    Y
18601.420    X

2543247.000  B
2. 54 32 47  B
162.360      D
10502.584    Y
18733.676    X

2550119.000  B
2. 55 01 19  B
148.780      D
10417.294    Y
18855.582    X

2541535.000  B
2. 54 15 35  B
139.230      D
10335.968    Y
18968.591    X

2555244.000  B
2. 55 52 44  B
156.410      D
10248.230    Y
19098.076    X
-------------------

.1380        Y
-.2860       X
1910.35      E
4. 64 14 05  B
.317         D
10596.760    Y
18601.419    X
```

```
10596.760    Y
18601.419    X

2. 54 32 29  B
162.276      D
10502.621    Y
18733.599    X

2. 55 01 02  B
148.703      D
10417.364    Y
18855.435    X

2. 54 15 17  B
139.158      D
10336.071    Y
18968.379    X

2. 55 52 27  B
156.328      D
10248.369    Y
19097.789    X

5869558.33   F
134.746518   A
```

FIG. 15.24.6. TRANSIT RULE ADJUSTMENT ON WANG 500 PT CALCULATOR.

15.25. CALCULATION OF SIDES OF A TRIANGLE.—In a triangulation system the base-line is the only line whose length is known at the start. The sides of any triangle are found from the law of sines, as follows:

$$\frac{\sin A}{\sin B} = \frac{a}{b} \qquad \frac{a \sin B}{\sin A} = b$$

$$\frac{\sin A}{\sin C} = \frac{a}{c} \qquad \frac{a \sin C}{\sin A} = c$$

Find the coördinates of Point C.
Coördinates of A are:
N 71,642.492
E 37,002.169
Bearing of AB is:
S 27-14-32 W
Base Line AB measured: 487.812 ft.
Angles measured and adjusted:

At A:	97-39-03
B:	40-31-42
C:	41-49-15
Sum:	180-00-00

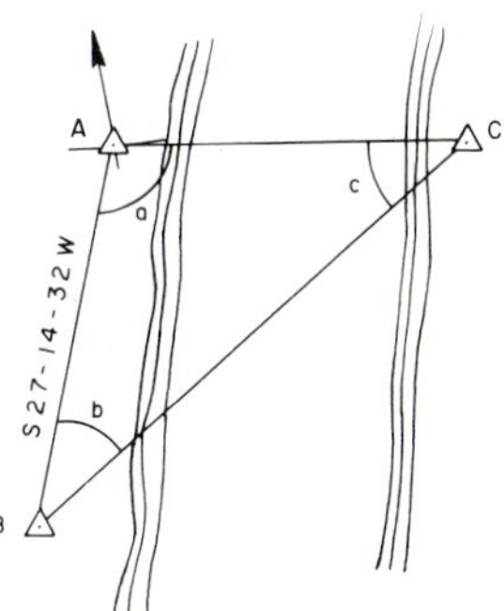

FIG. 15.25.1. SINE-RULE CALCULATION.

Solution:

$$AC = \frac{AB \sin b}{\sin c}$$

$$= \frac{487.812\ (0.6498240)}{(0.6668035)}$$

$$= 475.390$$

Bearing of AB:	S 27-14-32 W
at A:	97-39-03
Difference:	70-24-31

Bearing of AC: S 70-24-31 E

$\Delta y = 475.390 \cos$ bearing	$\Delta x = 475.390 \sin$ bearing
$\Delta y = 475.390\ (0.3353100)$	$\Delta x = 475.390\ (0.9421079)$
$\Delta y = -159.403$	$\Delta x = +447.869$
N71,642.492 (A)	E37,002.169 (A)
N71,483.089 (C)	E37,450.038 (C)

15.26. AVAILABILITY OF COMPUTERS AND PROGRAMS.—Computing costs are diminishing with advancing technology. The number of installations are increasing due to the widespread application of computers to all phases of business and industry. Most machines, because of large capital outlay, high maintenance, obsolescence and costs, are leased from the manufacturer, rather than purchased.

Computing facilities of all sizes and configuration are available throughout the country, in small cities as well as large, through Service Companies. The cost of computer use is usually a function of time used and will vary with the size of machine, the time of day and the total time per month. The larger the machine, the greater its capacity, flexibility and speed and the higher the unit hourly cost. The normal work day hours are in greatest demand and therefore command the highest price. Most Service Companies seek a guarantee of a minimum amount of computer time per month, and as more time is guaranteed, the rate is reduced accordingly.

Computer programs are available from Service Companies, Consultants, User Groups and the Computer Manufacturers. User Groups are mostly affiliated with a profession or field. Many programs are free while others are available on a fee basis.

Most of the problems described in this book can be solved on the smaller machines, but because of the savings in computer time, the larger ones may be more economical to use if the volume of computations warrants.

15.27. APPLICATIONS OF COMPUTER TO SPECIFIC SURVEYING PROBLEMS.—The application of computers to the surveying problems described in various chapters of this book will be examined. They include traverse closure and area computations, boundary calculations, horizontal and vertical geometry and earthwork volumes. The traverse calculations will be carried out in some detail; the other applications will be outlined.

Before proceeding with the traverse solution, the following computer terminology will be explained. The information fed to a computer is called *input*, and the answers returned by the computer are called *output*. The input and output usually go to or come from the memory of the computer. The *memory* is that portion of the computer in which information can be introduced and at a later time retrieved. A single item of input or output is a *word*. The media of input and output are the console or remote typewriter, punched cards, punched or magnetic tape and high speed printers. Perhaps the most common medium is punched cards. These are of uniform size and shape normally having 80 columns in each of which any 12 positions may be punched (See Fig. 15.27.1).

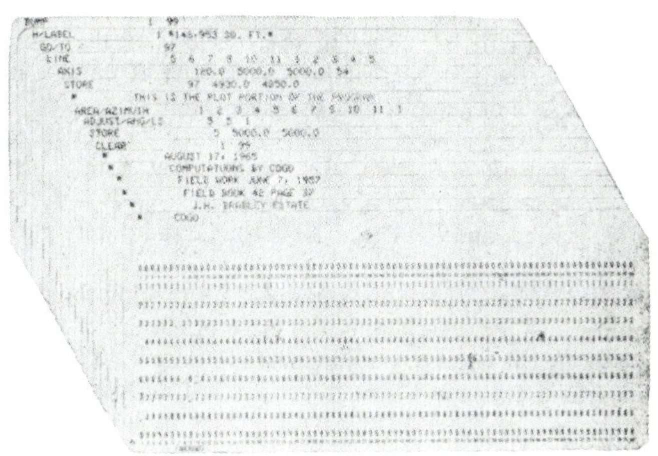

FIG. 15.27.1. PUNCHED CARD INPUT FOR TRAVERSE PROBLEM. (Excluding Data Cards)

15.28. Program Systems for Electronic Computers.—There are large program systems commercially available which are capable of solving any surveying problem, however they do require access to an electronic computer. Systems which are most useful for the surveyor include ICES COGO (Integrated Civil Engineering System—Coördinate Geometry), ICES ROAD (Roadway Analysis and Design System), SAPGO (Simultaneous Adjustment of Photogrammetric and Geodetic Observations) and many others.

The manufacturers of programmable desk calculators also have programs available for most of the common surveying problems, such as traversing, which are either provided with the machine or can be purchased.

To illustrate the use of computer program systems, the COGO system is considered in more detail, while for others the reader is referred to the proper user's manuals.

15.29. THE COGO SYSTEM.*—The basis for the COGO system is the location of points on a plane surface, identified by a point number and its associated X and Y coördinates. A program in this system consists of a series of "*commands*" and the data required to solve a geometric problem. No knowledge of the intricacies of the computer is necessary. Typical commands in the COGO system are shown in Fig. 15.29.1. Some of these will be used in solving a traverse closure problem, Art. 15.30.

Referring to Fig. 15.29.1, the words with a slash between them comprise the command, such as LOCATE/line, and the following row of letters, such as J K N D, is the data. In applying the method, the points are given numbers keyed to their coördinates, and other data are quantified, as shown by the example of input given for each command and as illustrated in accompanying graphs. Note that the notations of degrees, minutes and seconds are omitted from angles. Also, in expressing bearings, quadrants are designated by numbers rather than by compass directions: NE = 1, SE = 2, SW = 3, and NW = 4. When azimuth is used, it is measured clockwise from the north.

The commands listed in Fig. 15.29.1 illustrate but a few of those in the COGO system. Others are available for solving many geometric, area and intersection problems. Plotting commands instruct plotting machines to make drawings which include straight and curved lines, with or without annotation and titles.

The COGO System has extensive use in both engineering and surveying operations.

15.30. TRAVERSE CALCULATIONS.—The problem to be solved is that of the Bradley Estate which is carried through the book, with field notes on Fig. 5.4.1, balancing angles in Art. 15.4, coördinates for plotting in Fig. 17.9.1 and completed land plan on Fig. 18.1.1.

---

*"COGO-90: Engineering User's Manual", Daniel Roos and C. L. Miller Civil Engineering Systems Laboratory, M.I.T., April 1964.

```
LOCATE/LINE    J K N D

   From J in the direction of K, Locate N
at a distance D from J.  Minus D would
locate N 180° from K.

Example:  LOCATE/LINE 20 8 81 156.23

   OUTPUT:  Coordinates of N. (81)
```

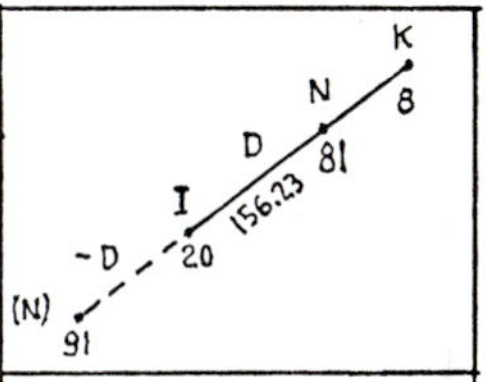

```
LOCATE/ANGLE   K J N D A

   Backsight on K, turn angle at J to locate
N at a distance D and angle A (degrees,
minutes, seconds).  The angle may be clockwise
(plus A) or counterclockwise (minus A).

Example:  LOCATE/ANGLE 8 20 81 156.23 61 23 32

   OUTPUT:  Coordinates of N. (81)
```

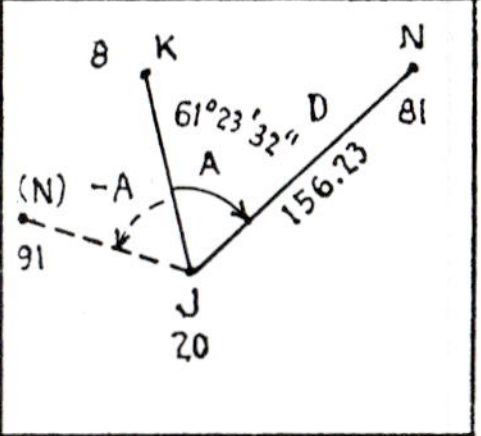

```
LOCATE/DEFLECTION   K J N D A

   Backsight on K, turn deflection angle at
J to locate N at distance of D and deflection
angle A (degrees, minutes, seconds).  Angle
may be clockwise or counterclockwise (minus A).

Example:
 LOCATE/DEFLECTION 8 20 81 156.23 61 23 32

   OUTPUT:  Coordinates of N. (81)
```

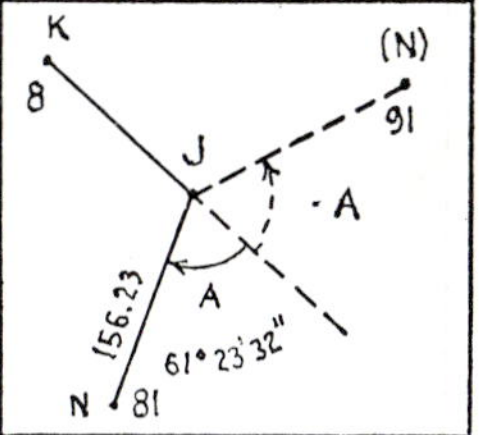

```
LOCATE/BEARING   J N D B

   From J locate N at a distance D and
bearing B (quadrant, degrees, minutes,
seconds).

Example:
 LOCATE/BEARING 20 81 156.23 1 62 41 20
   OUTPUT:  Locate coordinates of N.  (81)
```

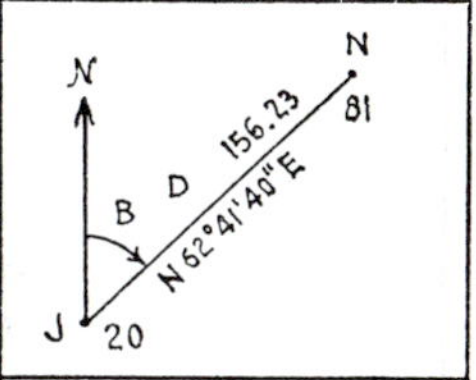

```
PARALLEL/LINE   J N D K L

   Locate a line parallel to the line
through points J and N at an offset distance
D with points K and L opposite J and N.
(-distance locates K and L to the left of
J and N ).

Example:  PARALLEL/LINE 23 9 41.23 2 63

   OUTPUT:  Coordinates of K and L (2 and 63).
```

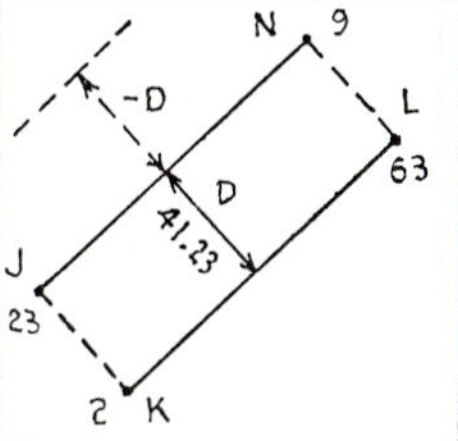

FIG. 15.29.1. TYPICAL COMMANDS IN THE COGO SYSTEM.

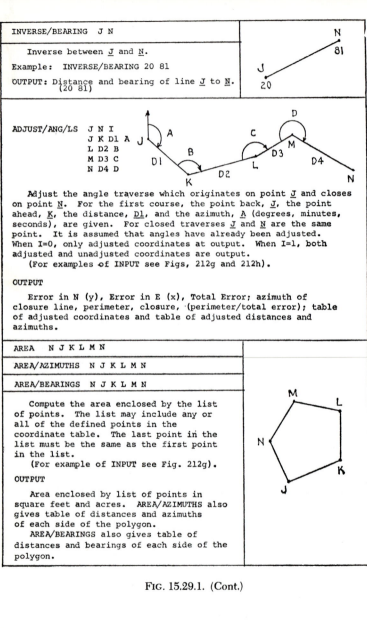

INVERSE/BEARING J N

Inverse between J and N.

Example: INVERSE/BEARING 20 81

OUTPUT: Distance and bearing of line J to N. (20 81)

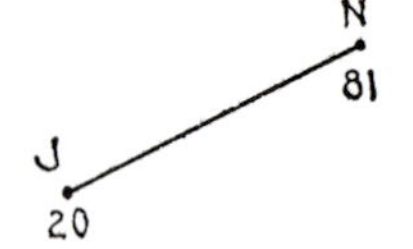

ADJUST/ANG/LS J N I
J K D1 A
L D2 B
M D3 C
N D4 D

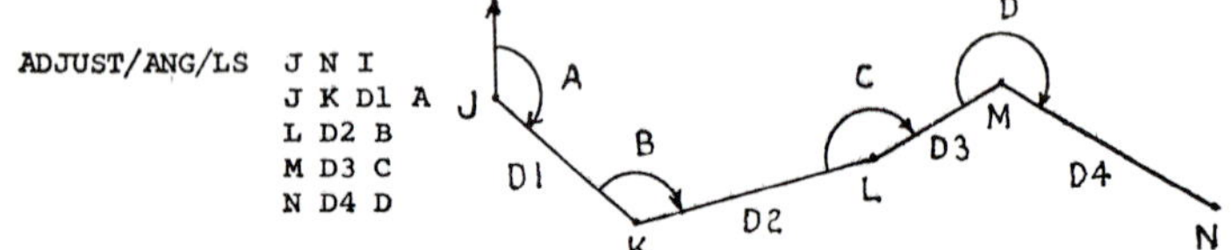

Adjust the angle traverse which originates on point J and closes on point N. For the first course, the point back, J, the point ahead, K, the distance, D1, and the azimuth, A (degrees, minutes, seconds), are given. For closed traverses J and N are the same point. It is assumed that angles have already been adjusted. When I=0, only adjusted coordinates at output. When I=1, both adjusted and unadjusted coordinates are output.

(For examples of INPUT see Figs, 212g and 212h).

OUTPUT

Error in N (y), Error in E (x), Total Error; azimuth of closure line, perimeter, closure, (perimeter/total error); table of adjusted coordinates and table of adjusted distances and azimuths.

AREA N J K L M N

AREA/AZIMUTHS N J K L M N

AREA/BEARINGS N J K L M N

Compute the area enclosed by the list of points. The list may include any or all of the defined points in the coordinate table. The last point in the list must be the same as the first point in the list.

(For example of INPUT see Fig. 212g).

OUTPUT

Area enclosed by list of points in square feet and acres. AREA/AZIMUTHS also gives table of distances and azimuths of each side of the polygon.

AREA/BEARINGS also gives table of distances and bearings of each side of the polygon.

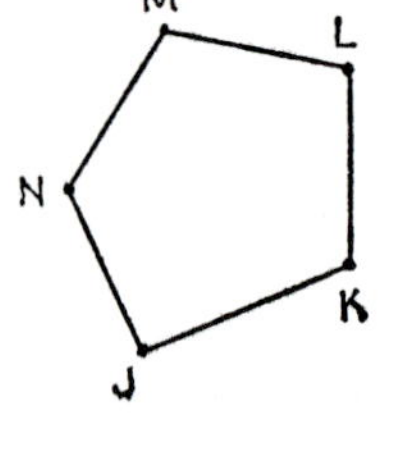

FIG. 15.29.1. (Cont.)

15.31. Input to Traverse Program.—The deck of input command cards is shown in Fig. 15.27.1. To reduce the size of illustration, the data cards following the command ADJUST/ANG/LS, have been omitted. They do appear in Fig. 15.31.1

```
*      COGO
*           J.H. BRADLEY ESTATE
*           FIELD BOOK 42 PAGE 37
*           FIELD WORK JUNE 7, 1957
*           COMPUTATIONS BY COGO
*           AUGUST 17, 1965
CLEAR                   1  99
STORE                   5  5000.0  5000.0
ADJUST/ANG/LS           5  5  1
                        5  6  68.62   323 22 00.0
                        7  95.10  157 09 00.0
                        8  207.41   111  09 45.0
                        10  103.75  90 12 00.0
                        11  96.75  124 59 45.0
                        1  420.77  233 49 00.0
                        2  208.64  101 50 00.0
                        3  436.79  91 32 00.0
                        4  56.48  112 57 45.0
                        5  98.80  146 20 45.0
AREA/AZIMUTH            1  2  3  4  5  6  7  8  10  11  1
*          THIS IS THE PLOT PORTION OF THE PROGRAM
STORE                   97  4930.0  4850.0
AXIS                      120.0  5000.0  5000.0  54
LINE                    5  6  7  8  10  11  1  2  3  4  5
GO/TO                   97
H/LABEL                 1 *148,953 SQ. FT.*
DUMP                    1  99
```

Fig. 15.31.1. Computer Input to Traverse Problem.

which is the input to the COGO program for adjusting the closed traverse shown on Fig. 15.4.1. This program also computes the area bounded by the traverse, and will operate a plotter to draw and label the traverse.

The items preceded by an asterisk (*) are "comments" which are read into the computer to identify the problem. CLEAR is the usual first command. It clears the coördinate table in the computer much as a surveyor would start a new job with a clean sheet of paper. The numbers 1-99 indicate the range of point numbers to be cleared.

The command STORE causes the starting point number and its coördinates to be stored in the memory of the computer. In this case, the traverse computations start at point 5 having a Y coördinate of 5000 ft. and a X coördinate of 5000 ft.

In the COGO system the survey points are designated by numbers instead of letters. In the Bradley Estate problem, point A = 1, B = 2, C = 3, D = 4, E = 5, F = 6, G = 7, H = 8, J = 10, and K = 11. (Note that I and 9 were purposely omitted.)

ADJUST/ANG/LS is the command for making the necessary computations for determining the error of closure of a traverse, and making the necessary adjustments to close it. The command applies to a traverse where directions are derived from clockwise angles, and the adjustments are made by least squares. As described in Art. 15.7, the commonly used methods for adjusting a closed traverse are the compass rule, the transit rule and arbitrary adjustment based on field party chief's knowledge of the conditions of the survey. The COGO system employs a more precise method based on principle of least squares.* This method is more complicated than the other "rules." However, the computational speed of the computer makes its use quite feasible. In each case, angles are assumed less susceptible to accidental errors than are distances. The angles are adjusted (balanced) before entering into closure calculations. The final adjustments are applied only to distances.

An explanation of the command ADJUST/ANG/LS is given in Fig. 15.29.1, and is further illustrated in Fig. 15.31.2. The commands are generally written to apply to traverse notation in a clockwise direction. However they apply just as well to counter-clockwise notation as used in the Bradley traverse.

Referring to the input data in Fig. 15.31.1, and to the sketch Fig. 15.31.2, the numbers 5 5 1 (opposite the command ADJUST/ANG/LS) indicate that the traverse begins and ends on point 5 and is thus a closed one (J J), and that the unadjusted as well as the adjusted coördinates are to be included in

*The Crandell method was used. See "Surveying-Theory and Practice" by R. E. Davis and F. S. Foote, McGraw-Hill Book Company, 4th Ed., p. 449.

the output (1). The following lines are the individual courses. From point 5 (J) to 6 (K) the distance is 68.62 ft. (D1) having an azimuth of 323° 22′ 00″ (A). The next line is the second course to point 7 (L) with a distance 95.10 (D2) and the clockwise angle at 6 is 157° 09′ 00″ (B). The following lines are succeeding courses to point 5.

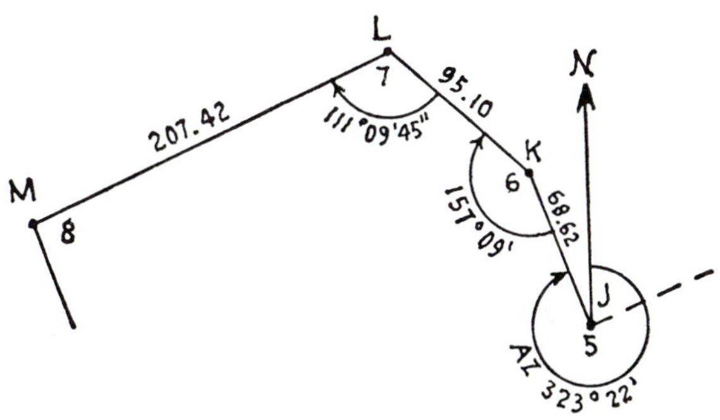

FIG. 15.31.2. EXAMPLE OF ADJUST/ANG/LS COMMAND.

The command AREA/AZIMUTH (Fig. 15.31.1) will compute the area enclosed by the designated points, 1 2 3 4 5 6 7 8 10 11 1. The last point in the list must be the same as the first. The area is computed by the coördinate method (Art. 14.17), using adjusted coördinates. This method is more convenient in the COGO system than the Double-Meridian-Distance method.

In the plot portion of the program, the STORE command is used only to position the drawing tool to plot the area. AXIS controls scale and location of drawing on paper. For example, the scale is 120 ft. per in., the coördinates of the center of plotting surface are Y = 5000 ft. and X = 5000 ft., and the width of paper is 54 in. LINE means to mark points and draw lines between them in the sequence indicated. GO/TO directs the drafting tool to commence drawing from its present position (97), the characters enclosed by asterisks, using the size of character specified by the number preceding the first asterisk.

DUMP causes all defined points to be punched out in a form that can be read back at some later date.

15.32. Output of Traverse Computation.—The output from the Bradley Estate program is shown in Fig. 15.32.1, just as it came from the computer. The output preceded by an asterisk (*) identifies the problem. The numbers to the extreme right of the page sequence number the lines of output. The first line of all input is reproduced in the output and is not sequence numbered. Output, lines 1-10, is the unadjusted coördinates Y and X, reading from the left. The system with which this problem was run automatically prints out 8 digits regardless of the significance of figures. In this computation only two places to the right of the decimal point are significant.

Line 11 gives the error in the Y and X coördinates and the total error (square root of the sum of the squares of Y and X errors). The computer notation E-02 indicates that the decimal point is to be moved 2 places to the left. Thus, the errors are 0.0192 ft. in Y, 0.0435 ft. in X and 0.0475 ft. in total.

Line 12 gives the azimuth in degrees and minutes (66° 11′) of the closure line, the perimeter of the traverse (1793.11 ft.), and the denominator of the error fraction (1/37710). The latter result might differ from manual results. The difference is apparently due to the rounding of numbers in the manual latitude and departure calculations. The computer in this case carries out calculations to two places beyond significance, the rounding being left to the final error result. Since the computer is operating outside the limits of significance, it cannot be assumed that its results are more accurate than manual ones. However, the comparison does bring out the sensitivity of the error ratio to the rounding of numbers.

Lines 13-21 in Fig. 15.32.1 give the Y and X coördinates after adjustment by the least squares method (Art. 15.31).

Lines 22 through 41 give the corrected courses derived from the adjusted coördinates, two lines per course. The first output line of each pair gives the point numbers and the second the distances and the azimuth in degrees, minutes and seconds. The numbers beyond two decimals or closer than 15 seconds

```
*           COGO
*           J.H. BRADLEY ESTATE
*           FIELD BOOK 42 PAGE 37
*           FIELD WORK JUNE 7, 1957
*           COMPUTATIONS BY COGO
*           AUGUST 17, 1965
CLEAR                    1   99
STORE                    5   5000.0    5000.0
ADJUST/ANG/LS            5   5   1
          6          5055.0654         4959.0550
          7          5103.3560         4877.1281
          8          4974.7482         4714.4047
         10          4893.1272         4778.4520
         11          4898.4017         4875.0581
          1          4572.8326         5141.6141
          2          4669.0917         5326.7217
          3          5061.8699         5135.6439
          4          5058.9343         5079.2403
          5          4999.9808         4999.9565
  1.9200000E-02      4.3500000E-02     4.7548818E-02
         66                11          1793.1100         37710.926
          6          5055.0652         4959.0551
          7          5103.3543         4877.1309
          8          4974.7555         4714.4190
         10          4893.1340         4778.4667
         11          4898.4088         4875.0788
          1          4572.8371         5141.6369
          2          4669.1031         5326.7577
          3          5061.8847         5135.6782
          4          5058.9493         5079.2781
          5                 6
 68.619681                323                21         59.688000
          6                 7
 95.096854                300                30         59.940000
          7                 8
207.39527                 231                40         44.544000
          8                10
103.75055                 141                52         44.508000
         10                11
 96.755985                 86                52         29.571600
         11                 1
420.77320                 140                41         29.400000
          1                 2
208.65485                  62                31         29.236800
          2                 3
436.79369                 334                 3         29.232000
          3                 4
 56.476435                267                 1         14.412000
          4                 5
 98.792894                233                21         59.004000
AREA/AZIMUTH             1 2 3 4 5 6 7 8 10 11 1
          1                 2
208.65485                  62                31         29.236800
          2                 3
436.79369                 334                 3         29.232000
          3                 4
 56.476435                267                 1         14.412000
          4                 5
 98.792894                233                21         59.004000
          5                 6
 68.619681                323                21         59.688000
          6                 7
 95.096854                300                30         59.940000
          7                 8
207.39527                 231                40         44.544000
          8                10
103.75055                 141                52         44.508000
         10                11
 96.755985                 86                52         29.571600
         11                 1
420.77320                 140                41         29.400000
148952.65            3.4194823
DUMP                     1   99
 10.000000                  1          4572.8371          5141.6369
 10.000000                  2          4669.1031          5326.7577
 10.000000                  3          5061.8847          5135.6782
 10.000000                  4          5058.9493          5079.2781
 10.000000                  5          5000.0000          5000.0000
 10.000000                  6          5055.0652          4959.0551
 10.000000                  7          5103.3543          4877.1309
 10.000000                  8          4974.7555          4714.4190
 10.000000                 10          4893.1340          4778.4667
 10.000000                 11          4898.4088          4875.0788
```

FIG. 15.32.1. COMPUTER OUTPUT OF TRAVERSE PROBLEM.

are not significant in this problem. This step of adjusting distances and azimuths can easily be had from the computer. A comparison of measured and adjusted distances show small differences. In other traverses of larger size and greater closing error, the differences could be quite significant.

Lines 42 through 61 are a repeat of lines 22-41 since the traverse follows the actual property line. When this is not so, a different set of data would appear.

Line 62 gives the area in square feet (148,953) and in acres (3.42948). These values are in close agreement with the results of the manual calculation (148,947).

Line 63 through 72 are the Y and X coördinates of all established points.

The COGO system also provides for the computing of the area from surveys run on offset traverses, similar to the plots in Figs. 5.2 and 5.4. In addition, the command SEGMENT com-

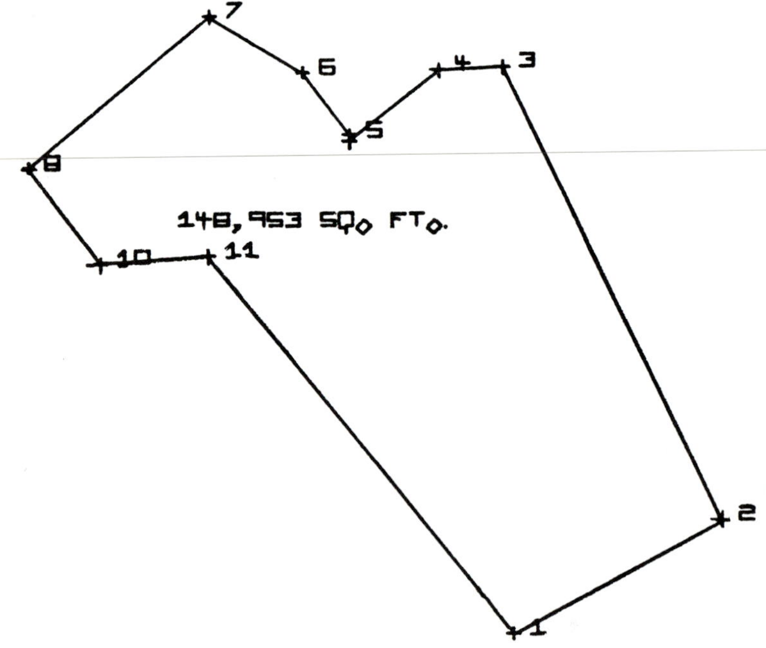

FIG. 15.33.1. COMPUTER-CONTROLLED PLOT OF TRAVERSE.

putes the area between a chord and an arc and is therefore useful in obtaining the area within curved boundaries. See Art. 14.10.

15.33. Automated Plotting from Computer Output.—The COGO system provides for plotting traverses and other geometric features on commercial plotters of both the paper roll and drawing board type.

As described in Art. 15.32, commands are given to establish the scale of the drawing and its location on the paper with respect to the coördinates of the points to be plotted. The plotter is instructed to start plotting points at a specified coördinate point and to proceed in a prescribed sequence from point to point until the list of point numbers is exhausted. Fig. 15.16 is a plot of the Bradley Estate produced by a computer-controlled plotter. Titles and labels may be automatically drawn either parallel to the X (as shown) or the Y axis. At the present state of the art, programming and plotting and annotations is rather detailed and time consuming.

PART IV

PLOTTING

CHAPTER XVI

DRAFTING INSTRUMENTS AND MATERIALS—REPRODUCING PLANS

It is assumed in this section that the reader is familiar with the ordinary drawing instruments such as the T-square, triangles, dividers, compasses, and scales, as well as with their use.

ENGINEERING DRAFTING INSTRUMENTS

16.1. Drafting Instruments.—There are several drafting instruments which are used by engineers and surveyors but which are not so generally employed in other kinds of drafting work. The most common of these are described briefly in the following articles.

16.2. Straight-Edge.—Engineering drawings are made with greater accuracy than much of the drafting work of other professions. In fact many engineering drawings are limited in precision only by the eyesight of the draftsman. It is evident, then, that to use a T-square which is run up and down the more or less uneven edge of a drawing board will not produce drawings of sufficient accuracy. For this reason in many classes of engineering work the edge of the drawing board is not relied upon. Furthermore, in most plots of surveying work the lines are not parallel or perpendicular to each other except by chance, but run at any angle which the notes require; and there is therefore not so much call for the use of a T-square as there is in architectural, machine, or structural drawings. Drawings are usually laid out by starting from some straight line drawn on the paper by means of a straight-edge, which is simply a flat piece of steel or wood like the blade of a

T-square. Steel straight-edges are more accurate and are more commonly used by engineering draftsmen than the wooden ones, the edges of which are likely to nick or warp and become untrue. They can be obtained of almost any length and of any desired weight, the common length being about 3 ft. Wooden straight-edges and blades of T-squares are made with a plastic edge about ¼ inch wide, which, on account of its transparency, is of great advantage to the draftsman.

16.3. Engineer's Scale.—Engineer's plans usually are made on scales of 10, 20, 30, etc. feet to an inch. In the engineer's scale, therefore, the inch is divided into 10, 20, 30, etc. parts, instead of into eighths and sixteenths. Engineer's scales are made 3, 6, 12, 18, and 24 inches long. One form is the flat wooden rule with both edges beveled and a scale marked on each bevel. Some flat rules are beveled on both faces and on both edges of each face, thereby giving four scales on one rule. Still another common form is the triangular scale having six different scales, one on each edge of the three faces. In such rules the scales are usually 20, 30, 40, 50, 60, and 80 ft. or 10, 20, 30, 40, 50, and 60 ft. to an inch. Scales are made having the inch divided into 100 parts, but in plotting a map which is on a scale of 100 ft. to an inch the work is probably done more easily and quite as accurately by using a scale of 10, 20, or 50 divisions to an inch and estimating the fractional part of a division. A 20-ft. or 50-ft. scale is more satisfactory for precision than a 10-ft. scale when it is desired to plot on a scale of 100 ft. to the inch. A plan on a 200-ft. scale is always plotted by using a 20-ft. scale, a 300-ft. plan by using a 30-ft. scale, etc.

A map covering a large area, like that of a state, for example, must be plotted to a very small scale, and this is usually given in the form of a ratio such as 1 to 20 000, 1 to 62 500, etc., meaning that one unit on the map is 1/20000, 1/62500, etc. of the corresponding distance on the ground; this is sometimes called the *natural* or *fractional** scale. For plotting such maps specially constructed scales with decimal subdivisions are used.

---

*Called also the representative fraction, *R.F.*

16.4. PROTRACTOR.—A *protractor* is a graduated arc made of plastic, metal, or paper, and is used in plotting angles. There are many varieties of protractor, most of them being either circular or semicircular.

16.5. Semicircular Protractor.—Probably the most common is the semicircular protractor, which is usually divided into degrees, half-degress, and sometimes into quarter-degrees. Fig. 16.5.1 represents a semicircular protractor divided into

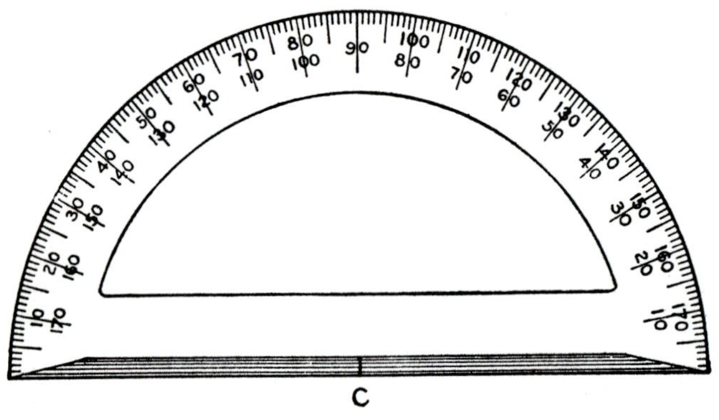

FIG. 16.5.1. SEMICIRCULAR PROTRACTOR.

degrees. In plotting an angle with this protractor the bottom line of the instrument is made to coincide with the line from which the angle is to be laid off, and the center of the protractor, point C, is made to coincide with the point on the line. On the outside of the arc a mark is made on the drawing at the desired reading. The protractor is then removed from the drawing and the line drawn on the plan.

Instead of having the 0° and 180° of the protractor on its lower edge some instruments are made as shown in Fig. 16.5.2. This form is claimed by some draftsmen to be more convenient, because in handling the protractor by placing the fingers on the base neither the graduations nor the line on the plan are covered by the hand.

16.6. Full-Circle Protractor.—The full-circle protractor is of use particularly in stadia work or in plotting any notes where azimuths over 180° have been taken. For such work as stadia

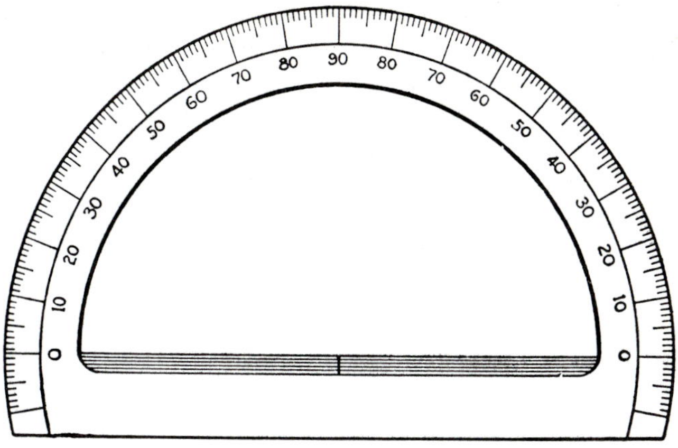

FIG. 16.5.2. SEMICIRCULAR PROTRACTOR.

plotting, a protractor 8 to 12 inches in diameter is required for accurate work. Paper protractors of this size will yield good results. The more durable plastic protractors are available in 6-, 8- and 10-inch diameters.

16.7. Some of the metal protractors are provided with an arm and vernier attachment. These, while giving more precise results, require more time for manipulation, and a full-circle protractor with a diameter of, say, 8 inches will give sufficiently close results for all ordinary work. As a matter of fact a protractor with a vernier reading to minutes can be set much closer than the line can be drawn, and it is therefore a waste of time to attempt to lay off the angles on a drawing with any such accuracy. There is, however, a protractor of this type with a vernier reading to about 5 minutes which may be of use in precise plotting.

16.8. Three-Armed Protractor.—The three-armed protractor is used for plotting two angles which have been taken with an instrument (usually a sextant) between three known points (in pairs), for the purpose of locating the position of the observer (the vertex of the two angles). The protractor has three arms, the beveled edges of which are radial lines. The middle arm is fixed at the 0° mark and the other two arms, which are movable, can be laid off at any desired angle from the fixed arm by means of the graduations on the circle, which number each way from the fixed arm. The two movable arms having been set at the desired angles and clamped, the protractor is laid on the plan and shifted about until each of the three known points (which have already been plotted on the plan) lies on a beveled edge of one of the three arms of the protractor. When the protractor is in this position its center locates the point desired, which is then marked by a needle point. Only one location of this center point can be obtained except when the three known points lie in the circumference of a circle that passes also through the vertex. In these circumstances the location becomes indeterminate.

16.9. PANTOGRAPH.—This instrument is composed of several flat pieces of metal or wood joined in such a way as to form a parallelogram. In Fig. 16.9.1 the point *A* is fixed and points *B* and *C* are movable. The supports at *A* and *B* may be interchangeably fitted with a pencil point or a tracing point. If, in Fig. 16.9.1, an enlargement is being made, the tracing point

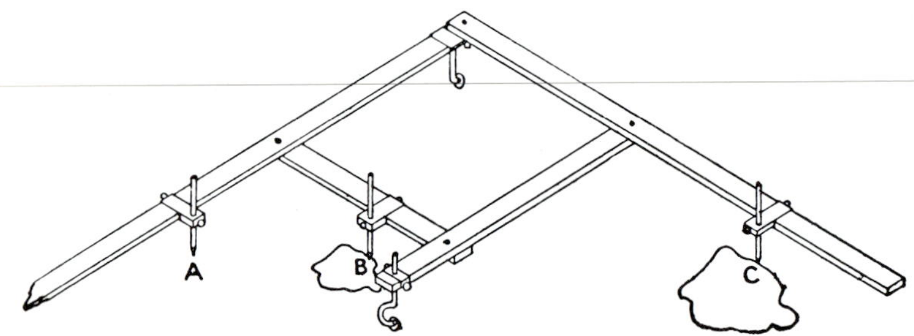

FIG. 16.9.1. THE PANTOGRAPH.

would be at *B* and the pencil point at *C*. If a reduction was desired the points would be interchanged. The other bearing points shown in the figure are only for the purpose of supporting the instrument. The two movable points at *B* and *C* are so attached to the instrument that they will trace out exactly similar figures. The essential condition is that all three points *A*, *B*, and *C* must lie in a straight line and each point must be on one of the three different sides (or sides produced) of a jointed parallelogram. It is evident then that by changing the relative positions of these points, by moving them up or down the arms of the parallelogram, but always keeping the points on a straight line, the scale of the copy can be made to bear any desired relation to the scale of the original drawing. These instruments are usually provided with scales marked on the arms indicating the proper settings for various reductions or enlargements. The simpler types have reducing and enlarging ratios ranging from 8:1 to 1⅛:1 or vice versa. The more complicated types have all ratios from 1:1 to 1:20 or vice versa. Very accurate results cannot as a rule be obtained with a pantograph because there is lost motion or play in the several joints of the instrument. Some of the expensive metal pantographs of the suspended type, however, will give fairly good results.

16.10. Parallel Ruler.—This is a beveled rule made of metal and mounted on two rollers of exactly the same diameter. It is used for drawing parallel lines. This instrument can be made to do accurate work, but it must be handled with a great deal of care to prevent the rollers from slipping. It is especially useful in drafting diagrams of graphic statics in connection with structural design, in drawing the parallel sides of buildings, section lining, blocking out for titles, and in drafting large titles that require mechanical lettering.

16.11. Beam Compass.—This is an instrument used for drawing the arcs of circles whose radii are longer than can be set out with the ordinary compass drafting instrument. It is composed of a strip of wood or metal with two metal attachments which can be fastened to it. One of these attachments carries a needle point, and the other, which is usually provided with a slow-motion screw for exact settings, carries a pencil or a pen. This instrument is particularly useful in laying out large rectangles such as are called for when surveys are plotted by coördinates (Art. 17.6).

16.12. Contour Pen.—This pen is constructed very much like an ordinary right-line ruling pen except that it has a metal shaft, running through the entire length of the holder, to which the pen is attached. The shaft revolves inside of the holder, and the pen is so shaped that it drags behind taking a position in the direction in which it is being moved. It is used for drawing irregular curved lines such as contours or shore lines. Not a little practice is required before one can use a pen of this type accurately. When skill in its use is once acquired, however, a plan can be easily made on which the contours all have a uniform weight of line giving a very satisfactory appearance. The purpose of a contour line is to show the facts as to the land surface, and this pen should not be used unless it is found by trial that it does the work in hand properly. Accuracy is more important than appearance.

16.13. PROPORTIONAL DIVIDERS.—Proportional dividers are virtually an ordinary pair of dividers with both legs prolonged through the pivot-point thereby forming another pair of legs above the pivot. The pivot can be pushed up and down in a slot in the legs and clamped in any desired position, thereby altering the relative lengths of the two pairs of legs. The sliding is accomplished in some dividers by a rack-and-pinion motion. There are marks on the legs showing the proper settings for the pivot so that the space between one pair of points will bear any desired ratio to the space between the other pair. The marks on the legs should not be accepted as correct, but should be tested by actual trial. By means of this instrument a drawing can be enlarged or reduced to a definite scale without the use of the engineer's scale.

16.14. The Section Liner.—This is an instrument used for drawing cross-section lines, or hatching lines, which are equidistant and parallel. One kind consists of a triangle which slides in contact with a short straight-edge, the amount of motion being regulated by an adjustable stop. In drawing the lines the triangle and the straight-edge are shifted alternately along the plan. In another kind the ruling edge is on a pivoted arm and by varying the angle between the ruling edge and the straight-edge different spacings of the lines may be obtained. Another type of the movable arm pattern has a device by means of which the spacing of the lines may be made double or treble the usual spacing.

16.15. Curve Templates.—For drawing arcs of curves of long radii, such as occur on railroad and highway plans and on plans of curved city streets, templates or circular curves made of metal or plastic are used; these come in sets of 30, 50 or 100, with radii varying from about 2 inches to 160 or 300 inches depending upon the number of curves in the set. Curves are also available in radii corresponding to degrees of curves used in highway or railroad practice. The inside and outside edges are of the same radii, and are beveled so that when ink lines are drawn the beveled edges may be turned up, thus preventing the ink from running in under the curve on to the paper. Some curves for railroad work are made with a short straight edge tangent to the curve at one end and with the point where the curve begins marked by a line across it.

16.16. Irregular curves, called *French Curves*, are of a variety of shapes. They are usually made of plastic and are used to guide the pencil or pen in tracing out irregular curved lines on the map. For a larger choice of shapes, *ship curves* may be used.

16.17. A *Flexible Curve* consists of a strip of rubber fastened to a flexible metal back. This curve can be twisted to conform to any irregular curved line on the map and can then be used as a guide against which the pencil or pen is held in tracing out the curve.

16.18. A *Spline* is a long thin flexible piece of wood, hard rubber, plastic, or metal which can be bent so as to conform to a curve. It is usually held in position by specially designed weights with light metal arms which fit into a thin groove in the top edge of the spline. This instrument is used by naval architects for drawing long flat irregular curves such as occur in ship designs. In engineering drafting it is used in drawing the lines of arches, which frequently are not circular.

DRAWING PAPERS

16.19. Classification.—The drawing papers used by surveyors may be divided into four general classes: (1) those used for plotting plans, (2) tracing paper or tracing cloth which is used for copying drawings, (3) cross-section and profile papers, and (4) process papers.

16.20. DRAWING PAPER FOR PLANS.—There are numerous grades of drawing paper ranging from very cheap "detail" to heavy paper mounted on cloth, called "mounted paper." For rough plots which are to be copied later or which are for temporary use only, a manila detail paper is frequently used; but where the drawing is to be of a more permanent character a heavy white or manila paper is used. Still more permanent plans, such as the plan of a survey of a city, should be plotted on heavy mounted paper. In order to be satisfactory a paper should have a surface that is not too porous to take ink nicely, and a fiber such that after scratching with a knife or rubbing with an ink eraser, the surface will still take ink effectively. No paper, however, after scratching can be expected to take bottle red ink, which permeates the fiber with extraordinary ease. Mounting cloth is available which can be easily applied to paper plans, prints or photographs to stiffen and preserve them.

16.21. TRACING PAPER.—In making copies of drawings, a thin transparent paper called *tracing paper* is often used. Some brands are thin and easily torn and are only suitable for sketches or drawings of a temporary character. Tracing papers of the vellum type, however, are quite durable and will withstand considerable hard usage.

16.22. TRACING CLOTH.—For more permanent drawings a *tracing cloth* is used, made of a very uniform quality of linen coated with a preparation to render it transparent. Most tracing cloth as it comes from the manufacturer will not take the ink readily, and it is necessary to rub powdered chalk or talc powder over the entire surface of the cloth before inking the drawing. After the surface chalk is brushed off, the tracing cloth is ready for use. Tracing linen generally has one side glazed and the other dull. Pencil lines can be drawn on the rough side, but the smooth side will not take even a very soft pencil; either side may be used, however, for ink drawings. Some draftsmen prefer to use the glazed side but the dull side is more commonly used. A tracing inked on the glazed side may be tinted on the dull side either by crayons or by a wash; the latter will cockle the cloth unless it is put on quite "dry." It is easier to erase from the glazed than from the dull side, but the dull side will stand more erasing and gives more uniform lines. Pencil tracing cloth is also available upon which the drawing may be made in either pencil or ink. Erasures can be made easily and clear prints obtained from a pencil drawing.

Recently developed drafting materials of plastic, mylar, or other film types are easier to work with albeit more costly. They are dimensionally stable, however, and prove very economical where highly stable drawing sheets are required for precision work.

Erasure of ink lines from a tracing, as well as from any drawing paper, is a delicate undertaking. Success will result if the following suggestions are observed carefully: with a smooth sharp knife shave off the ink from the paper; this can be done without touching the paper. When nearly all of the ink is off, rub the line with a pencil eraser. This will take off the rest of the line except perhaps a few specks of ink which can be removed readily by a sharp knife. This method of erasing takes more time than the ordinary method of rubbing with an ink eraser until the line has disappeared, but it leaves the paper in much better condition to take another line. It is impossible to obtain good results by this method unless the knife has an edge which is both smooth and sharp. Where the surface of the tracing cloth has been damaged the application of a thin coating of collodion on the damaged portion will produce a surface that will take ink. Electric erasing machines are available which will quickly remove ink lines; these must be used cautiously in order not to damage the surface of the drawing or tracing.

In making a tracing of another tracing it will be found that the lines can be seen more readily if a white paper is put under the lower tracing also, in tracing from a blueprint or negative photostat it is difficult to see the white lines through tracing paper or tracing cloth. A tracing table is used for this purpose in such cases. It consists of a piece of plate glass set into and flush with the top of a drawing table, with a light mounted underneath the glass. The light shining up through the print will make the white lines easily visible for copying. This device is also useful for tracing detail from aerial photographs.

While the survey is in progress, or after substantial portions have been completed, field notes are plotted accurately on detail paper. Later the completed plot is traced in neat form to produce the final record drawing of the survey.

From these tracing drawings any number of process prints can be made, the tracing taking the place of the negative used in photographic printing. (Arts. 16.26-16.28.)

16.23. CROSS-SECTION.—Paper divided into square inches which, in turn, are divided into small subdivisions is used to plot cross-sections of earthwork and for many other purposes. The inch squares are usually divided into 1/4", 1/5", 1/8", 1/10", 1/12", or 1/16". Cross-section paper can also be obtained divided according to the metric system, or with logarithmic divisions. Cross-section paper comes in sheets or in rolls.

16.24. Profile Paper which, as the name implies, is used for plotting profiles comes in rolls of 10 yds. or more. The vertical divisions are usually much smaller than the horizontal divisions, which makes it easier to plot the elevations accurately. The horizontal distances to be plotted occur mostly at full station points; these are represented on the profile by the vertical rulings on the paper.

Both the cross-section and the profile papers come in colors (usually red, green, blue, orange, or burnt sienna), so that a black or a red ink line (the two most commonly used) will show up distinctly on the paper. These papers can be obtained also of very thin transparent material or in tracing cloth form, suitable for use in making process prints. Profile papers usually come in long rolls 20 inches wide. Special papers and cloth for highway plans are available with profile rulings on the lower half of the sheet only. These conform to the standards of the Bureau of Public Roads and are called "Federal Aid Sheets."

REPRODUCING PLANS

16.25. Blueprints.—Plans may be reproduced from transparent tracings by the blueprint process. In this method light is passed through the tracing onto a sheet of sensitized paper. The light affects the sensitized coating chemically causing it to turn blue after washing. The portions of the sheet that were under the lines and lettering of the plan are shielded from the light and turn white in the washing; the result is a reproduction of the original plan in white on a blue background.

Most blueprints are made on paper. They may be made directly upon cloth, if so desired. The advantages of the cloth prints are that they do not shrink so badly as the paper prints and they are more durable. Prints which are to be used on construction work where they are sure to get rough usage are sometimes made on cloth. Both cloth and paper shrink so much, however, that scaled dimensions are unreliable.

Commercial blueprint companies equipped to make the common types of reproductions will be found in most cities and large towns. The larger companies offer mail-order service. Engineering firms which have much reproduction work to do usually purchase and operate their own machines. In remote regions where commercial reproduction is not available, it is possible to make blueprints by hand using a simple printing frame and sunlight as the light source. Sensitized bluprint paper will be required for this process.

16.26. Vandyke Prints.—Positive prints (i.e., those with a white background) may be obtained by the blueprint process by first making a Vandyke negative. The tracing is printed with the ink lines in contact with the Vandyke paper. Where the light penetrates the tracing the coating on the Vandyke paper is affected chemically so that after washing and fixing in a solution of hyposulphate of soda the background turns a dark brown and the lines appear white. When a print is made from the Vandyke by the blueprint process, the light only penetrates through the white lines producing blue lines on a white background.

Another method of producing positive prints from either a tracing or an opaque original is to photograph it on to a

Kodalith negative film, from which blue-line prints may be obtained by the blueprint process. The original may be enlarged or reduced in the Kodalith negative, whereas the Vandyke print must be the same size as the original.

One of the advantages of making a Vandyke or a photographic negative is that the original tracing may be filed away and kept in excellent condition, while prints are made from the Vandyke or negative. Positive prints on a white background also have the advantage that corrections or notations may easily be made on these prints in pencil or ink, whereas on blueprints corrections must be made with a bleaching fluid or in colored crayon or water color in order to be easily read.

16.27. Ozalid Prints.—Positive prints on a white background may be produced directly from tracings by a "dry" process employing ammonia fumes as a developing agent. These prints are very close to the scale of the tracing, and do not have the shrinkage characteristics of blueprints. Ozalid prints have superseded blueprints for many purposes where readability is important. They are subject to fading over a period of years especially if exposed to sunlight. With the ammonia process it is possible to obtain positive prints with blue or red lines if these colors are preferred to black. Transparent ozalids called "sepia transparencies" may be made from tracings and used in place of the original tracing.

16.28. Reproducing Tracings.—Tracings may be reproduced on sensitized tracing cloth by first making a brown print negative, which is placed over the sensitized cloth and printed by the usual blueprint process. The cloth is given a waterproofing coating so that is it unaffected by the washing process.

Tracings may also be reproduced photographically on a sensitized cloth. By this method, it is possible to change the scale of the map, and it is also possible to assemble plans of the same scale into one tracing, likewise material may be blanked out or added to the original tracing merely by placing inserts over the portions to be blanked out or changed. This process is expensive in relation to other processes, but it affords a means for altering and assembling plans which would be very costly if done by drafting.

There is a process of reproducing tracing cloth drawings on various kinds of material, such as tracing cloth, tracing paper, drawing paper, cardboard, water-color paper, or even on ordinary cotton cloth; this process of reproduction has a trade mark of "Gelitho." These "Gelitho" prints are made from a gelatin plate instead of from an expensive stone necessary in ordinary lithography work. The advantages of this method are that the prints are true to scale. They do not curl as they are not treated with a fixative or waterproofing solution, and since printer's ink is used they make a permanent record. Constant use in making blueprints or handling in the drafting room does not seem to affect their surface.

16.29. Transparencies from Opaque Drawings.—Drawings on bond paper or photostats which are too opaque for blueprinting may be rendered transparent by soaking them in a solution of carbon tetrachloride and mineral oil; but this treatment leaves the drawings slightly greasy. They may stain other drawings that come in contact with them, and therefore this process is an expedient rather than recommended practice.

A better method is to make photographic negatives from opaque drawings. These serve as transparencies or "intermediates" from which prints may be made by the blueprint process giving blue lines on a white background. The photographic process also makes it possible to produce prints at a different scale from that of the original.

An "autopositive" transparency may be made to the same scale directly on paper, film or tracing cloth by a contact or "reflection" process. Lines may be added or deleted from the autopositive, and prints may be made by the blueprint or ozalid processes. By this process well-drawn pencil drawings may be reproduced to appear as though they were drawn in ink.

Positive prints may be made directly from opaque originals by the "Xerox" process which employs electrical charges to reproduce a fused-powder image of the original. The reproductions can be either opaque or transparent.

16.30. Photostats.—The photostat method is commonly used to reproduce documents, plans, or pictures by photography. For this work a special camera is used which contains a roll of prepared sensitized paper. The first print made from a white plan with black lines is in the form of a negative with white lines on a black background. For many purposes these negatives are satisfactory, but if positive prints are desired, the negatives are rephotographed. Photostats of blueprints give black lines on white background directly. By adjusting the camera and lenses, enlargements or reductions can be obtained.

Photostat prints are unreliable for scaling because they shrink so much and so unevenly. It is always advisable to have a graphical scale on the plan that is to be photostated. Then if the reduction or enlargement is not exactly correct, the graphical scale will, nevertheless, fit the plan.

Plans can be made up from prints of several plans pasted together, forming a large compiled map, and then photostated to any desired scale, thus producing at little cost an assembly of several plans. Similarly a photograph can be attached to the face of a drawing before being photostated thus giving a plan and photograph on the single sheet.

The photostat and Xerox processes are useful, because at small cost they give faithful copies of any document or plan.

There are several xerography processes, and recent developments make copying more and more efficient and simple.

Reduced size photographic masters with blow-up to full size can work well for engineering drawings.

The photostat process may also be used to produce transparent negatives which are printed by the blueprint process giving prints with blue lines on a white background. The original does not have to be transparent; the scale can be changed as desired. The prints from transparent negatives are very attractive in appearance and look well in engineering reports.

16.31. Offset Printing Process.—Where a hundred or more positive prints are required on paper, the offset printing method will prove quick and economical. The original need not be transparent since the plate used for printing is produced photographically. The scale of the plan may be changed as desired. Best results are obtained from ink drawings, although pencil drawings with firm black lines can also be reproduced satisfactorily. The offset process has several trade names, such as photo-offset, photo-litho, planograph, etc. This method is particularly valuable for reproducing land subdivision plans for use in sales promotion, and also for reducing construction drawings to a scale suitable for binding with specifications. Revisions may be made on the original drawing by covering certain portions and adding new data in the form of patches. Offset prints may be made in more than one color, such as red and black, thus accentuating certain features of the plan.

16.32. Inks and Water-Colors.—Bottled ink, prepared in various colors, is sometimes used on engineering drawings. The so-called "waterproof" inks differ from other inks in that a water-color wash can be put over the lines without causing them to "run." Bottled black inks are quite satisfactory for all drawings. Colored inks, however, particularly the red inks, are very unsatisfactory. They will sometimes run on paper where only very slight erasures have been made; in fact, on some of the cheaper papers red ink will always run. For tracing purposes red ink is wholly unsatisfactory, as it is impossible to obtain a good reproduction of a red ink line by any of the process prints. Where red lines are needed the use of *scarlet vermilion water-color* will be found to give not only a brilliant red line on the tracing, but also "body" enough in the color so that the lines will print fully as well as the black ink lines. Scarlet vermilion water-color will give much better lines on any paper than the bottled red inks. Only enough water should be used to make the water-color flow well in the pen. Other water-colors are used in the place of the bottled colored inks, such as *Prussian blue* instead of bottled blue ink, or *burnt* *sienna* instead of brown ink, and these give much better results.

Frequently it is necessary to make additions on blueprints in white, red, or yellow. A white line can be put on easily by using *Chinese white* water-color; but sometimes a bleaching fluid is used, which bleaches out the blue leaving the white paper visible. The best color for a red line on blueprints is scarlet vermilion water-color; and for a yellow line none of the ordinary yellow water-colors gives as brilliant lines as *middle chrome yellow.*

For tinting drawings, water-colors and dilute inks are used. Effective tinting may be done on tracings by using colored pencils on the rough side of the linen. Prints may be tinted with colored pencil, and the coloring spread uniformly by rubbing it with a piece of cloth, a blotter or with stumps. The latter are small, pointed tubes of tightly rolled cardboard and resemble pencils. When the pencil coloring is dampened, it becomes "fixed" and will not rub off or erase easily.

CHAPTER XVII

METHODS OF PLOTTING*

17.1. Laying Out a Plan.—Laying out a plan requires careful work. If a good-looking plan is to be obtained this part of the work must be done with a little judgment. Besides the plan of the survey or property the drawing must have a title, and sometimes notes and an arrow to show the direction of the meridian. These must all be arranged so that the entire drawing will have a symmetrical appearance. Often the plot is of such awkward shape that it is difficult to lay out the drawing so that it will look well, and even then the draftsman's artistic instincts are taxed to produce a satisfactory result.

17.2. Scale.—In many cases the scale of the plan as well as the general arrangement of its parts must be chosen by the engineer. Surveys of large extent which do not contain many details, such as the preliminary survey for a highway, may be drawn to a scale of 200 ft. to an inch. A plan of a large piece of woodland or a topographical map of a section of a town may be represented on a scale of from 100 ft. to 400 ft. to an inch. A plan of a city lot for a deed is represented on a 20-ft. to 80-ft. scale; and city streets, such as sewer plans and the like, are frequently drawn to a scale of 20 ft. to 40 ft. to an inch. Highway construction plans are usually plotted to a scale of 50 ft. to an inch. On some construction plans details are shown to different scales; the drawing of a conduit may be to a scale of 80 ft. to an inch, while on the same sheet the conduit cross-section may be shown to a scale of 4 ft. to the inch.

The field maps of the U. S. Coast and Geodetic Survey are plotted on scales of 1/20000 or 1/10000, but some special maps

---

*For a brief description of different projections for maps of large areas, such as states or counties, see Volume II, Chapter 9.

are made on scales as large as 1/2000. The field maps of the U. S. Geological Survey are plotted to various scales and reduced on the lithograph sheets to 1/24000, 1/21680, 1/62500 or 1/125000.

These remarks in regard to scales are not to be considered in any sense as hard and fast rules to govern all conditions. They are suggested simply to give some idea of the existing practice in this matter.

METHODS OF PLOTTING TRAVERSES

17.3. PLOTTING BY PROTRACTOR AND SCALE.—The most common method of plotting angles is by use of the protractor (Art. 16.4), and of plotting distances, by use of the engineer's scale. Every traverse consists of a series of straight lines and angles, which can be plotted by a protractor in the following manner: First, the survey to be mapped should be sketched out roughly to scale, in order to ascertain its extent and shape so as to decide the size of paper necessary for any given scale of drawing and to determine its general position on the sheet; this will fix the direction of the first line of the traverse, to be used as a starting line for the entire drawing. After this has been done, the first line is drawn in the proper place on the paper, its length is scaled off by using the proper scale, and its two extremities accurately marked by pencil dots or by means of a needle point, and surrounded by a light penciled circle. The line should be drawn so that it will extend beyond the next angle point a distance greater than the radius of the protractor, this extension of line being used when the next angle is laid off with the protractor.

The protractor is placed so that its center is exactly on the second angle point and so that **both** the 0° and 180° marks of the protractor exactly coincide with the line. The traverse angle taken from the field-notes is plotted, the protractor removed, the line drawn, and the length of the second course carefully scaled. Then the protractor is placed along this new line and opposite the third point, the angle at that point is laid off, the next line drawn, and the distance scaled. By this process the entire traverse is plotted.

17.4. Checks.—On all plotting work, just as on all fieldwork and computations, frequent checks should be applied to insure accuracy.

If the traverse is a closed traverse the plot, of course, should close on the paper.* If it does not and the error of closure is in a direction parallel to any one of the lines, there is probably a mistake in plotting the length of that line. If there is no indication of this sort the mistake may be either in scaling, in laying off the angles, or in both. In such a case the entire plot should be checked unless there is some reason to think that a certain line may have been laid off at the wrong angle, in which event that questionable angle should be replotted. The bearings of all the lines of the traverse can be computed with reference to the magnetic or to any assumed meridian; any line can be produced to meet the meridian line, and this angle measured and checked. Similarly, the bearing of the last line of a traverse which does not close can be computed and the angle the last line makes with the meridian measured. If it checks the computed angle it is evident that no error has been made in the angles unless mistakes were made that exactly balance each other, which is not probable. In this way, by "cutting into" the drawing here and there, the angular error, if there is one, can be quickly "run down," without laying out all of the angles again and so possibly repeating the mistake that was originally made. The angles measured in applying this check have different values from the ones first laid out, and the chance of repeating the original mistake is thereby eliminated. If no error is found to exist in the angles, the distances should next be checked. This can be done in two ways, and in some drawings both of these checks should be applied.

First, scale each line separately setting down the results

*Instead of plotting every line of the traverse from its preceding line and returning, in the case of a closed traverse, to the beginning of the starting line, it may be well to plot half the traverse from one end of the starting line and the other half from the other end; the check will then come at a point about half-way around the traverse. The advantage of this method lies in the fact that accumulative errors are to some extent avoided since they are carried through only half as many courses.

independently upon a sheet of paper. After these are all recorded (and not before), compare the lengths with the lengths of lines as taken from the field-notes. No error should be allowed to pass if it is large enough to be readily plotted by the use of the scale.

Second, take a long straight piece of paper, lay this on the drawing, and mark off the length of the first line on the edge of the paper; then mark off the length of the second line starting from the mark which denotes the end of the first line, and proceed in a similar way to the end of the traverse. Apply the scale to the strip of paper and read the station of each mark; record each of these independently and afterwards compare them with the field-notes. The entire length of line should check within a reasonable amount depending upon the scale; the allowable error can be determined as explained in Art. 1.10.

By checking angles and distances by the above methods errors of any consequence can be avoided; in any case **a draftsman should not allow a drawing to leave his hands which has not been properly checked and is not known to be correct.**

When the traverse is not closed, such checks as have been described above must **always** be applied and also the checks mentioned in the last paragraph of Art. 17.13; otherwise there is no assurance whatever that the plan is correct. It is especially necessary to check the bearings of lines frequently, so that the accumulation of small errors may not become appreciable.

17.5. Protractor and T-square.—While the ordinary T-square is not much used in plotting engineering plans, there are some occasions where it is convenient to use it. Where a traverse has been run by bearings or by deflection angles the T-square with a shifting head can be conveniently used in connection with a protractor for plotting the angles by bearings.

The paper is fastened to a drawing board having a metal edge, which insures one straight edge to the board. A meridian line is drawn on the paper, and the shifting head of the T-square is fastened so that the blade coincides with the meridian line. Then as the T-square is slid up and down the edge of the drawing board its blade always takes a direction parallel to the meridian. By means of the protractor shown in Fig. 16.5.2 the bearing of each line can be readily laid off or checked as illustrated by Fig. 17.5.1 and the distances laid off with the scale. In order to secure a satisfactory check, the de-

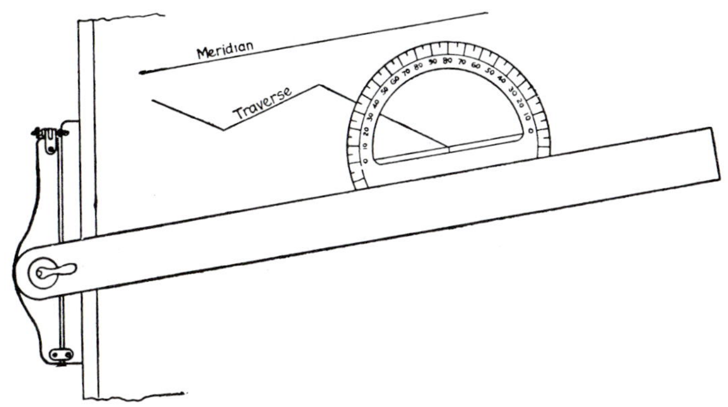

FIG. 17.5.1. LAYING OFF BEARINGS BY USE OF T-SQUARE AND PROTRACTOR.

flection angles should be laid off directly from the previous line, and the bearings checked by means of the T-square and protractor.

It is evident that the bearings of the lines may be computed just as well from any assumed meridian as from the magnetic or true meridian; and that the drawing can be fastened to the board in such a way that the T-square can be conveniently used. This method is especially applicable to compass surveys as it obviates the need for a meridian line through each angle point.

A *universal drafting machine* is useful for plotting survey data. This machine has a linkage system by means of which graduated scales may be maintained in parallel directions over all parts of the drawing. It combines the work of a T-square, triangles, protractor and scale.

17.6. PLOTTING BY RECTANGULAR COÖRDINATES.—In plotting by this system all points in the traverse are referred to a pair of coördinate axes. For convenience these axes are often the same as those used in calculating the area enclosed by the traverse. The advantages of this method are, (1) that all measurements are made by means of the scale only and (2) that the plotting may be readily checked.

To plot a survey of a field by rectangular coördinates, first calculate the *total latitude* and the *total departure*, that is, the ordinate and the abscissa, of each point in the survey. If the meridian through the most westerly point and the perpendicular through the most southerly point are chosen as the axes negative signs in the coördinates will be avoided. The coördinates of the transit points are computed by beginning with the most westerly point, whose total departure is zero, and adding successively the departure of each of the courses around the traverse. *East* departures are called *positive* and *West* departures *negative*. The total departure of the starting point as computed from that of the preceding point will be zero if no mistake is made in the computations. The total latitudes may be computed in a similar manner beginning, preferably, with the most southerly point as zero.

17.7. Rectangle Construction.—For plotting the traverse points on the plan, a convenient method of procedure is to construct a rectangle whose height equals the difference in latitude of the most northerly and the most southerly points and whose width equals the difference in departure of the most westerly and the most easterly points. If the most westerly and the most southerly points are taken as zero then the greatest ordinate and the greatest abscissa give the dimensions of the rectangle (Fig. 17.9.1). The right angles should be laid off either by the use of a reliable straight-edge and a triangle or by the beam compass.

17.8. Perpendiculars with Straight Edge and Triangle.—The better method, however, is to construct the perpendiculars by means of a straight-edge and a triangle. It is not at all necessary, although it is always desirable, that the triangle shall be accurate. It should be used in the following manner: It is first placed against the straight-edge, as shown by the full lines in Fig. 17.8.1, and a point *A*, marked on the paper. Point *C* is also marked opposite a certain definite part of the triangle.

Then the triangle is reversed to the dotted position and brought so that its edge coincides with point *A*, and then point *B* is marked opposite point *C*, as nearly as can be judged. A point *D* is plotted midway between *B* and *C* and the line *AD* is

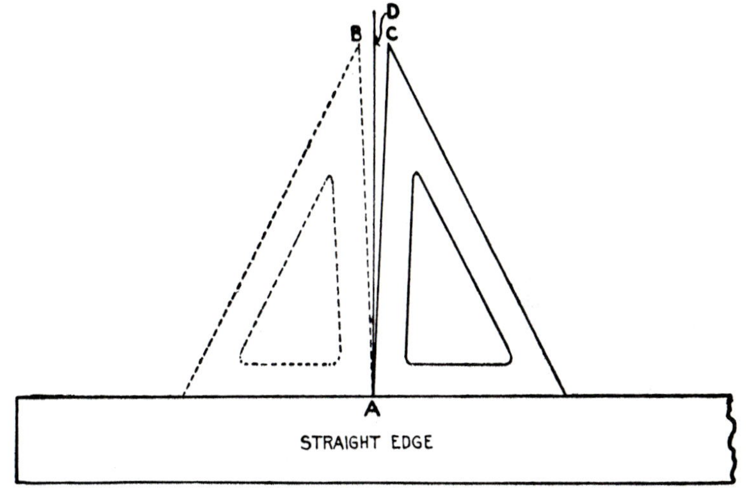

FIG. 17.8.1. ERECTING A PERPENDICULAR WITH A STRAIGHT-EDGE AND AN INACCURATE TRIANGLE.

then drawn which is perpendicular to the straight-edge. If the triangle is accurate point *B* will fall on point *C*, so that this is a method of testing the accuracy of the right angle of any triangle. If it is found to be inaccurate it should be sent to an instrument maker and be "trued up." A few cents spent in keeping drafting instruments in shape will save hours of time trying to locate small errors, which are often due to the inaccuracy of the instruments used.

If the compass is used the right angle may be laid off by geometric construction. On account of the difficulty of judging the points of intersection of the arcs, very careful work is required to obtain good results with the compass.

Since the accuracy of all of the subsequent work of a coördinate plot depends upon the accuracy with which the rectangle is constructed, great care should be taken to check this part of the work. The opposite sides of the rectangle should be equal and the two diagonals should be equal, and these conditions should be tested by scaling or with a beam compass before continuing with the plot.

17.9. Plotting of Points.—After the rectangle has been constructed, all points in the survey can be plotted by use of the scale and straight-edge. To plot any point, lay off its total

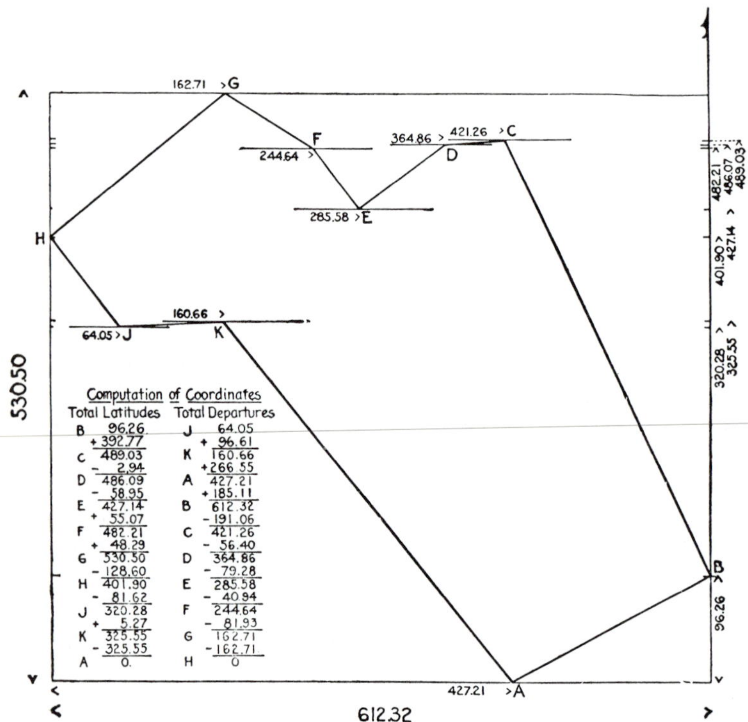

FIG. 17.9.1.
COMPUTATIONS AND PLOTTING BY RECTANGULAR COÖRDINATES.

latitude on both the easterly and the westerly of the two meridian lines of the rectangle, beginning at the southerly line of the rectangle. Draw a line through both of these points by means of a straight-edge. Then lay off along this line the total departure, beginning at the westerly side of the rectangle, thus obtaining the desired position of the point.

The computations of the total latitudes and departures and the method of plotting a traverse by the Coördinate Method are shown in Fig. 17.9.1. This is the survey which is shown in Fig. 5.4.1.

In Art. 15.33, this traverse is plotted automatically by a plotting machine "on line" with an electronic computer which has computed and stored the X and Y coördinates of each traverse point.

17.10. Checks.—When the transit points have been plotted, the scaled distance between consecutive points should equal the distance measured in the field. It sometimes happens that some of the transit lines run so nearly parallel to one of the axes that the distances will scale the right amount even though a mistake has been made in laying off one of the coördinates. If so any appreciable error can be detected by testing the bearings of the lines by means of a protractor. These two tests, together with the scaled distances of any cut-off lines which may have been measured in the field (Art. 5.17), form a good check on the accuracy of the plotting. Since all of the points are plotted independently errors cannot accumulate. If it is found that any scaled distance fails to check with the measured distance it is probable that one of the two adjacent lines will also fail to check and that the point common to the two erroneous lines is in the wrong position.

It should be remembered that everything depends upon the accuracy of the rectangle and that nothing should be plotted until it is certain that the right angles have been laid off accurately.

17.11. Most Accurate Method of Plotting.—Plotting by rectangular coördinates is the most accurate of all the methods usually employed. For plans of closed traverses, where the latitudes and departures have been computed in connection with the calculation of its area, this coördinate method of plotting is frequently used. It can be applied equally well to a traverse which does not close, but such traverses may also be plotted by the Tangent Method, or the Chord Method, as explained in the following articles.

17.12. PLOTTING BY TANGENTS.—The traverse should first be plotted approximately on some convenient small scale by use of the protractor and scale, to ascertain its extent and shape. The importance of this little plot is often overlooked, with the result that when the plan is completed it is found to be too close to one edge of the paper or otherwise awkwardly located on the sheet. It takes only a few moments to draw such a sketch, and unless the draftsman is sure of the shape and extent of the plot he should always determine it in some such manner before the plan is started.

The directions of all the lines are referred to some meridian and the bearings determined with an accuracy consistent with the measured angles. From the auxiliary plot it can be decided where to start the first course of the traverse on the paper and in what direction to draw the meridian, so that the lines of the completed traverse will be well balanced with the edges of the sheet, and so that the needle will be pointing, in a general way, toward the top of the drawing rather than toward the bottom.

The bearing of the first line is plotted as follows (Fig. 17.12.1): Lay off on the meridian line a length *Aa* of at least 10 inches and erect a perpendicular at *a* on the right-hand side of the meridian if the bearing of the first course is east, and on the left-hand side if it is west. Look up in the table of natural functions the tangent of the bearing of the first course and scale off this distance *ab* on the perpendicular.* Draw *Ab* which is the direction of the first course. On this line scale off

---

*These distances and also the 10-inch base-lines are all laid off by use of the engineer's scale. By using the 10-ft. or 100-ft. scale the tangents can be laid off without any computation, whereas with the other scales the tangent must be multiplied by some number, e.g., by 2 if the 20-ft. scale is used, by 3 if the 30-ft. scale is used, etc., taking care in the pointing off.

If it is deemed unnecessary to use a base as long as 10 inches, one can be laid off at the "10" mark on any engineer's scale and the tangent distances laid off by using the same scale, e.g., if a 20-ft. scale is used the "10" mark will give a base-line 5 inches long.

AB, the length of the first course. On this line produced lay off *Bg* equal to 10 inches and erect a perpendicular, scaling off on the perpendicular the length *gd* equal to the tangent of the **deflection angle** at *B*. This determines the direction of *BC* from the first course. The remaining lines of the traverse are plotted in the same manner, using each time the deflection angle.

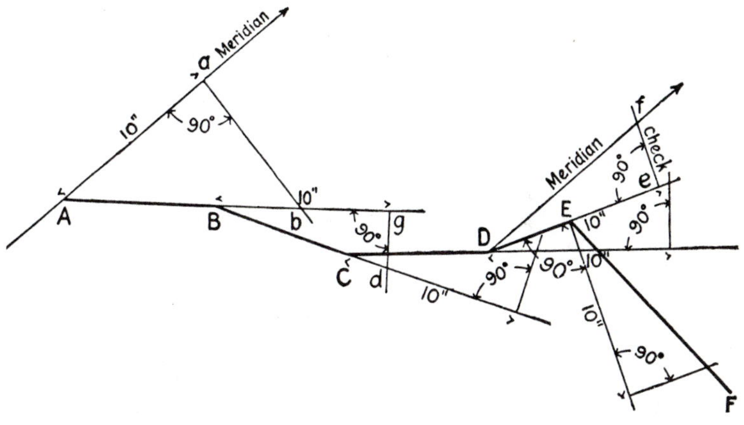

FIG. 17.12.1. PLOTTING BY TANGENT OFFSETS.

17.13. Checks.—Unless the survey is a closed traverse checks must be occasionally applied. Every third or fourth course should be checked by finding the angle between it and the meridian line. This angle should be found by the same method (tangent offset method) and by using a base of 10 inches as in plotting the angles. In checking the course *De*, for example, a meridian is drawn through *D* parallel to *Aa*, *De* is scaled off 10 inches, and a perpendicular *ef* erected. The distance *ef* is scaled and from the table of tangents the angle *fDe* is obtained. If the angle that the course makes with the meridian line disagrees with the calculated bearing of that course by any large amount, say, 10 minutes of angle or more, the previous courses should be replotted. If the error is less than 10 minutes the course which is being checked should be drawn in the correct direction so that even the slight error discovered may not be carried further along in the plot. Then after the plotting has proceeded for three or four more courses the check is again applied.

The bearings of the lines can be checked by use of the protractor and this will detect errors of any large size, but this method will not disclose small errors; moreover, if it is desired to have the plot, when completed, as accurate as could be expected from the precise method employed, it is entirely inconsistent to check by use of a method far less accurate than the one used in making the plot. For this reason the checks on the direction of the lines are applied with the same care and by the same method as was used in the original layout of the angles.

Occasionally it is more convenient to plot the complement of an angle rather than the angle itself, as was done in plotting the line *EF*. When doing this the right angle erected at *E* must be laid off with great care, preferably by the method explained in Art. 17.8.

It is evident that the direction of each course could have been plotted by drawing a meridian line through the transit points and by laying off the **bearings** by the tangent method. But if such a method were used there would be no single check applied that would check all the previous courses; this check is an important feature of the method explained above.

If the traverse is not closed the lengths of the lines of the traverse should **always** be checked by the methods explained in Art. 17.4.

When plotting a traverse which does not close a series of checks may be obtained as follows: compute the coördinates of all of the transit stations, and the length and bearing of the closing line. Plot the initial and final points by coördinates, as well as a few of the intermediate stations. Then plot the traverse by the Tangent Method, checking upon the plotted points as they are reached.

17.14. Plotting by Chords.—This method, which is employed by some draftsmen in plotting traverse lines, is fairly good although probably not so accurate as the Coördinate or so convenient as the Tangent Methods.

Fig. 17.14.1 represents the traverse *ABCDEF* which has been plotted by chords. It is the same traverse that is shown in Fig. 17.12.1.

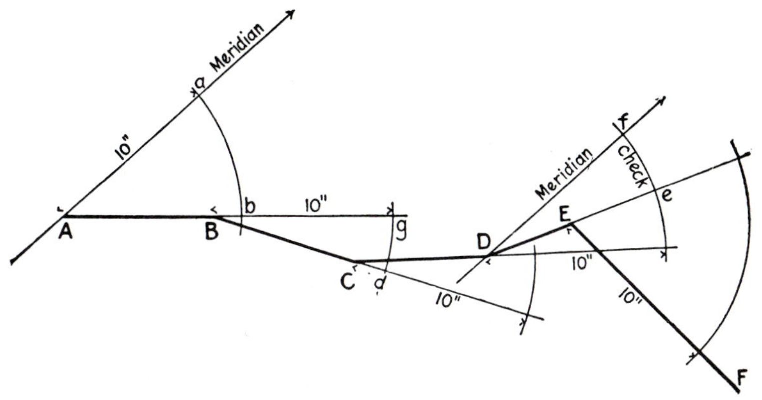

FIG. 17.14.1. PLOTTING BY CHORDS.

On the meridian line the distance *Aa* is scaled off equal to 10 inches and the arc *ab* swung from A as a center by use of the ordinary pencil compass. Then from a table of chords* the length of the chord *ab* is found for the angle *aAb*. The point *b* is sometimes located by setting the dividers at the distances *ab* and with *a* as a center intersecting the arc *ab* at *b;* but the more accurate method is to scale from point *a* the chord distance and mark the point *b* on the arc. Then the line *Ab* is drawn and *AB* scaled off on it. With *B* as a center the arc *gd* is drawn and the chord *gd*, corresponding to the deflection angle at *B*, is scaled off. *Bd* is then drawn and *BC* scaled off on it. In the same way the entire traverse is plotted.

17.15. Plotting by Use of the Sine.—It is evident that the chord

$$ab = 2 \times 10 \times \sin\frac{A}{2} \text{ (see Fig. 17.15.1)}$$

*Tables of chords can be found in Trautwine's "Civil Engineer's Pocket Book," published by John Wiley & Sons, New York.

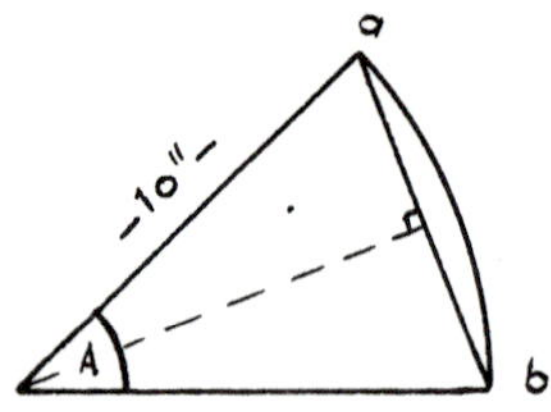

FIG. 17.15.1. CHORD LENGTH USING SINE.

hence, if a table of chords is not available, a table of sines (always easily obtainable) can be used. The sine of half the angle can be taken from the tables and multiplied by 20 mentally. Some draftsmen use the table of sines and a radius of 5 inches to avoid the multiplication. This is not recommended because a base of 5 inches is not long enough to insure an accurate drawing. The necessity of multiplying by 2 can easily be done away with by laying off the radius with a 20-ft. scale and scaling off the sine of the angle with a 10-ft. scale.

With dividers of the ordinary size it is impossible to lay out an arc with a 10-inch radius. Either beam compasses must be used or the radius employed must be shorter, so short, in fact, that it will frequently be better to resort to the Tangent Method.

17.16. Checks.—Since this method is usually applied to traverses which do not close it is desirable to check every fourth or fifth course so that a mistake will not be carried too far before it is discovered and thereby cause a waste of time. In Fig. 17.5 it is desired to check the calculated bearing of *De*. The meridian *Df* is drawn through *D* parallel to *Aa*, the arc *fe* is swung with *D* as a center and with a radius of 10 inches, and the chord *ef* is scaled. From the table of chords (or sines) the angle *fDe* (the bearing) can be found. It should agree reasonably well with the calculated bearing. The degree of precision to be expected when plotting by chords is a little less than that suggested for the Tangent Method in Art. 17.12, unless the beam compass is used. The Tangent Method, especially if the right angles are laid off by reversing the triangle, gives more accurate results than the Chord Method, for the use of the ordinary compass in the Chord Method is a fruitful source of error unless it is handled with the utmost care.

METHOD OF PLOTTING DETAILS

17.17. Buildings, Fences, Streams, etc.—The previous articles have dealt with the plotting of the traverse lines only, and these in many instances form merely the skeleton of the final plan. In the field the details of the survey are located from the transit line; and, in a similar manner, the details are located on the plan from the traverse line which has been plotted already.

Buildings, fences, shore-lines, streams, etc. are all plotted by means of the scale for distances and the protractor for the angles. Often a smaller protractor is used for this sort of work than for the traverse lines. This is permissible, for the lines which locate the details are usually short in comparison with the traverse lines and the resulting error is small; furthermore any slight error in the location of a detail will not as a rule affect the rest of the drawing, whereas an error in a transit line will have an effect on all of the rest of the drawing. The plotting of buildings has been taken up in connection with their location. (See Chapter VI.)

In plotting a set of notes where several angles have been taken at one point, such as in stadia surveying, it is well to plot all of the angles first, marking them by number or by their value, and then to plot the distances with the scale.

17.18. Contours.—Where contours are located by the cross-section method (Art. 7.10), this cross-section system is laid out in soft penciled lines on the drawing. The elevations which were taken are written at their respective points on the plan and then the contours desired are sketched. The ground is assumed to slope uniformly between adjacent elevations, and, by interpolation between these points, the location of the contours on the plan can be made. When the contours have been located, the cross-section lines and elevations are erased unless the plan is intended to be used as a working drawing. As a rule all useful data, such as construction lines and dimensions, are left on a working drawing.

When the contours are located by any other means the principle is the same. The points whose elevations have been determined are plotted by scale and protractor, and the contours are interpolated between the elevations and sketched on the plan.

17.19. CROSS-SECTIONS.—In plotting on cross-section paper, the rulings of the paper are used as the scale, and all the dimensions of the cross-section, which are to be plotted, are laid off by counting the number of squares on the cross-section paper.

Cross-sections for earthwork computations are plotted to a natural scale of either 10 ft. or 5 ft. to the inch (the latter scale being adopted usually only when the amount of cut or fill is small). These cross-sections should always be plotted **UP** the sheet.

Progress records of construction, such as for a highway or a dam, are often kept by plotting the cross-section at each station and marking on each section in colored ink the progress of the work over certain periods, such as a week or a month. In this way monthly estimates can be readily made, each month being represented by a different color or a different style of line.

Where a series of cross-sections like this are to be plotted the station number and the elevation of the finished grade are recorded just under or over the section. To avoid mistakes in numbering the sections this should be done at the time of plotting the section. The areas of the sections and quantities of earthwork are usually recorded on the sections, together with other data which may be of use in calculating volumes. (See Fig. 14.22.1.)

17.20. PROFILES.—Profiles are almost always plotted on profile paper, although occasionally they are plotted on the same sheet with the plan so that the two can be readily compared, the plan being shown at the top of the sheet on plain paper and the profile at the bottom on profile paper. Sometimes the plan is placed below the profile.

The profile is intended to show (graphically) relative elevations. In most surveys the differences in elevation are so small

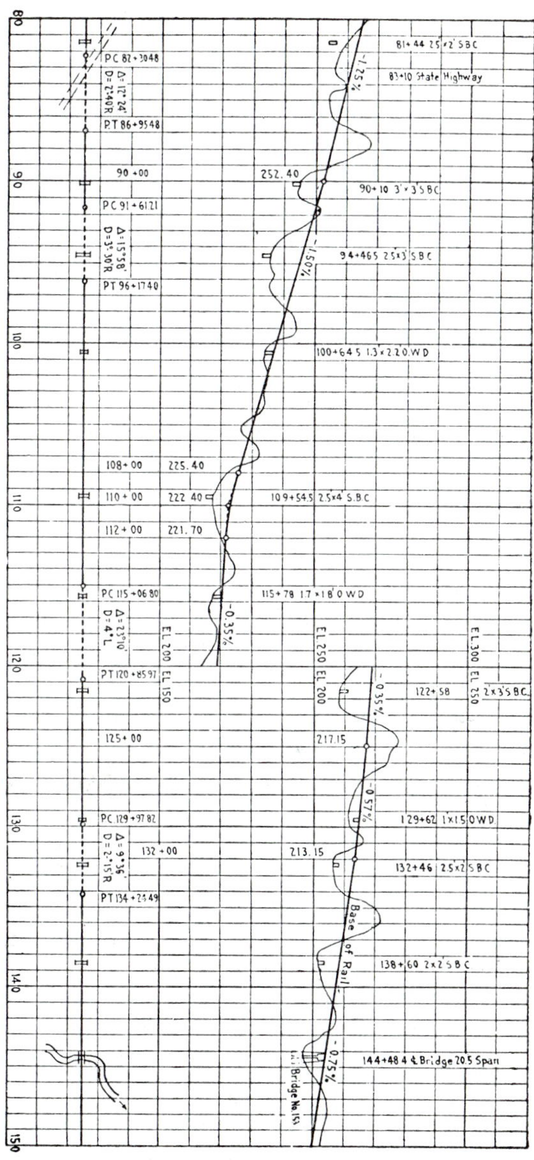

Fig. 17.20.1. Profile of Railway.

in comparison with the horizontal distances that it is necessary to exaggerate the vertical scale of the profile so that the elevations can be read from the profile with a reasonable degree of accuracy. The horizontal scale of profile should be the same as the scale of the plan, but the vertical scale should be exaggerated, say, 5 to 20 times the horizontal scale, depending upon how close it is desired to read the elevations from the drawing. If the horizontal scale of the profile is 80 ft. to an inch its vertical scale should probably be 20, 10, or 8 ft. to an inch. (See Fig. 17.20.1.)

17.21. Proper Layout.—In plotting any profile the first step is to lay it out properly on the paper, i.e., to decide, from an examination of the range of the elevations, where to start it on the paper so that it will look well when completed, and so that any additions or studies which may be drawn on it subsequently will come within the limits of the paper. Station 0 of the profile should come on one of the heavy vertical lines at the left of the paper, and the heavy horizontal lines should represent some even elevation such as 100, 125, 150, etc.

The profile is plotted by using the rulings of the profile paper as a scale; it is drawn in pencil first and afterward inked in. It will be found, if these profile papers are carefully measured with a scale, that owing to the shrinkage of the paper the divisions frequently do not scale as long as they should. In plotting a profile or section on such paper no attempt is made to use a scale; the scale of the paper is assumed to be correct and the intermediate points are plotted by estimation, which can almost always be done accurately since the rulings of the paper are quite close together.

The data for a profile of the ground generally consist of levels taken in the field at such points that the ground may be assumed to run straight between adjacent elevations. For this reason, in drawing the profile, the points where the slope of the ground changes should not be rounded off. On the other hand the ground probably does not come to a sharp angle at that point. The profile should be plotted therefore as a series of free-hand straight lines drawn so that the angles are not emphasized. When a profile is made from a contour map, the line should be a smooth, rather than an angular line.

17.22. Profile Plotting.—Profiles of the surface of the ground are generally made for the purpose of studying some proposed construction which is represented on the profile by a grade line, consisting usually of a series of straight lines. The points where the gradient changes are plotted and connected by straight ruled lines unless the proposed grade should happen to be a vertical curve (Art. 10.30). Vertical lines are also drawn from the bottom of the profile to the grade line at these points. Notes of alignment are recorded at the bottom of the profile as shown in Fig. 17.20.1.

17.23. "Breaking" the Profile.—When the difference of elevation is such that the profile line, if continued, would run off the top or the bottom of the paper, the profile line may be stopped at some vertical line, preferably a heavy line, and resumed at the same vertical line, say 20 ft. or 50 ft. lower or higher, as the case may require. The numbering of horizontal lines to the right of this point should then be changed to correspond (see Fig. 17.20.1).

17.24. Checks.—After plotting the surface and grade elevations in pencil, read off from the profile the station and elevation of each point as plotted and record both the station and elevation on a piece of paper. Compare these readings with the data given and make the necessary corrections. Time can be saved if one man reads off the station and elevation from the profile while a second man compares the readings with the note-book. A quick method of **plotting** profiles is to have one man read the notes while the other man plots them, but when the profile is being checked this method should not be used; the man, preferably the one who did not do the plotting, should **read from the profile as plotted** and these readings should be compared with the note-book.

CHAPTER XVIII

FINISHING AND FILING DRAWINGS*

18.1. What Should Appear on a Drawing.—Drawings are made for a great variety of purposes, so that the data which a plan should contain depend entirely upon the use to which it is to be put. There are, however, several important things which should appear on every engineering drawing. In the first place, it should have a complete title, and this should be a brief description of the drawing (see Figures 18.7.1 and 18.7.2). The title should state whether the drawing is a plan, cross-section, profile, etc.; what it represents—a lot of land, a sewer, highway, airport, etc.; the name of the owner; the place; the date; the scale; and the name of the surveyor and his address. Besides the title, some plans, such as land plans, always require the names of owners of abutting property, and a meridian (needle). Notes are frequently added giving such information as is necessary to interpret the plan. All essential dimensions are lettered in their proper places.

Besides these it is well to insert in some inconspicuous place (preferably near the border) the number of the notebook and the page from which the notes were plotted, and also the initials of the draftsman who made the drawing and of the man who checked it.

Fig. 18.1.1 represents a land plan which contains all of the essentials; it is a plot of the land shown in the form of notes in Fig. 5.4.1; and its working plot is in Fig. 17.9.1; and computer plot in Fig. 15.33.1. In Fig. 18.1.2 a planimetric map with coördinate grid is illustrated. It also contains a title box and an indication of magnetic north.

---

*For methods of finishing topographic and hydrographic maps see Volume II, Chapter 9. For plotting with electronic computer, see Art. 15.33.

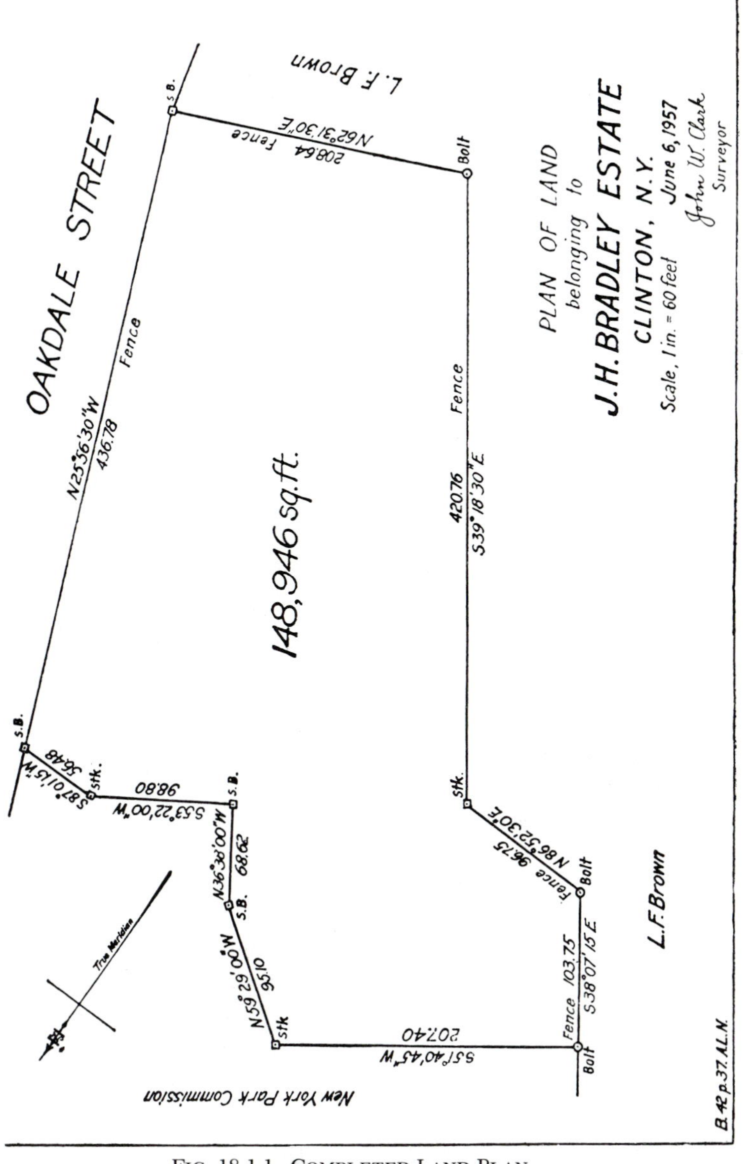

Fig. 18.1.1. Completed Land Plan.
(Scale of Reduced Drawing, 1 inch = 120 feet.)

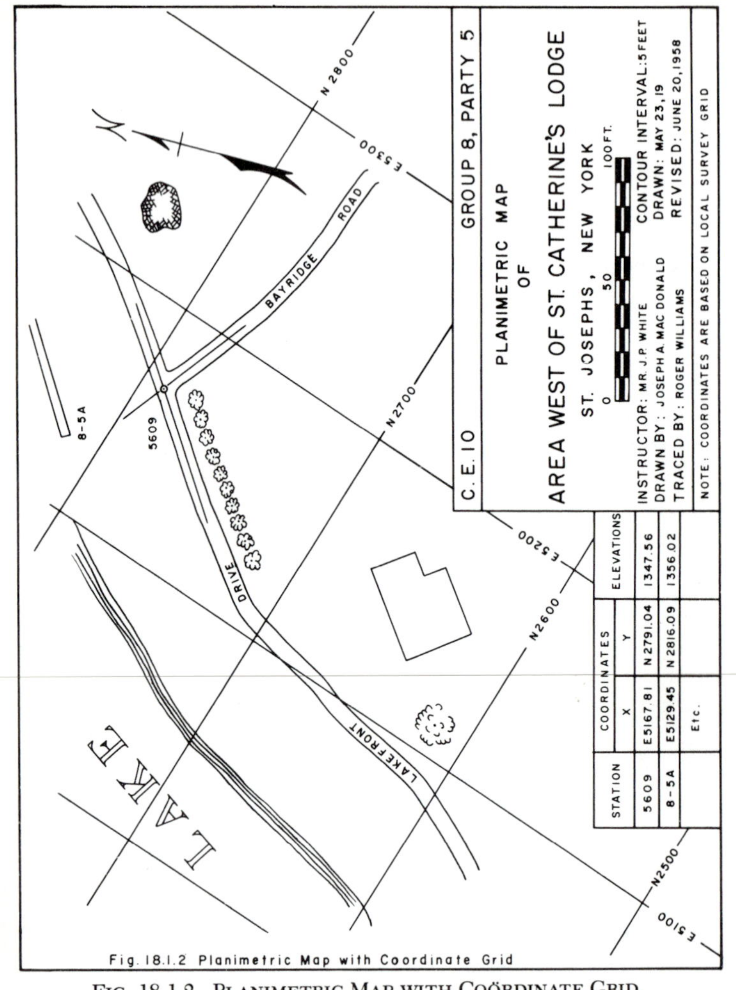

STATION	COORDINATES X	COORDINATES Y	ELEVATIONS
5609	E 5167.81	N 2791.04	1347.56
8-5A	E 5129.45	N 2816.09	1356.02
	Etc.		

Fig. 18.1.2 Planimetric Map with Coordinate Grid

FIG. 18.1.2. PLANIMETRIC MAP WITH COÖRDINATE GRID.

18.2. TRAVERSE LINES.—The convenient use of a plan sometimes requires the traverse line to be shown on the completed drawing. It is sometimes shown as a full colored line, each of the angle points being represented by a very small circle of the same color, the center of which marks the angle point. Sometimes the lines of the traverse are drawn to the angle points and these are marked by very short lines bisecting the angles. Fig. 18.2.1 illustrates these two methods of marking transit points.

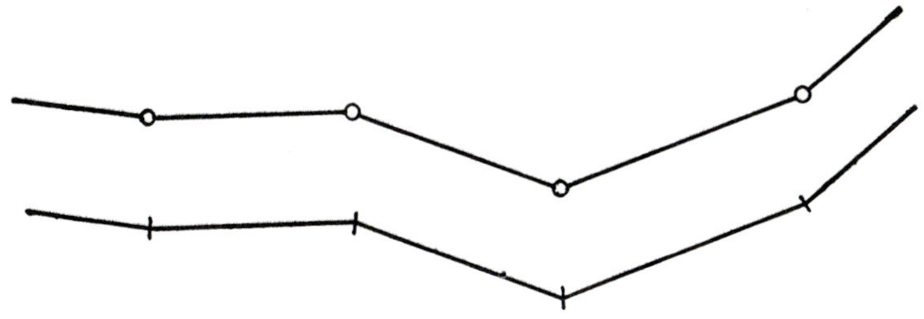

FIG. 18.2.1. METHODS OF MARKING ANGLE POINTS ON TRAVERSE LINES.

18.3. PHYSICAL FEATURES.—The boundaries of property and the physical features which are represented on a plan, such as streets and buildings, are usually drawn in black ink. Any additions or proposed changes may be drawn in colored ink, usually in red, although water-color is much better, for the reasons stated in Art. 16.33.

Shore lines and brooks are represented in black unless a colored map is desired, in which case they are shown in Prussian blue. The shore line may be emphasized by water-lining as shown in topographical signs in Fig. 18.4.1, or by a light blue band of shading applied along the water side of the shore line.

18.4. TOPOGRAPHIC CONVENTIONAL SIGNS.—Standard conventional symbols for use on small-scale maps, such as those issued by various U.S. Government agencies, have been adopted by the U.S. Geological Survey. Their "Topographic Maps—Descriptive Folder" shows standard symbols and colors used, as well as types of maps available. These may be had by writing to Map Information Office, United States Geological Survey, Washington, D.C. Fig. 18.4.1 shows some of the conventional signs formerly used. These are useful for showing details of culture on large scale maps, such as landscaping or park development plans. In recent years, the trend is to colors and color tints in place of symbols. For example, on Geological Survey maps, black is used for names, boundaries,

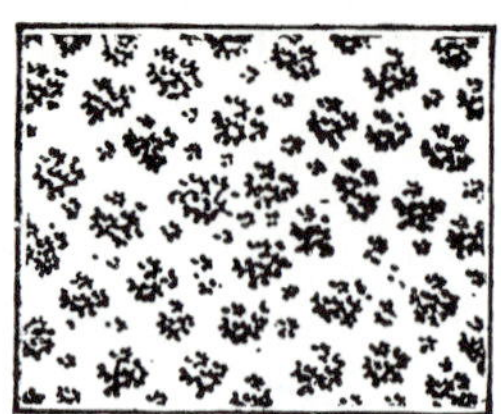

DECIDUOUS TREES (OAK).

DECIDUOUS TREES (ROUND LEAF.)

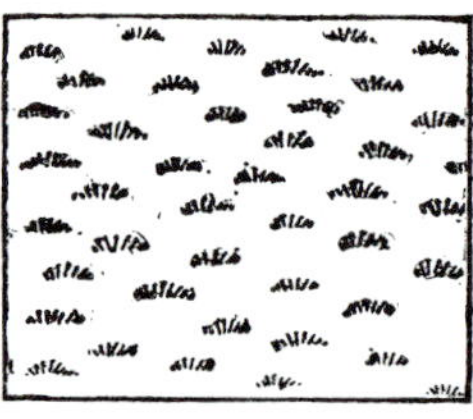

GRASS.

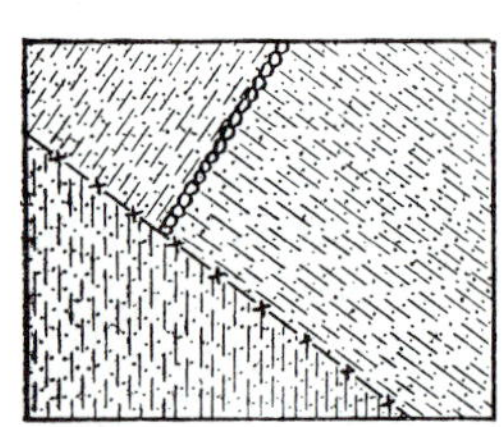

CULTIVATED LAND.

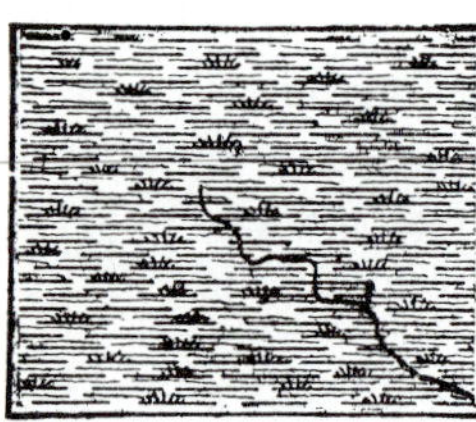

FRESH MARSH.

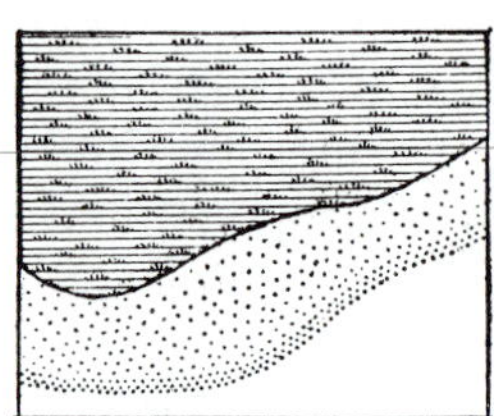

SALT MARSH — SAND.

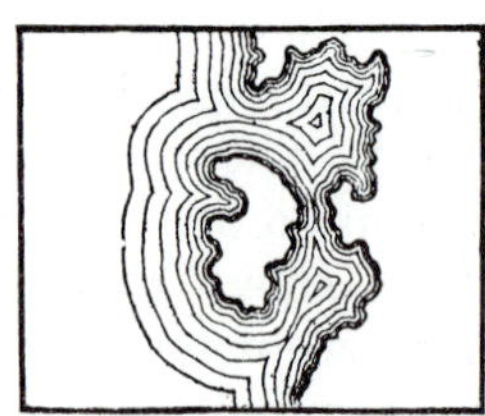

WATERLINING.

LEDGES — EVERGREEN TREES.

FIG. 18.4.1. TOPOGRAPHIC CONVENTIONAL SIGNS.

secondary roads, trails, buildings, railroads and beach sand (dots); deep red is used for primary roads such as numbered routes; green shading represents wooded areas; a light red tint denotes thickly settled areas; blue is used for water surfaces, marsh symbols and under-water contours; and brown for land contours.

Referring to Fig. 18.4.1, "cultivated land" and the horizontal lines of "Salt Marsh" are ruled; the rest are drawn freehand with a fine pen. In the symbol for "grass," the individual lines of a group radiate from a center below the group. The bottom of grass and marsh lines are always parallel to bottom margin of the map.

In executing "water-lining" the first line outside the shore line should be a full line drawn just as close to the shore line as possible, and should follow very carefully every irregularity of the shore line. The next water-line should be drawn parallel to the first but with a little more space between them than was left between the shore line and the first water-line. Then the third water-line should be spaced a little farther out, and so on; five to ten lines are sufficient to represent this symbol properly. As successive lines are added, farther and farther from the shore line, little irregularities of the shore gradually disappear.

On colored maps waterlining, fresh marsh and salt marsh are drawn in Prussian blue. Contour lines, Chapter VII, are usually drawn in burnt sienna (brown) water color. On tracings of engineering plans, they are usually drawn in black since the brown color will not reproduce. Every fifth or tenth contour is represented by a slightly heavier line.

A contour pen, which has adjustable nibs like a ruling pen but swivels about the holder, will give uniform lines when it is skillfully handled and there are not too many sharp turns to be made. Every fifth or tenth contour is usually numbered with its elevation by breaking the line and inserting a small number or by placing the number on top of the line. Sufficient contours should be numbered so that the elevation of any contour may be found without much effort. On any plan which shows elevations, the datum to which the elevations refer should be clearly stated, such as mean sea level, U. S. C. & G. S. datum, city base, etc. See Art. 9.1.

18.5. Conventional Signs for Physical Features.—Features such as highways, railways, walls, buildings, and boundaries are usually represented in black by the symbols in Fig. 18.5.1.

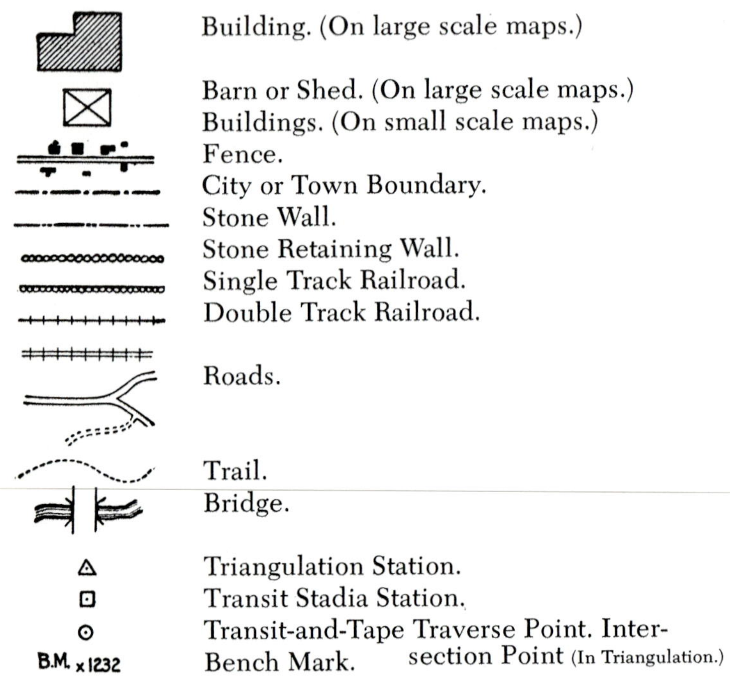

FIG. 18.5.1. CONVENTIONAL SIGNS FOR PHYSICAL FEATURES.

18.6. LETTERING.*—The lettering on a drawing probably has more to do with its appearance than any other feature. To be able to do good lettering at first is a gift which but few men possess. It is an art that can be acquired by most draftsmen,

*For a complete discussion and illustrations of lettering see any of the following publications: "Technical Drawing," by Giesecke, Mitchell and Spencer, published by The Macmillan Co.; "Manual for Engineering Drawing for Students and Draftsmen," by Thomas E. French and Charles F. Vierck, and "Lessons in Lettering," by Thomas E. French and William D. Turnbull, both published by McGraw-Hill Book Co., New York.

however, if they will study it carefully and devote a little time to systematic practice.

Several different styles of lettering are shown in Figs. 18.6.1 and 18.6.2. The general style to use in any given case depends on the kind of drawing and on the use to which it is to be put. On most plans a simple style of lettering is appropriate such as obtained with Reinhardt letters. Roman or Gothic styles may be used in titles, particularly on plans that are prepared for public display.

The use of lettering guides has superseded much of the free-hand lettering formerly required on maps and engineering drawings. These guides produce a simple clean style of lettering, as illustrated in middle title of Fig. 18.7.1 and bottom title of Fig. 18.7.2.

On plans which are to be reproduced photographically, such as prints to a reduced scale for enclosure in engineering reports, the lettering can be done on strips of paper and pasted on the plan. Sheets of letters of different sizes and styles are available. These can be cut out and assembled into titles or captions. A special kind of typewriter has been developed which prints in letters similar to those obtained with lettering guides.

Some plans are not inked by the draftsman, but are prepared as neat pencil drawings executed in a fairly soft but sharp pencil. From these pencil drawings, transparencies can be produced which will give prints that are as satisfactory for many purposes as those printed from ink tracings. See Art. 16.29.

All plans should be lettered so as to read from the bottom. Unless a draftsman exercises care he will find, when the plan is completed, that some of the lettering is upside down. Fig. 18.6.3 illustrates the proper lettering of lines of various slopes.

Many drafting pens are available in several weights of line, and are easily operated as either line pens or lettering pens.

The appearance of freehand lettering is entirely dependent upon the skill and experience of the letterer. Draftsmen who cannot letter well freehand can produce a neat-looking plan with the aid of lettering guides. These guides are of two types:

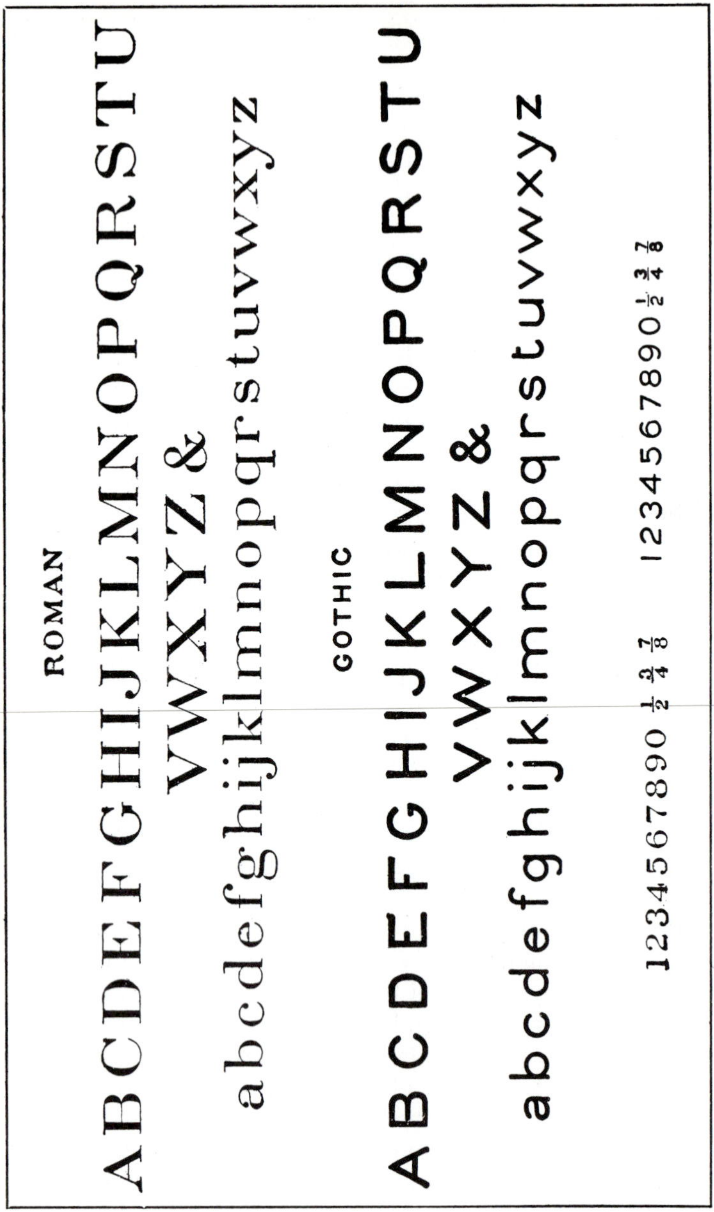

ROMAN

ABCDEFGHIJKLMNOPQRSTU
VWXYZ&
abcdefghijklmnopqrstuvwxyz

GOTHIC

ABCDEFGHIJKLMNOPQRSTU
VWXYZ&
abcdefghijklmnopqrstuvwxyz

1234567890 $\frac{1}{2}$ $\frac{3}{4}$ $\frac{7}{8}$ 1234567890 $\frac{1}{2}$ $\frac{3}{4}$ $\frac{7}{8}$

FIG. 18.6.1.

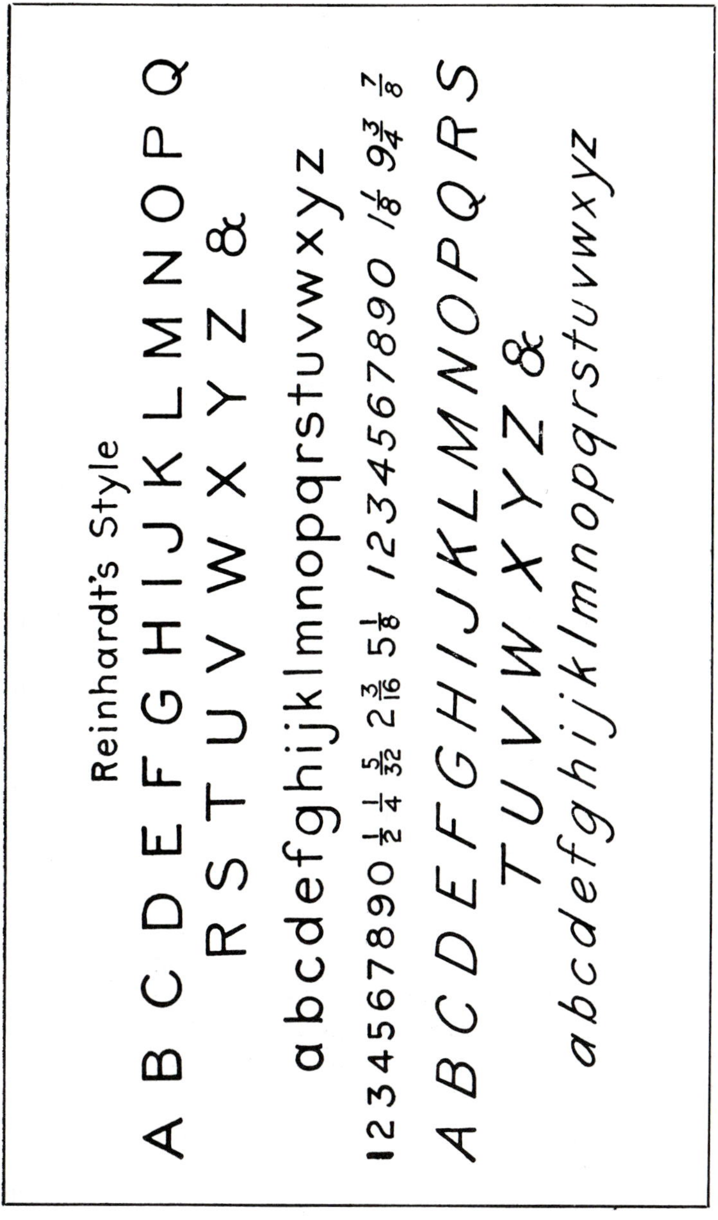

Reinhardt's Style

A B C D E F G H I J K L M N O P Q

R S T U V W X Y Z &

abcdefghijklmnopqrstuvwxyz

1234567890 $\frac{1}{2}$ $\frac{1}{4}$ $\frac{5}{32}$ $2\frac{3}{16}$ $5\frac{1}{8}$

1234567890 *$\frac{1}{8}$* *$9\frac{3}{4}$* *$\frac{7}{8}$*

ABCDEFGHIJKLMNOPQRS

TUVWXYZ &

abcdefghijklmnopqrstuvwxyz

FIG. 18.6.2.

in one type the lettering pen is moved in contact with the characters cut in the guide; in the other type a tracing point is moved in the grooved letters of the guide as the pen describes the letters on the drawing on a line parallel to and a short distance above the guide.

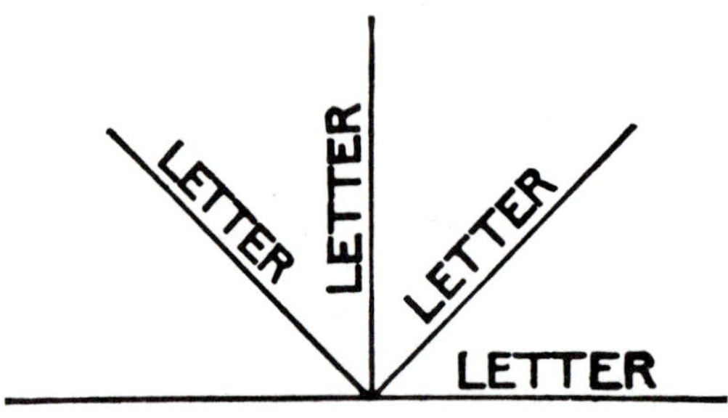

FIG. 18.6.3. LETTERING ON SLOPES.

Lettering guides can also be used for pencil lettering. One disadvantage of the guides is that the size and spacing of letters cannot always be adjusted to the space available on the plan, and recourse has to be made to freehand lettering. A good draftsman can do faster and just as presentable lettering freehand.

Automatic Plotters have lettering capabilities as illustrated in Fig. 17.9.1.

18.7. Titles.—The design of the title of a plan gives the draftsman an opportunity to exercise good taste. It should be so arranged and the size of the letters so chosen that the most important part of the title strikes the eye first. In general, each line of lettering should be centered, and the spacing between the lines should be so arranged that no part will either appear crowded or seem to be floating away from the rest of the title. The general outline of the title should be pleasing to the eye.

Where a number of sheets are used covering the same work, it is cheaper to have the title set up in type and stamped on. This also applies to notes on construction plans that appear on several sheets.

Three typical titles are shown in Fig. 18.7.1. The upper title is sloping Reinhardt style, the middle title is an example of the

Plan of Land
belonging to
James O. Farrington
Somerville....Mass.

Scale, 1 in. = 40 ft. *January 8, 1952*

HILTON AIRPORT COMMISSION
HILTON, OHIO

MASTER PLAN FOR HILTON AIRPORT
PROPOSED DEVELOPMENT

SCALE 1 IN. = 200 FT. JUNE 1952
EASTMAN & WESTMAN ENGINEERS
COLUMBUS, OHIO

UNITED STATES–EAST COAST

NEW BEDFORD HARBOR
AND APPROACHES
MASSACHUSETTS

Scale $\frac{1}{20000}$

FIG. 18.7.1. TITLES OF PLANS.

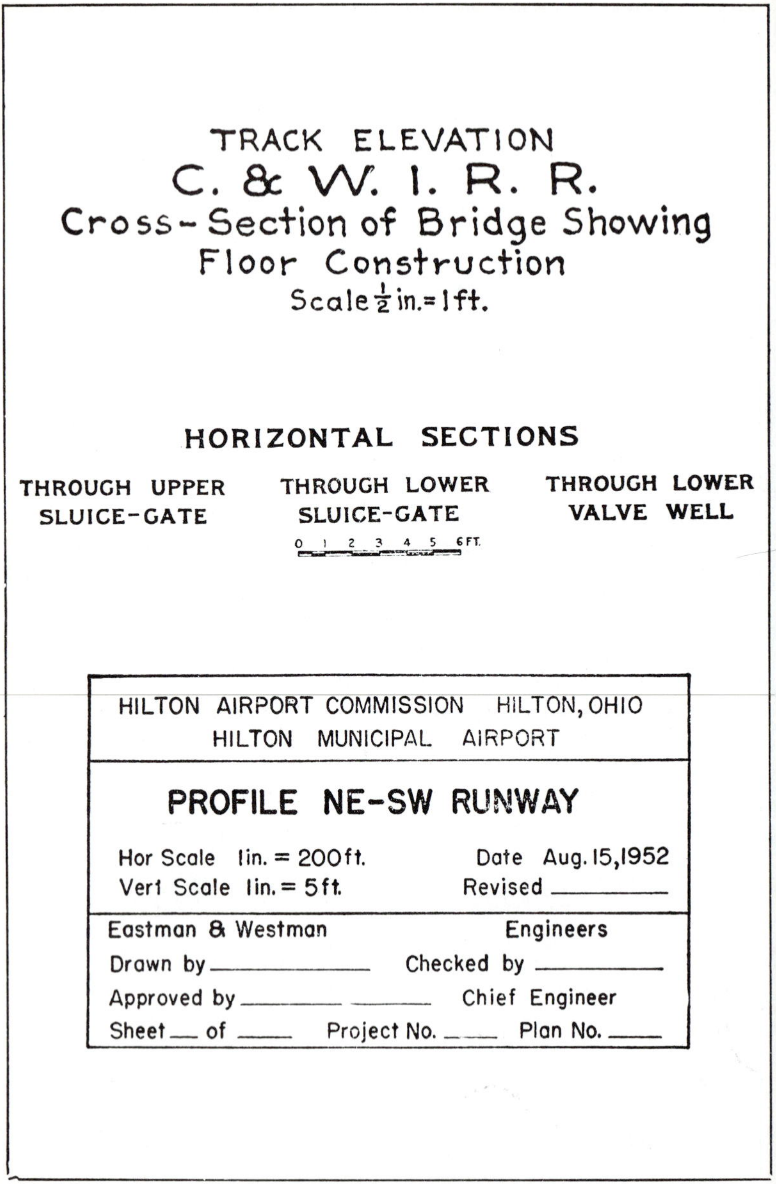

TRACK ELEVATION
C. & W. I. R. R.
Cross-Section of Bridge Showing
Floor Construction
Scale $\frac{1}{2}$ in. = 1 ft.

HORIZONTAL SECTIONS

THROUGH UPPER SLUICE-GATE

THROUGH LOWER SLUICE-GATE

0 1 2 3 4 5 6 FT.

THROUGH LOWER VALVE WELL

HILTON AIRPORT COMMISSION HILTON, OHIO
HILTON MUNICIPAL AIRPORT

PROFILE NE-SW RUNWAY

Hor Scale 1 in. = 200 ft. Date Aug. 15, 1952
Vert Scale 1 in. = 5 ft. Revised ______

Eastman & Westman Engineers
Drawn by ______ Checked by ______
Approved by ______ ______ Chief Engineer
Sheet __ of ___ Project No. ___ Plan No. ___

FIG. 18.7.2. TITLES OF PLANS AND PROFILES.

use of lettering guides, and the lower title is in the Roman style. Fig. 18.7.2 shows typical forms of titles for a cross-section, construction details and a profile. The upper two titles are in freehand Gothic style; the lower title was made with lettering guides. The box type of title is commonly used on sets of plans pertaining to one project. The upper and lower portions of the box are the same for all drawings of the set. The appropriate title and scales for a particular plan are lettered in the middle portion of the box.

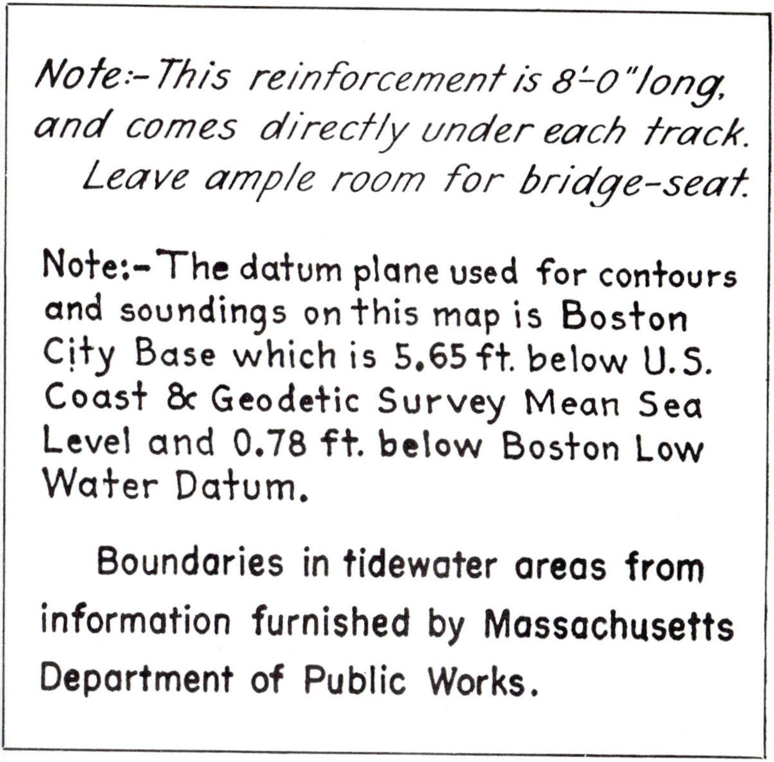

Note:-This reinforcement is 8'-0" long, and comes directly under each track. Leave ample room for bridge-seat.

Note:- The datum plane used for contours and soundings on this map is Boston City Base which is 5.65 ft. below U.S. Coast & Geodetic Survey Mean Sea Level and 0.78 ft. below Boston Low Water Datum.

Boundaries in tidewater areas from information furnished by Massachusetts Department of Public Works.

FIG. 18.8.1. SAMPLES OF NOTES.

18.8. Notes.—Most drawings require notes of some sort. These are usually lettered in the plain Reinhardt style or with lettering guides. Three samples of notes are shown in Fig.

18.8.1. The upper two notes were lettered freehand; the lower one was lettered with a lettering guide.

18.9. Colors on Plans.—Colors are sometimes used on plans to distinguish certain lines from topographical features; such as the use of red for transit lines or grade lines and blue for shore lines. Colors may also be used effectively on construction plans to emphasize new work, or on land taking plans to designate the parcels of land required for a project. If such plans are to be reproduced, however, it should be remembered that the colors will not be distinguishable on the prints. Red lines print as black; other colors, such as blue, print with less intensity than black. Therefore the colored lines should be of such weight or convention that they will convey their meaning on the reproduction. When only a few copies are to be made and colors are desirable, they may be added to the prints in crayon or watercolor.

18.10. Border Lines.—The border line of a drawing should consist of a heavy single line or of double lines closely spaced. It should neither be so heavy nor of such fancy design as to be conspicuous. Plain clear drawings are the practice of today, and the border line should be in keeping with the rest of the drawing.

For drawings up to 2 by 3 feet in size, borders of ½″ to ¾″ are appropriate on the top, bottom and right-hand edge with a 2″ border on the left for binding. For larger drawings the borders may be increased to 1″ with 2½″ at the binding edge. Some offices have standard size sheets prepared for their use with borders and title box printed on them.

18.11. Meridians.—On all land plans it is customary to draw either the true or the magnetic meridian; often both of them are represented. To be in keeping with the rest of the drawing the needle should be simple in design. It should not be too ornate or too prominent. The simple meridians shown in Fig. 18.11.1 are suggested as suitable for land plans.

The plan should always be drawn, if possible, so that the meridian will point, in general, toward the top of the drawing rather than toward the bottom. The line of the meridian should not cross any of the lines of the drawing, and it should be located far enough away from the body of the drawing so that it will not be mistaken for part of the drawing.

If the plan is referred to an established coördinate system, the grid may be represented by short intersecting lines representing the E–W and N–S coördinates at, say, 1000 ft. intervals.

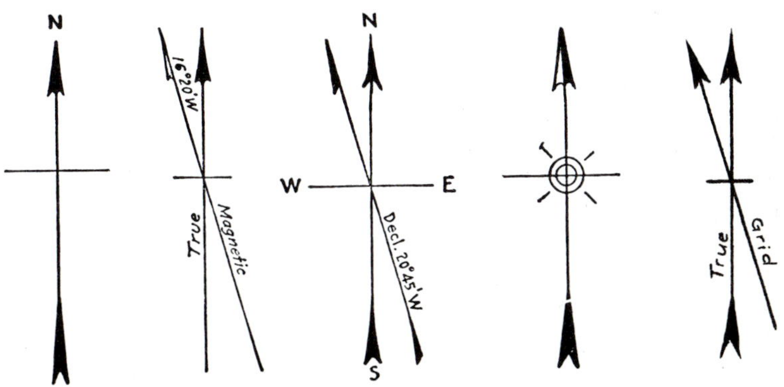

FIG. 18.11.1. MERIDIANS.

18.12. Scales.—The scale of a plan may be stated in words, given by a representative fraction or shown graphically, as indicated in Fig. 18.12.1. The graphical scale has the advantage that it remains a true scale if the plan is enlarged or

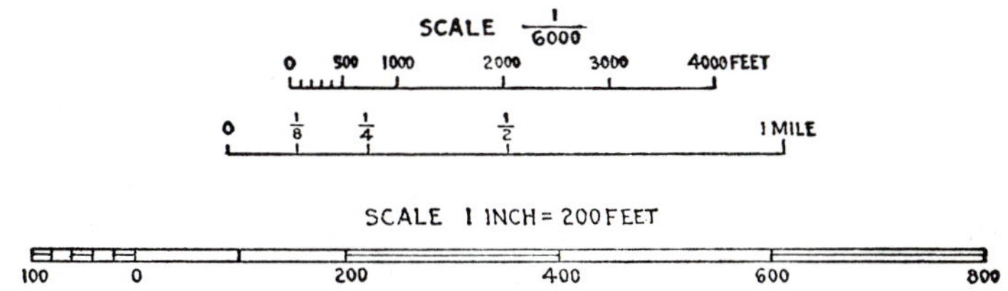

FIG. 18.12.1. SCALES.

reduced. It also serves to detect shrinkage of the drawing paper, since the plotted scale can at any time be checked against the draftsman's rule. To detect shrinkage it is well to plot a fairly long scale (3 to 10 inches depending upon the size of the drawing). A graphical scale will only detect shrinkage in the area and in the direction in which it is plotted. Such scales are usually placed directly under the title or in the lower central portion of the map.

In plotting a coördinate survey, the intersections of the north and south with the east and west lines should be marked on the finished drawing, as these are of great assistance in plotting additions. Moreover the distances between these points give a reliable measure of the change in scale of the map (at that place and in that direction) due to shrinkage of paper.

18.13. SRINKAGE OF DRAWING PAPERS.—All of the papers in use will shrink and swell more or less with variations of weather conditions. The heavy mounted papers are affected the least, but large drawings even on such paper will be found on examination to change in size perceptibly. The fact that they do not always shrink the same amount in different directions makes it difficult to estimate the amount of the change and to allow for it. This effect can be estimated quite closely, however, by testing the drawing by measuring accurately a few lines running in different directions **when it is plotted** and scaling the same lines at any other time and making allowance for the change. Scaled distances on tracing cloth are quite unreliable if it is not kept in a dry place, and blueprints generally shrink in washing so that scale measurements taken from them usually contain large errors.

18.14. MAPS OF LARGE EXTENT.—Some maps, like the location map of a highway or the map of a city, are so large that they must be made in sections. Two slightly different methods are employed for their construction. One method is to plot the several sheets so that the drawing on one will extend to but not include any of the drawing on the adjacent sheet, the limits of the drawings being defined by straight lines. The other method is to have the drawing on each sheet lap over the drawings on the adjacent sheets a little. Identifying marks are made on all drawings which make it possible to fit them to the corresponding marks on the adjacent drawings when they are being assembled at the match lines.

In attempting to arrange the sheets of adjacent drawings

after they have been in use for a long time, it is often found that they do not fit well on account of the unequal shrinking and swelling of the paper. Moreover, in plotting lines on separate sheets so that they will match exactly, there are mechanical difficulties which can only be appreciated by the draftsman who has had experience with them. If separate sheets are used the first sheet should be a "key plan" or "site plan" showing the relation of the different sheets and giving a comprehensive view of the whole project. Station zero on the first detail sheet should be at the *left-hand* side. Separate sheets are convenient for use in the field.

18.15. Inking a Profile.—The surface line is usually shown as a full firm black line and the grade line as a full red or black line (Art. 16.33). The gradients are lettered in red or black on the grade line (Fig. 17.20.1). A horizontal base-line is sometimes drawn a short distance above the bottom of the paper and vertical lines are drawn from this line to the grade line at every change of gradient, and at both ends of the profile. On these vertical lines are recorded the grade elevations at these points and the "plus" if the place where the gradient changes is not at a full station. The gradients are sometimes lettered on the base-line instead of on the grade line. The notes of alignment are recorded on the base-line, and under these the stations are marked at every heavy vertical ruling of the profile paper.

Information, such as the names of streets and brooks, is lettered vertically above the profile and at the proper station. A title and the scale are placed on the face of the profile; sometimes these are put on the back of the profile at one end of it (or both in the case of a long profile), so that the title can be read when it is rolled up.

18.16. Cleaning Drawings.—Every drawing, during its construction, collects more or less dirt. Often construction lines are drawn which must be erased when the plan is completed. A soft pencil eraser or art-gum may be used to remove construction lines.

18.17. Filing Drawings.—While the particular method of filing plans varies considerably in different offices, there are a few general ideas carried out by all drafting offices in regard to the preservation as well as the systematic filing of drawings. There is no doubt that the best method of filing plans is to keep them flat, but this is not practicable with large plans, which must usually be filed rolled. In all systems of plan filing there appears to be a proper use of both flat and rolled plans.

In large offices plans are, as a rule, made in several standard sizes prescribed by the rules at the office, and are filed flat in shallow drawers to fit the different sizes of drawings, or they may be clamped together at the binding edge and suspended vertically in a plan rack. In some offices the adherence to standard sizes is very rigid, and time is often spent to bring drawings within the limits of one of these sizes. When these sizes are exceeded the plans are either made in sections of standard size, as explained in Art. 18.14, or they are made as large plans which are rolled and filed away in cardboard tubes.

Plans filed flat are marked each with its proper index number in one corner, preferably the lower right-hand corner, so that as the drawer is opened the numbers can be readily examined. In some offices it is required that in returning a drawing it shall be placed in its proper order in the drawer as well as in the proper drawer, while in other offices the plan drawers are made very shallow, so as to contain only about 15 or 20 drawings, and when a plan is returned no attempt is made to put it in any particular place in the drawer, there being, at the most, only a very few drawings to handle to obtain the one desired.

Rolled drawings are marked on the side of the rolls at each end so as to be easily read by one standing in front of the shelf on which the plans are stored. Another style of roll is closed at one end with a white label on the outside of the closed end. When the plan has been put into the tube it is so placed on the shelf that the label on which the plan number is marked is at the front edge of the shelf where it can be read conveniently.

Large plans which are made in sections are often filed in large folios or books in such a way that they can be taken out readily and used separately.

18.18. Indexing Drawings.—A system of numbering plans should be such that one can tell from its number whether the drawing is a sketch, a working drawing, a finished drawing, a tracing, or a process print. The numbering also should suggest the kind of drawing, as a land plan, or a construction plan, and the project.

For offices where few plans are on file an index book may suffice for recording the plans, but in large drafting offices the card catalog system is used extensively. By a judicious use of "markers" a card catalog system can be so devised that it will be necessary to examine only a very few cards to find the one corresponding to any plan.

18.19. Filing and Indexing Notes.—To prevent loss from accident or fire, notebooks and original plans should be filed in a safe place. Reproductions should be made for general use by photostat or Xerox process. Notebooks can be recorded on microfilm. Care should be taken to preserve documents of value for future reference and contingencies, such as a lawsuit.

Field notes and their contents may be indexed in a book or card catalog, classified by date and job number.

PROBLEMS

PROBLEMS FOR CHAPTER I

1. Convert the following values from the English system into SI-units:

   (a) 267.3 miles
   (b) 121 yards
   (c) 2 ft 7½ inches
   (d) ⅛ inch
   (e) 6 square feet
   (f) 5½ acres

2. Convert the following metric values into the English system:

   (a) 16.31 km
   (b) 239.47 m
   (c) 8.34 cm
   (d) 3.6 mm
   (e) 1.25 m²
   (f) 2.6 ha

3. A quantity was measured 6 times with the following results: 214.30, 214.36, 214.29, 214.33, 214.36, 214.34.

(a) Determine the most probable value of the quantity and its standard deviation.

(b) Earlier measurements had resulted in 214.36 ± 0.01 and 214.32 ± 0.06 respectively. Compute now the most probable value and its standard deviation.

See also problems 16 to 18 in chapter 3.

PROBLEMS FOR CHAPTER II

1. Compute the angles *AOB*, *COD*, *EOF* and *GOH* from the given magnetic bearings.

(a) *OA*, N 39°¼ E.
OB, N 76°¾ E.

(b) *OC*, N 35° 15′ E.
OD, S 88° 00′ W.

(c) *OE*, N 15° E.
OF, S 36° E.

(d) *OG*, N 40° 15′ E.
OH, N 66° 45′ W.

2. The bearing of one side of a field in the shape of a regular hexagon is S 10°¼ F proceeding around the field in the left-hand (counter-clockwise) direction. Find the bearings of the other sides taken around the field in order.

3. (a) In 1859 a certain line had a magnetic bearing of N 21° W. The declination of the needle at that place in 1859 was 8° 39′ W. In 1902 the declination was 10° 58′ W. What was the magnetic bearing of the line in 1902?

(b) In 1877 a line has a magnetic bearing of N 89° 30′ E. The declination was 0° 13′ E. In 1902 the declination was 1° 39′ W. Find the magnetic bearing of the line in 1902.

4. Magnetic bearings of a closed field are as follows. Short side is *DE*.

Sta.	Forward Bearing	Reverse Bearing	Sta.	Forward Bearing	Reverse Bearing
A	due N.	S 85° ¾ W.	*D*	N 86° ½ W.	N 2° E.
B	N 87° ½ W.	due S.	*E*	S 10° E.	S 89° E.
C	S 1° ¼ W.	S 87° ¼ E.	*F*	N 82° ¾ E.	N 12° W.

Bearing error may be caused by local magnetic attraction or by errors in observation.

(a) Will the former affect the interior angles computed from the bearings?

(b) In the above survey of a closed figure first compute the interior angles from the forward and reverse bearing at each station and adjust the interior angles, i.e., make them equal to the proper number of right angles, placing the error adjacent to the short side. Then start at a station free from local attraction and compute the correct bearings thus producing a closed figure in which the local attraction has been eliminated from the bearings and the other errors have been adjusted.

5. Magnetic bearing of line *AB* is N 48° 15′ E. Angle *ABC* is 99° 50′, *C* being south of *AB*. Compute the bearing of *BC*. The deflection angle at *C*, i.e., angle between *BC* produced and line *CD* is 31° 10′ Right. Compute the bearing of *CD*.

6. Determine from the Isogonic Chart the approximate declination in 1968 at a place in latitude 30° N, longitude 100° W.

7. In 1969 a land surveyor near Boston, Massachusetts, wished to rerun a compass line 1530 feet long originally surveyed in 1775. (a) Find the amount and direction of the resulting divergence in the magnetic meridian between these dates. (b) If the bearing of the line in 1775 had been S 54° 30′ W, what would the bearing be in 1969?

8. A 4-sided piece of land in Western Pennsylvania in latitude 40° N and longitude 80° W was surveyed in 1840, and the bearings recorded as follows: N 30° 15′ E, N 89° 30′ E, S 5° 45′ E, and S 88° 00′ W. Magnetic declination in 1840 = 0° 17′ E. What bearings should be used to retrace the lines in 1969 (to nearest ¼°)?

9. Is it necessary that the adjustments of the transit should be made in the order given in this chapter? Give your reasons.

10. A transit is sighting toward *B* from a point *A*. In setting up the transit at *A* it was carelessly set 0.01 ft. directly to one side of *A*, as at *A′*. What would be the resulting error, i.e., the difference in direction (in seconds) between *AB* and *A′B* (1) when *AB* = 40 ft., (2) when *AB* = 1000 ft.?

11. An angle of 90° is laid off with a "one minute" transit, and the angle then determined by six repetitions, the final reading being 179° 58′ + 360°. The point sighted is 185 ft. from the transit. Compute the offset to be laid off in

order to correct the first angle. Express the result in feet and also in inches.

12. An angle measured with a transit is 10° 15′ 40″. The telescope of a leveling instrument is placed in front of the transit (with its objective toward the transit) and the angle again measured and found to be 0° 18′ 20″. What is the magnifying power of this level telescope?

13. If a transit is set up so that the horizontal axis is inclined one minute with the true horizontal direction, what will be the angular error in sighting on a point on a hill, vertical angle 20°, and then setting a point in line and on the same level as the instrument?

14. Design a vernier to read to 30″ for a circle divided into 15′ spaces.

15. Design a vernier to read to 5′ when the circle is divided into degrees.

16. In the transit having a graduated circle five inches in diameter, the centers do not coincide by 0.001 of an inch.

(a) What is the maximum error in seconds (on one vernier) due to this eccentricity?

(b) What would be the maximum error in a circle seven inches in diameter?

PROBLEMS FOR CHAPTER III

1. A distance is measured with an engineer's chain and found to be 796.4 ft. The chain when compared with a standard is found to be 0.27 ft. too long. What is the actual length of the line?

2. A metallic tape which was originally 50 ft. is found to be 50.14 ft. long. A house 26 ft. × 30 ft. is to be laid out. What measurements must be made, using this tape, in order that the house shall have the desired dimensions? Using the same tape what should the diagonals read?

3. A steel tape is known to be 100.000 ft. long at 68° F. with a pull of 12 lbs. and supported its entire length. Its coefficient of expansion is 0.000 006 4 for 1° F. A line was measured and found to be 142.67 ft. when the temperature was 8° below zero. What is the true length of the line?

4. In chaining down a hill with a surveyor's chain the head-chainman held his end of the chain 1.5 ft. too low. What error per chain-length would this produce?

5. In measuring a line with a 100-ft. tape the forward end is held 3 ft. to the side of the line. What is the error in one tape-length?

6. A certain 100-ft. steel tape was tested by the Bureau of Standards and its length given as 100.015 ft. at 68° F. when supported throughout its length and a 10 lb. tension applied. The tape was afterward tested with a 10 and then a 20 lb. pull and found to stretch 0.016 ft. (or 0.0016 per lb.). To obtain the length when supported at the ends only, a 12 lb. pull was used and the full length between end graduations was marked off on tripods, the tape being supported throughout its length. The intermediate supports were then removed. The tape was found to sag 0.77 ft. and was 0.017 ft. shorter than when supported full length, the tension remaining at 12 lbs. Compute the true (chord) length of the tape for 12 lb. pull, 68° F. supported at ends only.

7. A distance measured with a 50-ft. steel tape is recorded as 696.41 ft. The tape is known to be 49.985 ft. long. What is the correct length of the line?

8. Two measurements of a line were made with the same tape on different days. The first length was 510.02 ft., the temperature being 65° F. The second length was 510.06, the temperature being 50° F. The tape is standard at 60°; coefficient = .000 006 45. Compute the length of the line from each of the two measurements and explain any difference in the reduced readings.

9. A distance is measured on slope with a 300-ft. tape and found to be 299.79 ft. If the difference of level of the two points is 15.10 ft. what is the horizontal distance?

10. If the slope distance is 201.61 ft. and the vertical angle is 4° 21′ what is the horizontal distance?

11. With transit set up at B, a slope distance of 182.42 ft. is measured with a 200-ft. tape from the horizontal axis of the transit to a tack in the top of stake A; similarly to tack in stake C (on the other side of B), the slope distance was 191.16 ft. Vertical angle to point A was +7° 12′ and to point C, −6° 47′. The tape used was 0.01 ft. too long at temperature 68° F. While this work was being done the tape temperature was 80° to 86° F. The suspended tape was given sufficient pull in both cases so as to offset the shortening due to sag.

(a) What is the corrected horizontal distance ABC to 1/100 ft.?

(b) Had the vertical angles been +7° 13′ and −6° 46′, how much difference (to the nearest 0.01 ft.) would this have made in the length of ABC?

12. A slope distance of 172.62 ft. is measured from the horizontal axis of a transit to a drill hole in a ledge, vertical angle +4° 33′.

(a) How many minutes of error must there be in this vertical angle to affect the computed horizontal length by 0.005 ft.?

(b) Had the angle been −1° 27′, how much error in the angle would cause an error of 0.005 ft. in the horizontal distance?

13. With a 300-ft. tape the slope distance measured from the horizontal axis of the transit to a point on a stake was 283.73 ft.; the vertical angle was read +6° 15′ (to the nearest 05′ only).

(a) What error may there be in the horizontal distance due to the fact that the angle was read to the nearest 05′?

(b) Had the vertical angle been +2° 25′ (to the nearest 05′), how large an error in horizontal distance may have been introduced by the approximation in reading the vertical angle?

(c) The maximum error computed in (a) might be caused by how much change in temperature of this tape from its standard temperature of 68° F.? Use result of (a) to nearest 0.01 ft.

(d) If the tape temperature was 77° F. and the vertical angle +6° 15′ (to nearest 05′) can you be certain of the exact horizontal distance within 0.01 ft.?

14. A 100-ft. steel tape is 0.005 ft. too short at 68° F. with 12 lbs. pull while supported. A rectangle is measured with this tape (supported). The sides were recorded as 601.72 ft. and 437.15 ft. The temperature of the tape was 81° F. at the start and 85° F. at the end of the job, with pull 12 lbs.

(a) If the recorded measurements are used will the square feet in the field be too great or too small?

(b) What is the error (in sq. ft.)?

(c) If the land is worth 80¢ per sq. ft., how does this error in dollars and cents compare with the $100 per day cost of a surveying party?

15. Using a 300-ft. tape a distance is measured as 286.42 ft. but the tape was bent out of line to the right 0.7 ft. at about the 60-ft. point and to the left of the line 1.0 ft. at about the 200-ft. point.

(a) What was the correct length of the line?

(b) What would have been the correct length if both the 0.7 ft. and 1.0 ft. were to the right of the line?

(c) What would be the correct length had it been bent 0.7 ft. to the right of the line at the 20-ft. point and 1.0 ft. to the left at the 270-ft. point?

(d) Explain fully why (a), (b) and (c) give different values.

16. A line 4500 ft. in length was measured with a 100-ft. steel tape. The uncertainty in the length of the tape was ±0.005 ft. per tape-length. The uncertainty in determining a representative value for the temperature of the tape caused an error per tape-length of ±0.004 ft. The error in marking each tape-length was ±0.005 ft. The uncertainty introduced in each tape-length by not holding the tape exactly horizontal was 0.01 ft. Determine the probable error in the length of the line introduced by compensating errors. Determine the error due to the constant or accumulative error.

17. A base-line is measured in one operation and found to be 5282.76 ± 0.10 ft. It is then measured in two sections, the resulting lengths being 2640.35 ± 0.07 and 2642.36 ± 0.07 ft. It is then remeasured in three sections, the resulting values being 1761.27 ± 0.06, 1761.39 ± 0.05, and 1760.15 ± 0.06 ft. All the measurements are made under the same conditions.

(a) What is the standard dev. of the base-line as computed from each of the three determinations and which is most precise?

(b) Computing a simple mean from the separate measurements, what is the final expression for the length of the base-line?

(c) If the tape used in determining these lengths were 0.01 ft. too long in 100 ft., what would be the effect of disregarding this accumulative error on the recorded measurement of the base-line?

18. A length from Sta. 1 to Sta. 2 is measured with the following results:

1. 3254.65 feet	3. 3254.72 feet	5. 3254.70 feet
2. 3254.69 feet	4. 3254.66 feet	

(a) What is the standard dev. of an observation?

(b) What is the standard dev. of the mean?

(c) What is the final expression for the most probable length from Sta. 1 to Sta. 2?

19. A distance was measured with a 100-ft. steel tape and found to be 400.00 ft. The tape weighed 2.2 pounds. It was of standard length under a 15 pound pull, when fully supported, at a temperature of 70° F. The line was measured with the tape supported on stakes at the ends only, the applied tension being 15 pounds. All stakes for tape supports were set at full tape intervals. The difference in elevation between the top of stakes 1 and 2 was 1.5 ft; between 2 and 3, 3.2 ft.; between 3 and 4, 2.8 ft.; between 4 and 5, 1.9 ft. The temperature at the time of measurement was 34° F. The coefficient of the tape was 0.000 006 45 per degree F. Compute the true length of the line.

20. A 300-ft. steel tape is used to set a point on a pier from a point on the shore, both points being at the same elevation. The tape is compared with a standard and found to be 300.003 ft. long when supported throughout at the standard conditions (temperature, 68° F.; tension, 20 lbs.). The weight of the tape is 3.6 lbs. and the area of the cross-section is 0.0026 square inches. The modulus of elasticity is 30,000,000 lbs. per square inch.

(a) What is the true horizontal distance between the end divisions when the tape is supported only at the ends and is at the standard conditions?

(b) What would the horizontal distance be if the tension were increased to 40 lbs., the temperature remaining at 68° F.?

(c) What effect would an increase of temperature from 68° to 78° F. have on (a) and (b)?

21. For the purpose of establishing survey monuments in a city, a line along a paved sidewalk is measured on the slope with the tape fully supported. Observations of the temperature are made each tape length. The measured slope distance is 1122.78 ft., the sidewalk grade is +2.5%, the applied tension is 12 lbs. and the average observed temperature is 89.5° F. The 100-ft. steel tape used is standardized at 70° F. and is found to be 0.008 ft. too short under a tension of 12 lbs. with the tape fully supported. What is the horizontal length of this line?

22. In order to locate the off-shore pier of a bridge, measurements were taken between two points at the same elevation with a 300-ft. steel tape. The recorded distance to a reference point on the pier from the point defining the bridge seat on the abutment is 282.07 ft. The tape was subjected to a 30-lb. tension and the temperature at the time of measurement was 35° F. The tape is 0.027 ft. too long when fully supported throughout its entire length and proportionately less for shorter distances. The weight of the entire tape is 6 lbs. and its cross section is 0.025 × 5/16 inch. The modulus of elasticity of steel is 30,000,000 lbs. per sq. in. What is the corrected distance?.

23. What would be the result in Problem 22 if the off-shore pier was 7 feet lower than the abutment?

24. Using the tape described in Problem 22, it is desired to secure an accurate vertical distance within a tall building. The tape is fastened in a vertical position by means of two brackets and turnbuckles in a stairwell. A spring balance is inserted between the lower end of the tape and the lower turnbuckle. By means of the turnbuckles, tension is increased until the spring balance reads 30 lbs. An engineer's level is set up near the top of the tape and another near the bottom by means of which readings are taken on the tape. The line of sight through the top level intersects the tape at 298.88 ft. and the line of sight through the lower at 0.31 ft. If the observations were taken when the temperature was 75° F., what is the correct difference of elevation between the two levels?

PROBLEMS FOR CHAPTER IV

1. A level was tested for the sensitiveness of the bubble, as follows: the rod was held on a point 150 ft. away; the bubble was moved over 6.8 divisions of the scale; the rod-readings at the two extreme positions of the bubble were 6.964 and 7.005. Compute the average angular value of one division of the level.

2. A dumpy level was tested by the peg method with the following results:

Instrument at *A*:—	Instrument at *B*:—
+S. on *A*, 4.139	+S. on *B*, 3.900
−S. on *B*, 4.589	−S. on *A*, 3.250

Find the rod-reading on *A* to give a level line of sight, the instrument remaining 3.900 above *B*. Was the line of sight inclined upward or downward? How much?

3. In running a line of levels between two distant points, a survey party encounters a stream 200 ft. wide. How can all of them cross the stream together, with their level and rod, just once and still carry elevations forward?

4. A rod is found to be 0.002 ft. short, due to wear on the brass foot-plate. Explain what effect this will have in finding the difference in elevation between two points.

5. (a) A level is set up and a +S. of 6.037 is taken on a point 500 ft. away, then a −S. of 0.492 is taken on a point 750 ft. away. What is the net curvature and refraction correction? What is the difference in elevation of the two points?

(b) In another case a +S. of 5.098 was taken on a point 200 ft. away and a −S. of 6.405 taken on a point 900 ft. away. What is the net curvature and refraction correction? What is the difference in elevation of the two points?

6. A level is set up near a hydrant and a backsight taken on the hydrant, the reading being 3.17. A foresight is taken to a stone bound and equals 5.29. The level is next set up close to the bound, the rod-reading upon it being 4.82. The second reading upon the hydrant is 2.66. What is the error of the sight line and which way must the cross-hairs be moved to correct this error?

7. In testing a level by the "peg method" the following rod-readings are taken: Instrument at B.M.$_1$, + sight on B.M.$_1$ = 5.32, − sight on B.M.$_2$ = 9.43: Instrument at B.M.$_2$, + sight on B.M.$_2$ = 5.32; − sight on B.M.$_1$ = 1.19. Which way should the cross-hairs be moved and how much?

8. The level bubble of a certain wye level is centered and the telescope then reversed in the wye, the bubble moving 4 divisions toward the eyepiece. If the angular value of 1 division of the level scale corresponds to 0.02 ft. on the rod per 100 ft. of distance, what is the error from this source alone in the elevation of B.M.$_2$ above B.M.$_1$ when the sum of the backsight (horizontal) distances taken with this level is 200 ft. greater than the sum of the foresight distances?

9. The line of sight of a level is found by the "peg adjustment" test to be inclined downward 0.014 ft. per 100 ft. in distance. What is the allowable difference in the backsight and foresight distances if readings are to be correct within 0.003 ft.?

10. Levels are run between B.M.$_1$ and B.M.$_2$ with a recorded difference of elevation of +84.19 feet. The error in reading the rod is ±0.003 ft. and the error in the rod-reading due to the bubble not being in the middle of the tube at the instant of sighting is ±0.005 ft. 16 set-ups were required to run the levels.

(a) What is the probable error of the difference of elevation?

(b) Assuming that a 12-ft. rod is used and that the rod is 0.008 ft. too long, what is the corrected difference of elevation?

(c) If the difference of elevation had been −84.19 ft., what would be the corrected difference?

11. A rod which is 2 inches square on its base is held on the flat top of a stone step which is used as a bench-mark. If the rod is held on a rounded surface and waved toward and from the levelman, and the levelman records the shortest reading, this is the vertical rod-reading. But if the rodman waves the rod on a flat surface obviously as the rod is waved backward from the vertical position it is supported on the back edge of its base and when in that position the least rod-reading is obtained. This reading was 0.38 ft. (a very short reading). What is the correct rod-reading?

12. A line of differential levels was run between two bench-marks 9 miles apart, sights averaging 250 feet in length. The measured difference in elevation was 746.38 feet. On comparison with a standard it was found that the rod, having a nominal length of 12 feet, was actually 12.000 feet long, a shortening of 0.003 feet because of wear on the brass foot plate being offset by a 0.003 foot lengthening distributed over the entire length due to a slight swelling of the wood during damp weather. Determine the true difference in elevation between the two bench-marks.

13. If, in running levels between two points the top of the rod is held 0.3 foot out of plumb on the average, what error would be introduced per set-up when backsight readings average 1 foot and foresight readings average 10 feet? A 12-foot rod is used for the work.

14. If the sights average 250 feet in length and the uncertainty in a single observation is ±0.002 foot, what is the probable error in running a line of levels 13 miles long?

15. Two level parties starting from the same bench *A* at Elev. 19.760 at the foot of a hill are to determine the elevation of bench *B* at the top of the hill. One party got 1740.65 ft. for the elevation of *B* and the other party got 1740.90. The aggregate backsights and foresights were equal and errors due to personal equation were compensating. In order to check the work each party closed the level circuit by running back to *A* where both checked on the original elevation.

How do you account for the differences in elevation the parties got at the top of the hill?

16. It is desired to set a bench-mark on the roof of a building about 325 feet high. The levels are carried up the stair-well substituting a 100-foot steel tape for the usual leveling rod. A bench-mark ($B.M._1$) is established on the first floor (elevation = 24.57 feet above mean sea level). The tape was tested and found to have a length of 100.013 feet when under a tension of 10 lbs. and supported throughout, at a temperature of 68° F. The weight of the tape is 2 lbs. It is known that the length of a vertically suspended tape of weight w with a weight W attached to the lower end is equal to the length of the same tape supported horizontally throughout its length and when under a tension of $W + \frac{1}{2}w$. The top of the tape is held suspended by a bracket fixed to the handrail of the stairs. A spring balance is inserted between the bottom of the tape and a second bracket. Tension is applied by means of a turnbuckle until the reading of the spring balance is 9 lbs. Instrument set-ups are established on landings in the stair-well at about every tape length. The work was done at an average tape temperature of 74° F, In the following readings, the numbers in italics are readings on the tape. No correction is necessary for the reading on the level rod.

	B.S.	F.S.	Elev.
$B.M._1$	3.29 ft.		24.57
Tape 1.	98.23	0.62 ft.	
Tape 2.	92.33	0.04	
Tape 3.	94.31	0.19	
Tape 4.	43.29	0.31	
$B.M._{roof}$		3.74	

Compute the correct elevation of the bench-mark on the roof to 0.01 ft.

PROBLEMS FOR CHAPTER V

1. At sta. q the meridian is established by an observation on the pole-star. (Art. 8.3, p. 325.) A traverse is run northerly and westerly to sta. 120. The total difference in latitude is 2400 ft. N and the total departure is 61,000 ft. W. The latitude of sta. 1 is 43° 10′ N. At sta. 120 an observation is made on the sun for azimuth of line 120-121. (Art. 8.9, p. 336). What correction should be applied to the azimuth of 120-121 to refer it to the meridian of sta. 1? (One minute of latitude equals about 6080 ft.)

2. Test the angles of the traverses shown in Probs. 11, 12, 13, and 14, Chapter XIV.

3. What is the difference in the lengths of the northern and southern boundaries of a Township (Art. 5.28, p. 179), the latitude of the southern boundary being 43° 00′ N? (One minute of latitude equals 6080 ft., nearly.)

4. A field that is approximately square is surveyed by setting the transit up in its approximate center and measuring the four angles between lines running to its corners. Assuming that the diagonal distances from the transit to the two ends of one side AB are 842.17 and 837.86, respectively, and that the error in measuring each one of these lines does not exceed 0.03 ft., and assuming also that the angle between the two diagonals is measured to the nearest minute, what is the greatest error in the computed length of line AB that may occur because of uncertainty in measurements?

5. From a base line 250.00 ft. long (on a street line monumented at both ends), the back corners of a rectangular lot 400.00 ft. deep are to be set with transit and tape. Assume that the uncertainty in measuring a 100-ft. tape-length is ±0.01 ft. and the uncertainty in turning the right angles after repetitions with a one-minute instrument is ±15 seconds of arc.

(a) What is the linear standard dev. in setting a back corner because of error in measuring with tape?

(b) What is the likely standard dev. in setting a back corner because of error in turning the angle?

(c) What is the linear standard dev. in setting a back corner from these combined causes?

(d) What is the greatest error in the back length of the lot that can occur within limits of these standard deviations?

(e) Assuming the worst possible case, what is the maximum error in the area?

6. The angles of a five-sided traverse are being measured by the deflection angle method. All the deflection angles are to the right and the uncertainty in measuring each deflection angle is ±10 seconds. What is the probable sum of the deflection angles?

7. A transit point X was referenced by means of three tie measurements radiating at angles of about 120° from X. $AX = 43.26$, $BX = 42.98$, $CX = 39.21$. The measurements were carelessly made; the tape was 1 ft. higher at A than at X, 2 ft. lower at B than at X, and 6 in. higher at C than at X. Later the point X was lost and had to be reproduced by means of ties AX and BX. The point C was also lost so that CX could not be used. In reproducing the point care was taken to hold the tape horizontal as it should be held. How far was the second determination of X away from the original position?

8. Angle AOB is to be determined by tape measurements. Along OA and OB a point is placed exactly 200 ft. from O at C and D. CD measures 80.66 ft.

(a) Compute the angle *AOB* to the nearest ½ minute.

(b) Assume that in measuring *CD* an error of 0.2 ft. was made and it was actually 80.86 ft. What error did this mistake introduce in the angle *AOB*?

9. Referring to Fig. 5.4.2, a line *EF* is to be established parallel to *AD*. As a first trial the instrument is set up at *E* and sighted on a swing offset *FD* = 4.00 ft. The distance *AH* is measured and found to be 3.08 ft. How far should points *H* and *F* be moved to the left or right in order to establish line *EF* truly parallel to *AD*?

Give results to nearest half-hundredth. The distance *EA* = 6.55 ft., *AD* = 268.72 ft.

10. The boundaries of a triangular lot of land are: *AB* = 400.00 ft., *BC* = 300.00 ft., *CA* = 500.00 ft.

(a) What is the angle at *A*?

(b) If the point *B* is set by laying off the angle *CAB* and measuring out 400.00 ft. from A, what is the greatest discrepancy that may be expected in the length of *BC* if the angle is laid out to the nearest minute only? Assume *AC* is exactly 500.00 ft.

(c) How closely should the angle at *A* be laid off so that no error greater than .01 ft. will be introduced in line *BC* due to this cause?

(d) Explain how you would obtain this precision in laying out the angle.

11. Compute the azimuths of all the lines in the traverse shown in Fig. 5.5.2 assuming line 1–6 is due South. Show check on last azimuth.

PROBLEMS FOR CHAPTER VI

1. In the quadrilateral in Fig. 6.39.1, angle *ACD* = 73° 01′, *ACB* = 37° 30′, *BDC* = 58° 20′, and *BDA* = 31° 00′. No distance is known. Find the angle *BAD*.

2. In Fig. 6.32.1. suppose angle *ABC* = 121° 52′, *BC* = 91.27 ft., angle *C* = 59° 30′; what will be the required distance *CD* to place *D* on *AB* prolonged? What will be the angle to lay off at *D* to place *E* on *ABD* prolonged? What is the length of *BD*?

3. In Fig. 6.26.1 *AX* = 391.24 ft.; swing offset *BZ* = 2.04 ft.; *YB* = 8.91 ft. What will be the offset *ME* if *AM* = 190.10 ft.?

4. In Fig. 6.[illegible].1 indicate complete and accurate location for each building shown.

5. The horizontal tape distance from the center of the base of an inclined pole to the transit set at A is 214.72 ft. When the telescope is horizontal the line of sight cuts the pole at a point 5.14 ft. vertically above its base. A vertical angle of 21° 16′ is measured to the top of the pole, and as this is done it is found that pole lines up with the vertical cross-hair (i.e., it appears vertical in that direction). The transit is then set up at point *B* on a line running from the base of the pole at right angles to the first line. This time the horizontal tape distance from the center of pole to transit is 210.16 ft., and when the transit telescope is horizontal the line of sight cuts the pole 5.32 ft. vertically above base of pole. The pole is not vertical when viewed from point *B* through the transit; it is straight but it inclines to the left toward A. A sight is then taken bisecting the top of the pole, the telescope lowered and an offset taken horizontally from the center of the pole at its base to this transit line. It measures 4.66 ft.

(a) What is the actual length of the pole (to 0.01 ft.) if the 100-ft. tape used was 0.015 ft. too long?

(b) What should the vertical angle at *B* read to the top of the pole?

(c) Suppose the vertical angle at *B* to top of pole read 21° 43′, what would this indicate to you?

6. Two church steeples, *A* and *B*, have been located by triangulation; they are approximately in latitude 40° North. The distance between them is 9347.9 feet. The bearing of *A* from *B* is N 2° 53′ W. It is desired to use this known distance and bearing for surveys in the vicinity. The steeples are visible from two intervisible points, *C* and *D*, on the ground. The line *CD* lies to the eastward of *AB*, point *C* being to the northward of *D*. Angles are measured as follows: *DCB* = 27° 54′; *ADC* = 32° 47′; *BDA* = 68° 21′; *ACB* = 57° 58′. Compute the distance *CD* and the bearing of *C* from *D*.

7. Referring to Fig. 6.34.1, the computed value of *AC* is 1000.000 ft. The angle *A* is 30° 00′, and the angle *B* is exactly 90° 00′. Now, suppose angle *A* was measured only to the nearest minute, but angle *B* is exactly 90°. What is the maximum error in the line *AC* due to the precision adopted for angle *A*?

8. Plot the buildings in Fig. 6.24.1 to scale: 1 in. = 50 ft.

9. A church steeple is over the center of a square masonry base. The south, east and west faces of the base measure 32.00 ft.; the north side is inaccessible but can be assumed to be 32.00 ft. The east side is ranged out for a distance of 100.00 ft. to *A* at which the transit is set up and vertical angle measured to top of steeple is 40° 34′. The telescope of the transit is leveled; it cuts the steeple base 4.72 ft. above the base.

(a) Are there sufficient data to compute to 0.01 ft. the height of top of steeple above the base course? If not, what additional data are required?

(b) If there are sufficient data, what is the height of the top of steeple above the base?

(c) Had the vertical angle been 30 seconds more, how much greater would the vertical distance be?

10. Indicate on a sketch drawn to scale how you would lay out with transit and tape the front corners (*E* and *F*) of the proposed building *EFGH* (Fig. 10a) so that they will be exactly on line with the front corners (*A* and *B*) of the existing building *ABCD*; also *E* must be precisely 8.00 ft. from *B*. The front of building *ABCD* is slightly irregular so that *EF* cannot be ranged out from *AB*.

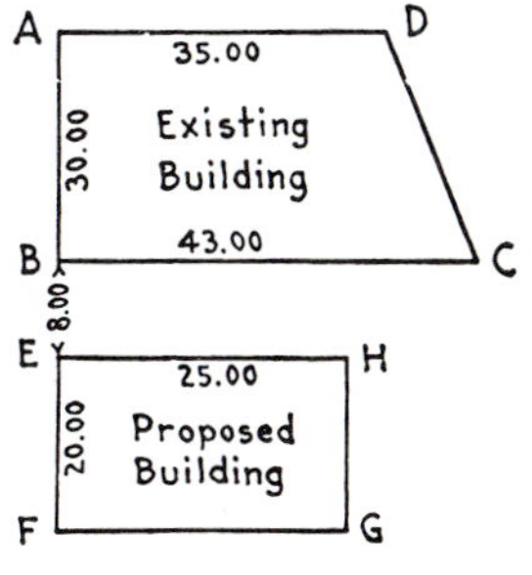

FIG. 10a.

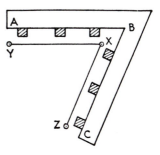

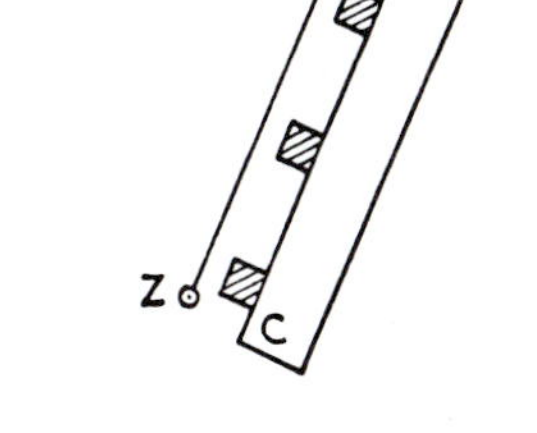

FIG. 10b.

11. Referring to Fig. 10b, explain explicitly with the help of a sketch how you would establish lines XY and XZ precisely parallel to lines AB and BC respectively, for the purpose of measuring the angle $YXZ = ABC$. The lines AB and BC are straight between corners of the building; the building is faced with pilasters as shown in Fig. 10b.

12. From a corner post at A a random line was run in a westerly direction towards another corner B. The perpendicular offset to B in a northerly direction from the random line at X was 2.64 ft. The distance AX on the random line was 1052.40 ft. Another random line was run southerly from A towards another corner C. The perpendicular offset to C in an easterly direction from a point Y on the random line was 1.98 ft. and the distance AY was 1292.50 ft. The angle at A between random lines was 87° 49′ 20″.

(a) Find the angle between the property lines AB and AC.

(b) Find the length of the property lines AB and AC.

13. If, in Fig. 6.30.1, the angles $B'AB$ and $B''AB$ were laid off to the nearest minute only, what maximum displacement may occur in the position of point B from its correct position on the prolongation of the line, and what maximum angular divergence from the true line may occur in line BC? Distances $AB' = AB'' = B'C' = B''C'' = 200.00$ ft.

14. The following measurements were taken to check the perpendicularity of a tapering factory chimney of circular cross-section. Point A was established on the ground about 150 ft. from the base of the chimney and the shortest dimension to the base measured carefully and found to be 150.32 ft. The instrument was set up at A and sighted to as to bisect the top of the chimney; the telescope was then lowered and a point B set on the circumference at the base, which was level with the telescope. The instrument was reversed and the sighting repeated, and, the instrument being in adjustment, the line of sight again fell at B. With the transit at A an angle was measured from B to a line tangent to the right side of the chimney at the base and found by repetition to be 2° 12′ 20″. Similarly an angle was measured from AB to the extreme left side of the chimney and found to be 2° 58′ 40″.

(a) How much is the chimney out of plumb at the top in a direction at right angles to AB?

(b) What is the radius of the chimney at the base to 0.01 ft.?

15. A line is being measured along the side of a straight highway pavement. Between transit points A and C a river intervenes, and a point B is chosen off line on the floor of a bridge. AB measures 538.21 ft. and BC. 762.18 ft. By means of a rod, the perpendicular distance from B to the transit line AC measures 7.5 ft.

(a) Compute distance AC to nearest 0.01 ft.

(b) If the offset at B were 1 ft. instead of 7.5 ft., what would be the tape distances AB and BC consistent with the computed value of AC?

16. Make a paper tracing of Fig. 6.25.2 and show on it the best method of locating the fences and buildings by tape measurements.

PROBLEMS FOR CHAPTER VII

1. Compute the Horizontal Distances and the Elevations in the Following Stadia Survey (F + c = 1, not included).

Pt.	Dist.	Azimuth	Observed Bearing	Vert. Angle	Diff. El.	Eleva.
		Instr. at ⊡ 4		H.I. = 4.29		133.1
81	61	168° 20′		− 13° 40′		
82	90	84° 30′		− 5° 45′		
83	140	218° 10′		− 5° 40′		
84	101	343° 10′		− 2° 16′ on 9.3		
85	185	310° 05′		− 4° 56′ on 5.3		
86	276	301° 25′		− 5° 26′ on 2.3		
87	373	277° 05′		− 4° 05′		
88	280	261° 45′		− 4° 44′		
89	210	233° 15′		− 5° 13′		
90	220	216° 50′		− 3° 55′		
91	201	202° 30′		− 3° 25′		
⊡ 5	255	202° 19′	N 22° E	− 3° 08′		
		Instr. at ⊡ 5		H.I. = 5.01		
92	88	85° 30′		− 1° 55′		
93	157	113° 20′		− 2° 55′		
94	183	169° 10′		− 1° 36′		
95	90	205° 20′		+ 0° 09′		
96	193	194° 25′		− 0° 06′		
97	218	222° 15′		+ 2° 06′		
98	115	228° 40′		+ 2° 30′		
99	39	230°		+ 2° 56′		
100	90	283° 40′		+ 1° 32′		
101	122	255° 20′		+ 3° 18′		
102	185	279° 00′		+ 2° 10′		
103	213	261° 30′		+ 3° 14′		
104	308	272° 00′		+ 0° 55′		
105	353	279° 10′		+ 0° 05′		
106	225	284° 50′		+ 0° 46′		
107	288	290° 30′		− 0° 20′		

2. A plane table is located by the three-point method and the elevation of the alidade found by measuring vertical angles to the following three stations.

Sta.	Vertical Angle	Scaled Horizontal Distance	Elevation of Point Sighted
A	+ 3° 52′	1530 ft.	510.6 ft.
B	− 2° 33′	4340 ft.	212.0 ft.
C	− 1° 23′	7420 ft.	227.9 ft.

Compute the mean elevation of the alidade. The ground is 3.6 ft. lower.

3. In running a stadia traverse 6000 ft. long the transit points were taken at about every 300 ft., care was exerted in holding the rod vertical, the vertical angles were closely read, and the rod interval was read to the nearest 0.01 ft. on both the backsight and foresight. The average uncompensated error in reading a distance between transit points is ±0.25 ft.

(a) What error in the distance would be caused by neglecting the constant of the instrument ($F + c = 1$ ft.)?

(b) What is the probable error in the distance of the entire traverse due to the average uncompensated error in reading a distance between transit points?

4. Stadia levels are being run up an approximately constant grade from station *A* to station *K*. The average stadia distance ($F + c$ included) is 250 ft. and the average vertical angle is +3° (read to the nearest 05°). If the levels are run for a distance of 2500 ft., what probable error may be expected in the difference of elevation between *A* and *K*? Assume that the average uncompensated error in each reading is 1'.25.

5. In a closed stadia traverse approximately 1200 ft. long, and with 6 nearly equal sides, the elevation of each station was obtained by stadia distances and vertical angles read to the nearest minute and taken backward and forward from each point. The forward angles from *A* to *D* were from +1° to +1° 30' and from *D* to *A* were from −1° to −1° 30'.

(a) How close would you expect the elevation of *A* obtained by going around the traverse to agree with the elevation originally used?

(b) If instead the elevations had been obtained by using the transit as a level and the same points occupied as in (a), how close would you expect the first and last elevations to agree? Rod-readings were taken to nearest .01 ft. and the instrument was free from errors of adjustment.

6. The alidade on a plane table is sighted on the top of the mast of a signal 3500 ft. away and a vertical angle of −2° 58' measured to the nearest minute. The height of the mast is 18.5 ft. and the height of the alidade above the point over which it is set is 3.7 ft. The elevation of the base of signal is 356.79 ft.

(a) Compute the elevation of point under the alidade.

(b) What is the likely error in the elevation determined in (a) introduced by measuring the vertical angle to the nearest minute?

7. Compute the Horizontal Distances and the Elevations in the following stadia survey ($F + c = 1$, not included below).

Pt.	Dist.	Azimuth	Observed Bearing	Vert. Angle	Diff. El.	Eleva.
Instr. at ⊡ *A*		B.S. 0° on ⊡ *X*		H.I. = 4.34		
B.M.				0°00' on 2.73		82.68
1	254	36° 25'		−2° 29'		
2	125	72° 20'		−4° 43'		
3	311	112° 10'		+1° 42'		
4	262	226° 05'		+5°32' on 6.5		
5	74	278° 55'		+3° 21'		
⊡ *B*	421	178° 26'	N 16° ¼ W	+2° 58'		
Instr. at ⊡ *B*		B.S. 0° on ⊡ *A*		H.I. = 4.12		
⊡ *A*	420			−2° 57'		
6	56	24° 50'		−1° 23'		
7	178	78° 40'		−0° 20'		
8	212	234° 45'		+3° 17'		
9	317	178° 15'		+5°14' on 2.00		
⊡ *C*	389	156° 24'		+4° 37'		

8. (a) From the map shown in Fig. 7.6.1 draw a profile along the line indicated by the lower border. Vertical scale, 1 inch = 200 ft.

(b) Draw a profile along a line drawn from the upper left to the lower right corner. For the vertical scale let 1/20 in. = one contour interval.

(c) Assume a straight line drawn from the lower left corner of the contour map to a point midway between *A* and *B*, and from that point to the upper right corner of the map.

Draw a profile of this broken line. Use vertical scale of 1 inch = 200 ft.

9. Assume that the side borders on the map shown in Fig. 7.7.1 are meridians with North toward the top of the page. Draw the side elevation (or projection) looking from the North toward the South. Vertical scale, 1 inch = 200 ft.

10. If the scale of the map shown in Fig. 7.8.1 is 1 inch = 1 mile, determine the number of square miles in the drainage area by means of the method explained in the second paragraph of Art. 14.11.

11. In Fig. 7.13.3, if the road had been level and at Elev. 62 ft., plot the intersection of the side slopes and the surface.

12. In Fig. 7.13.1, if the elevation of *A* is 5 feet and the line *BC* is level at Elev. 25 ft., draw the intersection of the plane defined by *ABC* and the surface.

PROBLEMS FOR CHAPTER VIII

1. (a) What was the azimuth of Polaris at its greatest western elongation at Chicago when the polar distance of the star was 0° 53′ 35″? The latitude of Chicago is 41° 23′ N.

(b) In making an observation for meridian two stakes were set 329 ft. apart, marking the direction of the star at elongation. Compute the length of the perpendicular offset to be laid off at one end of the line to obtain the true meridian.

2. What is the approximate Eastern Standard Time of the eastern elongation of Polaris on March 5, 1976, at a place in longitude 75° 36′ West and latitude 38° 23′ North?

3. Compute the azimuth of the mark from the following on the sun July 5, 1976. Horizontal angles measured to the right.

	VERNIER A	ALTITUDE	WATCH (E.S.T.)
Mark	0° 00′		
Left and	151° 12′	52° 10′	$3^h\,39^m\,58^s$ P.M.
Lower Limbs	151° 31′	51° 54′	$3^h\,41^m\,24^s$
Telescope reversed			
Upper and	151° 15′	50° 53′	$3^h\,44^m\,07^s$
Right Limbs	151° 32′	50° 38′	$3^h\,45^m\,27^s$
Mark	0° 00′		

Latitude = 40° 43′.3 N; longitude = 74° 17′.8 W; IC = +01′.

Declination at 0^h Greenwich Civil Time = +22° 51′ 15″.0 (North); tabular difference for 24^h = −333″.4 (sun going south).

4. Observation for latitude. The observed altitude of Polaris at upper culmination was 43° 27′. The polar distance of the star was 0° 55′. What was the latitude of the place?

5. Observed maximum altitude of the sun's lower limb June 28, 1966 = 76° 58′ (bearing north). Index Correction = +30″. Greenwich Civil Time is 8^h 08^m A.M. Sun's declination June 28 at 0^h Gr. Civ. Time = +23° 18′ 55″.2; tabular difference for 24^h = −164 ″.3; semi-diameter = 15′ 46″. Compute the latitude.

6. A mark was set roughly in the meridian (to the north of the transit) and 3 repetitions taken from Polaris to the mark (mark east of star), the third reading being 6° 39′ 00″. The observed watch readings were 8^h 24^m 10^s, 8^h 26^m 25^s, and 8^h 28^m 20^s. The watch was 20″ slow of Pacific Standard Time. Observed altitude of Polaris, 42° 02′, date April 7, 1967, latitude 42° 00′.0 N; longitude 117° 01′ W. Compute the true bearing of the mark. Watch readings are P.M.

7. With the transit on Station 12 on June 7, 1966, in latitude 42° 29′.5 N, longitude 71° 07′.5 W, a sun observation is made to determine the bearing 12–13. The corrected altitude is 42° 03′, the corrected declination is +22° 46′.1, the mean horizontal angle from 13, clockwise, to the sun is 156° 22′ and the mean watch reading is 2^h 31^m 50^s.2 P.M. E.S.T. The deflection angle at station 13 is 5° 26′ R; at 14 it is 7° 36′ L; at 15 it is 2° 13′ R. At station 15 an observation is taken on Polaris; 0° on station 16; first angle, at 7^h 02^m 05^s P.M. E.S.T., 252° 44′ (clockwise); second repetition, at 7^h 04^m 25^s, 145° 29′; third repetition, at 7^h 06^m 40^s, 38° 14′. Corrected mean altitude is 41° 35′.5. Lat. sta. 15, 42° 29′.4, long. 71° 06′.9. Assuming no error in the observations, compute the error in the traverse angles.

PROBLEMS FOR CHAPTER IX

1. Compute the following set of level notes and show check.

Sta.	+ S	H. I.	− S	Elev.
B. M.$_1$	4.702			16.427
B. M.$_2$	11.846		6.727	
T. P.$_1$	7.276		9.689	
B. M.$_3$	8.760		4.726	
T. P.$_2$	0.687		11.000	
B. M.$_4$			8.496	

2. Compute the following set of level notes and show check.

Sta.	B. S.	H. I.	F. S.	Elev.
B. M.$_{12}$	6.427			62.473
20			6.2	
21			7.4	
+ 42			5.2	
22			4.7	
T. P.$_{27}$	4.724		9.976	
23			11.2	
+ 63			10.4	
B. M.$_{22}$	0.409		7.482	
24			11.2	

3. Compute the following set of level notes and show check.

Sta.	+ S	H. I.	− S	Elev.
B. M.$_{24}$	6.214			84.238
T. P.$_1$ L.	3.515		9.280	
T. P.$_1$H.	2.152		7.919	
T. P.$_2$ L.	2.971		8.263	
B. M.$_{25}$ H.	2.338		7.629	
T. P.$_3$ L.	4.278		7.529	
T. P$_3$ H.	2.646		5.894	
B. M.$_{26}$ L.	5.721		6.072	
T. P.$_4$ H.	4.837		5.187	
B. M.$_{27}$			5.817	

4. Write cross-section notes from Sta. 0 to 3 inclusive for road in cut consistent with the following: road 50 ft. wide, slopes $1\frac{1}{2}$ to 1; grade elev. at Sta. 0 = 107.20; gradient +1.4%. Sta. 0 is level section; Sta. 1, a three-level section; Sta. 2, a five-level section; Sta. 3, an irregular section. Except at Sta. 0, ground slopes down transversely from left to right.

5. Compute the following set of level notes and show check.

Sta.	+ S	H. I.	− S	Elev.
B. M.1	8.21			47.19
T. P.	11.01		3.07	
T. P.	9.61		4.19	
0			9.0	
+ 50			8.1	
1			6.0	
+ 50			4.5	
2			1.0	
T. P.	12.00		0.17	
+ 50			11.0	
3			8.8	
B. M.2			7.91	

6. Compute the following set of level notes and show check.

Sta.	+ S	H. I.	− S	Elev.
B. M.6	8.21			207.33
0			4.0	
1			8.1	
2			11.2	
+ 50			11.9	
T. P.	4.01		12.19	
3			6.9	
+ 10.90			1.0	
4			2.6	
5			2.9	
B. M.			5.27	

7. Compute the following set of level notes and show check.

Sta.	+ S	H. I.	− S	Elev.
B. M.21	0.27			1164.20
T. P.	1.16		12.41	
T. P.	1.01		10.91	
T. P.	2.16		7.99	
T. P.	0.79		11.32	
B. M.22	4.71		4.90	
T. P.	3.02		8.00	
T. P.	0.64		9.69	
T. P.	2.26		11.49	
B. M.23			10.20	

8. The elevation of B.M.$_1$ is 20.96 ft. By route 1, B.M.$_A$ is 60.17 above B.M.$_1$. B.M.$_2$ is 101.10 above B.M.$_A$. By route 2, B.M.$_2$ is 161.40 above B.M.$_1$. By route 3, B.M.$_2$ is 161.10 above B.M.$_1$. Route 1 is 6 miles long and B.M.$_A$ is 3 miles from B.M.$_1$. Route 2 is 4 miles long. Route 3 is 8 miles long. Compute the adjusted elevation of B.M.$_2$ and of B.M.$_A$. Compute the probable error in differences in elevation between B.M.$_1$ and B.M.$_2$.

9. By route 1 (3 miles long) point B is 111.20 ft. higher than point A. By route 2 (4 miles long) B is 110.97 ft. above A. By route 3 (6 miles long) B is 111.26 ft. above A. Compute the adjusted elevation of B above A. Compute the probable error in difference in elevation between A and B.

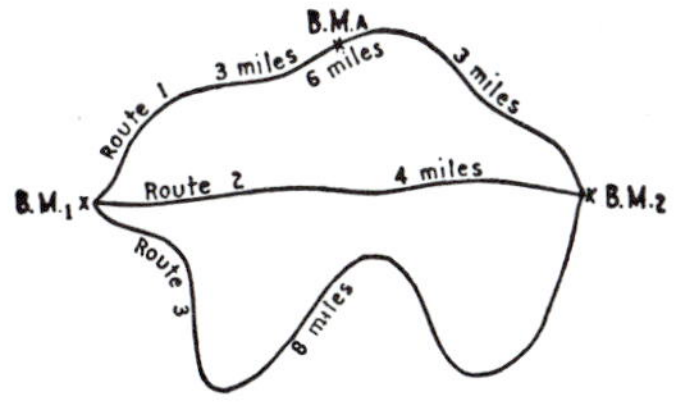

FIG. 8. DIAGRAM ILLUSTRATING PROBLEM 8.

10. The accepted elevation of B.M. 572 is 1928.73 feet. A fifteen-mile level circuit was run, closing back on B.M. 572 with an elevation, as determined from the notes of 1928.58 feet. The observed elevations of other bench marks which were established during the run, and their approximate distances from the initial bench, are given below. Adjust the elevations of the intermediate bench marks.

Bench Mark	Dist. from B.M. 572 (Miles)	Observed Elevation
B.M. 573	1.2	1983.21
B.M. 574	2.5	2091.76
B.M. 575	3.8	2243.19
B.M. 576	5.3	2339.70
B.M. 577	7.1	2302.95
B.M. 578	9.4	2441.62
B.M. 579	11.3	2117.37
B.M. 580	13.8	2003.49

11. Level notes taken in a tunnel read as follows:

Sta.	B.S.	H.I.	F.S.	Elev.	
B.M. 16	4.57			162.72	Top of bolt in floor of shaft 16th level
Sta. 161	−3.24		−2.35		In Roof of Tunnel
Sta. 162	−2.19		−3.62		" " " "
Sta. 163	4.07		3.33		" Floor " "
Sta. 164	5.12		4.01		" " " "
Sta. 165	−0.85		−1.12		" Roof " "
Sta. 166			5.16		" Floor " "

In the above set of level notes compute the elevations of stations 161 to 166 inclusive.

Note: Sights marked − (minus) indicate that the point on which the rod was held is above the H.I. The rod-reading on these points was obtained by inverting the rod.

PROBLEMS FOR CHAPTER X

1. The deflection angle between two lines is 46° 24′ to the right. The lines are to be connected with a curve of 300-foot radius.

(a) Find the tangent distance to .01 ft. and the total length of arc to .01 ft.

(b) If the curve is laid out in 10 equal parts, compute the deflection angles (to ½ minute) from the tangent at the P.C. to each point and show check on the deflection angle to the P.T. Also compute the length of the short chord (to .01 ft.) to be used in laying out these points.

(c) If, instead, the curve is laid out by a series of 25-foot chords starting at the P.C., compute all the deflection angles necessary to lay out this curve and the length of the last chord to close on the P.T.

2. Compute chords to .01 ft. and deflection angles to ½ minute for center line, right, and left edges of a 40-foot pavement, on a curve of 250-foot center-line radius to the left. The center-line curve is to be staked out in 25-foot arcs starting at the P.C. The points set on the left and right edges of the pavement are to be radially opposite those on the center line. The station of the P.C. = 20 + 00.00. I = 36° 00′.

3. The grade lines shown below are to be connected by a vertical parabolic curve 400 ft. long.

(a) Compute the elevations to .01 ft. at full stations and +50 points along the curve.

Sta. 50 + 00	El. 24.60
Sta. 65 + 00	El. 57.60
Sta. 85 + 00	El. 26.80

(b) What is the elevation at Sta. 65 + 20?

4. A pavement 30 ft. wide has a parabolic surface with a total crown of 6 in. If the pavement grade elevation at the center line is 96.20, what are the elevations at quarter points between the center line and the curb?

5. Design the cross-section of a residential street, 30 feet wide, on which an unsymmetrical crown is required. The highest point of the pavement is to

be elevation 656.62; it is located 12 ft. east of the face of the curb on the west side of the street. The elevation of the gutter on the west side of the street is to be 656.27 ft. The elevation of the gutter on the east side of the street is to be 655.98. Curbs are to be 7 inches in height on each side of the street. On a sheet of cross-section paper, plot the cross-section of the street. Use a horizontal scale of 8 ft = 1 in. and a vertical scale of 1 ft. = 1 in. Determine the elevation of the top of each curb and also the elevation for each 5 ft. point across the pavement. Record the elevations on the sketch.

6. A +4.0% grade line meets a −5.5% grade line at elevation 238.92 at Sta. 48 + 00.0. An 800-foot parabolic vertical curve is to be used to connect these grade lines.

(a) Compute and tabulate the elevations for points at every full station and +50 points along the curve.

(b) Determine the station and elevation of the point of highest elevation on the curve.

7. A −3 per cent grade meets a +5 per cent grade at a vertex which lies directly under the center line of an overpass bridge which carries another road across these grades at right angles. The underside of the bridge is level at elevation 760.00 ft.

(a) What is the longest vertical (parabolic) curve that can be used (to nearest 50 ft.) to connect the −3 per cent and +5 per cent grades, and provide at least 14.50 ft. of clearance under the bridge at its center line? The station of the intersection of the straight grades is 10 + 00 and the elevation of this intersection is 740.50 ft.

(b) If the underside of the bridge is 40 ft. wide, find the actual clearances (to .01 ft.) at the two edges and at the center line of the bridge for the vertical curve found in (a).

(c) Find the elevations of each 50-ft. station on the vertical curve to .01 ft.

8. (a) Compute the lengths to 0.01 ft. not given on the sketch (Fig. 8a) of all straight and curved boundaries on the lots shown; also compute the area of each lot to nearest 10 s.f.

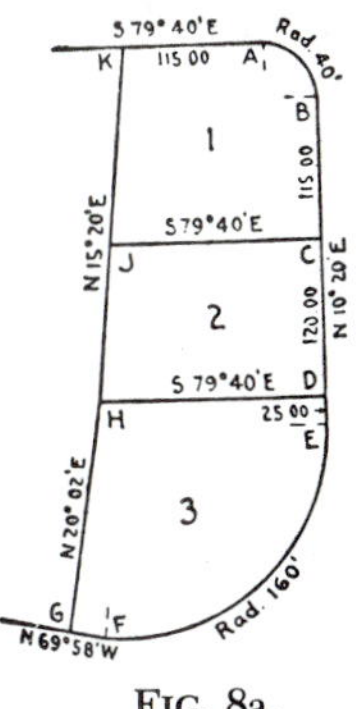

FIG. 8a.

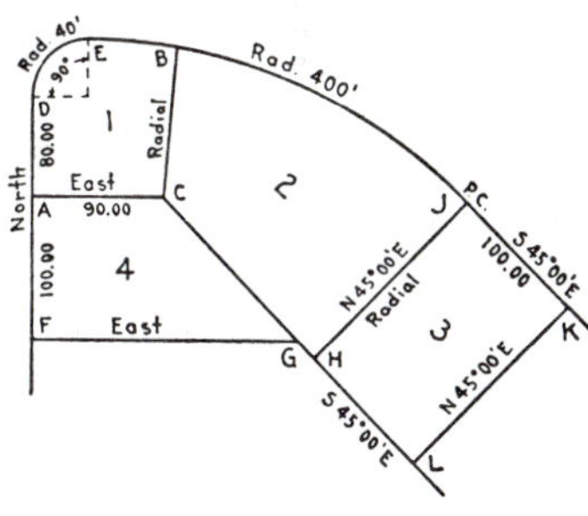

FIG. 9a.

(b) Plot the lots to a scale of 50 ft = 1 in. and letter all dimensions, bearings and areas on the plot.

9. (a) Compute the lengths to 0.01 ft. not given on the sketch (Fig. 9a) of all straight and curved boundaries of the lots shown.

(b) Compute the area of each lot to nearest 10 s.f.

10. Referring to Fig. 10.28.1, the distance AD = 102.58 ft., and the angle α = 48° 24′. If the radius is 300 ft., find distance CB and arc AB to nearest 0.01 ft.

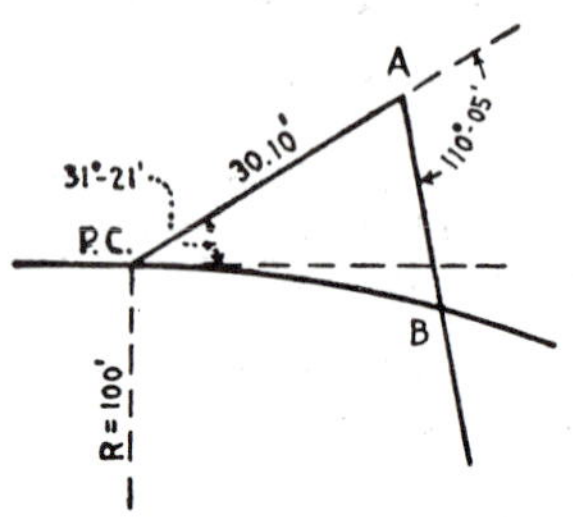

FIG. 11.

11. In Fig. 11 compute the distances from B to A and to $P.C.$ (See Art. 10.28.)

12. Data for a circular curve are as follows:

I = 63° 17′ R = 300 ft. V = 27 + 36.23 *Arc Stationing*

(a) Find station of *P.C.* and *P.T.* to 1/100 ft.

(b) Find deflection angles (to ½ minute) for setting all full stations on the curve with the transit at the *P.C.*

(c) Find chord lengths (to 0.01 ft.) for sub-chords at each end of curve and chord length for a full arc station (100-ft. arc).

13. (a) A 5-degree curve is to connect two tangents which intersect at an angle of 51° 23′. If the station of the *P.C.* = 12 + 42.71, find the station of the *P.T.* and the deflection angles (to ½ minute) required to set all the stations on the curve by the chord-station method.

(b) An obstacle prevents running the curve beyond Sta. 17 with the transit at the *P.C.* With the transit at Sta. 17, what angle should be set on the plate when backsighting on the *P.C.* and what are the deflection angles required to set the remaining stations on the curve and the *P.T.*?

14. A circular curve has an angle I = 41° 34′, radius 700 ft. and *P.C.* = 21 + 50.00 *Arc Stationing.*

(a) Find right-angle offset from the *P.C.–V* tangent to STa. 23 + 00 on the curve.

(b) Find also the distance from the vertex to the point on the tangent from which the offset is to be laid off. Distances to 1/100 ft.

15. Two center-line tangents intersect with a deflection angle of 35° 22′ to the right. It is desired to connect these tangents with a circular curve of a full degree that will come nearest to having an external distance of 50 ft.

(a) Find the degree of curve by arc stationing method which will come nearest to fulfilling the above condition.

(b) Assuming that the *P.C.* falls on Sta. 10 + 00, prepare complete notes for laying out stations on the curve by deflection angles from the *P.C.* Include in notes the external distance, middle ordinate, tangent distance and length of curve computed to nearest 0.01 ft.

16. Solve Problem 15, using chord stationing method.

PROBLEMS FOR CHAPTER XI

1. From a station at the mouth of a tunnel a line is run in the tunnel, azimuth 37° 24′, slope distance 424.34, vertical angle + 2° 10′, H.I. + 4.62, H.Pt. 0.0; thence azimuth 62° 42′, slope distance 278.53, vertical angle + 2° 18′, H.I. + 4.21, H.Pt. + 3.12, to breast. From the same station at tunnel a line is run on the surface, azimuth 98° 33′, slope distance 318.57, vertical angle − 3° 22′, H.I. + 4.87, H.Pt. 0.0; thence azimuth 38° 02′, slope distance 647.82, vertical angle + 14° 13′, H.I. + 4.73, H.Pt. 0.0, to the center of a vertical shaft. How deep must the shaft be to meet a connecting drift run on a grade of + 2.4% from a point in the floor at the breast of tunnel which is 7.0 ft. vertically under the roof point, and what is the slope length and azimuth of this drift? (Breast is face of heading where work is going on).

2. The strike of a certain vein at point of outcrop is N 43° E and the dip is 71° 50′ to S.E. From this point of outcrop a surface line is run, N 83° 15′ E, slope distance 247.12, vertical angle − 12° 34′; thence S 2° 54′ E, slope distance 208.52, vertical angle − 14° 34′, to a point from which the tunnel is to be driven in the direction N 71° W and with a grade of + 3.8% until it intersects the vein. All vertical angles were taken to points above the station sighted equal to the distance the horizontal axis of the transit was above the station point under the instrument.

(a) What would be the slope length of such a tunnel?

(b) What would be the slope length and bearing of the shortest possible tunnel run on a + 1.3% grade to intersect the vein?

3. A vein has a strike of S 67° W and its dip is 55°. What is the bearing of a line lying in the plane of the vein and having a vertical angle of − 44°?

4. From station M at the mouth of a tunnel a traverse is run in the tunnel, azimuth 20° 35′, distance 352.16, vertical angle + 1° 02′, H.I. + 4.71, H.Pt. + 3.42 to point A; thence azimuth 61° 07′, distance 528.24, vertical angle + 0° 40′, H.I. − 3.62, H.Pt. + 4.07, to point B at the breast of the tunnel. From M a surface traverse is run, azimuth 25° 10′, distance 578.34, vertical angle + 4° 25′ to point C; thence azimuth 11° 15′, distance 407.62, vertical angle + 14° 20′ to point D, which is the center of a vertical shaft 120 ft. deep. In sighting surface points the vertical angles were in all cases read to point above the stations sighted equal to the distance the horizontal axis of the telescope was above the station point under the instrument.

Find the azimuth, vertical angle and slope distance of the line from the point at the bottom of the shaft to the roof point at breast of the tunnel.

PROBLEMS FOR CHAPTER XII

1. Vertical aerial photographs of an area are to be taken to a scale of 1/24000 with a camera having a focal length of 8.25 inches. Average elevation of ground above sea level is 500 ft.

(a) At what altitude above sea level should the exposures be made?

(b) If the altitude actually flown is 5% higher than found in (a), what will be the scale of prints?

2. The size of an area to be photographed is 100 sq. mi., the prints are 9 by 9 inches, end lap 60% and side lap 25%. Scale of prints 1/10000. Find the net number of exposures necessary to cover the area.

3. Find the displacement on the print of the top of a chimney 100 ft. high above datum, if the top of the chimney appears in the print at a point 3.98 inches from the center of the print. Altitude of airplane above datum (H) = 8000 ft. Focal length = 6 in.

4. Upon photographs supplied by the instructor, identify by means of coördinates parallel to and at right angles to the bottom of the print physical and topographical features designated by the instructor.

5. Plot a graph showing the relationship of scale, flight height and focal length for focal lengths of 4″, 6″, 8.25″, 12″ and 24″. The limits of the graph are to be from 0 to 20,000 for flight heights and from 0 to 30,000 for scale number. (For scale of 1:30,000 the "scale number" is 30,000, etc.) Also indicate equivalent scales in units of feet per inch. A plotting scale of 4,000 = 1 inch is suggested for both flight heights and scale numbers.

6. Plot a graph showing the relation between H, f and area in square miles covered by a 9″ × 9″ negative for the focal lengths and the limits of flight height stated in Problem 5, and for area limits from 0 to 15 square miles.

7. A vertical aerial photograph taken with a camera of 6″ focal length has a scale of 1:7,000 in a valley area.

(a) If the image of a hill top appears in the photograph 3.58 inches from the principal point and the height of the hill is 200 ft. above the valley, find the amount and direction of the displacement caused by relief with respect to the valley floor.

(b) What is the scale of the photograph at the top of the hill?

PROBLEMS FOR CHAPTER XIV

1. A series of perpendicular offsets are taken from a straight line to a curved boundary line. The offsets are 15 ft. apart and were taken in the following order: 6.8, 7.2, 4.6, 5.7, 7.1, 6.3, and 6.8.

(a) Find the area between the straight and curved lines by the Trapezoidal Rule.

(b) Find the same area by Simpson's One-Third Rule.

2. It is desired to substitute for a curved boundary line a straight line which shall part off the same areas as the curved line. A trial straight line AB has been run; its bearing is S 10° 15′ W, its length is 418.5 ft., and point B is on a boundary line CD which has a bearing S 80° W. The sum of the areas between the trial line and the crooked boundary on the easterly side is 2657 sq. ft.; on the westerly side it is 7891 sq. ft. It is required to determine the distance BX along CD such that AX shall be the straight boundary line desired. Also find the length of the line AX, and bearing of AX.

3. In the quadrilateral $ACBD$ the distances and angles which were taken in the field are as follows:

$$
\begin{array}{llll}
AB = 50.63 & & ABC = 105^\circ\,39'\,00'' \\
BC = 163.78 & & BAD = 89^\circ\,37'\,30'' \\
CD = 93.80 & & & \\
DA = 160.24 & & & \\
DB = 167.73 & & &
\end{array}
$$

Check the fieldwork by computations, and figure the area of the quadrilateral by using right triangles entirely.

4. Two street lines intersect at an angle (deflection angle) of 48° 17′ 30″. The corner lot is rounded off by a circular curve of 40-ft. radius.

(a) Find the length of this curve to the nearest 1/100 ft.

(b) Find the area of the land included between the curve and the two tangents to the curve (the two street lines produced).

5. Find the quantity in cubic yards, in the borrow-pit shown in Fig. 5; the squares are 20 ft. on a side. Bank is vertical 4 ft. down from the top of slope; then slopes uniformly to "foot of slope," except where the depths are marked in figures. Here the slope is to be taken as uniform from top to bottom.

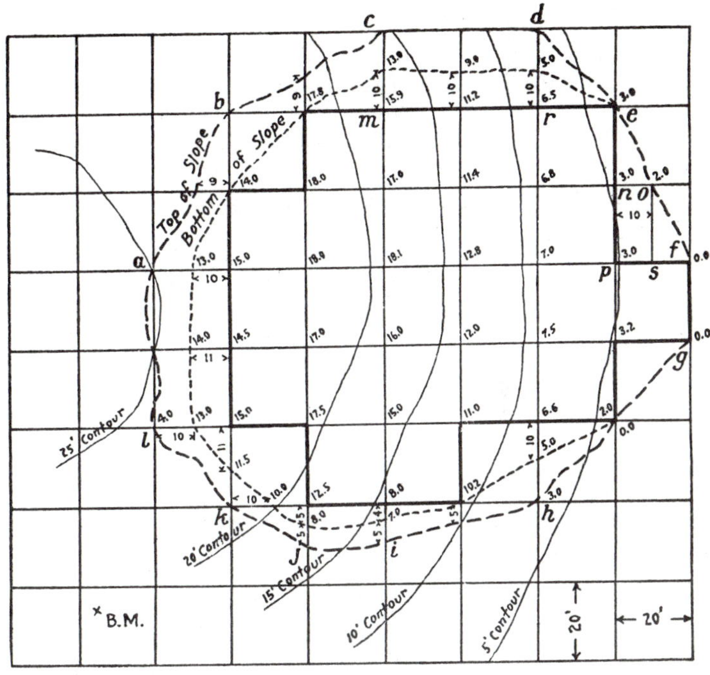

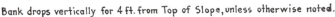

Bank drops vertically for 4 ft. from Top of Slope, unless otherwise noted.

FIG. 5. PLAN OF PORTION OF A BORROW-PIT.

6. At station 6 a rectangular trench was measured and found to be 3 ft. wide and 4 ft. deep. At stations 6 + 70 it was found to be 3.2 ft. wide and 8.6 ft. deep.

(a) Find by use of the Prismoidal Formula the quantity of earthwork between stations 6 and 6 + 70. Result in cubic yards.

(b) Find the volume of the same by End-Area Method.

7. The following is a set of notes of the earthwork of a road excavation.

12	$\frac{27.0}{+8.0}$		+4.2		$\frac{23.4}{+5.6}$
11 +60	$\frac{30.0}{+10.0}$	$\frac{15.0}{+4.5}$	+4.0	$\frac{15.0}{+7.5}$	$\frac{24.0}{+6.0}$
Sta. 11	$\frac{21.0}{+4.0}$		+6.0		$\frac{25.8}{+7.2}$

The base of the road is 30 ft. and the slopes are 1½ to 1.

Find by the End-Area Method the quantity of earthwork from station 11 to station 12. Result in cubic yards.

8. A pyramid has a square base 8 ft. on a side, the height of the pyramid being 12 ft. What is the error (in cubic feet) in the volume computed by the "end-area method"? What is the percentage error?

9. Referring to Fig. 7.13.3, compute cu. yds. of cut for this road.

10. Following is a "closed" compass and chain survey. Calculate the area.

Station.	Bearing.	Distance.
1	S 40° W	17.50 chains
2	N 45° W	22.25
3	N 36°¼ E	31.25
4	North	13.50
5	S 81° E	46.50
6	S 8°½ W	34.25
7	West	32.50

11. Following is a closed transit and tape traverse. Calculate the area.

Station.	Deflection Angle.	Distance.
A	52° 49′ R	400.9 ft.
B	72 12 R	526.1
C	90 22 R	390.8
D	85 11 L	339.6
E	123 17 R	816.7
F	106 30 R	829.5

12. Calculate the area.

Station.	Deflection Angle.	Distance.
1	124° 31′ R	118.17 ft.
2	69 07 R	106.18
3	20 56 L	91.39
4	126 23 R	156.55
5	100 54 R	82.19
6	39 59 L	121.67

13. Calculate the area.

Station.	Deflection Angle.	Bearing.	Distance.
1	74° 43′½ R	N 22° 28′ E	229.15
2	125 22′½ R		105.29
3	15 29′¼ R		148.75
4	34 24′ L		276.68
5	9 53′½ L		247.67
6	172 11′ R		457.68
7	16 31′ R		242.02

14. Calculate the area.

Station.	Deflection.	Bearing.	Distance.
1	91° 41′ 30″ R	N 65° 51′ W	319.15
2	120 03′ 30″ R		168.40
3	11 14′ L		498.80
4	151 23′ R		394.18
5	8 07′ R		229.35

In this problem 100-ft tape is 0.067 ft. short.

15. Calculate area

Station.	Deflecting Angle.	Bearing.	Distance.
1	40° 08′ R	N 25° 43′ 30″ W	140.74
2	2° 34′ L		100.97
3	0° 58′½ R		145.74
4	51° 51′ R		84.87
5	7° 49′½ R		301.32
6	82° 39′½ R		527.70
7	107° 57′ R		498.90
8	11° 14′½ R		168.37
9	59° 56′½ R		24.35

100-ft. tape is 0.089 ft. short.

16. Compute the area.

Station.	Deflection Angle.	Bearing.	Distance (Feet).
A	78° 10′ 00″ L		208.64
B	88° 28′ 00″ L		436.79
C	67° 02′ 15″ L		56.48
D	33° 39′ 15″ L		98.80
E	90° 00′ 00″ R		68.62
F	22° 51′ 00″ L		95.10
G	68° 50′ 15″ L	N 36° 14′ 00″ W	207.41
H	89° 48′ 00″ L		103.75
I	55° 00′ 15″ L		96.75
J	53° 49′ 00″ R		420.77

17. Compute the area enclosed by the following traverse.

Station	Deflection Angle	Bearing	Distance
1	79° 13′ *R*	North	1278.2 ft.
2	76 21 *R*		230.6
3	17 03 *R*		383.4
4	22 00 *L*		500.2
5	113 38 *R*		850.3
6	18 58 *L*		881.1
7	114 45 *R*		1241.5

18. Compute the area enclosed by the following traverse.

Station	Deflection Angle	Bearing	Distance
1	99° 49′ *R*		854.1 ft.
2	100 06 *R*		203.9
3	19 47 *L*		250.7
4	22 10 *R*		479.6
5	32 54 *L*	S 1° 39′ W	454.0
6	112 06 *R*		687.3
7	78 32 *R*		1341.6

PROBLEMS FOR CHAPTER XV

1. Referring to Fig. 5.5.1, calculate the coördinates of all points of the traverse (using balanced Lat. and Dep.) referred to the meridian through the most westerly point and the perpendicular through the most southerly point.

2. From the coördinates of problem 1 calculate bearing and length of *AD*.

3. Using the coördinates of problem 1 and the data of Fig. 5.5.1, calculate the coördinates of the corners near points *C* and *B*; from the coördinates find the length and bearing of the property line itself.

4. In Fig. 5.5.1, calculate the new bearings and distances consistent with the balanced latitudes and departures.

5. Referring to Fig. 5.4.1, if lines *ED* and *BC* are prolonged to intersect at *C′*, a property corner, what is the area of the triangle that must be added to the area of the traverse? Referring to the same figure, if the property corner is on *JH* prolonged to *H′*, 50.00 ft. from *H*, and another corner is at *G′* 65.10 ft. from *G*, defl. angle from *FG* = 31° 10′ *left*, what is the area that must be added to the area of the traverse? What is the length and bearing of *H′G′*?

6. In the following traverse there are two mistakes. Find where they occur and determine their amounts.

Station.	Observed Bearing.	Deflection Angle.	Distance (Feet).	Calculated Bearings.	Remarks.
A	N 34° E	164° 14′ R	240.2	N 34° 00′ E	
B	S 73°½ E	62° 16′ R	163.7		
C	S 10°¼ W	84° 22′ R	207.6		*CE* = 188.1
D	N 26°½ W	142° 49′ R	273.1		*BCE* = 34° 14′
E	S 52° W	103° 41′ L	147.4		*DEC* = 81° 25′

7. The following is a set of notes of an irregular boundary of a lot of land. It is desired to straighten this crooked boundary line by substituting a straight line running from *B* to the line *EF*. Find the bearing of the new boundary line and its length; also the distance along *EF* from *E* to where the new line cuts *EF*.

Station.	Bearing.	Distance (Feet).
A	S 89° 14′ E	373.62
B	N 13° 10′ E	100.27
C	N 0° 17′ W	91.26
D	N 27° 39′ E	112.48
E	N 72° 12′ W	346.07
F	S 5° 07′ W	272.42
	etc.	etc.

8. (a) In the lot of land, *ABCD*, the lines *AB* and *DC* both have a bearing of N 23° E; the bearing of *AD* is due East; *AD* is 600 ft., *AB* is 272.7 ft., and *DC* is 484.6 ft. Find the length of a line *EF* parallel to *AB* which will cut off an area *ABFE* equal to half an acre. Also find the length of the lines *AE*, and *BF*. (b) What is the area of *EFCD*?

9. Given the notes of a traverse, which does not close, as follows:—

Station	Deflection Angle	
0		
6 + 40	6° 17′ L	Find the length of a straight line from 0 to 20 + 64 and the angle it makes with the line from 0 to 6 + 40.
9 + 20	18° 43′ L	
14 + 55	12° 47′ R	
17 + 18	45° 24′ L	
20 + 64	68° 06′ R	

10. The line *Ax* has a bearing S 51° 10′ E, and line *Ex* has a bearing N 21° 04′ E. *A* and *E* are connected by the following traverse: *AB*, S 60° 51′ W, 102.69 ft.; *BC*, S 0° 59′ E, 298.65 ft.; *CD*, S 89° 01′ E, 101.20 ft; *DE*, N 80° 52′ E, 148.28 ft. Compute the distance *Ax* and *Ex*.

11. On a street line *ACE* whose bearing is S 89° 10′ E, the distance *AC* is 50.00 ft. and *CE* is 101.50 ft. The side line *AG* of a house lot has a bearing of S 2° 05′ W. Point *B* is on this side line and is 31.25 ft. from A. The opposite side of the lot, *CH* has a bearing S 1° 20′ W, and the side line of the next lot, *EI* has a bearing S 1° 10′ E. From *B* a line having a bearing N 89° 16′ E cuts these side lines, at points *B*, *D* and *F*. Compute distances *CD*, *BD*, *EF*, and *DF*.

12. A traverse run from *A* to *D* has the following bearings and distances: *AB*, S 26° 19′ E, 91.2 ft.; *BC*, S 89° 58′ E, 216.00 ft.; *CD*, N 1° 19′ E, 371.25 ft. Point *E* is 191.00 ft. from *A* and 212.10 ft. from *D*. Compute the bearings of *AE* and *DE*, and the angles *EAD*, *EDA* and *AED*.

13. A corner city lot *ABCDE* has the following interior angles. *A* = 7° 15′, *B* = 94° 50′, *C* = 91° 22′, *D* = 120° 27′, *E* = 226° 06′. The measured sides are *AB* = 583.58, *BC* = 102.88, *CD* = 153.01, *DE* = 86.85, *EA* = 391.42. Bearing of *AB* is

N 66° 50′ E. At the corner B the lot is rounded by a circular curve of 20.00 ft. radius; point B is at the vertex of this curve. Compute the area of the lot.

PROBLEMS FOR CHAPTER XVI

1. Plot the surveys given in Fig. 5.2.1, and in Fig. 5.5.1 by Protractor and Scale, Rectangular Coördinates, Tangents, or Chords.
2. Plot by use of Scale and Protractor the notes given in Fig. 7.16.1.

TABLES

TABLE I. GEOMETRIC FORMULAS.

Required.	Given.	Formulas.
Area of		
Circle	Radius $=r$	πr^2
Sector of Circle	Radius $=r$, Arc $=L_c$	$\frac{rL_c}{2}$
Segment of Circle	Chord $=C$, Middle Ordinate $=M$	$\frac{2}{3}CM$ (Approximate)
Ellipse	Semi-axes $=a$ and b	πab
Surface of		
Cone	Radius of Base $=r$; Slant Height $=s$	πrs
Cylinder	Radius $=r$, Height $=h$	$2\pi rh$
Sphere	Radius $=r$	$4\pi r^2$
Zone	Radius of Sphere $=r$, Height of Zone $=h$	$2\pi rh$
Volume of		
Prism or Cylinder	Area of Base $=b$; Height $=h$	bh
Pyramid or Cone	Area of Base $=b$; Height $=h$	$\frac{bh}{3}$
Frustum of Pyramid or Cone	Area of bases $=b$ and b'; Height $=h$	$\frac{h}{3}(b+b'+\sqrt{bb'})$
Sphere	Radius $=r$	$\frac{4}{3}\pi r^3$

GENERAL TRIGONOMETRIC FORMULAS.

$$\sin A = 2 \sin \tfrac{1}{2} A \cos \tfrac{1}{2} A = \sqrt{1-\cos^2 A} = \tan A \cos A = \sqrt{\tfrac{1}{2}(1-\cos 2A)}$$

$$\cos A = 2\cos^2 \tfrac{1}{2}A - 1 = 1 - 2\sin^2 \tfrac{1}{2}A = \cos^2 \tfrac{1}{2}A - \sin^2 \tfrac{1}{2}A = 1 - \text{vers } A$$

$$\tan A = \frac{\sin A}{\cos A} = \frac{\sqrt{1-\cos^2 A}}{\cos A} = \frac{\sin 2A}{1+\cos 2A}$$

$$\cot A = \frac{\cos A}{\sin A} = \frac{\sin 2A}{1-\cos 2A} = \frac{\sin 2A}{\text{vers } 2A}$$

$$\text{vers } A = 1 - \cos A = \sin A \tan \tfrac{1}{2} A = 2\sin^2 \tfrac{1}{2} A$$

$$\text{exsec } A = \sec A - 1 = \tan A \tan \tfrac{1}{2} A = \frac{\text{vers } A}{\cos A}$$

$$\sin 2A = 2\sin A \cos A$$

$$\cos 2A = 2\cos^2 A - 1 = \cos^2 A - \sin^2 A = 1 - 2\sin^2 A$$

$$\tan 2A = \frac{2\tan A}{1-\tan^2 A}$$

$$\cot 2A = \frac{\cot^2 A - 1}{2\cot A}$$

$$\text{vers } 2A = 2\sin^2 A = 2\sin A \cos A \tan A$$

$$\text{exsec } 2A = \frac{2\tan^2 A}{1-\tan^2 A}$$

$$\sin^2 A + \cos^2 A = 1$$

$$\sin(A \pm B) = \sin A \cos B \pm \sin B \cos A$$

$$\cos(A \pm B) = \cos A \cos B \mp \sin A \sin B$$

$$\sin A + \sin B = 2\sin \tfrac{1}{2}(A+B)\cos \tfrac{1}{2}(A-B)$$

$$\sin A - \sin B = 2\cos \tfrac{1}{2}(A+B)\sin \tfrac{1}{2}(A-B)$$

$$\cos A + \cos B = 2\cos \tfrac{1}{2}(A+B)\cos \tfrac{1}{2}(A-B)$$

$$\cos B - \cos A = 2\sin \tfrac{1}{2}(A+B)\sin \tfrac{1}{2}(A-B)$$

$$\sin^2 A - \sin^2 B = \cos^2 B - \cos^2 A = \sin(A+B)\sin(A-B)$$

$$\cos^2 A - \sin^2 B = \cos(A+B)\cos(A-B)$$

$$\tan A + \tan B = \frac{\sin(A+B)}{\cos A \cos B}$$

$$\tan A - \tan B = \frac{\sin(A-B)}{\cos A \cos B}$$

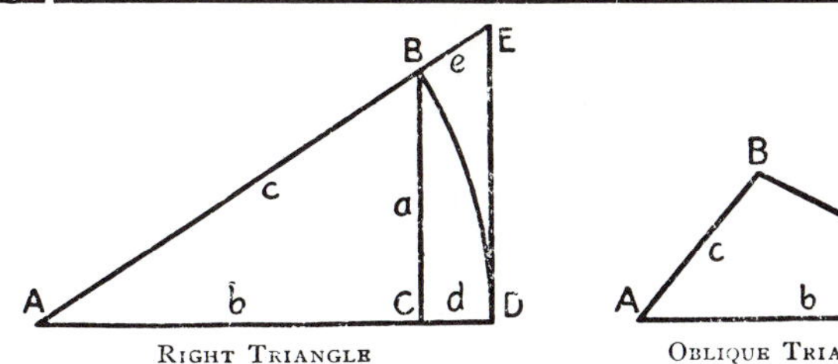

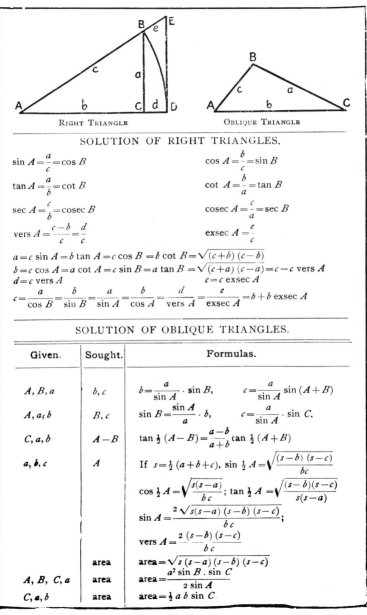

RIGHT TRIANGLE   OBLIQUE TRIANGLE

SOLUTION OF RIGHT TRIANGLES.

$\sin A = \frac{a}{c} = \cos B$ $\cos A = \frac{b}{c} = \sin B$

$\tan A = \frac{a}{b} = \cot B$ $\cot A = \frac{b}{a} = \tan B$

$\sec A = \frac{c}{b} = \operatorname{cosec} B$ $\operatorname{cosec} A = \frac{c}{a} = \sec B$

$\operatorname{vers} A = \frac{c-b}{c} = \frac{d}{c}$ $\operatorname{exsec} A = \frac{e}{c}$

$a = c \sin A = b \tan A = c \cos B = b \cot B = \sqrt{(c+b)(c-b)}$

$b = c \cos A = a \cot A = c \sin B = a \tan B = \sqrt{(c+a)(c-a)} = c - c \operatorname{vers} A$

$d = c \operatorname{vers} A$ $e = c \operatorname{exsec} A$

$c = \frac{a}{\cos B} = \frac{b}{\sin B} = \frac{a}{\sin A} = \frac{b}{\cos A} = \frac{d}{\operatorname{vers} A} = \frac{e}{\operatorname{exsec} A} = b + b \operatorname{exsec} A$

SOLUTION OF OBLIQUE TRIANGLES.

Given.	Sought.	Formulas.
A, B, a	b, c	$b = \frac{a}{\sin A} \cdot \sin B$, $c = \frac{a}{\sin A} \sin(A+B)$
A, a, b	B, c	$\sin B = \frac{\sin A}{a} \cdot b$, $c = \frac{a}{\sin A} \cdot \sin C$.
C, a, b	$A - B$	$\tan \frac{1}{2}(A-B) = \frac{a-b}{a+b} \tan \frac{1}{2}(A+B)$
a, b, c	A	If $s = \frac{1}{2}(a+b+c)$, $\sin \frac{1}{2} A = \sqrt{\frac{(s-b)(s-c)}{bc}}$
		$\cos \frac{1}{2} A = \sqrt{\frac{s(s-a)}{bc}}$; $\tan \frac{1}{2} A = \sqrt{\frac{(s-b)(s-c)}{s(s-a)}}$
		$\sin A = \frac{2\sqrt{s(s-a)(s-b)(s-c)}}{bc}$;
		$\operatorname{vers} A = \frac{2(s-b)(s-c)}{bc}$
	area	$\text{area} = \sqrt{s(s-a)(s-b)(s-c)}$
A, B, C, a	area	$\text{area} = \frac{a^2 \sin B \cdot \sin C}{2 \sin A}$
C, a, b	area	$\text{area} = \frac{1}{2} a b \sin C$

Greek Alphabet.

LETTERS	NAME
Α, α,	Alpha
Β, β,	Beta
Γ, γ,	Gamma
Δ, δ,	Delta
Ε, ϵ,	Epsilon
Ζ, ζ,	Zeta
Η, η,	Eta
Θ, θ,	Theta
Ι, ι,	Iota
Κ, κ,	Kappa
Λ, λ,	Lambda
Μ, μ,	Mu
Ν, ν,	Nu
Ξ, ξ,	Xi
Ο, ο,	Omicron
Π, π,	Pi
Ρ, ρ,	Rho
Σ, σ, ς,	Sigma
Τ, τ,	Tau
Υ. υ,	Upsilon
Φ, φ,	Phi
Χ, χ,	Chi
Ψ, ψ,	Psi
Ω, ω,	Omega

INDEX